TOXICOEPIGEN

TOXICOEPIGENETICS
Core Principles and Applications

Edited by

SHAUN D. MCCULLOUGH

DANA C. DOLINOY

ACADEMIC PRESS

An imprint of Elsevier

Academic Press is an imprint of Elsevier
125 London Wall, London EC2Y 5AS, United Kingdom
525 B Street, Suite 1650, San Diego, CA 92101, United States
50 Hampshire Street, 5th Floor, Cambridge, MA 02139, United States
The Boulevard, Langford Lane, Kidlington, Oxford OX5 1GB, United Kingdom

Notices
Knowledge and best practice in this field are constantly changing. As new research and experience broaden our understanding, changes in research methods, professional practices, or medical treatment may become necessary.

Practitioners and researchers must always rely on their own experience and knowledge in evaluating and using any information, methods, compounds, or experiments described herein. In using such information or methods they should be mindful of their own safety and the safety of others, including parties for whom they have a professional responsibility.

To the fullest extent of the law, neither the Publisher nor the authors, contributors, or editors, assume any liability for any injury and/or damage to persons or property as a matter of products liability, negligence or otherwise, or from any use or operation of any methods, products, instructions, or ideas contained in the material herein.

Library of Congress Cataloging-in-Publication Data
A catalog record for this book is available from the Library of Congress

British Library Cataloguing-in-Publication Data
A catalogue record for this book is available from the British Library

ISBN 978-0-12-812433-8

For information on all Academic Press publications
visit our website at https://www.elsevier.com/books-and-journals

Working together
to grow libraries in
developing countries

www.elsevier.com • www.bookaid.org

Publisher: John Fedor
Acquisition Editor: Kattie Washington
Editorial Project Manager: Jennifer Horigan
Production Project Manager: Swapna Srinivasan
Cover Designer: Mark Rogers

Typeset by SPi Global, India

Contents

2
DNA METHYLATION

3
NONCODING RNAs

5-3. Methods for Analyzing miRNA Expression

VALENTINA BOLLATI, LAURA DIONI

Contributors

Patrick Allard Institute for Society and Genetics, University of California, Los Angeles, Los Angeles, CA, United States

Michelle M. Angrish U.S. Environmental Protection Agency, National Center for Environmental Assessment, Research Triangle Park, NC, United States

Andrea A. Baccarelli Environmental Precision Biosciences Laboratory, Columbia University, Mailman School of Public Health, New York, NY, United States

Valentina Bollati Center of Molecular and Genetic Epidemiology, Department of Clinical Sciences and Community Health, Università degli Studi di Milano; EPIGET (Epidemiology, Epigenetics and Toxicology) Lab, Department of Clinical Sciences and Community Health, University of Milan, Milan, Italy

Paige A. Bommarito Department of Environmental Sciences and Engineering, Gillings School of Global Public Health, University of North Carolina at Chapel Hill, Chapel Hill, NC, United States

Emma C. Bowers Curriculum in Toxicology, University of North Carolina—Chapel Hill, Chapel Hill, NC, United States

Jessica A. Camacho Molecular Toxicology Interdepartmental Program, University of California, Los Angeles, Los Angeles, CA, United States

Raymond G. Cavalcante Epigenomics Core, Biomedical Research Core Facility, University of Michigan, Ann Arbor, MI, United States

Robert Y.S. Cheng Cancer and Inflammation Program, Center for Cancer Research, National Cancer Institute, Frederick, MD, United States

Brian N. Chorley National Health and Environmental Effects Research Laboratory, US Environmental Protection Agency, RTP, NC, United States

Ila Cote Cote and Associates, Boulder, Colorado, United States

Julia Yue Cui Department of Environmental and Occupational Health Sciences, University of Washington, Seattle, WA, United States

Ian J. Davis Lineberger Comprehensive Cancer Center; Departments of Genetics and Pediatrics, The University of North Carolina, Chapel Hill, NC, United States

Joseph Dempsey Department of Environmental and Occupational Health Sciences, University of Washington, Seattle, WA, United States

Maya A. Deyssenroth Department of Environmental Medicine and Public Health, Icahn School of Medicine at Mount Sinai, New York, NY, United States

Laura Dioni EPIGET (Epidemiology, Epigenetics and Toxicology) Lab, Department of Clinical Sciences and Community Health, University of Milan, Milan, Italy

Ingrid L. Druwe U.S. Environmental Protection Agency, National Center for Environmental Assessment, Research Triangle Park, NC, United States

Christopher Faulk Department of Animal Science, College of Food, Agricultural, and Natural Resource Sciences, University of Minnesota, Saint Paul, MN, United States

Rebecca C. Fry Department of Environmental Sciences and Engineering, Gillings School of Global Public Health; Curriculum in Toxicology, School of Medicine, University of

North Carolina at Chapel Hill, Chapel Hill, NC, United States

Zidong Donna Fu Department of Environmental and Occupational Health Sciences, University of Washington, Seattle, WA, United States

Jaclyn M. Goodrich University of Michigan School of Public Health, Ann Arbor, MI, United States

Patrick A. Grant Department of Biochemistry and Molecular Genetics, University of Virginia Medical School, Charlottesville, VA, United States

Tiffany A. Katz Center for Precision Environmental Health, Department of Molecular and Cellular Biology, Baylor College of Medicine, Houston, TX, United States

Christina Y. Lee Department of Biochemistry and Molecular Genetics, University of Virginia Medical School, Charlottesville, VA, United States

Ronit Machtinger Sheba Medical Center, Ramat-Gan and Tel-Aviv University, Tel Aviv, Israel

Elizabeth M. Martin National Health and Environmental Effects Research Laboratory, US Environmental Protection Agency Research Triangle Park; Curriculum in Toxicology, University of North Carolina —Chapel Hill, Chapel Hill, NC, United States

Shaun D. McCullough National Health and Environmental Effects Research Laboratory, US Environmental Protection Agency, Research Triangle Park, NC, United States

Doan M. On National Health and Environmental Effects Research Laboratory, US Environmental Protection Agency Research Triangle Park; Department of Pharmacology and Toxicology, Medical College of Virginia, Richmond, VA, United States

Samantha G. Pattenden Lineberger Comprehensive Cancer Center; Division of Chemical Biology and Medicinal Chemistry, Center for Integrative Chemical Biology and Drug Discovery, Eshelman School of Pharmacy, The University of North Carolina, Chapel Hill, NC, United States

Jairus Pulczinski Department of Environmental Health and Engineering, Johns Hopkins Bloomberg School of Public Health, Baltimore, MD, United States

Tingting Qin Department of Computational Medicine and Bioinformatics, University of Michigan, Ann Arbor, MI, United States

Karilyn E. Sant San Diego State University School of Public Health, San Diego, CA, United States

Maureen A. Sartor Epigenomics Core, Biomedical Research Core Facility; Department of Computational Medicine and Bioinformatics; Biostatistics Department, University of Michigan, Ann Arbor, MI, United States

Wan-yee Tang Department of Environmental Health and Engineering, Johns Hopkins Bloomberg School of Public Health, Baltimore, MD, United States

John J. Vandenberg U.S. Environmental Protection Agency, National Center for Environmental Assessment, Research Triangle Park, NC, United States

Cheryl L. Walker Center for Precision Environmental Health, Department of Molecular and Cellular Biology, Baylor College of Medicine, Houston, TX, United States

Emily Woolard Oak Ridge Institute for Science and Education at US Environmental Protection Agency, RTP, NC, United States

Robert O. Wright Department of Environmental Medicine and Public Health, Icahn School of Medicine at Mount Sinai, New York, NY, United States

Qian Wu Department of Environmental Health and Engineering, Johns Hopkins Bloomberg School of Public Health, Baltimore, MD, United States; Department of Hygienic Analysis and Detection and Ministry of Education Key Lab for Modern Toxicology, School of Public Health, Nanjing Medical University, Nanjing, China

Bonnie H.Y. Yeung Department of Environmental Health and Engineering, Johns Hopkins Bloomberg School of Public Health, Baltimore, MD, United States

Editors' Biography

Dr. Shaun D. McCullough is a principal investigator in the National Health and Environmental Effects Research Laboratory of the US Environmental Protection Agency (EPA) where he leads a research laboratory that focuses on exploring the molecular and epigenetic mechanisms of interindividual variability in adverse health effects of inhaled toxicants. Dr. McCullough holds a PhD in biochemistry and molecular genetics from the University of Virginia School of Medicine and currently serves on the editorial boards of *Environmental Epigenetics* and the *Journal of Toxicology and Environmental Health* and as a subject matter expert in epigenetics for the EPA's review of Organisation for Economic Co-operation and Development (OECD) guidelines. He has also been an active contributor to the field of toxicoepigenetics as an invited speaker in a variety of meetings and has chaired  scientific sessions on toxicoepigenetics at a broad range of conferences, including the Society of Toxicology (SOT) annual meeting, the American Association for the Advancement of Science (AAAS) annual meeting, and the Gordon Research Conference in Cellular and Molecular Mechanisms of Toxicity. Dr. McCullough has also served as the president of SOT's Molecular and Systems Biology Specialty Section, as the chair of SOT's Career Resources and Development Committee, and as a member of the board of directors for the American Society for Cellular and Computational Toxicology. He has also received EPA's Superior Accomplishment Award and SOT's Gabriel L. Plaa Education Award, and his transition into the field of epigenetics was featured in the journal *Science*. Drs. McCullough and Dolinoy also served as co-chairs of the 2016 Contemporary Concepts in Toxicology meeting *Toxicoepigenetics: The Interface of Epigenetics and Risk Assessment*.

Dr. Dana C. Dolinoy serves as NSF International chair of Environmental Health Sciences and professor of Environmental Health Sciences and Nutritional Sciences at the University of Michigan School of Public Health as well as Faculty Director of the Epigenomics Core at Michigan Medicine. She leads the Environmental Epigenetics and Nutrition Laboratory, which investigates how nutritional and environmental factors interact with epigenetic gene regulation to shape health and disease. Dr. Dolinoy holds a PhD in genetics and genomics and integrated toxicology from Duke University and MSc in public health from Harvard University and serves as associate editor of *Environmental Health Perspectives*, *Environmental Epigenetics*, and *Toxicological Sciences* and served as chair of the Gordon Research Conference in Cellular and Molecular Mechanisms of  Toxicity. She has been an invited speaker at numerous national and international meetings and authored more than 85 peer-reviewed scientific manuscripts and 10 book chapters. In 2011, Dr. Dolinoy received the Norman Kretchmer Memorial Award from the American Society for Nutrition and the Classic Paper of the Year Award from *Environmental Health Perspectives*. In 2012, she was the recipient of the Association of Schools of Public Health (ASPH)/Pfizer Research Award for the article "An expression microarray approach for the identification of metastable epialleles in the mouse genome." This work was cited as a model approach that may allow for directly assessing the role of early environmental exposures in human adult disease. Dolinoy received the 2015 NIH Director's Transformative Research Award to develop novel epigenetic editing tools to reduce disease risk and served with Dr. McCullough as chair of the 2016 Contemporary Concepts in Toxicology meeting *Toxicoepigenetics: The Interface of Epigenetics and Risk Assessment*.

Introduction to the Role of the Epigenome in Health and Disease

Shaun D. McCullough*, Dana C. Dolinoy†

*National Health and Environmental Effects Research Laboratory, US Environmental Protection Agency, Research Triangle Park, NC, United States †Department of Environmental Health Sciences, University of Michigan School of Public Health, Ann Arbor, MI, United States

The genetic material of every organism exists within the context of regulatory networks that govern gene expression collectively called the epigenome. Epigenetics have taken center stage in the study of diseases such as cancer, diabetes, and neurodegeneration; however, its integration into the field of toxicology is in its infancy. Increasing the presence of epigenetics in toxicological research (e.g., toxicoepigenetics) will allow for a more in-depth understanding of important aspects of toxicology, including (1) the mechanisms underlying toxicant-mediated health effects, (2) the role of the environment and lifestyle in modulating the individual susceptibility to these effects, (3) the multi- and transgenerational transmission of these health effects and susceptibilities, and ultimately (4) the integration of epigenetic information into the next risk assessment framework. There is a rapidly growing appetite for the integration of epigenetics into the field of toxicology; however, the recent emergence of the field means that many toxicologists have not yet had the opportunity to gain formal training in epigenetics to effectively incorporate toxicoepigenetics into their research or risk assessment programs. This textbook provides the fundamental principles and practices critical to understanding the role of the epigenome in regulating gene expression and thus the response of cells, tissues, and individuals to toxicant exposures.

This book covers the core aspects of epigenetics, including chromatin biology and histone modifications, DNA methylation, and noncoding RNA (Sections 1–3), as well as special considerations for studying each of these mechanisms of epigenetic regulation (Section 4). The editors and authors have designed this text to serve as an introduction to epigenetics for toxicologists and scientists trained and experienced in a wide range of biomedical subspecialties who want to incorporate epigenetics into their research programs. Each of the first three sections describes the basic biology of epigenetic marks followed by implications for studying each epigenetic mark in the context of toxicology. Special considerations for incorporating toxicoepigenetics into research programs including germ line and transgenerational responses to toxicant exposures (Chapter 4-1), the development of novel bioinformatics tools for toxicoepigenetics (Chapter 4-2), and the incorporation of toxicoepigenetic data into the risk assessment framework (Chapter 4-3) are also explored. Further, given the diverse nature of available protocols for epigenetic analysis, this text also includes unified, practical, and easy-to-follow protocols that will make it a useful resource in both the classroom and the laboratory (Section 5). These include protocols for evaluating histone modifications via chromatin immunoprecipitation (Chapter 5-1), assays for both targeted and genome-wide DNA methylation (Chapter 5-2), and methods for analyzing small noncoding RNAs (Chapter 5-3).

Section 1 on histone modifications and chromatin structure contains four chapters evaluating the role of histone proteins and chromatin accessibility in toxicological sciences. Chapter 1-1 by Grant and Lee, "The Role of Histone Acetylation and Acetyltransferases in Gene Regulation," provides a current review of the role of histone acetylation and acetyltransferases in the regulation of transcriptional initiation, elongation, and gene expression. The goal of this chapter is to improve the reader's ability to determine whether specific histone acetylation targets and their writers, readers, and erasers are pertinent to their studies. Chapter 1-2 by Cui and colleagues evaluates "The Role of Histone Methylation and Methyltransferases in Gene Regulation," expanding on the role of histone methylation and methyltransferases as mediators of both transcriptional activation and silencing. Through a deeper understanding of the different mechanisms responsible for methylation-dependent activation and repression of gene expression, the reader will be able to determine whether specific histone methylation targets and associated proteins are pertinent to their studies. Chapter 1-3 by Davis and Pattenden on "Chromatin Accessibility as a Strategy to Detect Changes Associated with Development, Disease, and Exposure and Susceptibility to Chemical Toxins" provides background on the implications of chromatin accessibility in gene regulation and describes how techniques such as formaldehyde-assisted isolation of regulatory elements (FAIRE) and assay for transposase-accessible chromatin using sequencing (ATAC-seq) data can be used to explore the mechanisms of disease. To conclude this section, Chapter 1-4 on "Implications for Chromatin Biology in Toxicology," by Katz and Walker, ties together the information presented in Section 1 by discussing direct applications of chromatin biology to toxicological studies, which will give the reader a clear representation of how toxicant exposures alter the chromatin landscape. Further, the reader will gain an understanding of how changes in chromatin modification states are important in the developmental origins of health and disease and in aging.

Section 2 covers the epigenetic mark DNA methylation, with attention to its application in both animal and human studies. Chapter 2-1 on "The Role of DNA Methylation in Gene Regulation," by Bommarito and Fry, reviews the mechanisms underlying the writing and reading of DNA methylation as modulators of gene expression. Emphasis is given to the interactions between DNA methylation state and associated chromatin structure in the context of transcriptional regulation. In Chapter 2-2, Faulk discusses "Implications of DNA Methylation in Toxicology," including the investigation of DNA methylation in current toxicology studies with an emphasis on the modulation of the DNA methylation state by environmental exposures and modifiable risk factors. Chapter 2-3 focuses on human studies with "DNA Methylation as a Biomarker in Environmental Epidemiology," by Deyssenroth and Wright. This chapter highlights the use of DNA methylation as a marker for a host of disease states and its predictive value. To close out the DNA methylation section, Chapter 2-4 on "DNA Hydroxymethylation: Implications for Toxicology and Epigenetic Epidemiology," by Tang and colleagues, covers the role of DNA hydroxymethylation as a biomarker of exposure and effects and as a risk-modifying factor in epidemiological toxicology.

Section 3 focuses on the expanding role of long and small noncoding RNAs as biomarkers and mechanistic links to disease outcomes. Chapter 3-1 by Woolard and Chorley discusses "The Role of Noncoding RNAs in Gene Regulation" with a focus on microRNAs (miRNAs), which are thought to regulate many expressed human genes. Chapter 3-2 by Machtinger, Bollati, and Baccarelli introduces "miRNAs and lncRNAs as Biomarkers of Toxicant Exposure" and discusses the use of noncoding RNAs in cells, tissues, and bodily fluids as markers of toxicant exposure and the way in which these biomarkers are beneficial tools in toxicological studies.

Section 4 builds upon the concepts presented in Sections 1–3 by examining special considerations in toxicoepigenetics research. Chapter 4-1 on "Germline and Transgenerational Impacts of Toxicant Exposures" by Camacho and Allard discusses how in utero exposure to an epigenotoxicant may directly impact not only her epigenome but also the epigenome of her offspring and grand offspring, commonly referred to as inter- or multigenerational effects. Much attention has been given to G0 exposure and F1 effects. Much less attention, however, has been given to direct effects of exposures on the germ line, the eventual F2 (grand offspring) generation. This may be due to the intense focus over the last decade on the potential for exposures to influence transgenerational effects (F3 and beyond). Chapter 4-2 by Cavalcante, Qin, and Sartor describes "Novel Bioinformatics Methods for Toxicoepigenetics." The bioinformatics analysis of epigenomic data is a rapidly expanding scientific discipline, and this chapter evaluates the complexity and evolution of the field of environmental epigenomics and describes novel bioinformatics methods and tools for analyzing high-throughput toxicoepigenomic data. Examples of bioinformatics databases and resources of special interest to toxicoepigenetics researchers are described, including recent consortia projects such as the Epigenome Roadmap and Toxicant Exposures and Responses by Genomic and Epigenomic Regulators of Transcription (TaRGET). Our special consideration section concludes with Chapter 4-3 by Angrish and colleagues on "Incorporating Epigenomes Into a Risk Assessment Framework." As described in previous chapters, the field of toxicoepigenetics is rapidly evolving to provide novel insights into the mechanisms underlying exposure-related susceptibility and disease; however, the utility and practicality of using epigenetic data in public health and risk assessment remains unclear. The overall goal of this chapter is to address some of the barriers the incorporation of toxicoepigenetic data into risk assessment will face.

Our book concludes with three chapters on protocols for toxicoepigenetics research. Chapter 5-1 by McCullough and colleagues on "Chromatin Immunoprecipitation: An Introduction, Overview, and Protocol" provides general protocols for both native and formaldehyde-based chromatin immunoprecipitation (ChIP), including variations that accommodate variable sample size and the preparation of samples for genome-wide analysis of histone modifications and transcription factor binding by ChIP-seq. This chapter includes a support protocol for the isolation of leukocytes from whole blood in preparation for ChIP and discusses troubleshooting and technical limitations of the ChIP assay. Chapter 5-2 by Sant and Goodrich on "Methods for Analysis of DNA Methylation" details methods for the analysis of DNA methylation from global approaches (e.g., luminometric methylation assay (LUMA) and long interspersed nuclear element-1 (LINE-1)), to candidate gene assays (e.g., pyrosequencing), to genome-wide technologies (e.g., Illumina BeadArray and next-generation sequencing). Finally, Chapter 5-3 on "Methods for Analyzing miRNA Expression," by Bollati and Dioni, describes methods for selective purification and analysis of small RNA, with an emphasis on miRNA, from readily accessible biological matrices, tissue samples, and cell culture.

HISTONE MODIFICATIONS AND CHROMATIN STRUCTURE

Role of Histone Acetylation and Acetyltransferases in Gene Regulation

Christina Y. Lee, Patrick A. Grant

Department of Biochemistry and Molecular Genetics, University of Virginia Medical School, Charlottesville, VA, United States

O U T L I N E

INTRODUCTION

History and Overview

Within the eukaryotic nucleus, DNA is compacted 10,000–20,000-fold, in part by being wound around octamers of core histone proteins that form nucleosomes. Nucleosomes each consist of 147 bp of DNA wrapped approximately twice around the protein core that contains two copies of each histone, H2A, H2B, H3, and H4 (Luger et al., 1997). The interaction of DNA and histones is critical to regulating transcription, replication, and repair of the genome and is influenced by various chromatin-modifying enzymes. Over 100 distinct histone modifications have been discovered (Zhao and Garcia, 2015). Histone proteins, the sites of posttranslational modification and modifying enzymes, are highly conserved across species from yeast to humans. This level of conservation has allowed great insights into the functions of histone modifications being garnered from model organisms such as the budding yeast or fruit flies. The N-terminal tails of the eight core histones are exposed to the nucleosome surface, allowing them to be modified by various mechanisms such as phosphorylation, methylation, ubiquitination, and acetylation (see review, Suganuma and Workman, 2011). Histone acetylation is associated with a variety of functions including regulation of nucleosome assembly, folding and decondensation of chromatin, heterochromatin silencing, and gene transcription (Fig. 1). Histone acetylation is conducted by histone acetyltransferases (HATs) and deacetylation by histone deacetylases (HDACs). In addition to the direct structural changes that lead to reorganization of chromatin as a consequence of histone acetylation,

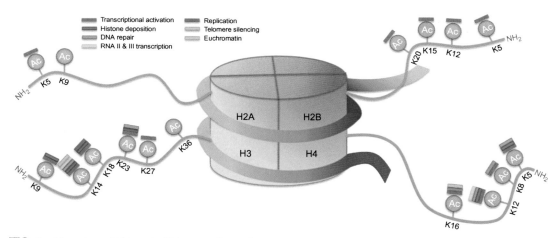

FIG. 1 Overview of N-terminal histone tail acetylations in mammals. The known functionality of each acetylation is color coded above the acetylation mark. The majority of marks are associated with transcriptional activation, and other functions include histone deposition for chromatin remodeling, DNA repair, replication, telomere silencing, and modulating euchromatin structure for chromatin folding. Histone N-terminal tails are in *green*; DNA wrapping around the nucleosome is in *blue*. In general, the overall acetylation level of histone proteins is more important than acetylation at specific lysines as there is functional redundancy in transcription (Shia et al., 2006). Key acetylations not only are noted on the histone tails but also can appear in the globular domain such as H3K122ac and H364ac, which mark gene enhancers (Pradeepa et al., 2016).

the histone modifications are also involved in the recruitment of specific "reader" proteins with binding affinity for specific marks. Acetylated lysines are read primarily by bromodomains, discussed in more detail below.

Although protein phosphorylation and acetylation were discovered within 4 years of each other in the 1960s, the understanding of histone acetylation continued to be relatively unknown until the 1990s when Allis and coworkers purified a HAT from *Tetrahymena thermophila*. This led to a clear link between histone acetylation and transcription regulation when it was revealed to be an enzyme orthologous to a yeast transcription regulator, Gcn5 (Brownell et al., 1996). Concurrently, Schreiber and colleagues discovered a human HDAC orthologous to a yeast transcription regulator, Rpd3 (Taunton et al., 1996). These discoveries were followed by a flurry of identifications of additional HATs associated with transcriptional regulation like CREB-binding protein (CBP), E1a-binding protein p300, TAF(II)250/ TAF1 subunit of transcription factor IID, and members of the MYST family (Table 1; Ogryzko et al., 1996; Borrow et al., 1996; Smith et al., 1998; Mizzen et al., 1996). Prior to these discoveries, Vincent Allfrey was the leading acetylation pioneer who from his research proposed that there must be a "dynamic and reversible mechanism for activation and repression of RNA synthesis" via histone acetylation (Allfrey et al., 1964).

A major hurdle in the transcription of genes and replication of the genome is the constitutively condensed chromatin structure referred to as heterochromatin. Nucleosomes are about 11 nm in diameter and form a configuration commonly referred to as "beads on a string." The interactions between adjacent nucleosomes create denser structures of 30 nm fibers (Schalch et al., 2005). This packaging of DNA alone reduces basal transcription levels, which hinders the access of polymerases and transcription factors. Acetylation neutralizes the positive charge of lysine residues, weakening the charge-dependent interactions between DNA and histones allowing accessibility to transcriptional machinery. Acetyl-lysines also serve to recruit acetyl-binding proteins, which are frequently transcriptional regulators (Marmorstein and Zhou, 2014). The chromatin dynamics that allow for the transcription and replication are heavily dependent on histone acetylation, including nucleosome assembly, chromatin structure and folding, and DNA damage repair.

TABLE 1 Overview of Histone Tail Acetylation Marks and Known Enzymatic Modifiers

H2A		H2B		H3		H4	
Mark	*Modifiers*	*Mark*	*Modifiers*	*Mark*	*Modifiers*	*Mark*	*Modifiers*
Lys5	Tip60, p300/CBP	Lys5	p300, ATF2	Lys9	Gcn5, SRC-1, unknown	Lys5	Hat1, Esa1, Tip60, ATF2, Hpa2, p300
Lys9		Lys12	p300/CBP, ATF2	Lys14	Unknown, Gcn5, PCAF, Esa1, Tip60, SRC-1, Elp3, Hpa2, hTFIIIC90, TAF1, Sas2, Sas3	Lys8	Gcn5, PCAF, Esa1, Tip60, ATF2, Elp3, p300
		Lys15	p300/CBP, ATF2	Lys18	Gcn5, p300/CBP	Lys12	Hat1, Esa1, Tip60, Hpa2, p300
		Lys20	P300	Lys23	Unknown, Gcn5, Sas3, p300/CBP	Lys16	Gcn5, Esa1, Tip60, ATF2, Sas2
				Lys27	Gcn5		

Nucleosome Assembly

Proper nucleosome assembly is key for efficient replication and particularly at sites of highly transcribed genes, as nucleosomes are evicted in these processes. During S phase, the entire genomic DNA is replicated as well as the underlying chromatin structure prior to cell division (Lucchini and Sogo, 1995). The nucleosomes that are in place preceding DNA replication can contribute to the reestablishment of chromatin on both the original and newly synthesized DNAs; however, new histone synthesis is also required (Sogo et al., 1986). The newly synthesized histones are acetylated transiently; once deposited on DNA, they are rapidly deacetylated suggesting the marks are critical to assembly, but not further structural stability. The ordered process of histone assembly requires histone chaperones (Verreault, 2000). The acetylation marks are important for chaperone recognition of histones and deposition onto replicated DNA. The chaperone, CAF-1, mediates the assembly of the H3/H4 tetramer and deposits them solely on replicated DNA and not mature chromatin, most likely due to different modification patterns (Smith and Stillman, 1991). Not only do histone tail domain acetylations play a key role in nucleosome assembly, but also sites within the globular domain may be important. For example, the histone H3K56 acetyl modification peaks during S phase and is deacetylated after histone deposition (Masumoto et al., 2005). This modification is recognized by Asf1, which works synergistically with CAF-1 (Recht et al., 2006). Additionally, there is also an acetylation that is important to the interaction of the H3/H4 tetramer, H4K91. Replication-independent roles of histone chaperones are still relatively unclear, as the histones that are released during transcription of active genes are not necessarily used to reassemble chromatin structure (see review, Burgess and Zhang, 2010).

Chromatin Folding

The degree of chromatin folding is significantly regulated by histone acetylation. Neutralization of positive charges on the lysine residues by acetylation disrupts the electrostatic interactions between histones and the phosphate groups in DNA, leading to a looser configuration. For example, hyperacetylation can prevent chromatin from compacting into a 30 nm fiber (Tse et al., 1998; Annunziato et al., 1988). The histone H4 tail is the most influential site of acetylation for chromatin folding. The structure includes a stretch of basic residues that form hydrogen bonds and salt bridges with the acidic region of the H2A-H2B dimer of the next nucleosome. Histone H4 residues 14–19 must be deacetylated for chromatin folding to occur (Dorigo et al., 2003). Moreover, H4K16 acetylation in vitro reduces the ability of nucleosome arrays to self-associate in a manner that characterizes higher-order chromatin structures (Shogren-Knaak et al., 2006).

Gene Expression

The acetylation of histones within nucleosomes is generally associated with the generation of an "open" and transcriptionally permissive chromatin structure (Fig. 2). Indeed, a number of acetyltransferases were initially identified as transcriptional coactivators prior to the discovery of their enzymatic activity. Likewise, a number of HATs were subsequently found to function as activators of gene expression (Torok and Grant, 2004). The opening of

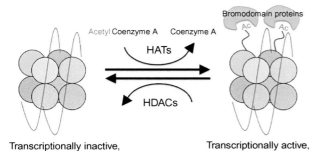

Acetyl Coenzyme A   Coenzyme A

HATs

HDACs

Bromodomain proteins

Transcriptionally inactive,
"closed" chromatin

Transcriptionally active,
"open" chromatin

FIG. 2 Regulation of histone acetylation by HATs and HDACs. HATs catalyze the acetylation (Ac) of specific lysine amino acids within histones by transferring an acetyl moiety from the cofactor acetyl coenzyme A. This leads to a more open and transcriptionally active chromatin conformation and provides binding sites for bromodomain-containing proteins. HDACs catalyze the reverse reaction and are generally associated with gene repression.

chromatin through acetylation allows the transcription machinery to more effectively access DNA leading to increased gene expression. The action of HDAC enzymes reverses this reaction to effectively silence gene transcription. A second mechanism by which histone acetylation regulates gene expression is through the recruitment of acetyl-binding proteins. Most frequently, proteins containing one or more bromodomains, discussed below, display specificity for binding to specific acetylated peptide sequences. In this manner, acetyl marks provide sites for the recruitment of transcription regulatory factors that increase the rate of transcription.

DNA Damage Repair

Acetylation is also key in DNA damage repair, particularly in double-stranded breaks. There are two major pathways for fixing DSBs, which include nonhomologous end joining (NHEJ) and homologous recombination. Histones are acetylated at DSB sites and are critical for repair, in particular H3 and H4 recruit proteins involved in DSB repair such as chromatin-remodeling complexes of the SWI2/SNF2 superfamily. The chromatin-remodeling complexes allow for the "loosening" of chromatin, so DNA repair proteins can access the point of damage. In particular, the HATs CBP and p300 acetylate histones H3 and H4 at DSB sites to recruit SWI/SNF and NHEJ factors (Ogiwara et al., 2011). Acetylation on histone H3K56 drives reassembly of chromatin after DSB repair and is a signal of repair completion and coordinates the function of H3/H4 chaperones in nucleosome assembly (Li and Heyer, 2008). Other modifiers of histone acetylation involved in efficient NHEJ include the HAT Esa1 and the HDAC Rpd3 with its binding partner Sin3 (Bird et al., 2002; Jazayeri et al., 2004). The requirement for both acetylation and deacetylation during this process could function to relax chromatin to allow access of repair proteins and stabilize chromatin structure around the break for the rejoining of DNA ends, respectively (Jazayeri et al., 2004). HATs also aid in nucleotide excision repair (NER). For example, the HAT p300 induces a transcription-independent chromatin-remodeling process and also interacts with p21 and PCNA to shift from transcription to DNA repair at sites of DNA damage (Cazzalini et al., 2008; de Boer and Hoeijmakers, 2000; see review, Reed, 2011). Altogether, understanding the epigenetic mechanisms of DNA damage repair can aid in discovering possible methods of radiotherapy and chemotherapy sensitization since NHEJ is preferentially employed when

DSBs are caused by ionizing radiation and when there is treatment with anticancer drugs like etoposide (topoisomerase II inhibitor), making the suppression of this pathway of interest in sensitizing cancer cells to these types of drugs (Adachi et al., 2003).

Toxicoepigenetic Relevance

Histone acetylation plays a diverse role in normal cellular functions and a range of disease etiologies. Epigenetic changes can be caused by environmental factors and are able to persist even in the absence of the establishing factors (Anway et al., 2006; Dolinoy, 2008). These factors can include toxicants such as pesticides, herbicides, metals, plastics, resins, and addictive drugs such as opiates. DNA damage and mutations are typically a major landmark in determining the toxicity and risk due to exposure; however, there is increasing evidence that the effects of toxic exposures are mediated by epigenetic changes from DNA methylation, histone methylation/acetylation, and microRNA expression. Histone acetylation also influences memory formation, which is associated with an increase in acetylation within cells in different areas of the brain leading to distinct epigenetic signatures that are thought to influence brain function (Levenson et al., 2004). HDACs also contribute to manipulate memory traces that form long-term memory in the brain (Haggarty and Tsai, 2011). Additionally, aging leads to a decrease of H4K12ac that leads to disruption of memory-associated behavior in mice due to loss of expression of several memory-related genes (Peleg et al., 2010). The imbalance of histone acetylation/deacetylation or misregulation of the enzymatic complexes that control these modifications are thus associated with several neurodevelopmental, neuropsychiatric, and neurodegenerative disease including Rubinstein-Taybi syndrome (RTS), Alzheimer's disease (AD), spinocerebellar Ataxia type 7 (SCA7), and Parkinson's disease (Peleg et al., 2010; Baker and Grant, 2007).

HISTONE ACETYLTRANSFERASES

Families and Structures

Several families of enzymes that modify histones have been identified, the majority of which are conserved throughout evolution. A common nomenclature was introduced in 2007 to group these enzymes based on their enzymatic activities, sequence and domain structure/organization, and sequence homology or substrate specificity of the enzyme catalytic domain (Allis et al., 2007). As a consequence, HAT enzymes are also referred to as lysine acetyltransferases (KATs; Table 2). HATs can be grouped into at least five different subfamilies based on sequence divergence within the catalytic HAT domain. These major HAT subfamilies include HAT1 and Gcn5/PCAF of the GNAT family, MYST, p300/CBP, and Rtt109. MYST is named for its founding members, MOZ, Ybf2/Sas3, Sas2, and TIP60. All of these families have alternative nomenclature that can be located in Table 2. Rtt109 and p300/CBP are the only two subfamilies that do not have yeast to human homologues; they are fungal- and metazoan-specific, respectively. X-ray crystallography on each of the five subfamilies reveals the molecular characteristics of their enzymatic domains and insight into the catalysis and substrate acetylation (Neuwald and Landsman, 1997). Despite extremely

TABLE 2 Summary of HAT Families and Properties

HAT Family	HATs	Key Structural and Biochemical Properties
GNAT	Gcn5/KAT2A	Ternary complex catalytic mechanism, amino- and carboxy-terminal used for histone substrate binding
	PCAF/KAT2B	
	Hat1/KAT1	
	Elp3/KAT9	
	Hpa2/KAT10	
MYST	Esa1/yKAT5	Ping-pong catalytic mechanism
	MOF/KAT8/MYST1	Requires autoacetylation at a particular lysine in the active site for cognate histone acetylation
	Sas2/KAT8	
	Sas3/KAT6	
	MORF/MYST4/ KAT6B	
	TIP60/hKAT5	
	HBO11/MYST2/ KAT7	
p300/CBP (metazoan-specific)	P300/KAT3B	Hit-and-run mechanism, contains substrate-binding loop that participates in AcCoA and lysine binding, contains autoacetylation loop that is required for maximal activation
	CBP/KAT3A	
Rtt109 (fungal)	yRtt109/KAT11	Structural homology to p300, shares all key structural and biochemical properties except catalytic mechanism

limited sequence conservation, each of the protein families contain a structurally conserved core region that is flanked by alpha/beta segments that are structurally different between the different HAT families—together that form a cleft for substrate binding and catalysis. Notably, p300/CBP and yRtt109 show structural homology in the flanking regions despite the lack of sequence conservation (see review, Yuan and Marmorstein, 2013). Remarkably, each HAT subfamily has a unique catalytic strategy to acetyl transfer (Table 2), most likely due to the fact that transfer of an acetyl group from a thioester to an amine is not a thermodynamically challenging reaction allowing many pathways for catalysis.

Regulation of HATs

The activity and specificity of histone acetyltransferases can be modulated by a variety of mechanisms: protein-protein interactions, protein cofactors, and autoacetylation. Multiprotein complexes influence catalytic activity and specificity of the HAT domain. Gcn5/PCAF is exclusively found in multiprotein complexes in vivo and exhibits different behavior than their recombinant counterparts, which are active on free histones and histone peptides and much less active on nucleosomes than when they are in complexes such as SAGA/SLIK and TFTC/STAGA

TABLE 3 Summary of HAT, HAT Complexes, and Substrates

	HAT	HAT Complex	Histone Substrates (Recombinant HAT)	Histone Substrates (HAT Complex)
GNAT family	Gcn5	SAGA (*Sc*)	H3, H2B	H2B/H3/H4
		SLIK (*Sc*)		H2B/H3/H4
		ADA (*Sc*)		H3
	GCNL	STAGA (*Hs*)		H3/H4
		TFTC (*Hs*)		H3/H4
		PCAF (*Hs*)	H3 ≫ H4	H3/H4
	Elp3	Elongator (*Sc*)	H2A, H2B, H3, H4	H3
	Hat1	HATB (*Sc*)	H4 ≫ H2A	H2A/H4
MYST family	Esa1	NuA4 (*Sc*)	H4 ≫ H3, H2A	H2A/H4
	TIP60	TIP60 (*Dm/Hs*)	H4 ≫ H3, H2A	H2A/H4
	Sas3	NuA3 (*Sc*)	H3, H4 ≫ H2A	H3
	HBO1	HBO1 (*Hs*)		H3/H4
	MOZ/MORF	MOZ/MORF (*Hs*)	H4 >H3	H3
	MOF	MSL (*Dm*)	H4 ≫ H3, H2A	H4

Sc, Saccharomyces cerevisiae; Hs, Homo sapiens; Dm, Drosophila melanogaster.

(Carrozza et al., 2003; Lee and Workman, 2007; Table 3). Incorporation of Gcn5/PCAF into these complexes both facilitates nucleosomal acetylation and conveys target lysine substrate specificity. MYST family HATs are also often assembled in multiprotein complexes such as yEsa1 in NuA4 and piccolo/NuA4 for chromatin acetylation (Sapountzi and Côté, 2011). Different subunits within each complex contribute to specificity, as the GNAT family SAGA complex preferentially modifies H3K9, but not H3K14, and the MYST family NuA3 complex preferentially modifies H3K14. Furthermore, even within the same family with overlapping histone substrate specificity, the complexes can be specific for different regions such as promoters (SAGA) or gene-coding regions (Elongator complex).

Surprisingly, in many instances, neither the binding partners of HATs nor other domains within the enzyme proteins themselves cause conformational changes that alter activity. Regulation appears to employ autoinhibition, substrate delivery, and localization. HAT proteins can also be regulated by other HAT-associated domains, their binding partners, and autoacetylation. Binding partners include the dynamics between Hat1 and Hat2, where Hat2 increases the substrate-binding specificity of Hat1 and therefore increases the catalytic efficiency for H4K12ac (Li et al., 2014). Furthermore, Esa1 can use different catalytic mechanisms depending on context and its binding partners. Additionally, sometimes, protein cofactors are necessary for full or increased activity of HATs. Rtt109, which acetylates the internal K56 residue of newly synthesized histone H3 prior to incorporation onto DNA, also requires at least one of the two binding partners for full activity, either Asf1 or Vps75. Vps75 delivers substrate to the reactive core of Rtt109 independent of Rtt109 activation by Asf1 (Tsubota et al., 2007; Han et al., 2007). Another example of dependence on cofactor binding is the MYST HAT, Sas2. Sas2 requires the binding of Sas4 and Sas5 for catalytic activity (Sutton et al., 2003). Furthermore, yeast Hat2 and Hif1 increase Hat1 activity by 10-fold when they are constituents in the NuB4 complex (Parthun et al., 1996).

Autoacetylation provides another mechanism of regulation for acetyltransferases such as p300, the MYST family, Rtt109, and PCAF. The p300 family has a unique structure that contains a bromodomain, a discontinuous PHD domain, and a RING domain that form a compact module. The RING domain performs regulatory function, inhibiting p300 HAT activity (Delvecchio et al., 2013). In p300, acetylation of its autoinhibitory loop exposes the substrate-binding region of the HAT activating the enzyme (Thompson et al., 2004). Autoacetylation of Rtt109 K290 increases the affinity for AcCoA and increases the rate of catalytic turnover (Albaugh et al., 2011). Autoacetylation also activates MOF (K274) and PCAF (K416, 428, 430, 431, and 442) that increases catalytic activity (Sun et al., 2011; Santos-Rosa et al., 2003). Additionally, there are complexes that can regulate HAT activity by "masking" histones. For example, inhibitor of acetyltransferases (INHAT) binds to histones and inhibits p300/CBP and PCAF acetylation (Seo et al., 2002).

TRANSCRIPTIONAL ACTIVATION

HAT Complexes

Below, we discuss select HAT complexes that are prevalent in transcriptional activation and serve as archetypal examples of regulators of aspects of gene activation. Notably, most HATs identified to date are not catalytically active in vivo in the absence of other proteins. These complexes often consist of HATs that associate with other HATs or histone-modifying enzymes, as well as coactivators, introducing more complexity in elucidating specific functions of each subunit in the context of their complex. The complexes also can have more than one chromatin-binding domain, as observed in yeast, especially Spt-Ada-Gcn5 acetyltransferase (SAGA), which contains bromo, chromo, Tudor, SANT, SWIRM, WD40, and PHD domains. The presence of such a broad range of chromatin-binding domains within a complex is not unique to SAGA, as it is also observed in yeasts NuA4 and NuA3 (Lee and Workman, 2007). The GNAT family-containing complexes of ADA (Gcn5), SAGA (Gcn5), PCAF, and Elongator (Elp3) and MYST family-containing complexes of NuA3 (Sas3), NuA4 (Esa1), and MSL (MOF) are typical examples of HAT complexes; however, a more extensive list can be found in Table 3 and reviewed in Lee and Workman (2007).

SAGA Transcription Regulatory Complex

SAGA and the highly related SLIK (SAGA-like) complexes are roughly 1.8 MDa complexes that modify chromatin with the ability to acetylate and deubiquitinate both histone and nonhistone substrates (Grant et al., 1997; Pray-Grant et al., 2002). SAGA is conserved between yeast and humans as well. It is modular in structure and has many distinct functional units such as the recruitment module (Tra1), the acetylation module (Gcn5, Ada2, Ada3, and Sgf29), a TBP interaction unit (Spt3 and Spt8), and a deubiquitination module (Ubp8, Sus1, Sgf11, and Sgf73), as well as an architecture unit (Baker and Grant, 2007). The complex aids in transcription and is recruited to gene loci via specific transcriptional activators interacting with Tra1 and the bromodomain of Gcn5 if histone H3 or H4 is acetylated, positively reinforcing further acetylation by Gcn5. In humans, hSAGA is recruited by activators such as c-Myc and E2F to target specific genes for activation of transcription (McMahon et al., 2000; Lang et al., 2001). Additionally, the Spt3 subunit recruits TBP to aid in the preinitiation complex (PIC) formation and transcriptional activation. Elongation is also aided by SAGA's deubiquitination module that then allows for Ctk1 kinase and further Ser2 phosphorylation of the RNA polymerase II (RNAP II) C-terminal repeat domain (CTD) (for more details about SAGA, see review, Koutelou et al., 2010; Baker and Grant, 2007). SAGA functions as a transcriptional coactivator at inducible genes and is required for RNAP II transcription (Bonnet et al., 2014). SAGA targets a different subset of genes compared with TFIID, and in yeast, the genes are mainly stress-induced genes (Basehoar et al., 2004). However, genome-wide studies analyzing H3K9ac and H2Bub densities revealed that SAGA modifies the promoter and transcribed region of all expressed genes in both yeast and human cells (Bonnet et al., 2014). The difference in expression mechanisms between TFIID/TATA-like promoters associated with housekeeping/constitutively active genes and SAGA/TATA-box promoters associated with environmental responsive expression is still relatively unclear. SAGA and TFIID promoters respond differently to the presence of activators, with SAGA being more responsive than TFIID (de Jonge et al., 2017). This may be explained in part by the fact that the SAGA/TATA-box-dominated promoters are susceptible to Mot1 eviction of TBP, which increases the turnover of TBP and the disruption of the PIC decreasing transcription (de Jonge et al., 2017). SAGA is implicated in a polyglutamine expansion of SAGA-associated protein ataxia 7 that leads to the dominant disorder SCA7, a progressive neurodegenerative disease of the cerebellum and retina (McCullough and Grant, 2010). These expansions lead to the sequestration of the deubiquitinase activity and alteration of HAT activity leading to aberrant regulation of transcription (Lan et al., 2015; Burke et al., 2013; McCullough et al., 2012).

NuA4 Transcription Regulatory Complex

Nucleosomal acetyltransferase of H4 (NuA4) is a 1.3 MDa complex consisting of 13 subunits that primarily modifies not only histone H4 but also H2A to a lesser extent via the HAT subunit, Esa1 (Allard et al., 1999). Esa1 is also present in a smaller complex known as piccolo NuA4 (picNuA4). The other subunits of NuA4 provide the recruitment module and allow NuA4 to participate in numerous roles such as DNA repair, transcription initiation, and elongation. These subunits include Tra1, Epl1, Arp4, Yaf9, Act1, Eaf1, and Eaf9. NuA4 often can work in tandem with other complexes such as SAGA and can be corecruited to promoter regions via the Tra1 subunit, and additionally, both can be recruited to the phosphorylated CTD of elongating RNAP II. Additionally, it can facilitate binding of chromatin remodelers such as RSC and SWI/SNF. Subunits of NuA4 contain both chromodomains (CHDs) and PHDs. Esa1 contains a CHD, and Yng2 contains a PHD, which binds trimethylated H3K4 at the sites of actively transcribed genes in vitro (Shi et al., 2007). H3 methylation can stimulate NuA4 interaction that leads to NuA4 acetylation of H4 that subsequently leads to SAGA acetylation of histone H3 (Ginsburg et al., 2014). NuA4 can facilitate transcriptional initiation of TFIID and is targeted to promoters of ribosomal protein genes and is dependent on its subunit, Eaf1. In the absence of Eaf1 or NuA4, SAGA becomes involved in targeting recruitment of TBP to promoters of ribosomal protein genes (Ginsburg et al., 2014). Homologues of the yeast NuA4 complex include TIP60, a complex of at least 16 subunits, and are a key regulator of cell homeostasis, stress response, stem cell renewal, DNA repair, and transcriptional activation (Avvakumov and Côté, 2007).

Elongator Complex

Elp3 is the catalytic subunit of the Elongator complex. It makes up one of the six subunits of the complex that is composed of the core Elongator components Elp1, Elp2, and Elp3 and the smaller subunits of Elp4, Elp5, and Elp6. The elongating and hyperphosphorylated RNAP II in yeast is associated with the Elongator complex (Otero et al., 1999). Elp3 acetylates histone H3 at K14 and histone H4 at K8 to churn the chromatin in front of Pol III to facilitate polymerase movement along the body of the gene (Shilatifard et al., 2003; Pokholok et al., 2002). The function of Elongator ranges from organism to organism, but it generally contributes to transcription of inducible genes rather than global transcription (Chen et al., 2006; Nelissen et al., 2010). In the context of chromatin, the Elongator complex is similar to the FACT complex, which aids procession of RNAP II through nucleosomes. Altogether, the Elongator complex acetylates histone H3 in the transcriptional start site (TSS) distal gene body of stress-inducible genes (Creppe and Buschbeck, 2011).

Chromatin Remodeling Complexes

Nucleosome-depleted regions (NDRs) often contain promoter elements and upstream activator binding sites (UAS elements) and are flanked by the −1 and +1 positioned nucleosome (Cui et al., 2012). NDRs are modulated by a variety of factors in combination of poly(dA/dT) tracts, bindings sites for general transcription factors and recruitment of chromatin-remodeling complexes by gene-specific activators bound to UAS elements. Acetylation of promoters by HAT transcriptional coactivator complexes such as SAGA and NuA4 recruits chromatin-remodeling complexes that "loosen" chromatin structure to facilitate transcription. These complexes use ATP hydrolysis to alter the contacts between DNA and histones, facilitating the movement and even eviction of histones/nucleosomes. SWI/SNF can bind chromatin acetylated by SAGA or NuA4 via the bromodomain of Snf2, aiding in transcriptional activation. The SWI/SNF complex is a 2MDa multisubunit DNA-dependent ATPase that can bind DNA and nucleosomes with high affinity. SWI/SNF remains associated with RNAP II during elongation (Schwabish and Struhl, 2007). Gcn4 recruits SAGA and SWI/SNF to highly induced promoters in order to enhance nucleosome eviction (Sanz et al., 2016; Qiu et al., 2016).

The RSC complex is another chromatin-remodeling complex that shares two identical subunits with SWI/SNF and also travels with RNAP II during transcription. Overexpression of transcription activators can compensate for the loss of SWI/SNF but not for the loss of

SAGA in glucose-dependent genes (Biddick et al., 2008). However, some promoters require SWI/SNF for histone eviction, such as the *SUC2* promoter during induction (Schwabish and Struhl, 2007).

GLOBAL HISTONE ACETYLATION

Global histone acetylation refers to acetylation throughout the genome that is independent of recruitment by transcriptional activators. In yeast, bulk acetylation levels can be as high as 13 lysines per octamer (Waterborg, 2000). Examples of global histone modifiers include Gcn5 and Esa1 in yeast, which acetylate adjacent nucleosomes that include coding and intragenic gene regions (Vogelauer et al., 2000). This rivals local, targeted histone acetylation that is observed at the sites of specific promoter and enhancer elements. Characteristic broad acetylation patterns of global histone acetylation can be observed at the beta-globin loci and the male X chromosome in *Drosophila*. Dosage compensation leads to increased transcriptional activity on many genes throughout the X chromosomes via H4K16ac by MOF (Bone et al., 1994). Global histone acetylation provides a baseline for chromatin structure and gene expression; alterations to the balance between acetylation and deacetylation likewise affect the transcriptome globally. Acetyl-CoA carboxylase additionally regulates global histone acetylation, and thus, histone acetylation competes with metabolic processes such as fatty acid biosynthesis (Galdieri and Vancura, 2012).

ROLE OF HISTONE ACETYLTRANSFERASES IN GENE ACTIVATION

Recruitment of Transcriptional Machinery

Histone acetylation provides a substrate for the binding of effector proteins via bromodomains and less commonly by other protein modules such as PHD domains (Marmorstein and Zhou, 2014). These "reader" proteins can range from chromatin-remodeling complexes SWI/SNF and RSC (see review, Becker and Workman, 2013) to transcription factors or even HAT complexes like SAGA. Complexes such as SAGA and Esa1 are recruited to promoters via their Tra1 subunit (TRAAP in humans). Bromodomain-extraterminal (BET) family proteins localize to acetylated promoters and are able to recruit specific and general transcription factors, some of which even can contribute to chromatin remodeling. In yeast, there are two BET bromodomain proteins, Bdf1 and Bdf2, which preferentially acetylate complimentary histones, H3 and H4 versus H2B and H3, respectively (Matangkasombut et al., 2000; Matangkasombut and Buratowski, 2003). Bdf1 and Bdf2 associate with the general transcription factor complex TFIID. The mammalian Brd2 BET protein is important for cell-cycle gene expression and is responsible for recruiting the general transcription factor TATA-binding protein (TBP) to the E2F complex (Peng et al., 2007). Brd4 is integral to transcription through its interaction with some forms of the mediator coactivator complex that are necessary for the transcription of various genes (Jiang et al., 1998; Houzelstein et al., 2002). Additionally, due to the association of bromodomains with disease and their ability to read acetyl marks, several selective inhibitors of BrD have been developed for the treatment of cancers, HIV, and bacterial sepsis (Marmorstein and Zhou, 2014).

Active Genes

Active genes are typically enriched with acetylated histones H3 and H4 on gene bodies and regulatory elements. In particular, H3K4ac demonstrates a genome-wide localization pattern at the promoters of active genes, as well as H3K14ac and H3K18ac (Guillemette et al., 2011). In yeast, global acetylation is required for increased levels of basal transcription (Vogelauer et al., 2000). For highly active genes, the transcription start sites are often marked by competing H4K5 and K8 acetylation and butyrylation (Goudarzi et al., 2016). Together, H3K4me3 and H3K9ac colocalize on active gene promoters as well. The functional role of H3K9ac is not fully understood, but it is believed to promote the release of paused RNAP II by recruiting super elongation complex (SEC) to chromatin (Gates et al., 2017). The dependence of gene expression on HAT activity varies by gene. Some genes (e.g., *HIS3*) require acetylation for transcription, while the expression of others (e.g., *PHO5*) is only delayed in its absence (Kuo et al., 1998; Barbaric et al., 2001). The varying role of HATs in gene expression is most likely due to the resident chromatin structure, stability of the promoter, and overlapping contributions of other histone-modifying enzymes. A plethora of histone modifications mark active and repressed genes, with unique localization according to their association with gene promoters, enhancers, and gene bodies (for more, see review, Miller and Grant, 2013).

Inducible/Repressed Genes

Histone acetylation plays a role in activating repressed genes or inducing gene expression based on external factors. The NuA4 complex is implicated in priming activation of such genes like *PHO5*, which has served as a model for the mechanism of activation by acetylation, which in turn recruits SAGA via interactions with the transcription factor Pho4 (Fig. 3; Nourani et al., 2004). The induction of heat-shock-associated gene offers another example. Heat-shock factor I (HSFI) recruits multiple acetyltransferases GCN5, TIP60, and p300 to pericentric heterochromatin leading to hyperacetylation and subsequently recruitment of bromodomain and extraterminal (BET) proteins that are required for transcription by RNAP II for satellite III DNA sequences (Col et al., 2017). The BET family members recruited (i.e., BRD2, BRD3, and BRD4) are major components of transcription elongation factor b (P-TEFb) complex, which is required for the initiation of transcription elongation (Brès et al., 2008; Gaucher et al., 2012). Gene induction via acetylation extends to many other processes, from immune responses and developmental cues to metabolic stress. Acetylation at histones H3 and H4 at the HLA-DRA promoter can be induced by IFNγ and rapidly returns to baseline after IFNγ removal (Beresford and Boss, 2001). Hormone-dependent transcriptional activation also relies on histone acetylation, particularly for the progesterone receptor (PR). Acetyltransferase activity of SAGA and NuA4 is also important for glucocorticoid receptor (GR) activation in a glucocorticoid-dependent manner to modulate chromatin structure (Wallberg et al., 1999; Guo et al., 2017).

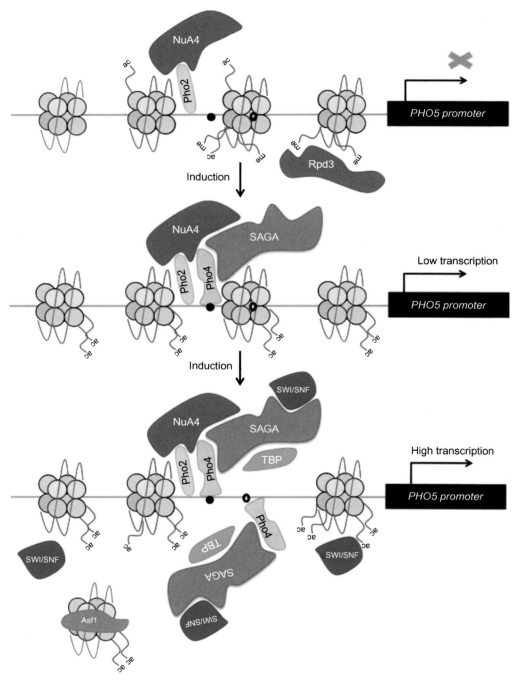

FIG. 3   Histone acetylation in the induction of *PHO5* promoter. The *PHO* genes in *S. cerevisiae* are regulated by the availability of phosphate. When inorganic phosphate is abundant, the *PHO* genes are repressed. When intracellular inorganic phosphate is low, induction of the *PHO5* gene is initiated when the *PHO* regulator and transactivator, Pho4,

ENVIRONMENTAL EXPOSURE

A wide variety of environmental conditions can influence histone acetylation and thus gene expression on both gene-specific and global scales. Notably, the influences of these toxicants can continue even when the toxicant is no longer present. These environmental conditions can range from toxic metals (e.g., nickel and arsenic) to pollutants (e.g., pesticides, BPA, and cigarette smoke) and even dietary conditions such as hyperglycemia (Cantone et al., 2011; Arita et al., 2012; Song et al., 2010; Kumar and Thakur, 2017; Chen et al., 2011; Ding et al., 2017; Pirola et al., 2011). The adverse health effects of nickel exposure, such as carcinogenicity and cardiorespiratory disease, have largely uncharacterized etiologies. Along with nickel-induced hypermethylation, several studies also demonstrate that exposure to $NiCl_2$ reduces histone acetylation, increases dimethylation H3K9, and increases monoubiquitination of H2A and H2B in vitro. Loss of acetylation of H2A, H2B, H3, and H4 is exhibited in human lung cells exposed to soluble nickel (Broday et al., 2000; Golebiowski and Kasprzak, 2005). Arsenic also influences epigenetic marks and has been shown to cause a global reduction of H3K9 acetylation in peripheral mononuclear cells of individuals exposed to arsenic in their drinking water (Arita et al., 2012; Arita and Costa, 2009). Evidence suggests that As_2O_3 directly binds HAT, hMOF (MYST1), reducing histone acetylation (Liu et al., 2015). Another example of environmental exposure is to pesticides/herbicides, which are linked to the pathogenesis of Parkinson's disease and Alzheimer's disease (Collotta et al., 2013). The herbicide paraquat, an organochlorine insecticide dieldrin, induces histone H3 and H4 core acetylation in a time-dependent manner and decreased total HDAC activity (Song et al., 2010). Hyperacetylation is attributed to dieldrin-induced proteasomal dysfunction, resulting in an accumulation of a key HAT, CREB-binding protein (CBP), and leads to caspase-3 activation and ultimately apoptosis of dopaminergic neuronal cells.

Bisphenol A (BPA), another common environmental toxicant that is used to create polycarbonate plastics and epoxy resins, is found in plastic-containing materials ranging from water bottles to food packaging. Perinatal long-term exposure to BPA in mice causes increased acetylation of histone H3K9 and H3K14 in both the cerebral cortex and hippocampus (Kumar and Thakur, 2017; Gao et al., 2016). The alterations of epigenetic marks persist even after the withdrawal of exposure to BPA. The use of both pharmaceutical and illicit drugs also influences histone acetylation. Cocaine-induced alterations in neural plasticity are linked to changes in acetylation at the promoters of genes predicted to have an underlying role in addiction-like behaviors. Nicotine also induces robust acetylation of histones H3 and H4 in the striatum, and this is believed to contribute to nicotine's role as a gateway drug, amplifying the effects of other drugs such as cocaine (Kandel and Kandel, 2014). Furthermore, the condition of

---

translocate into the nucleus. Pho2 binds near the constitutively accessible UASp1 *(closed circle)* and is critical in recruiting Pho4 during activation and additionally constitutively recruits NuA4. NuA4 acetylates histone H4 (Nourani et al., 2004). Pho4 recruits SAGA and SWI/SNF complex, and SAGA hyperacetylates other promoter nucleosome histones on histone H3. Acetylated histone H3 and H4 recruit various other modules, including chromatin remodelers binding bromodomain proteins like Bdf1. The chromatin remodelers disrupt nucleosome stability and reveal other UAS sites *(open circle)* for Pho4 to bind and further recruit SAGA, SWI/SNF, and TBP (Chandy et al., 2006). When phosphate is abundant, Rpd3 is recruited to the repressed genes via histone methylations by Set1 at H3K4 that then inhibit turnover of histones.

hyperglycemia can lead to genome-wide H3K9/K14 hyperacetylation and unique CpG methylation signatures, significantly influencing vascular chromatin, and leads to the upregulation of genes involved in metabolic and cardiovascular disease (Pirola et al., 2011). Valproate is a known antiepileptic and mood-stabilizing drug that was believed to act on GABAergic neurons but had an unknown mechanism of action when first introduced. However, it was found that it is actually a histone deacetylase inhibitor (Phiel et al., 2001). Chronic treatment with fluoxetine (Prozac) has been reported to cause gene expression changes that are a result of altered DNA structure caused by chromatin remodeling. These changes include increased DNA methylation and reduced histone acetylation that accompany persistent desensitization of serotonin 5HT1A receptors in the brain, even after the removal of fluoxetine (Csoka and Szyf, 2009).

HISTONE DEACETYLASES

Families

HDACs are classified into two families and four classes (Table 4). The two families belong to either the histone deacetylase family or the Sir2 regulatory family. Human HDACs are split into families based on sequence similarities (Seto and Yoshida, 2014). Class I HDACs have similarity to yeast Rpd4 and include HDAC1, HDAC2, HDAC3, and HDAC8. Class II HDACs are similar to yeast Hda1 and include HDAC4–7, 9, and 10 (Yang and Grégoire, 2005). Both Class I and Class II are related to the yeast Hos proteins, as Rpd3 and Hda1 share significant similarity to Hos1, Hos2, and Hos3 (Gregoretti et al., 2004). The yeast Sir2 protein is similar to Class III human HDACs, which consist of SIRT1-7 and Class IV protein; HDAC11 shares sequence similarity with both Class I and Class II proteins. Class I, II, and IV HDACs are numbered according to order or discovery, and all belong to the arginase/deacetylase superfamily of proteins. Surprisingly, these HDACs predate the evolution of histone proteins, and thus, their primary substrate may not be histones. The Class III proteins of the sirtuins/Sir2 regulatory family belong to the deoxyhypusine synthase like NAD/FAD-binding

TABLE 4 Classification of HDACs

Family	Class	Protein (Yeast)	Subclass	Protein (Human)
Histone deacetylase	Class I	Rpd3, Hos1, Hos2, Hos3		HDAC1, HDAC2, HDAC3, HDAC8
	Class II	Hda1	Class IIa	HDAC4, HDAC5, HDAC7, HDAC9
			Class IIb	HDAC6, HDAC10
	Class IV			HDAC11
Sirtuin	Class III	Sir2, Hst1	I	SIRT1, SIRT2, SIRT3
		Hst2, Hst3	II	SIRT4
		Hst4	III	SIRT5
			IV	SIRT6, SIRT7

domain superfamily (Frye, 2000). Sirtuins have two enzymatic modules including mono-ADP-ribosyltransferase and histone deacetylase (see review, Seto and Yoshida, 2014).

Catalytic Mechanisms and Structures

Class I, II, and IV HDACs share a common catalytic mechanism that requires a zinc ion to facilitate the hydrolysis of the acetamide bond of the acetylated lysine. A study analyzing HDAC8 proposed that there is a key tyrosine and two key histidines in the active site. One of the histidine residues acts as a general base (H143), which accepts a proton from a zinc-bound water molecule, while another histidine (H142) acts as a general electrostatic catalyst that leads to cleavage of the amide bond (Wu et al., 2010). The exception to this mechanism are Class IIa enzymes (HDAC4, 5, 7, and 9), in which the tyrosine that is important for stabilizing the tetrahedral intermediate is replaced with a histidine residue and may contribute to their low catalytic activity (Lahm et al., 2007). Class III HDACs are dependent on NAD^+ as a cofactor for enzyme activity. The catalytic domain of sirtuins resides inside a cleft created by a large domain with a Rossmann fold and a small zinc-binding domain that creates a protein tunnel where the substrate interacts with NAD^+ (Finnin et al., 2001). The mechanism for catalysis involves a nucleophilic addition and release of nicotinamide as a by-product. Histone specificity for Class I, II, and IV is still ambiguous, as many HDACs serve redundant functions and can compensate for the loss of other HDACs. Additionally, similar to HAT enzymes, the substrate specificity of HDACs is also influenced by their interaction with other proteins in multiprotein complexes. In contrast, sirtuin substrate specificity is less ambiguous, where preferential substrates and specific target lysines have been relatively well characterized in general (Seto and Yoshida, 2014).

Regulation of HDAC Activity

HDACs are regulated from the transcriptional to posttranslational level. The most thoroughly characterized of these regulatory mechanisms occurs at the protein-protein level within multiprotein complexes. HDAC1 and HDAC2 exist in at least three distinct multiprotein complexes: Sin3, NuRD, and CoREST. Other protein-protein interactions that regulate HDAC activity can be observed by SMRT and nuclear receptor corepressor (NCoR) recruitment of HDACs. For Class III HDACs, SIRT1 negatively interacts with DBC1 but is enhanced by another nuclear protein, AROS (Kim et al., 2007). A variety of posttranslational modifications are observed on HDACs, but the regulation of their activity by phosphorylation is the most extensively studied. Phosphorylation can both activate and inhibit HDAC activity. HDAC1 phosphorylation promotes enzymatic activity and dictates the protein complex formation with Sin3 and Mi2. Additionally, HDAC3, which is phosphorylated by CK2 and DNA-PKcs, enhances enzymatic activity (Zhang et al., 2005). In contrast to other Class I HDACs, phosphorylation of HDAC8 downregulates enzymatic activity (Lee et al., 2004). Regulation of HDAC activity can also be influenced by subcellular localization, which is controlled by phosphorylation in Class IIa HDACs. Phosphorylation of Class IIa HDACs provides binding sites for 14-3-3 proteins, which mask the nuclear localization signal on HDAC4/5, leading to cytoplasmic sequestration (Grozinger and Schreiber, 2000). A variety of protein kinases regulate the phosphorylation binding sites of 14-3-3 proteins such as Ca^{2+}/calmodulin-dependent kinases, protein kinase D, microtubule affinity-regulating kinases, salt-inducible kinases, checkpoint kinase-1, and AMP-activated protein kinases (AMPK). Class III HDACs, such as SIRT1, are also phosphorylated, which leads to an increase in deacetylation rate and substrate-binding affinity (Kang et al., 2009).

Role of HDACs at Active Genes

The regulation of transcription does rely on not only elevated levels of acetylation but also rapid deacetylation. The fastest rates of acetylation turnover are associated with transcriptionally active regions (Waterborg, 2002). Additionally, turnover of acetylation is important to ensuring proper transcriptional initiation. Deletion of the yeast gene, *RPD3*, leads to a downregulation of greater than 250 genes despite deacetylation being associated with repression of genes (Bernstein et al., 2000). Furthermore, it is significant that more genes are downregulated rather than upregulated when *RPD3* is deleted. The turnover of acetylation at active genes most likely is mediated by the global balance of HAT and HDAC activity (Katan-Khaykovich and Struhl, 2002). Further, the specificity of HDACs may be fundamental to whether a deacetylation event activates or represses transcription. Specific combinations of hyperacetylated and hypoacetylated residues could be recognized by particular transcription factors, as these patterns correlate with specific expression profiles (Kurdistani et al., 2004).

Deacetylation and Gene Repression

HDACs are also recruited to genes by sequence-specific DNA-binding factors, with similar mechanisms as HATs. For example, Rpd3 deacetylase (yeast) is recruited to promoters of sporulation genes by the transcriptional regulator Ume6, and the histone deacetylase 1 (Hda1) can be recruited to genes by Tup1, a general repressor that is also targeted to specific genes. Deletion of either leads to hyperacetylation at the genes that are targeted by DNA-binding factors. Deacetylation by HDACs may promote the binding of repressor proteins through SANT domains, which are found in a number of corepressor proteins that are believed to recognize unmodified histone tails (de la Cruz et al., 2005). While there are multiple examples of SANT domain-containing proteins associating with HDACs and other repressors, the contribution of SANT domain-containing proteins is yet to fully be understood as SANT domains are also critical for the functions of the SAGA, RSC, and SWI/SNF complexes associated with transcriptional activation (Boyer et al., 2002). Outside of deacetylases ensuring proper transcription by preventing aberrant initiation within the coding region, they can be recruited to active genes for rapid repression. The transient induction of *c-fos* in response to growth factor stimulation of the MAPK pathway is reduced to basal levels within 2h by Elk-1 recruitment of HDAC-1 (Yang et al., 2001).

HDAC Complexes

The Sin3 and NuRD complexes both contain similar core components, sharing HDAC1, HDAC2, RbAp48, and RbAp46 (Fig. 4). The Sin3 complex uniquely contains Sin3 and

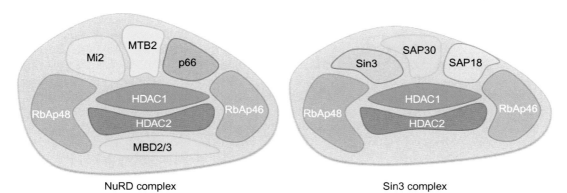

FIG. 4 Example of HDAC complexes: NuRD and Sin3 complex. The NuRD and Sin3 complex shares the core components of HDAC1, HDAC2, RbAp46, and RbAp48. RbAp46 is also known as Rbbp7, and RbAp48 is also known as Rbbp4. P66 (α and β) is also known as Gata2a and Gata2b. Mi-2 imparts chromatin-remodeling activity to the NuRD complex and contains both PHD and chromodomain binding domains and SWI/SNF2-type ATPase domain. In the Sin3 complex, there are two isoforms, Sin3a and Sin4b, which both recruit HDAC1 and HDAC2 (see review, McDonel et al., 2009). There are various other HDAC complexes such as but not limited to the CoREST complex and NCoR/SMRT.

SAP30 and is a gene-specific complex, while NuRD, which uniquely contains MBD3, MTA2, and Mi2, is both global- and gene-specific. Both the human and yeast Sin3 complexes have many proteins such as HDACs and sequence-specific DNA-binding proteins. Sin3 can directly interact with Ume6, a transcriptional repressor of meiotic genes (Kadosh and Struhl, 1997). Several proteins involved in transcriptional repression recruit Sin3, such as NoRC and MeCP2. NoRC is a nucleolar chromatin-remodeling complex that represses ribosomal gene transcription, and MeCP2 is a methyl-binding domain (MBD)-containing protein that represses promoters containing methylated CpG islands in DNA (Jones et al., 1998). The NCoR/SMRT corepressors also interact with the Sin3 complex both through Sin3 and Sap30, to repress transcription at receptor-targeted genes (Laherty et al., 1998). NuRD is recruited to specific genes by DNA-binding transcription factors and in a broader context to methylated DNA (Kehle et al., 1998; Zhang et al., 1999). NuRD can also be found in larger complexes, known as "Supra-NuRD" complexes that incorporate a large number of factors that impart various specific functions on the complex. Examples include the MeCP1 complex, cohesion complex, ALL-1 complex, and PYR (for a broader overview on HDAC complexes, see Verdin, 2006).

CONCLUSION AND PERSPECTIVES

Histone acetylation and deacetylation are critical to various cellular processes including nucleosome assembly, chromatin folding, DNA damage repair, and proper transcription. Furthermore, the regulation of histone acetylation and deacetylation plays a central role in disease etiology and adverse effects of chemical exposures. Many of these pathologies arise from misregulation of histone acetylation by HATs, but it also should be noted HATs are not only limited to histone substrates. Transcription factors are additional substrates for HAT acetylation, such as Smads, p53, and NF-κB (Tu and Luo, 2007; Gu and Roeder, 1997; Pejanovic et al., 2012). The acetylation of these transcription factors dictates DNA-binding ability, nuclear localization, protein stability, and interactions with other transcriptional regulators. The importance of histone acetylation and deacetylation in these various pathologies has led to a concerted focus on HATs, HDACs, and bromodomain-containing "reader" proteins as targets for therapeutic intervention. Currently, HDAC inhibitors are used to treat cancer, inflammation, and assorted neurological disorders (see review, Zwergel et al., 2016; Adcock, 2007; Didonna and Opal, 2015). Bromodomain proteins have also emerged as drug targets that can be inhibited with high specificity in human disorders such as HIV-1 and myocardial ischemia (Zeng et al., 2005; Mujtaba et al., 2006; Gerona-Navarro et al., 2011). Targeting HATs has been more difficult, as they are bisubstrate enzymes and more challenging to target due to the size, poor metabolic stability, and the lack of cell permeability of bisubstrate inhibitors (Wapenaar and Dekker, 2016). Small-molecule inhibitors of HATs also have largely exhibited poor target selectivity. Altogether, there is still much to learn on how histone acetylation/deacetylation modulates gene transcription, the response to environmental agents, and ultimately disease. A careful characterization of the HAT and HDAC enzymes will enable the continued development of potent and selective modulators of histone acetylation that will help to further advance our understanding of the mechanisms underlying chemical exposure effects and susceptibility and the development of novel therapeutic interventions.

References

Adachi, N., Suzuki, H., Iiizumi, S., Koyama, H., 2003. Hypersensitivity of nonhomologous DNA end-joining mutants to VP-16 and ICRF-193: implications for the repair of topoisomerase II-mediated DNA damage. J. Biol. Chem. 278, 35897–35902. https://doi.org/10.1074/jbc.M306500200.

Adcock, I.M., 2007. HDAC inhibitors as anti-inflammatory agents. Br. J. Pharmacol. 150, 829–831. https://doi.org/10.1038/sj.bjp.0707166.

Albaugh, B.N., Arnold, K.M., Lee, S., Denu, J.M., 2011. Autoacetylation of the histone acetyltransferase Rtt109. J. Biol. Chem. 286, 24694–24701. https://doi.org/10.1074/jbc.M111.251579.

Allard, S., Utley, R.T., Savard, J., Clarke, A., Grant, P., Brandl, C.J., Pillus, L., Workman, J.L., Côté, J., 1999. NuA4, an essential transcription adaptor/histone H4 acetyltransferase complex containing Esa1p and the ATM-related cofactor Tra1p. EMBO J. 18, 5108–5119. https://doi.org/10.1093/emboj/18.18.5108.

Allfrey, V.G., Faulkner, R., Mirsky, A.E., 1964. Acetylation and methylation of Histones and their possible role in the regulation of rna synthesis. Proc. Natl. Acad. Sci. U. S. A. 51, 786–794.

Allis, C.D., Berger, S.L., Cote, J., Dent, S., Jenuwien, T., Kouzarides, T., Pillus, L., Reinberg, D., Shi, Y., Shiekhattar, R., Shilatifard, A., Workman, J., Zhang, Y., 2007. New nomenclature for chromatin-modifying enzymes. Cell 131, 633–636. https://doi.org/10.1016/j.cell.2007.10.039.

Annunziato, A.T., Frado, L.L., Seale, R.L., Woodcock, C.L., 1988. Treatment with sodium butyrate inhibits the complete condensation of interphase chromatin. Chromosoma 96, 132–138.

Anway, M.D., Leathers, C., Skinner, M.K., 2006. Endocrine disruptor vinclozolin induced epigenetic transgenerational adult-onset disease. Endocrinology 147, 5515–5523. https://doi.org/10.1210/en.2006-0640.

Arita, A., Costa, M., 2009. Epigenetics in metal carcinogenesis: nickel, arsenic, chromium and cadmium. Metallomics 1, 222–228. https://doi.org/10.1039/b903049b.

Arita, A., Niu, J., Qu, Q., Zhao, N., Ruan, Y., Nadas, A., Chervona, Y., Wu, F., Sun, H., Hayes, R.B., Costa, M., 2012. Global levels of histone modifications in peripheral blood mononuclear cells of subjects with exposure to nickel. Environ. Health Perspect. 120, 198–203. https://doi.org/10.1289/ehp.1104140.

Avvakumov, N., Côté, J., 2007. The MYST family of histone acetyltransferases and their intimate links to cancer. Oncogene 26, 5395–5407. https://doi.org/10.1038/sj.onc.1210608.

Baker, S., Grant, P., 2007. The SAGA continues: expanding the cellular role of a transcriptional co-activator complex. Oncogene 26, 5329–5340. https://doi.org/10.1038/sj.onc.1210603.

Barbaric, S., Walker, J., Schmid, A., Svejstrup, J.Q., Hörz, W., 2001. Increasing the rate of chromatin remodeling and gene activation—a novel role for the histone acetyltransferase Gcn5. EMBO J. 20, 4944–4951. https://doi.org/10.1093/emboj/20.17.4944.

Basehoar, A.D., Zanton, S.J., Pugh, B.F., 2004. Identification and distinct regulation of yeast TATA box-containing genes. Cell 116, 699–709.

Becker, P.B., Workman, J.L., 2013. Nucleosome remodeling and epigenetics. Cold Spring Harb. Perspect. Biol. 5. https://doi.org/10.1101/cshperspect.a017905.

Beresford, G.W., Boss, J.M., 2001. CIITA coordinates multiple histone acetylation modifications at the HLA-DRA promoter. Nat. Immunol. 2, 652–657. https://doi.org/10.1038/89810.

Bernstein, B.E., Tong, J.K., Schreiber, S.L., 2000. Genomewide studies of histone deacetylase function in yeast. Proc. Natl. Acad. Sci. U. S. A. 97, 13708–13713. https://doi.org/10.1073/pnas.250477697.

Biddick, R.K., Law, G.L., Chin, K.K.B., Young, E.T., 2008. The transcriptional coactivators SAGA, SWI/SNF, and mediator make distinct contributions to activation of glucose-repressed genes. J. Biol. Chem. 283, 33101–33109. https://doi.org/10.1074/jbc.M805258200.

Bird, A.W., Yu, D.Y., Pray-Grant, M.G., Qiu, Q., Harmon, K.E., Megee, P.C., Grant, P.A., Smith, M.M., Christman, M.F., 2002. Acetylation of histone H4 by Esa1 is required for DNA double-strand break repair. Nature 419, 411–415. https://doi.org/10.1038/nature01035.

Bone, J.R., Lavender, J., Richman, R., Palmer, M.J., Turner, B.M., Kuroda, M.I., 1994. Acetylated histone H4 on the male X chromosome is associated with dosage compensation in Drosophila. Genes Dev. 8, 96–104.

Bonnet, J., Wang, C.-Y., Baptista, T., Vincent, S.D., Hsiao, W.-C., Stierle, M., Kao, C.-F., Tora, L., Devys, D., 2014. The SAGA coactivator complex acts on the whole transcribed genome and is required for RNA polymerase II transcription. Genes Dev. 28, 1999–2012. https://doi.org/10.1101/gad.250225.114.

Borrow Jr., J., Stanton, V.P., Andresen, J.M., Becher, R., Behm, F.G., Chaganti, R.S.K., Civin, C.I., Disteche, C., Dubé, I., Frischauf, A.M., Horsman, D., Mitelman, F., Volinia, S., Watmore, A.E., Housman, D.E., 1996. The translocation t (8;16)(p11;p13) of acute myeloid leukaemia fuses a putative acetyltransferase to the CREB–binding protein. Nat. Genet. 14, 33–41. https://doi.org/10.1038/ng0996-33.

Boyer, L.A., Langer, M.R., Crowley, K.A., Tan, S., Denu, J.M., Peterson, C.L., 2002. Essential role for the SANT domain in the functioning of multiple chromatin remodeling enzymes. Mol. Cell 10, 935–942.

Brès, V., Yoh, S.M., Jones, K.A., 2008. The multi-tasking P-TEFb complex. Curr. Opin. Cell Biol. 20, 334–340. https://doi.org/10.1016/j.ceb.2008.04.008.

Broday, L., Peng, W., Kuo, M.H., Salnikow, K., Zoroddu, M., Costa, M., 2000. Nickel compounds are novel inhibitors of histone H4 acetylation. Cancer Res. 60, 238–241.

Brownell, J.E., Zhou, J., Ranalli, T., Kobayashi, R., Edmondson, D.G., Roth, S.Y., Allis, C.D., 1996. Tetrahymena histone acetyltransferase a: a homolog to yeast gcn5p linking histone acetylation to gene activation. Cell 84, 843–851. https://doi.org/10.1016/S0092-8674(00)81063-6.

Burgess, R.J., Zhang, Z., 2010. Histones, histone chaperones and nucleosome assembly. Protein Cell 1, 607–612. https://doi.org/10.1007/s13238-010-0086-y.

Burke, T.L., Miller, J.L., Grant, P.A., 2013. Direct inhibition of Gcn5 protein catalytic activity by polyglutamine-expanded ataxin-7. J. Biol. Chem. 288, 34266–34275. https://doi.org/10.1074/jbc.M113.487538.

Cantone, L., Nordio, F., Hou, L., Apostoli, P., Bonzini, M., Tarantini, L., Angelici, L., Bollati, V., Zanobetti, A., Schwartz, J., Bertazzi, P.A., Baccarelli, A., 2011. Inhalable metal-rich air particles and histone H3K4 dimethylation and H3K9 acetylation in a cross-sectional study of steel workers. Environ. Health Perspect. 119, 964–969. https://doi.org/10.1289/ehp.1002955.

Carrozza, M.J., Utley, R.T., Workman, J.L., Côté, J., 2003. The diverse functions of histone acetyltransferase complexes. Trends Genet. 19, 321–329. https://doi.org/10.1016/S0168-9525(03)00115-X.

Cazzalini, O., Perucca, P., Savio, M., Necchi, D., Bianchi, L., Stivala, L.A., Ducommun, B., Scovassi, A.I., Prosperi, E., 2008. Interaction of p21(CDKN1A) with PCNA regulates the histone acetyltransferase activity of p300 in nucleotide excision repair. Nucleic Acids Res. 36, 1713–1722. https://doi.org/10.1093/nar/gkn014.

Chandy, M., Gutiérrez, J.L., Prochasson, P., Workman, J.L., 2006. SWI/SNF Displaces SAGA-Acetylated Nucleosomes. Eukaryot. Cell 5, 1738–1747. https://doi.org/10.1128/EC.00165-06.

Chen, R.-J., Chang, L.W., Lin, P., Wang, Y.-J., 2011. Epigenetic effects and molecular mechanisms of tumorigenesis induced by cigarette smoke: an overview [WWW Document]. J. Oncol. https://doi.org/10.1155/2011/654931.

Chen, Z., Zhang, H., Jablonowski, D., Zhou, X., Ren, X., Hong, X., Schaffrath, R., Zhu, J.-K., Gong, Z., 2006. Mutations in ABO1/ELO2, a subunit of holo-elongator, increase abscisic acid sensitivity and drought tolerance in *Arabidopsis thaliana*. Mol. Cell. Biol. 26, 6902–6912. https://doi.org/10.1128/MCB.00433-06.

Col, E., Hoghoughi, N., Dufour, S., Penin, J., Koskas, S., Faure, V., Ouzounova, M., Hernandez-Vargash, H., Reynoird, N., Daujat, S., Folco, E., Vigneron, M., Schneider, R., Verdel, A., Khochbin, S., Herceg, Z., Caron, C., Vourc'h, C., 2017. Bromodomain factors of BET family are new essential actors of pericentric heterochromatin transcriptional activation in response to heat shock. Sci. Rep. 7, 5418. https://doi.org/10.1038/s41598-017-05343-8.

Collotta, M., Bertazzi, P.A., Bollati, V., 2013. Epigenetics and pesticides. Toxicology 307, 35–41. https://doi.org/10.1016/j.tox.2013.01.017.

Creppe, C., Buschbeck, M., 2011. Elongator: an ancestral complex driving transcription and migration through protein acetylation [WWW Document]. Biomed. Res. Int. https://doi.org/10.1155/2011/924898.

Csoka, A.B., Szyf, M., 2009. Epigenetic side-effects of common pharmaceuticals: a potential new field in medicine and pharmacology. Med. Hypotheses 73, 770–780. https://doi.org/10.1016/j.mehy.2008.10.039.

Cui, F., Cole, H.A., Clark, D.J., Zhurkin, V.B., 2012. Transcriptional activation of yeast genes disrupts intragenic nucleosome phasing. Nucleic Acids Res. 40, 10753–10764. https://doi.org/10.1093/nar/gks870.

de Boer, J., Hoeijmakers, J.H., 2000. Nucleotide excision repair and human syndromes. Carcinogenesis 21, 453–460.

de Jonge, W.J., O'Duibhir, E., Lijnzaad, P., van Leenen, D., Groot Koerkamp, M.J., Kemmeren, P., Holstege, F.C., 2017. Molecular mechanisms that distinguish TFIID housekeeping from regulatable SAGA promoters. EMBO J. 36, 274–290. https://doi.org/10.15252/embj.201695621.

de la Cruz, X., Lois, S., Sánchez-Molina, S., Martínez-Balbás, M.A., 2005. Do protein motifs read the histone code? BioEssays 27, 164–175. https://doi.org/10.1002/bies.20176.

Delvecchio, M., Gaucher, J., Aguilar-Gurrieri, C., Ortega, E., Panne, D., 2013. Structure of the p300 catalytic core and implications for chromatin targeting and HAT regulation. Nat. Struct. Mol. Biol. 20, 1040–1046. https://doi.org/10.1038/nsmb.2642.

Didonna, A., Opal, P., 2015. The promise and perils of HDAC inhibitors in neurodegeneration. Ann. Clin. Transl. Neurol. 2, 79–101. https://doi.org/10.1002/acn3.147.

Ding, R., Jin, Y., Liu, X., Ye, H., Zhu, Z., Zhang, Y., Wang, T., Xu, Y., 2017. Dose- and time- effect responses of DNA methylation and histone H3K9 acetylation changes induced by traffic-related air pollution. Sci. Rep. 7, 43737. https://doi.org/10.1038/srep43737.

Dolinoy, D.C., 2008. The agouti mouse model: an epigenetic biosensor for nutritional and environmental alterations on the fetal epigenome. Nutr. Rev. 66, S7–S11. https://doi.org/10.1111/j.1753-4887.2008.00056.x.

Dorigo, B., Schalch, T., Bystricky, K., Richmond, T.J., 2003. Chromatin fiber folding: requirement for the histone H4 N-terminal tail. J. Mol. Biol. 327, 85–96.

Finnin, M.S., Donigian, J.R., Pavletich, N.P., 2001. Structure of the histone deacetylase SIRT2. Nat. Struct. Biol. 8, 621–625. https://doi.org/10.1038/89668.

Frye, R.A., 2000. Phylogenetic classification of prokaryotic and eukaryotic Sir2-like proteins. Biochem. Biophys. Res. Commun. 273, 793–798. https://doi.org/10.1006/bbrc.2000.3000.

Galdieri, L., Vancura, A., 2012. Acetyl-CoA carboxylase regulates global histone acetylation. J. Biol. Chem. 287, 23865–23876. https://doi.org/10.1074/jbc.M112.380519.

Gao, L., Wang, H.N., Zhang, L., Peng, F.Y., Jia, Y., Wei, W., Jia, L.H., 2016. Effect of perinatal bisphenol a exposure on serum lipids and lipid enzymes in offspring rats of different sex. Biomed. Environ. Sci. 29, 686–689. https://doi.org/10.3967/bes2016.092.

Gates, L.A., Shi, J., Rohira, A.D., Feng, Q., Zhu, B., Bedford, M.T., Sagum, C.A., Jung, S.Y., Qin, J., Tsai, M.-J., Tsai, S.Y., Li, W., Foulds, C.E., O'Malley, B.W., 2017. Acetylation on histone H3 lysine 9 mediates a switch from transcription initiation to elongation. J. Biol. Chem. 292, 14456–14472. https://doi.org/10.1074/jbc.M117.802074.

Gaucher, J., Boussouar, F., Montellier, E., Curtet, S., Buchou, T., Bertrand, S., Hery, P., Jounier, S., Depaux, A., Vitte, A.-L., Guardiola, P., Pernet, K., Debernardi, A., Lopez, F., Holota, H., Imbert, J., Wolgemuth, D.J., Gérard, M., Rousseaux, S., Khochbin, S., 2012. Bromodomain-dependent stage-specific male genome programming by Brdt. EMBO J. 31, 3809–3820. https://doi.org/10.1038/emboj.2012.233.

Gerona-Navarro, G., Yoel-Rodríguez, n., Mujtaba, S., Frasca, A., Patel, J., Zeng, L., Plotnikov, A.N., Osman, R., Zhou, M.-M., 2011. Rational design of cyclic peptide modulators of the transcriptional coactivator CBP: a new class of p53 inhibitors. J. Am. Chem. Soc. 133, 2040–2043. https://doi.org/10.1021/ja107761h.

Ginsburg, D.S., Anlembom, T.E., Wang, J., Patel, S.R., Li, B., Hinnebusch, A.G., 2014. NuA4 links methylation of histone H3 lysines 4 and 36 to acetylation of histones H4 and H3. J. Biol. Chem. 289, 32656–32670. https://doi.org/10.1074/jbc.M114.585588.

Golebiowski, F., Kasprzak, K.S., 2005. Inhibition of core histones acetylation by carcinogenic nickel(II). Mol. Cell. Biochem. 279, 133–139. https://doi.org/10.1007/s11010-005-8285-1.

Goudarzi, A., Zhang, D., Huang, H., Barral, S., Kwon, O.K., Qi, S., Tang, Z., Buchou, T., Vitte, A.-L., He, T., Cheng, Z., Montellier, E., Gaucher, J., Curtet, S., Debernardi, A., Charbonnier, G., Puthier, D., Petosa, C., Panne, D., Rousseaux, S., Roeder, R.G., Zhao, Y., Khochbin, S., 2016. Dynamic competing histone H4 K5K8 acetylation and butyrylation are hallmarks of highly active gene promoters. Mol. Cell 62, 169–180. https://doi.org/10.1016/j.molcel.2016.03.014.

Grant, P.A., Duggan, L., Côté, J., Roberts, S.M., Brownell, J.E., Candau, R., Ohba, R., Owen-Hughes, T., Allis, C.D., Winston, F., Berger, S.L., Workman, J.L., 1997. Yeast Gcn5 functions in two multisubunit complexes to acetylate nucleosomal histones: characterization of an Ada complex and the SAGA (Spt/Ada) complex. Genes Dev. 11, 1640–1650.

Gregoretti, I.V., Lee, Y.-M., Goodson, H.V., 2004. Molecular evolution of the histone deacetylase family: functional implications of phylogenetic analysis. J. Mol. Biol. 338, 17–31. https://doi.org/10.1016/j.jmb.2004.02.006.

Grozinger, C.M., Schreiber, S.L., 2000. Regulation of histone deacetylase 4 and 5 and transcriptional activity by 14-3-3-dependent cellular localization. Proc. Natl. Acad. Sci. U. S. A. 97, 7835–7840. https://doi.org/10.1073/pnas.140199597.

Gu, W., Roeder, R.G., 1997. Activation of p53 sequence-specific DNA binding by acetylation of the p53 C-terminal domain. Cell 90, 595–606.

Guillemette, B., Drogaris, P., Lin, H.-H.S., Armstrong, H., Hiragami-Hamada, K., Imhof, A., Bonneil, E., Thibault, P., Verreault, A., Festenstein, R.J., 2011. H3 lysine 4 is acetylated at active gene promoters and is regulated by H3 lysine 4 methylation. PLoS Genet. 7: e1001354. https://doi.org/10.1371/journal.pgen.1001354.

Guo, B., Huang, X., Cooper, S., Broxmeyer, H.E., 2017. Glucocorticoid hormone-induced chromatin remodeling enhances human hematopoietic stem cell homing and engraftment. Nat. Med. 23, 424–428. https://doi.org/10.1038/nm.4298.

Haggarty, S.J., Tsai, L.-H., 2011. Probing the role of HDACs and mechanisms of chromatin-mediated neuroplasticity. Neurobiol. Learn. Mem. 96, 41–52. https://doi.org/10.1016/j.nlm.2011.04.009.

Han, J., Zhou, H., Horazdovsky, B., Zhang, K., Xu, R.-M., Zhang, Z., 2007. Rtt109 acetylates histone H3 lysine 56 and functions in DNA replication. Science 315, 653–655. https://doi.org/10.1126/science.1133234.

Houzelstein, D., Bullock, S.L., Lynch, D.E., Grigorieva, E.F., Wilson, V.A., Beddington, R.S.P., 2002. Growth and early postimplantation defects in mice deficient for the bromodomain-containing protein Brd4. Mol. Cell. Biol. 22, 3794–3802.

Jazayeri, A., McAinsh, A.D., Jackson, S.P., 2004. Saccharomyces cerevisiae Sin3p facilitates DNA double-strand break repair. PNAS 101, 1644–1649. https://doi.org/10.1073/pnas.0304797101.

Jiang, Y.W., Veschambre, P., Erdjument-Bromage, H., Tempst, P., Conaway, J.W., Conaway, R.C., Kornberg, R.D., 1998. Mammalian mediator of transcriptional regulation and its possible role as an end-point of signal transduction pathways. Proc. Natl. Acad. Sci. U. S. A. 95, 8538–8543.

Jones, P.L., Veenstra, G.J., Wade, P.A., Vermaak, D., Kass, S.U., Landsberger, N., Strouboulis, J., Wolffe, A.P., 1998. Methylated DNA and MeCP2 recruit histone deacetylase to repress transcription. Nat. Genet. 19, 187–191. https://doi.org/10.1038/561.

Kadosh, D., Struhl, K., 1997. Repression by Ume6 involves recruitment of a complex containing Sin3 corepressor and Rpd3 histone deacetylase to target promoters. Cell 89, 365–371.

Kandel, E.R., Kandel, D.B., 2014. A molecular basis for nicotine as a gateway drug. N. Engl. J. Med. 371, 932–943. https://doi.org/10.1056/NEJMsa1405092.

Kang, H., Jung, J.-W., Kim, M.K., Chung, J.H., 2009. CK2 is the regulator of SIRT1 substrate-binding affinity, deacetylase activity and cellular response to DNA-damage. PLoS One 4, e6611. https://doi.org/10.1371/journal.pone.0006611.

Katan-Khaykovich, Y., Struhl, K., 2002. Dynamics of global histone acetylation and deacetylation in vivo: rapid restoration of normal histone acetylation status upon removal of activators and repressors. Genes Dev. 16, 743–752. https://doi.org/10.1101/gad.967302.

Kehle, J., Beuchle, D., Treuheit, S., Christen, B., Kennison, J.A., Bienz, M., Müller, J., 1998. dMi-2, a hunchback-interacting protein that functions in polycomb repression. Science 282, 1897–1900.

Kim, E.-J., Kho, J.-H., Kang, M.-R., Um, S.-J., 2007. Active regulator of SIRT1 cooperates with SIRT1 and facilitates suppression of p53 activity. Mol. Cell 28, 277–290. https://doi.org/10.1016/j.molcel.2007.08.030.

Koutelou, E., Hirsch, C.L., Dent, S.Y.R., 2010. Multiple faces of the SAGA complex. Curr. Opin. Cell Biol. 22, 374–382. https://doi.org/10.1016/j.ceb.2010.03.005.

Kumar, D., Thakur, M.K., 2017. Effect of perinatal exposure to Bisphenol-A on DNA methylation and histone acetylation in cerebral cortex and hippocampus of postnatal male mice. J. Toxicol. Sci. 42, 281–289. https://doi.org/10.2131/jts.42.281.

Kuo, M.-H., Zhou, J., Jambeck, P., Churchill, M.E.A., Allis, C.D., 1998. Histone acetyltransferase activity of yeast Gcn5p is required for the activation of target genes in vivo. Genes Dev. 12, 627–639.

Kurdistani, S.K., Tavazoie, S., Grunstein, M., 2004. Mapping global histone acetylation patterns to gene expression. Cell 117, 721–733. https://doi.org/10.1016/j.cell.2004.05.023.

Laherty, C.D., Billin, A.N., Lavinsky, R.M., Yochum, G.S., Bush, A.C., Sun, J.-M., Mullen, T.-M., Davie, J.R., Rose, D.W., Glass, C.K., Rosenfeld, M.G., Ayer, D.E., Eisenman, R.N., 1998. SAP30, a component of the mSin3 corepressor complex involved in N-CoR-mediated repression by specific transcription factors. Mol. Cell 2, 33–42. https://doi.org/10.1016/S1097-2765(00)80111-2.

Lahm, A., Paolini, C., Pallaoro, M., Nardi, M.C., Jones, P., Neddermann, P., Sambucini, S., Bottomley, M.J., Lo Surdo, P., Carfí, A., Koch, U., De Francesco, R., Steinkühler, C., Gallinari, P., 2007. Unraveling the hidden catalytic activity of vertebrate class IIa histone deacetylases. Proc. Natl. Acad. Sci. U. S. A. 104, 17335–17340. https://doi.org/10.1073/pnas.0706487104.

Lan, X., Koutelou, E., Schibler, A.C., Chen, Y.C., Grant, P.A., Dent, S.Y.R., 2015. Poly(Q) expansions in ATXN7 affect solubility but not activity of the SAGA deubiquitinating module. Mol. Cell. Biol. 35, 1777–1787. https://doi.org/10.1128/MCB.01454-14.

Lang, S.E., McMahon, S.B., Cole, M.D., Hearing, P., 2001. E2F transcriptional activation requires TRRAP and GCN5 cofactors. J. Biol. Chem. 276, 32627–32634. https://doi.org/10.1074/jbc.M102067200.

Lee, H., Rezai-Zadeh, N., Seto, E., 2004. Negative regulation of histone deacetylase 8 activity by cyclic AMP-dependent protein kinase A. Mol. Cell. Biol. 24, 765–773.

Lee, K.K., Workman, J.L., 2007. Histone acetyltransferase complexes: one size doesn't fit all. Nat. Rev. Mol. Cell Biol. 8, 284–295. https://doi.org/10.1038/nrm2145.

Levenson, J.M., O'Riordan, K.J., Brown, K.D., Trinh, M.A., Molfese, D.L., Sweatt, J.D., 2004. Regulation of histone acetylation during memory formation in the hippocampus. J. Biol. Chem. 279, 40545–40559. https://doi.org/10.1074/jbc.M402229200.

Li, X., Heyer, W.-D., 2008. Homologous recombination in DNA repair and DNA damage tolerance. Cell Res. 18, 99–113. https://doi.org/10.1038/cr.2008.1.

Li, Y., Zhang, L., Liu, T., Chai, C., Fang, Q., Wu, H., Garcia, P.A.A., Han, Z., Zong, S., Yu, Y., Zhang, X., Parthun, M.R., Chai, J., Xu, R.-M., Yang, M., 2014. Hat2p recognizes the histone H3 tail to specify the acetylation of the newly synthesized H3/H4 heterodimer by the Hat1p/Hat2p complex. Genes Dev. 28, 1217–1227. https://doi.org/10.1101/gad.240531.114.

Liu, D., Wu, D., Zhao, L., Yang, Y., Ding, J., Dong, L., Hu, L., Wang, F., Zhao, X., Cai, Y., Jin, J., 2015. Arsenic trioxide reduces global histone H4 acetylation at lysine 16 through direct binding to histone acetyltransferase hMOF in human cells. PLoS One 10, e0141014. https://doi.org/10.1371/journal.pone.0141014.

Lucchini, R., Sogo, J.M., 1995. Replication of transcriptionally active chromatin. Nature 374, 276–280. https://doi.org/10.1038/374276a0.

Luger, K., Mäder, A.W., Richmond, R.K., Sargent, D.F., Richmond, T.J., 1997. Crystal structure of the nucleosome core particle at 2.8 Å resolution. Nature 389, 251–260. https://doi.org/10.1038/38444.

Marmorstein, R., Zhou, M.-M., 2014. Writers and readers of histone acetylation: structure, mechanism, and inhibition. Cold Spring Harb. Perspect. Biol. 6, a018762. https://doi.org/10.1101/cshperspect.a018762.

Masumoto, H., Hawke, D., Kobayashi, R., Verreault, A., 2005. A role for cell-cycle-regulated histone H3 lysine 56 acetylation in the DNA damage response. Nature 436, 294–298. https://doi.org/10.1038/nature03714.

Matangkasombut, O., Buratowski, R.M., Swilling, N.W., Buratowski, S., 2000. Bromodomain factor 1 corresponds to a missing piece of yeast TFIID. Genes Dev. 14, 951–962.

Matangkasombut, O., Buratowski, S., 2003. Different sensitivities of bromodomain factors 1 and 2 to histone H4 acetylation. Mol. Cell 11, 353–363.

McCullough, S.D., Grant, P.A., 2010. Histone acetylation, acetyltransferases, and ataxia–alteration of histone acetylation and chromatin dynamics is implicated in the pathogenesis of polyglutamine-expansion disorders. Adv. Protein Chem. Struct. Biol. 79, 165–203. https://doi.org/10.1016/S1876-1623(10)79005-2.

McCullough, S.D., Xu, X., Dent, S.Y.R., Bekiranov, S., Roeder, R.G., Grant, P.A., 2012. Reelin is a target of polyglutamine expanded ataxin-7 in human spinocerebellar ataxia type 7 (SCA7) astrocytes. Proc. Natl. Acad. Sci. U. S. A. 109, 21319–21324. https://doi.org/10.1073/pnas.1218331110.

McDonel, P., Costello, I., Hendrich, B., 2009. Keeping things quiet: roles of NuRD and Sin3 co-repressor complexes during mammalian development. Int. J. Biochem. Cell Biol. 41, 108–116. https://doi.org/10.1016/j.biocel.2008.07.022.

McMahon, S.B., Wood, M.A., Cole, M.D., 2000. The essential cofactor TRRAP recruits the histone acetyltransferase hGCN5 to c-Myc. Mol. Cell. Biol. 20, 556–562.

Miller, J.L., Grant, P.A., 2013. The role of DNA methylation and histone modifications in transcriptional regulation in humans. Subcell. Biochem. 61, 289–317. https://doi.org/10.1007/978-94-007-4525-4_13.

Mizzen, C.A., Yang, X.-J., Kokubo, T., Brownell, J.E., Bannister, A.J., Owen-Hughes, T., Workman, J., Wang, L., Berger, S.L., Kouzarides, T., Nakatani, Y., Allis, C.D., 1996. The TAFII250 subunit of TFIID has histone acetyltransferase activity. Cell 87, 1261–1270. https://doi.org/10.1016/S0092-8674(00)81821-8.

Mujtaba, S., Zeng, L., Zhou, M.-M., 2006. Modulating molecular functions of p53 with small molecules. Cell Cycle 5, 2575–2578. https://doi.org/10.4161/cc.5.22.3464.

Nelissen, H., Groeve, S.D., Fleury, D., Neyt, P., Bruno, L., Bitonti, M.B., Vandenbussche, F., Straeten, D.V.D., Yamaguchi, T., Tsukaya, H., Witters, E., Jaeger, G.D., Houben, A., Lijsebettens, M.V., 2010. Plant Elongator regulates auxin-related genes during RNA polymerase II transcription elongation. PNAS 107, 1678–1683. https://doi.org/10.1073/pnas.0913559107.

Neuwald, A.F., Landsman, D., 1997. GCN5-related histone N-acetyltransferases belong to a diverse superfamily that includes the yeast SPT10 protein. Trends Biochem. Sci. 22, 154–155. https://doi.org/10.1016/S0968-0004(97)01034-7.

Nourani, A., Utley, R.T., Allard, S., Côté, J., 2004. Recruitment of the NuA4 complex poises the PHO5 promoter for chromatin remodeling and activation. EMBO J. 23, 2597–2607. https://doi.org/10.1038/sj.emboj.7600230.

Ogiwara, H., Ui, A., Otsuka, A., Satoh, H., Yokomi, I., Nakajima, S., Yasui, A., Yokota, J., Kohno, T., 2011. Histone acetylation by CBP and p300 at double-strand break sites facilitates SWI/SNF chromatin remodeling and the recruitment of non-homologous end joining factors. Oncogene 30, 2135–2146. https://doi.org/10.1038/onc.2010.592.

Ogryzko, V.V., Schiltz, R.L., Russanova, V., Howard, B.H., Nakatani, Y., 1996. The transcriptional coactivators p300 and CBP are histone acetyltransferases. Cell 87, 953–959. https://doi.org/10.1016/S0092-8674(00)82001-2.

Otero, G., Fellows, J., Li, Y., de Bizemont, T., Dirac, A.M., Gustafsson, C.M., Erdjument-Bromage, H., Tempst, P., Svejstrup, J.Q., 1999. Elongator, a multisubunit component of a novel RNA polymerase II holoenzyme for transcriptional elongation. Mol. Cell 3, 109–118.

Parthun, M.R., Widom, J., Gottschling, D.E., 1996. The major cytoplasmic histone acetyltransferase in yeast: links to chromatin replication and histone metabolism. Cell 87, 85–94. https://doi.org/10.1016/S0092-8674(00)81325-2.

Pejanovic, N., Hochrainer, K., Liu, T., Aerne, B.L., Soares, M.P., Anrather, J., 2012. Regulation of nuclear factor κB (NF-κB) transcriptional activity via p65 acetylation by the chaperonin containing TCP1 (CCT). PLoS One 7, e42020. https://doi.org/10.1371/journal.pone.0042020.

Peleg, S., Sananbenesi, F., Zovoilis, A., Burkhardt, S., Bahari-Javan, S., Agis-Balboa, R.C., Cota, P., Wittnam, J.L., Gogol-Doering, A., Opitz, L., Salinas-Riester, G., Dettenhofer, M., Kang, H., Farinelli, L., Chen, W., Fischer, A., 2010. Altered histone acetylation is associated with age-dependent memory impairment in mice. Science 328, 753–756. https://doi.org/10.1126/science.1186088.

Peng, J., Dong, W., Chen, L., Zou, T., Qi, Y., Liu, Y., 2007. Brd2 is a TBP-associated protein and recruits TBP into E2F-1 transcriptional complex in response to serum stimulation. Mol. Cell. Biochem. 294, 45–54. https://doi.org/10.1007/s11010-006-9223-6.

Phiel, C.J., Zhang, F., Huang, E.Y., Guenther, M.G., Lazar, M.A., Klein, P.S., 2001. Histone deacetylase is a direct target of valproic acid, a potent anticonvulsant, mood stabilizer, and teratogen. J. Biol. Chem. 276, 36734–36741. https://doi.org/10.1074/jbc.M101287200.

Pirola, L., Balcerczyk, A., Tothill, R.W., Haviv, I., Kaspi, A., Lunke, S., Ziemann, M., Karagiannis, T., Tonna, S., Kowalczyk, A., Beresford-Smith, B., Macintyre, G., Kelong, M., Hongyu, Z., Zhu, J., El-Osta, A., 2011. Genome-wide analysis distinguishes hyperglycemia regulated epigenetic signatures of primary vascular cells. Genome Res. 21, 1601–1615. https://doi.org/10.1101/gr.116095.110.

Pokholok, D.K., Hannett, N.M., Young, R.A., 2002. Exchange of RNA polymerase II initiation and elongation factors during gene expression in vivo. Mol. Cell 9, 799–809.

Pradeepa, M.M., Grimes, G.R., Kumar, Y., Olley, G., Taylor, G.C.A., Schneider, R., Bickmore, W.A., 2016. Histone H3 globular domain acetylation identifies a new class of enhancers. Nat. Genet. 48, 681–686. https://doi.org/10.1038/ng.3550.

Pray-Grant, M.G., Schieltz, D., McMahon, S.J., Wood, J.M., Kennedy, E.L., Cook, R.G., Workman, J.L., Yates III, J.R., Grant, P.A., 2002. The novel SLIK histone acetyltransferase complex functions in the yeast retrograde response pathway. Mol. Cell. Biol. 22, 8774–8786. https://doi.org/10.1128/MCB.22.24.8774-8786.2002.

Qiu, H., Chereji, R.V., Hu, C., Cole, H.A., Rawal, Y., Clark, D.J., Hinnebusch, A.G., 2016. Genome-wide cooperation by HAT Gcn5, remodeler SWI/SNF, and chaperone Ydj1 in promoter nucleosome eviction and transcriptional activation. Genome Res. 26, 211–225. https://doi.org/10.1101/gr.196337.115.

Recht, J., Tsubota, T., Tanny, J.C., Diaz, R.L., Berger, J.M., Zhang, X., Garcia, B.A., Shabanowitz, J., Burlingame, A.L., Hunt, D.F., Kaufman, P.D., Allis, C.D., 2006. Histone chaperone Asf1 is required for histone H3 lysine 56 acetylation, a modification associated with S phase in mitosis and meiosis. PNAS 103, 6988–6993. https://doi.org/10.1073/pnas.0601676103.

Reed, S.H., 2011. Nucleotide excision repair in chromatin: damage removal at the drop of a HAT. DNA Repair (Amst) 10, 734–742. https://doi.org/10.1016/j.dnarep.2011.04.029.

Santos-Rosa, H., Schneider, R., Bernstein, B.E., Karabetsou, N., Morillon, A., Weise, C., Schreiber, S.L., Mellor, J., Kouzarides, T., 2003. Methylation of histone H3 K4 mediates association of the Isw1p ATPase with chromatin. Mol. Cell 12, 1325–1332. https://doi.org/10.1016/S1097-2765(03)00438-6.

Sanz, A.B., García, R., Rodríguez-Peña, J.M., Nombela, C., Arroyo, J., 2016. Cooperation between SAGA and SWI/SNF complexes is required for efficient transcriptional responses regulated by the yeast MAPK Slt2. Nucleic Acids Res. 44, 7159–7172. https://doi.org/10.1093/nar/gkw324.

Sapountzi, V., Côté, J., 2011. MYST-family histone acetyltransferases: beyond chromatin. Cell. Mol. Life Sci. 68, 1147–1156. https://doi.org/10.1007/s00018-010-0599-9.

Schalch, T., Duda, S., Sargent, D.F., Richmond, T.J., 2005. X-ray structure of a tetranucleosome and its implications for the chromatin fibre. Nature 436, 138–141. https://doi.org/10.1038/nature03686.

Schwabish, M.A., Struhl, K., 2007. The Swi/Snf complex is important for histone eviction during transcriptional activation and RNA polymerase II elongation in vivo. Mol. Cell. Biol. 27, 6987–6995. https://doi.org/10.1128/MCB.00717-07.

Seo, S., Macfarlan, T., McNamara, P., Hong, R., Mukai, Y., Heo, S., Chakravarti, D., 2002. Regulation of histone acetylation and transcription by nuclear protein pp32, a subunit of the INHAT complex. J. Biol. Chem. 277, 14005–14010. https://doi.org/10.1074/jbc.M112455200.

Seto, E., Yoshida, M., 2014. Erasers of histone acetylation: the histone deacetylase enzymes. Cold Spring Harb. Perspect. Biol. 6, a018713. https://doi.org/10.1101/cshperspect.a018713.

Shi, X., Kachirskaia, I., Walter, K.L., Kuo, J.-H.A., Lake, A., Davrazou, F., Chan, S.M., Martin, D.G.E., Fingerman, I.M., Briggs, S.D., Howe, L., Utz, P.J., Kutateladze, T.G., Lugovskoy, A.A., Bedford, M.T., Gozani, O., 2007. Proteome-wide analysis in Saccharomyces cerevisiae identifies several PHD fingers as novel direct and selective binding modules of histone H3 methylated at either lysine 4 or lysine 36. J. Biol. Chem. 282, 2450–2455. https://doi.org/10.1074/jbc.C600286200.

Shia, W.-J., Li, B., Workman, J.L., 2006. SAS-mediated acetylation of histone H4 Lys 16 is required for H2A.Z incorporation at subtelomeric regions in Saccharomyces cerevisiae. Genes Dev. 20, 2507–2512. https://doi.org/10.1101/gad.1439206.

Shilatifard, A., Conaway, R.C., Conaway, J.W., 2003. The RNA polymerase II elongation complex. Annu. Rev. Biochem. 72, 693–715. https://doi.org/10.1146/annurev.biochem.72.121801.161551.

Shogren-Knaak, M., Ishii, H., Sun, J.-M., Pazin, M.J., Davie, J.R., Peterson, C.L., 2006. Histone H4-K16 acetylation controls chromatin structure and protein interactions. Science 311, 844–847. https://doi.org/10.1126/science.1124000.

Smith, E.R., Eisen, A., Gu, W., Sattah, M., Pannuti, A., Zhou, J., Cook, R.G., Lucchesi, J.C., Allis, C.D., 1998. ESA1 is a histone acetyltransferase that is essential for growth in yeast. PNAS 95, 3561–3565. https://doi.org/10.1073/pnas.95.7.3561.

Smith, S., Stillman, B., 1991. Stepwise assembly of chromatin during DNA replication in vitro. EMBO J. 10, 971–980.

Sogo, J.M., Stahl, H., Koller, T., Knippers, R., 1986. Structure of replicating simian virus 40 minichromosomes. The replication fork, core histone segregation and terminal structures. J. Mol. Biol. 189, 189–204.

Song, C., Kanthasamy, A., Anantharam, V., Sun, F., Kanthasamy, A.G., 2010. Environmental neurotoxic pesticide increases histone acetylation to promote apoptosis in dopaminergic neuronal cells: relevance to epigenetic mechanisms of neurodegeneration. Mol. Pharmacol. 77, 621–632. https://doi.org/10.1124/mol.109.062174.

Suganuma, T., Workman, J.L., 2011. Signals and combinatorial functions of histone modifications. Annu. Rev. Biochem. 80, 473–499. https://doi.org/10.1146/annurev-biochem-061809-175347.

Sun, B., Guo, S., Tang, Q., Li, C., Zeng, R., Xiong, Z., Zhong, C., Ding, J., 2011. Regulation of the histone acetyltransferase activity of hMOF via autoacetylation of Lys274. Cell Res. 21, 1262–1266. https://doi.org/10.1038/cr.2011.105.

Sutton, A., Shia, W.-J., Band, D., Kaufman, P.D., Osada, S., Workman, J.L., Sternglanz, R., 2003. Sas4 and Sas5 are required for the histone acetyltransferase activity of Sas2 in the SAS complex. J. Biol. Chem. 278, 16887–16892. https://doi.org/10.1074/jbc.M210709200.

Taunton, J., Hassig, C.A., Schreiber, S.L., 1996. A mammalian histone deacetylase related to the yeast transcriptional regulator Rpd3p. Science 272, 408–411. https://doi.org/10.1126/science.272.5260.408.

Thompson, P.R., Wang, D., Wang, L., Fulco, M., Pediconi, N., Zhang, D., An, W., Ge, Q., Roeder, R.G., Wong, J., Levrero, M., Sartorelli, V., Cotter, R.J., Cole, P.A., 2004. Regulation of the p300 HAT domain via a novel activation loop. Nat. Struct. Mol. Biol. 11, 308–315. https://doi.org/10.1038/nsmb740.

Torok, M.S., Grant, P.A., 2004. Histone acetyltransferase proteins contribute to transcriptional processes at multiple levels. In: Advances in Protein Chemistry, Proteins in Eukaryotic Transcription. Academic Press, San Diego, pp. 181–199. https://doi.org/10.1016/S0065-3233(04)67007-0.

Tse, C., Sera, T., Wolffe, A.P., Hansen, J.C., 1998. Disruption of higher-order folding by core histone acetylation dramatically enhances transcription of nucleosomal arrays by RNA polymerase III. Mol. Cell. Biol. 18, 4629–4638.

Tsubota, T., Berndsen, C.E., Erkmann, J.A., Smith, C.L., Yang, L., Freitas, M.A., Denu, J.M., Kaufman, P.D., 2007. Histone H3-K56 acetylation is catalyzed by histone chaperone-dependent complexes. Mol. Cell 25, 703–712. https://doi.org/10.1016/j.molcel.2007.02.006.

Tu, A.W., Luo, K., 2007. Acetylation of Smad2 by the co-activator p300 regulates activin and transforming growth factor beta response. J. Biol. Chem. 282, 21187–21196. https://doi.org/10.1074/jbc.M700085200.

Verdin, E. (Ed.), 2006. Histone Deacetylases: Transcriptional Regulation and Other Cellular Functions, Cancer Drug Discovery and Development. Humana Press, Totowa, NJ.

Verreault, A., 2000. De novo nucleosome assembly: new pieces in an old puzzle. Genes Dev. 14, 1430–1438. https://doi.org/10.1101/gad.14.12.1430.

Vogelauer, M., Wu, J., Suka, N., Grunstein, M., 2000. Global histone acetylation and deacetylation in yeast. Nature 408, 495–498. https://doi.org/10.1038/35044127.

Wallberg, A.E., Neely, K.E., Gustafsson, J.-Å., Workman, J.L., Wright, A.P.H., Grant, P.A., 1999. Histone acetyltransferase complexes can mediate transcriptional activation by the major glucocorticoid receptor activation domain. Mol. Cell. Biol. 19, 5952–5959.

Wapenaar, H., Dekker, F.J., 2016. Histone acetyltransferases: challenges in targeting bi-substrate enzymes. Clin. Epigenetics. 8 https://doi.org/10.1186/s13148-016-0225-2.

Waterborg, J.H., 2000. Steady-state levels of histone acetylation in *Saccharomyces cerevisiae*. J. Biol. Chem. 275, 13007–13011. https://doi.org/10.1074/jbc.275.17.13007.

Waterborg, J.H., 2002. Dynamics of histone acetylation in vivo. A function for acetylation turnover? Biochem. Cell Biol. 80, 363–378.

Wu, R., Wang, S., Zhou, N., Cao, Z., Zhang, Y., 2010. A proton-shuttle reaction mechanism for histone deacetylase 8 and the catalytic role of metal ions. J. Am. Chem. Soc. 132, 9471–9479. https://doi.org/10.1021/ja103932d.

Yang, S.-H., Vickers, E., Brehm, A., Kouzarides, T., Sharrocks, A.D., 2001. Temporal recruitment of the mSin3A-histone deacetylase corepressor complex to the ETS domain transcription factor Elk-1. Mol. Cell. Biol. 21, 2802–2814. https://doi.org/10.1128/MCB.21.8.2802-2814.2001.

Yang, X.-J., Grégoire, S., 2005. Class II histone deacetylases: from sequence to function, regulation, and clinical implication. Mol. Cell. Biol. 25, 2873–2884. https://doi.org/10.1128/MCB.25.8.2873-2884.2005.

Yuan, H., Marmorstein, R., 2013. Histone acetyltransferases: rising ancient counterparts to protein kinases. Biopolymers 99, 98–111. https://doi.org/10.1002/bip.22128.

Zeng, L., Li, J., Muller, M., Yan, S., Mujtaba, S., Pan, C., Wang, Z., Zhou, M.-M., 2005. Selective small molecules blocking HIV-1 Tat and coactivator PCAF association. J. Am. Chem. Soc. 127, 2376–2377. https://doi.org/10.1021/ja044885g.

Zhang, X., Ozawa, Y., Lee, H., Wen, Y.-D., Tan, T.-H., Wadzinski, B.E., Seto, E., 2005. Histone deacetylase 3 (HDAC3) activity is regulated by interaction with protein serine/threonine phosphatase 4. Genes Dev. 19, 827–839. https://doi.org/10.1101/gad.1286005.

Zhang, Y., Ng, H.-H., Erdjument-Bromage, H., Tempst, P., Bird, A., Reinberg, D., 1999. Analysis of the NuRD subunits reveals a histone deacetylase core complex and a connection with DNA methylation. Genes Dev. 13, 1924–1935.

Zhao, Y., Garcia, B.A., 2015. Comprehensive catalog of currently documented histone modifications. Cold Spring Harb. Perspect. Biol. 7, a025064. https://doi.org/10.1101/cshperspect.a025064.

Zwergel, C., Stazi, G., Valente, S., Mai, A., 2016. Histone deacetylase inhibitors: updated studies in various epigenetic-related diseases. J. Clin. Epigenet. 2.

The Role of Histone Methylation and Methyltransferases in Gene Regulation

Julia Yue Cui, Zidong Donna Fu, Joseph Dempsey

Department of Environmental and Occupational Health Sciences, University of Washington, Seattle, WA, United States

Abbreviations

AhR	aryl hydrocarbon receptor
APC	anaphase-promoting complex
APC^{CDH1}	anaphase-promoting complex/cadherin 1
ASH1L	ASH1-like histone lysine methyltransferase

ASH2L	absent, small, or homeotic discs 2-like histone lysine methyltransferase complex subunit
BaP	benzo[a]pyrene
BEAS-2B	human lung bronchial epithelial cell line
BPA	bisphenol A
BRCA-1	breast cancer 1
Bre2	Set1C complex subunit Bre2
CAR	constitutive androstane receptor
CARM1	coactivator-associated arginine methyltransferase 1
CBX1	chromobox homologue 1
ChIP-Seq	chromatin immunoprecipitation coupled with high-throughput sequencing
CITED2	Cbp/p300-interacting transactivator with Glu/Asp-rich carboxy-terminal domain 2
COMPASS	complex of proteins associated with Set1
COX	cyclooxygenase
CREB	cyclic AMP response element binding
CREBBP/CBP	CREB-binding protein
CRL4^{Cdt2}	Cullin-RING ubiquitin ligase 4 E3 ubiquitin ligase complex
CYP	cytochrome P450
DES	diethylstilbestrol
DOT1L	disruptor of telomeric silencing 1-like histone lysine methyltransferase
DPY30	histone methyltransferase complex regulatory subunit
DSB	DNA double-stranded break
EDCs	endocrine disruptors
EED	embryonic ectoderm development
EHMT	euchromatic histone lysine methyltransferase
ER	estrogen receptor
EZH	enhancer of zeste polycomb repressive complex subunit
FBL	fibrillarin
Fgf2	fibroblast growth factor 2
gmt	glutamic-pyruvic transaminase
gpt	glutamic-pyruvic transaminase
gst	glutathione *S*-transferase
H2AQ	histone H2AQ isoform
H2AR3	histone H2A arginine 3
H2AZ	histone H2AZ isoform
H2BK5	histone H2B lysine 5
H3K4	histone H3 lysine 4
H3K5	histone H3 lysine 5
H3K27	histone H3 lysine 27
H3K36	histone H3 lysine 36
H3K79	histone H3 lysine 79
H3K9	histone H3 lysine 9
H3R17	histone H3 arginine 17
H3R2	histone H3 arginine 2
H3R3	histone H3 arginine 3
H3R8	histone H3 arginine 8
H3R26	histone H3 arginine 26
H4K20	histone H4 lysine 20
H4R3	histone H4 arginine 3
HCC	hepatocellular carcinoma
HIF1α	hypoxia-inducible factor 1 α
HMOX	heme oxygenase
HOTAIR	HOX transcript antisense RNA
HOTTIP	HOXA distal transcript antisense RNA

1. HISTONE MODIFICATIONS AND CHROMATIN STRUCTURE

HOX	homeobox
HR	lysine demethylase and nuclear receptor corepressor
Igf2r	insulin-like growth factor 2 receptor
IL	interleukin
JARID1	jumonji AT-rich interactive domain 1
JMJD	jumonji domain-containing proteins
JNK	c-Jun N-terminal kinase
KDM	lysine demethylase
KDM6A/UTX	ubiquitously transcribed X-chromosome tetratricopeptide repeat protein
KMT	lysine methyltransferase
lncRNAs	long noncoding RNAs
LPS	lipopolysaccharide
LSD	lysine (K)-specific demethylase
me1	monomethylated
me2	dimethylated
me2a	asymmetrically dimethylated
me2s	symmetrically dimethylated
me3	trimethylated
MECOM	MDS1 and EVI1 complex locus
miRNAs	microRNAs
MLH1	mutL homologue 1
MLL	mixed-lineage leukemia
MSH6	mutS homologue 6
MYND	myeloid, Nervy, and DEAF-1
ncRNAs	noncoding RNAs
NNMT	nicotinamide N-methyltransferase
NOP1	rRNA methyltransferase NOP1
NSD	nuclear receptor-binding SET domain protein
PCNA	proliferating cell nuclear antigen
PCR2	polycomb repressive complex 2
PGR	progesterone receptor
PHF	PHD finger protein
PPAR	peroxisome proliferator activated receptor
PPARGC1A/PGC1α	PPARG coactivator 1 α
PRC2	polycomb repressive complex 2
PRDM	positive regulatory domain I-binding factor and retinoblastoma protein-interacting zinc-finger gene SET domain
PRMT	protein arginine methyltransferase
PSIP1	PC4 and SFRS1 interacting protein 1
PXR	pregnane X receptor
RAR	retinoic acid receptor
RBBP5	retinoblastoma-binding protein 5, histone lysine methyltransferase complex subunit
RBBP8/CTIP	RB-binding protein 8, endonuclease
RGMA	repulsive guidance molecule family member A
RIOX	ribosomal oxygenase
SAH	S-adenosylhomocysteine
SAMe	S-adenosylmethionine
Scd1	Dpy-30 domain protein Sdc1
SCF complex	Skp, Cullin, F-box containing complex
SET	Su(var)3–9, enhancer of zeste and trithorax
SETD	SET domain containing
SETMAR	SET domain and mariner transposase fusion gene
Slc22a3/Oct3	solute carrier family 22 member 3/organic cation transporter 3

1. HISTONE MODIFICATIONS AND CHROMATIN STRUCTURE

SMYD	SET and MYND domain containing
SUV39H	suppressor of variegation 3–9 homologue
SUZ12	SUZ12 polycomb repressive complex 2 subunit
Swd	Set1C complex subunit Swd
SYMD2	SET and MYND domain containing 2
TCDD	2,3,7,8-tetrachlorodibenzodioxin
TFF1/pS2	trefoil factor 1
TNF	tumor necrosis factor
Topo II2a	topoisomerase II2a
TSS	transcriptional start site
UCP	uncoupling protein
Ugt	UDP glucuronosyltransferase
UPR	unfolded protein response
UTY	ubiquitously transcribed tetratricopeptide repeat containing, Y-linked
WDR	WD repeat domain
ZNF	zinc finger

INTRODUCTION

In eukaryotic cells, DNA is packaged in the nucleus as sister chromatids into a highly organized structure called chromatin. This functional storage of DNA is necessary to appropriately induce, maintain, and repress gene transcription in response to developmental stimuli and xenobiotic exposures. The basic unit of chromatin, referred to as the nucleosome, is composed of 145–147 base pairs of DNA wrapped around an octamer of two copies each of the histone proteins H2A, H2B, H3, and H4 (Fig. 1). The pattern is continued throughout the genome, and chromatin is further condensed with the aid of additional proteins, such as histone H1. In what is referred to as the "histone code," posttranslational modifications to the

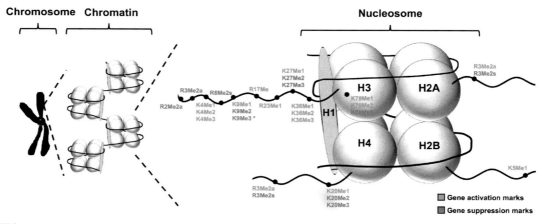

FIG. 1 Positions of various histone methylation marks. The histone octamer contains two molecules each of histones H2A, H2B, H3, and H4, whereas the linker histone (H1) and linker DNA connect the adjacent nucleosomes. Histone methylation occurs on H2A, H2B, H3, and H4. Methylation marks associated with active gene transcription are shown in *green*, whereas marks associated with transcriptional repression are shown in *red*. *Me*, methylation; *K*, lysine; *R*, arginine. *K9Me3 is usually a suppression mark, except that in cancer cells, it can lead to active gene transcription.

N-terminus of histone tails, including acetylation, glutathionylation, methylation, phosphorylation, sumoylation, and ubiquitination, determine gene transcription and influence cell identity through the regulation of transcription, replication, and chromosome maintenance (Strahl and Allis, 2000; Choudhuri et al., 2010; Migliori et al., 2010; Tan et al., 2011; Garcia-Gimenez et al., 2013). Histone modifications can differ among cell types leading to tissue-specific expression signatures of genes that define unique cellular functions (Kouzarides, 2007). Dysregulation of histone modifications may lead to various human diseases such as developmental disorders and cancers. The International Human Epigenome Consortium (http://ihec-epigenomes.org/) and the NIH Roadmap Epigenomics Consortium (http://www.roadmapepigenomics.org/) include a blueprint for epigenetic marks including histone methylation in major primary human cell types and human tissues (Roadmap Epigenomics et al., 2015; Stunnenberg et al., 2016). In this chapter, we will introduce various histone methylation marks and their functions on gene transcription, methyltransferases and demethylases that regulate specific sites of histone methylation patterns, cofactors and other regulators, and human diseases and toxicological responses associated with dysregulation of histone methylation.

Specific amino acid residues on histones are methylated, regulating chromatin architecture and accessibility of transcription factors to target genes. In 1964, methylation of the ε-amino group of lysine in histone proteins was first isolated in a study of histone biosynthesis using isolated calf thymus nuclei, as demonstrated by the transfer of C^{14}-labeled methyl groups from methionine to histones (Allfrey and Mirsky, 1964). Later that same year, lysine methylation was suggested to modulate RNA synthesis and gene expression (Murray, 1964). Over the next few decades, technological advances led to the identification of many methylation sites on the basic residues lysine and arginine (Byvoet et al., 1972). Lysines can be monomethylated (Me1), dimethylated (Me2), or trimethylated (Me3) on their ε-amino group; arginines can be monomethylated, symmetrically dimethylated (Me2s), or asymmetrically dimethylated (Me2a) on their guanidinyl group (Greer and Shi, 2012). Histidine residues have also been reported to be monomethylated in avian erythrocyte histone fractions and in HeLa cells during mitosis; however, the phenomenon appears to be rare (Gershey et al., 1969; Borun et al., 1972). There has been no further characterization of histidine methylation, effect of histidine marks, or identification of enzymes that methylate histidine residues in histones (Greer and Shi, 2012). A single glutamine residue on yeast histone H2A glutamine 104 (H2AQ105) is reported to be methylated by rRNA methyltransferase NOP1 (NOP1); in humans, this corresponds to the methylation of histone H2A glutamine 104 (H2AQ104) by fibrillarin (FBL; Tessarz et al., 2014). In yeast, methylation of Q105 disrupts the binding of the histone chaperone facilitator of chromatin transcription (FACT) complex, which is a complex that increases transcription permissibility and reduces RNA polymerase 1 activity in particular for the transcription of ribosomal components (Tessarz et al., 2014).

Posttranslational methylation of histone residues is catalyzed by methyltransferases and demethylases, also known as writers and erasers, respectively. S-Adenosylmethionine (SAMe) serves as the methyl donor in histone methylation reactions. In 2000, the first histone lysine methyltransferase (KMT) enzyme, namely, suppressor of variegation 3–9 homologue 1 (SUV39H1), was identified in humans and mice by Thomas Jenuwein and colleagues, and SUV38H1 was found to be conserved from yeast to humans (Rea et al., 2000). Afterward, many other KMT enzymes were identified based on gene homology through the conserved, catalytic Su(var)3–9, enhancer of zeste and trithorax (SET) domain (Jenuwein, 2006). A second class of KMTs was also identified that did not contain the SET domain (Table 1). The first coactivator-associated arginine methyltransferase 1 (CARM1) was identified in 1999 and provided the first evidence that arginine methylation can regulate transcription (Chen et al., 1999). Following the initial identification of lysine and arginine methyltransferases, a plethora of enzymes have been discovered with varying histone methylation site specificities (Kooistra and Helin, 2012).

TABLE 1   Human Histone Methyltransferases

Gene Symbol	Description	GeneID	Synonyms	Specificity (Product)	References
Lysine methyltransferases					
ASH1L	ASH1-like histone lysine methyltransferase	55870	ASH11, KMT2H	H3K36Me2, H3K36Me3, H3K4Me3	Gregory et al. (2007) and An et al. (2011)
DOT1L	DOT1-like histone lysine methyltransferase	84444	DOT1, KMT4	H3K79Me1, H3K79Me2, H3K79Me3	Feng et al. (2002) and Kim et al. (2014a,b)
EHMT1	Euchromatic histone lysine methyltransferase 1	79813	EHMT1-IT1, EUHMTASE1, Eu-HMTasel, FP13812, GLP, GLP1, KMT1D	H3K9Me1, H3K9Me2	Collins et al. (2008)
EHMT2	Euchromatic histone lysine methyltransferase 2	10919	BAT8, C6orf30, G9A, GAT8, KMT1C, NG36	H3K9Me1, H3K9Me2, H3K27Me1, H3K27Me2	Collins et al. (2008) and Wu et al. (2011)
EZH1	Enhancer of zeste 1 polycomb repressive complex 2 subunit	2145	KMT6B	H3K27Me1, H3K27Me2, H3K27Me3	Margueron et al. (2008)
EZH2	Enhancer of zeste 2 polycomb repressive complex 2 subunit	2146	ENX-1, ENX1b, KMT6, KMT6A, WVS, WVS2, EZH2	H3K27Me1, H3K27Me2, H3K27Me3	Margueron et al. (2008)
KMT2A	Lysine methyltransferase 2A	4297	ALL-1.CXXC7, HRX, HTRX1, MLL, MLL1, MLL1A, TRX1, WDSTS	H3K4Me1	Southall et al. (2009)
KMT2B	Lysine methyltransferase 2B	9757	CXXC10, DYT28, HRX2, MLL1B, MLL2, MLL4, TRX2, WBP-7, WBP7	H3K4Me3	Demers et al. (2007)

TABLE 1 Human Histone Methyltransferases—cont'd

Gene Symbol	Description	GeneID	Synonyms	Specificity (Product)	References
KMT2C	Lysine methyltransferase 2C	58508	HALR, MLL3	H3K4Me1	Chen et al. (2010)
KMT2D	Lysine methyltransferase 2D	8085	AAD10, ALR, CAGL114, KABUK1, KMS, MLL2, MLL4, TNRC21	H3K4Me1	Chen et al. (2010)
KMT5A	Lysine methyltransferase 5A	387893	PR-Set7, SET07, SET8, SETD8	H4K20Me2	Nishioka et al. (2002)
KMT5B	Lysine methyltransferase 5B	51111	CGI-85, CGI85, SUV420H1	H4K20Me3	Black et al. (2012) and Greer and Shi (2012)
KMT5C	Lysine methyltransferase 5C	84787	SUV420H2, Suv4-20h2	H4K20Me3	Black et al. (2012) and Green et al. (2012)
MECOM	MDS1 and EVI1 complex locus	2122	AML1-EVI-1, EVI1, KMT8E, MDS1, MDS1-EVI1, PRDM3, RUSAT2	H3K9Me1[a]	Black et al. (2012) and Pinheiro et al. (2012)
NSD1	Nuclear receptor-binding SET domain protein 1	64324	ARA267, KMT3B, SOTOS, SOTOS1, STO	H3K36Me1, H3K36Me2, H4K20Me2 (in vitro)	Li et al. (2009a,b), Qiao et al. (2011), and Wagner and Carpenter (2012)
NSD2	Nuclear receptor-binding SET domain protein 2	7468	KMT3F, KMT3G, MMSET, REIIBP, TRX5, WHS, WHSC1	H3K27Me, H3K20Me, H3K4Me, H3K27Me, H3K36Me1, H3K36Me2, H4K20Me3	Kim et al. (2008), Li et al. (2009a,b), and Wagner and Carpenter (2012)
NSD3	Nuclear receptor-binding SET domain protein 3	54904	KMT3F, KMT3G, WHISTLE, WHSC1L1, PP14328	H3K4Me2, H3K27Me2, H3K27Me3, H3K36Me1, H3K36Me2	Kim et al. (2006), Rahman et al. (2011), and Li et al. (2009a,b)
PRDM2	PR/SET domain 2	7799	HUMHOXY1, KMT8, KMT8A, MTB-ZF, RIZ, RIZ1, RIZ2	H3K9Me1	Wu et al. (2010a,b)
PRDM5	PR/SET domain 5	11107	BCS2, PFM2	H3K9Me2	Porter et al. (2015)
PRDM7	PR/SET domain 7	11105	PFM4, ZNF910	H3K4Me3	Blazer et al. (2016)

Continued

TABLE 1   Human Histone Methyltransferases—cont'd

Gene Symbol	Description	GeneID	Synonyms	Specificity (Product)	References
PRDM8	PR/SET domain 8	56978	EPM10, KMT8D, PFM5	H3K9Me2	Eom et al. (2009) and Yang and Shinkai (2013)
PRDM9	PR/SET domain 9	56979	KMT8B, MEISETZ, MSBP3, PFM6, ZNF899	H3K4Me3, H3K36Me1, H3K36Me2, H3K36Me3	Eram et al. (2014)
PRDM12	PR/SET domain 12	59335	PFM9, HSAN8	H3K9Me1, H3K9Me2	Yang and Shinkai (2013)
PRDM13	PR/SET domain 13	59336	MU-MB-20.220, PFM10	H3K9Me3, H3K27Me3	Hanotel et al. (2014)
PRDM16	PR/SET domain 16	63976	CMD1LL, KMT8F, LVNC8, MEL1, PFM13	H3K9Me1[a]	Nishikata et al. (2003), Pinheiro et al. (2012), and Yang and Shinkai (2013)
SETD1A	SET domain containing 1A	9739	KMT2F, Set1, Set1A	H3K4Me1[a]	Lee et al. (2007)
SETD1B	SET domain containing 1B	23067	KMT2G, Set1B	H3K4Me1	Lee et al. (2007)
SETD2	SET domain containing 2	29072	HBP231, HIF-1, HIP-1, HSPC069, HYPB, KMT3A, LLS, SET2, P231HBP	H3K36Me3	Sun et al. (2005) and Yuan et al. (2009)
SETD3	SET domain containing 3	84193	C14orf154	H3K36Me1, H3K36Me2	Eom et al. (2011)
SETD6	SET domain containing 6	79918		H2AZMe1	Levy et al. (2011)
SETD7	SET domain-containing lysine methyltransferase 7	80854	KMT7, SET7, SET7/9, SET9	H3K4Me1	Wang et al. (2001a)
SETDB1	SET domain bifurcated 1	9869	ESET, H3-K9-HMTase4, KG1T, KMT1E, TDRD21	H3K9Me3	Schultz et al. (2002)
SETDB2	SET domain bifurcated 2	83852	C13orf4, CLLD8, CLLL8, KMT1F	H3K9Me3	Falandry et al. (2010)
SETMAR	SET domain and mariner transposase fusion gene	6419	METNASE, Marl	H3K4Me1, H3K36Me2	Lee et al. (2005c) and Fnu et al. (2011)

TABLE 1 Human Histone Methyltransferases—cont'd

Gene Symbol	Description	GeneID	Synonyms	Specificity (Product)	References
SMYD1	SET and MYND domain containing 1	150572	BOP, KMT3D, ZMYND18, ZMYND22	H3K4Me1	Tan et al. (2006) and Nagandla et al. (2016)
SMYD2	SET and MYND domain containing 2	56950	HSKM-B, KMT3C, ZMYND14	H3K4Me (not specific), H3K36Me2	Abu-Farha et al. (2008)
SMYD3	SET and MYND domain containing 3	64754	KMT3E, ZMYND1, ZNFN3A1, bA74P14.1	H3K4Me2, H3K4Me3, H4K5Me1, H4K5Me2, H4K5Me3	Hamamoto et al. (2004), Van Aller et al. (2012), and Peserico et al. (2015)
SUV39H1	Suppressor of variegation 3–9 homologue 1	6839	H3-K9-HMTase1, KMT1A, MG44, SUV39H	H3K9Me2, H3K9Me3	Rea et al. (2000) and Wang et al. (2012)
SUV39H2	Suppressor of variegation 3–9 homologue 2	79723	KMT1B	H3K9Me3	Frontelo et al. (2004)
Arginine methyltransferases					
CARM1	Coactivator-associated arginine methyltransferase 1	10498	PRMT4	H3R17Me1, H3R17Me2, H3R26Me1, H3R26Me2	Li et al. (2002), Miao et al. (2006), and Lakowski and Frankel (2009)
METTL23	Methyltransferase-like 23	124512	C17orf95, MRT44	H3R17Me2a	Rinn et al. (2007)
PRDM4	PR/SET domain 4	11108	PFM1	H4R3Me2s	Chittka et al. (2012)
PRDM6	PR/SET domain 6	93166	KMT8C, PDA3, PRISM	H3R2Me1, H3R2Me2, H4K20Me1	Davis et al. (2006), Bedford (2007), Guccione et al. (2007), and Mozzetta et al. (2015)
PRMT1	Protein arginine methyltransferase 1	3276	ANM1, HCP1, HRMT1L2, IR1B4	H4R3Me1, H4R3Me2a	Wang et al. (2001b)
PRMT2	Protein arginine methyltransferase 2	3275	HRMT1L1	H3RMe2a, H4Me	El Messaoudi et al. (2006) and Lakowski and Frankel (2009)
PRMT5	Protein arginine methyltransferase 5	10419	HRMT1L5, IBP72, JBP1, SKB1, SKB1Hs	H2AR3Me2s, H3R8Me1, H3R8Me2 H4R3Me1, H4R3Me2s	Lacroix et al. (2008), Majumder et al. (2010), Tee et al. (2010), and Girardot et al. (2014)

Continued

1. HISTONE MODIFICATIONS AND CHROMATIN STRUCTURE

TABLE 1   Human Histone Methyltransferases—cont'd

Gene Symbol	Description	GeneID	Synonyms	Specificity (Product)	References
PRMT6	Protein arginine methyltransferase 6	55170	HRMT1L6	H2AR3Me1, H4R3Me1, H3R2Me2a, H4R3Me2a	Hyllus et al. (2007)
PRMT7	Protein arginine methyltransferase 7	54496	SBIDDS	H2A, H4R3Me2	Lee et al. (2005b)
PRMT8	Protein arginine methyltransferase 8	56341	HRMT1L3, HRMT1L4	H4R3Me2a	Lee et al. (2005a)
PRMT9	Protein arginine methyltransferase 9	90826	PRMT10	H4R3Me2	Niu et al. (2007)
Glutamine methyltransferase					
FBL	Fibrillarin (humans)	2091	FIB, FLRN, Nop1, RNU3IP1	H2AQ104Me (humans)	Tessarz et al. (2014)

[a] *Reaction occurs in cytoplasm.*

Initially, it was thought that methylation of histone residues was permanent, inherited, and irreversible (Byvoet et al., 1972; Kooistra and Helin, 2012). However, the identification of the histone H3 lysine 4 (H3K4) demethylase lysine demethylase 1A (KDM1A) demonstrated that histone methylation is reversible, and since then, more lysine demethylases have been identified (Table 2; Shi et al., 2004). Regarding arginine demethylases, only arginine demethylase and lysine hydroxylase (JMJD6) has been suggested to demethylate arginine residues (Chang et al., 2007). Although JMJD6 is a lysine hydroxylase (Webby et al., 2009; Mantri et al., 2010), some lysine demethylases have been shown to have arginine demethylase activity (Di Lorenzo and Bedford, 2011; Walport et al., 2016). Histone demethylases for histidine and glutamine have not been identified.

SITE SPECIFICITY OF HISTONE METHYLATION

Methylation of histones can occur on one or more residues of all histones, especially on lysine and arginine residues, which are the most characterized targets. Reported methylation targets include, but are not limited to, H2BK5, H3K4, H3K9, H3K27, H3K36, H3K79, H4K5, H4K20, H2A arginine 3 (H2AR3), H3R2, H3R3, H3R8, H3R17, H3R23, H3R26, H4R3, and H2AQ104 (Tables 1 and 3). Fig. 1 shows the sites where histone methylation occurs; marks associated with active gene transcription are shown in green, whereas marks associated with transcriptional silencing are shown in red. Interestingly, up to 40%–80% of mammalian cells are dimethylated at H3K9, H3K27, H3K36, or H4K20 (Schotta et al., 2008; Jung et al., 2010; Kooistra and Helin, 2012). Histone methylation marks can be either permissive or repressive for gene transcription (Table 3). For example, chromatin immunoprecipitation coupled with

TABLE 2 Human Histone Demethylases

Gene Symbol	Description	GeneID	Synonyms	Specificity (Substrate)	References
Lysine demethylases					
HR	HR, lysine demethylase and nuclear receptor corepressor	55806	ALUNC, AU, HSA277165, HYPT4, MUHH, MUHH1	H3K9Me1, H3K9Me2	Liu et al. (2014)
JMJD1C	Jumonji domain-containing 1C	221037	KDM3C, TRIP-8, TRIP8	H3K9Me1, H3K9Me2	Kim et al. (2010)
KDM1A	Lysine demethylase 1A	23028	AOF2, BHC110, CPRF, KDM1, LSD1	H3K4Me1, H3K4Me2, H3K9Me2	Forneris et al. (2005) and Laurent et al. (2015)
KDM1B	Lysine demethylase 1B	221656	AOF1, C6orf193, LSD2, bA204B7.3, dJ298J15.2	H3K4Me1, H3K4Me2	Ciccone et al. (2009)
KDM2A	Lysine demethylase 2A	22992	CXXC8, FBL11, FBL7, FBXL11, JHDM1A, LILINA	H3K36Me1, H3K36Me2	Blackledge et al. (2010)
KDM2B	Lysine demethylase 2B	84678	CXXC2, FBXL10, Fbl10, JHDM1B, PCCX2	H3K4Me3, H3K36Me1, H3K36Me2, H3K79Me2, H3K79Me3	Frescas et al. (2007) and Seo et al. (2017)
KDM3A	Lysine demethylase 3A	55818	JHDM2A, JHMD2A, JMJD1, JMJD1A, TSGA	H3K9Me1, H3K9Me2	Yamane et al. (2006)
KDM3B	Lysine demethylase 3B	51780	5qNCA, C5orf7, JMJD1B, NET22	H3K9Me1, H3K9Me2	Yamane et al. (2006)
KDM4A	Lysine demethylase 4A	9682	JHDM3A, JMJD2, JMJD2A, TDRD14A	H3K9Me3, H3K36Me3, H4K20Me3	Whetstine et al. (2006) and Lee et al. (2008)
KDM4B	Lysine demethylase 4B	23030	JMJD2B, TDRD14B	H3K9Me3	Whetstine et al. (2006)
KDM4C	Lysine demethylase 4C	23081	GASC1, JHDM3C, JMJD2C, TDRD14C	H3K9Me3, H3K36Me3	Whetstine et al. (2006)
KDM4D	Lysine demethylase 4D	55693	JMJD2D	H3K9Me3	Whetstine et al. (2006
KDM4E	Lysine demethylase 4E	390245	JMJD2E, KDM4DL, KDM5E	H3K9Me2, H3K9Me3	Hillringhaus et al. (2011)

Continued

1. HISTONE MODIFICATIONS AND CHROMATIN STRUCTURE

TABLE 2   Human Histone Demethylases—cont'd

Gene Symbol	Description	GeneID	Synonyms	Specificity (Substrate)	References
KDM5A	Lysine demethylase 5A	5927	RBBP-2, RBBP2, RBP2	H3K4Me3	Klose et al. (2007)
KDM5B	Lysine demethylase 5B	10765	CT31, JARID1B, PLU-1, PLU1, PPP1R98, PUT1, RBBP2H1A, RBP2-H1	H3K4Me3	Yamane et al. (2007)
KDM5C	Lysine demethylase 5C	8242	DXS1272E, JARID1C, MRX13, MRXJ, MRXSCJ, MRXSJ, SMCX, XE169	H3K4Me3	Iwase et al. (2007)
KDM5D	Lysine demethylase 5D	8284	HY, HYA, JARID1D, SMCY	H3K4Me3, H3K4Me2	Iwase et al. (2007)
KDM6A	Lysine demethylase 6A	7403	KABUK2, UTX, bA386N14.2	H3K27Me2, H3K27Me3	Lan et al. (2007a,b)
KDM6B	Lysine demethylase 6B	23135	JMJD3	H3K27Me2, H3K27Me3	Lan et al. (2007a,b)
KDM7A	Lysine demethylase 7A	80853	JHDM1D	H3K9Me2, H3K27Me1, H3K27Me2	Tsukada et al. (2010) and Horton et al. (2010)
KDM8	Lysine demethylase 8	79831	JMJD5	H3K36Me2	Hsia et al. (2010)
PHF2	PHD finger protein 2	5253	CENP-35, GRC5, JHDM1E, KDM7C	H3K9Me2	Baba et al. (2011)
PHF8	PHD finger protein 8	23133	JHDM1F, KDM7B, MRXSSD, ZNF422	H3K9Me2, H3K27Me2, H3K36Me2, H4K20Me1	Yu et al. (2011), Liu et al. (2010), Loenarz et al. (2010), and Horton et al. (2010)
RIOX1	Ribosomal oxygenase 1	79697	C14orf169, JMJD9, MAPJD, N066, ROX, URLC2, hsN066	H3K4Me3, H3K4Me1, H3K36Me3	Brien et al. (2012)
RIOX2	Ribosomal oxygenase 2	84864	JMJD10, MDIG, MINA, MINA53, N052, ROX	H3K9Me3	Lu et al. (2009)
UTY	Ubiquitously transcribed tetratricopeptide repeat containing, Y-linked	7404	KDM6AL, KDM6C, UTY1	H3K27Me3	Walport et al. (2014)

TABLE 2 Human Histone Demethylases—cont'd

Gene Symbol	Description	GeneID	Synonyms	Specificity (Substrate)	References
Arginine demethylases					
KDM3A	Lysine demethylase 3A	55818	JHDM2A, JHMD2A, JMJD1, JMJD1A, TSGA	H3K9RMe2a[a], H3K9RMe2s[a]	Walport et al. (2016)
KDM4E	Lysine demethylase 4E	390245	JMJD2E, KDM4DL, KDM5E	H3K9RMe2a[a], H3K9RMe2s[a], H3K9RMe1[a], H3R2Me2a[a], H4R3Me2a[a]	Walport et al. (2016)
KDM5C	Lysine demethylase 5C	8242	DXS1272E, JARID1C, MRX13, MRXJ, MRXSCJ, MRXSJ, SMCX, XE169	H3K4RMe2a[a], H3K4RMe2s[a], H3R2Me2s[a]	Walport et al. (2016)
KDM6B	Lysine demethylase 6B	23135	JMJD3	H3K27RMe2a[a]	Walport et al. (2016)
JMJD6	Arginine demethylase and lysine hydroxylase	23210	PSR, PTDSR, PTDSR1	H3R2Me2, H4R3Me2, H4R3Me2	Chang et al. (2007), Webby et al. (2009), Ng et al. (2009), and Hahn et al. (2010)

[a] *In vitro.*

high-throughput sequencing (ChIP-Seq)-based high-resolution profiling of histone methylation code of the human genome showed that the monomethylations of H2BK5Me1, H3K9me1, H3K27me1, H3K79me1, and H4K20me1 are linked to transcriptional activation whereas trimethylations of H3K27me3, H3K9me3, and H3K79Me3 are linked to transcriptional repression (Barski et al., 2007). In this section, we describe the biological functions of each histone methylation mark, the methyltransferases that catalyze the formation of these marks, and the demethylases that catalyze the removal of these marks (Tables 1–4).

TABLE 3 Site Specificity of Histone Methylation in Humans

Transcriptional Activation		Transcriptional Suppression	
H2	AR3Me2a	H2	AR3Me2s
			AQ104Me1
	BK5Me1		BK5Me3
H3	K4Me1, K4Me2, K4Me3, K9Me1, K9Me3 (cancer cells), K27Me1, K36Me1, K36Me2, K36Me3, K79Me1, K79Me2 R17Me2a	H3	K9Me2, K9Me3 (usually), K27Me2, K27Me3, K79Me3 R2Me2a, R3Me2a, R8Me2s
H4	K20Me1 R3Me2a	H4	K20Me2, K20Me3 R3Me2s

TABLE 4   A Summary of Major Histone Methylation Marks and Associated Enzymes for Their Methylation/Demethylation

Methyltransferases	Histone Methylation Marks	Demethylases
PRMT5	H2AR3Me2a	
	H2BK5Me1	
	H2AR3Me2s	
	H2AQ104Me1	
	H2BK5Me3	
KMT2A, KMT2C, KMT2D, SETD1A, SETD1B, SETD7, SETMAR, SMYD1	H3K4Me1	KDM1A, KDM1B, RI0X1
NSD3, SMYD3	H3K4Me2	KDM1A, KDM1B, KDM5D
ASH1L, KMT2B, PRDM7, PRDM9, SMYD3	H3K4Me3	KDM2B, KDM5A, KDM5B, KDM5C, KDM5D, RIOX1
EHMT1, EHMT2, MECOM, PRDM2, PRDM12, PRDM16	H3K9Me1	HR, JMJD1C, KDM3A, KDM3B
EHMT2, EZH1, EZH2	H3K27Me1	KDM7A
NSD1, NSD2, NSD3, PRDM9, SETD3	H3K36Me1	KDM2A, KDM2B
ASH1L, NSD1, NSD2, NSD3, PRDM9, SETD3, SETMAR, SMYD2	H3K36Me2	KDM2A, KDM2B, KDM8, PHF8
ASH1L, PRDM9, SETD2	H3K36Me3	KDM4A, KDM4C, RI0X1
DOT1L	H3K79Me1	
DOT1L	H3K79Me2	KDM2B
METTL23	H3R17Me2a	
EHMT1, EHMT2, PRDM5, PRDM8, PRDM12, SUV39H1	H3K9Me2	HR, JMJD1C, KDM1A, KDM3A, KDM3B, KDM4E, KDM7A, PHF2, PHF8
PRDM13, SETDB1, SETDB2, SUV39H1, SUV39H2	H3K9Me3	KDM4A, KDM4B, KDM4D, KDM4E, RI0X2
EHMT2, EZH1, EZH2, NSD3	H3K27Me2	KDM6A, KDM6B, KDM7A, PFH8
EZH1, EZH2, NSD3, PRDM13	H3K27Me3	KDM6A, KDM6B, UTY
DOT1L	H3K79Me3	KDM2B
PRMT6	H3R2Me2a	KDM4E
	H3R3Me2a	
	H3R8Me2s	
PRDM6	H4K20Me1	PHF8
PRMT1, PRMT6, PRMT8	H4R3Me2a	KDM4E
KMT5A, NSD1	H4K20Me2	
KMT5B, KMT5C, NSD2	H4K20Me3	KDM4A
PRDM4, PRMT5	H4R3Me2s	

Lysine Methylation

H2BK5 Methylation

Little is known regarding H2BK5 methylation. H2BK5me1, first reported in 2007, is associated with active promoter regions downstream of the transcriptional start site (TSS) near the 5′ end (Barski et al., 2007; Linghu et al., 2013). H2BK5me1 is enriched at exons and is positively correlated with gene expression levels, suggesting that H2BK5me1 is a permissive histone mark (Barski et al., 2007; Hon et al., 2009; Spies et al., 2009). No H2BK5 histone methyltransferases of demethylases have been identified.

H3K4 Methylation

H3K4 methylation is an evolutionarily conserved histone posttranscriptional modification pattern that marks active gene transcription. H3K4me1 is enriched at both promoters and distal enhancers, as well as introns; H3K4me2 is highly prevalent at actively transcribed gene bodies (immediately downstream of TSS), whereas H3K4me3 is present at promoters but not enhancers (Heintzman et al., 2007; Spies et al., 2009; Dhami et al., 2010; Tian et al., 2011; Kooistra and Helin, 2012; Shilatifard, 2012).

For H3K4me1, its prevalence at distal gene regulatory elements enables the identification of novel enhancer sequences (Heintzman et al., 2007; Hon et al., 2009; Rada-Iglesias et al., 2011). H3K4me1 usually colocalizes with E1A-binding protein p300, which also marks the enhancer region, whereas H3K4me3 does not show enrichment around p300 binding sites (Heintzman et al., 2007). An algorithm called REPTILE has been developed to identify the precise location of enhancers based on epigenetic marks including H3K4me1 (He et al., 2017). The presence of H3K4me1 is cell-type-specific and is highly correlated with cell-type-specific expression of the putative gene targets of these enhancers (Hon et al., 2009). For H3K4me2, it is present at both TSS and actively transcribed gene bodies, and its presence immediately downstream of TSS marks tissue-specific genes and is important for cell differentiation and tissue development (Pekowska et al., 2010; Zhang et al., 2012a). For H3K4me3, it marks the active gene promoters and is highly prevalent around RNA polymerase II binding sites, suggesting that these target genes are poised for transactivation or retain a memory for posttranscription (Barski et al., 2007; Barrera et al., 2008). Interestingly, H3K4me3 is also detected in inactive promoters, indicating that silent genes that acquire this permissive mark may become activated with the recruitment of additional transcriptional factors (Santos-Rosa et al., 2002; Barski et al., 2007; Mikkelsen et al., 2007). The presence of H3K4me3 is particularly important in cell-cycle regulation for genes that are silent at G0 but are active at G1 (Smith et al., 2009).

Regarding H3K4 methyltransferases, SET domain-containing lysine methyltransferase 7 (SETD7) was the first identified enzyme that monomethylates H3K4 (Wilson et al., 2002; Xiao et al., 2003). Its initial discovery in 2001 showed H3K4 methylation as a mark for active gene transcription that inhibited the repressive mark H3K9 produced by SUV39H1 (Wang et al., 2001a). Similarly, recombinant SET domain and mariner transposase fusion gene (SETMAR) also produces H3K4me1 and does not appear to have intrinsic activity for forming di- or trimethylation (Lee et al., 2005b). In yeast, histone lysine methyltransferase Set1 (Set1) performs all H3K4 methylations by forming a multisubunit complex of proteins associated with Set1 (COMPASS). The SET domain of the catalytic subunit of Set1 is assisted by five other subunits: Set1C complex subunit Swd1 (Swd1), Swd3, Set1C complex subunit Bre2 (Bre2), and Dpy-30 domain protein Sdc1 (Sdc1; Briggs et al., 2001; Roguev et al., 2003; Racine et al., 2012; Tang et al., 2013; Mikheyeva et al., 2014; Hyun et al., 2017). The loss of any subunit can affect the stability of Set1 activity, the global H3K4 methylation levels, and the distribution of H3K4 methylation along active genes (Dehe and Geli, 2006). Interestingly, Set1/COMPASS interacts with RNA polymerase II, and it has been suggested that the gradient for H3K4 methylation level transitions from H3K4me1 to H3K4me2 and H3K4me3 in a stepwise pattern that is dependent on the number of transcription events (Soares et al., 2017). This "time" model indicates that the same promoter may respond differently to transcriptional regulators depending on if the site is di- or trimethylated (Soares et al., 2017). Humans have six SET1 homologues: SETD1A, SETD1B, lysine methyltransferase 2A (KMT2A), KMT2B, KMT2C, and KMT2D. Each homologue is associated with four other subunits, each with distinct catalytic properties that are referred to as a myeloid/lymphoid or mixed-lineage leukemia (MLL) complex: WD repeat domain 5 (WDR5); retinoblastoma-binding protein 5, histone lysine methyltransferase complex subunit (RBBP5); absent, small, or homeotic discs 2-like histone lysine methyltransferase complex subunit (ASH2L); and histone methyltransferase complex regulatory subunit (DPY30; Kooistra and Helin, 2012; Shilatifard, 2012). For example, in the presence of WDR5, KMT2A and KMT2B core complexes catalyze H3K4me1 and H3K4me2 formation, whereas KMT2C and KMT2D are specific for H3K4me1; in the absence of WDR5, weak H3K4me1 is observed (Shinsky et al., 2015). The human enzyme ASH1L, the human homologue of the *Drosophila melanogaster* absent, small, or homeotic disc1 (Ash1), also contains a SET domain that catalyzes the product H3K4me3 in vivo and in vitro (Byrd and Shearn, 2003; Gregory et al., 2007) and has also been reported to methylate H3K36 in vitro (Tanaka et al., 2007; An et al., 2011). Due to in vitro experiments and no studies examining the full-length protein catalytic mechanism, some controversy persists regarding ASH1L site specificity and activity of ASH1L in vivo (Gregory et al., 2007; Tanaka et al., 2007; An et al., 2011). The enzyme nuclear receptor-binding SET domain protein 3 (NDS3) has increased RNA expression in bone marrow of leukemia patients, and it forms H3K4me2 and H3K27me2/3 in human transgenic NIH 3T3 mouse embryonic fibroblast cells (Kim et al., 2006) and has increased activity in the presence of heat shock protein 90 (HSP90; Kim et al., 2010). Initially identified in mice, the enzyme meiosis-induced factor containing a proline-rich/SET domain and zinc-finger motif (Meisetz) is essential for the progression of meiosis through H3K4 trimethylation (Fumasoni et al., 2007). The human ortholog, proline-rich domain-containing protein 9 (PRDM9), can form H3K4me1/2/3 (Eram et al., 2014), and a recent gene copy found only in primates, namely, PRDM7, can dimethylate H3K4 (Fumasoni et al., 2007; Blazer et al., 2016). Studied in mice and zebra fish muscle tissue and cells, SET and MYND domain containing 1 (Smyd1) also methylates H3K4, but methylation-state specificity is not known (Tan et al., 2006). Smyd1 interacts with histone deacetylases through the MYND domain to repress RNA transcription (Nagandla et al., 2016). Using an in vitro histone methylation assay and a nonspecific antibody for chromatin immunoprecipitation with 293 human embryonic kidney cells, SMYD2 was also shown to methylate H3K4, and SMYD2 activity was enhanced by the presence of HSP90A (Abu-Farha et al., 2008). Similarly, SMYD3, which is overexpressed in colorectal cancers and may regulate homologous recombination DNA repair, requires HSP90A for di- and trimethylation of H3K4 (Hamamoto et al., 2004; Chen et al., 2017).

Interestingly, recent studies have demonstrated that histone lysine methyltransferases also carry methylation activities toward nonhistone proteins (Steinmeyer and Kalbhen, 1991; Chuikov et al., 2004; Pradhan et al., 2009; Zhang et al., 2012b). For example, the KMT7-/SET7-mediated methylation of p53, which is the guardian of the genome that prevents gene mutation and carcinogenesis, was the first reported KMT-mediated methylation event among nonhistone proteins (Chuikov et al., 2004). It was later discovered that other KMTs and KMD1A also regulate p53 methylation or demethylation (Chuikov et al., 2004; Huang et al., 2006; Shi et al., 2007; Chen et al., 2010; Huang et al., 2010). The p53 methylation at the C-terminus plays a key role in the regulation of protein stability, transactivation of p53, and protein-protein interactions.

Regarding H3K4 demethylases, the histone demethylase lysine demethylase 1A (KDM1A/LSD1) is conserved from yeast to humans, and its family member KDM1B/LSD2 can demethylate H3K4me1 and H3K4me2 following their recruitment to target genes by repressive complexes (Shi et al., 2004; Forneris et al., 2005; Lee et al., 2005a; Lan et al., 2007b; Ciccone et al., 2009; Hyun et al., 2017). A second family of H3K4 demethylases referred to as the jumonji AT-rich interactive domain 1 (JARID1) protein family, namely, KDM5A, KDM5B, KDM5C, KDM5D, and ribosomal oxygenase 1 (RIOX1), was found to be specific for H3K4me2 and H3K4me3 demethylation and can function as transcriptional corepressors (Christensen et al., 2007; Iwase et al., 2007; Sinha et al., 2010; Hyun et al., 2017). The lysine demethylase KDM2B is a member of the Jmjc domain-containing histone demethylase family and, shown in human cell lines, specifically demethylates H3K4me3 in ribosomal DNA (Frescas et al., 2007).

The importance of broad H3K4me3 domains has been noted regarding the regulation of complex gene regulation in the third dimension (Chen et al., 2015). Stretched enhancers, such as superenhancers and broad H3K4me3 domains, have been associated with the maintenance of cell identity and cancer. In immortalized cell lines and primary normal human samples, superenhancers and broad H3K4me3 showed high association with chromatin interactions than their typical counterparts (Cao et al., 2017). Certain overlapping regions of superenhancers and H3K4me3 showed high association with cancer-related genes (Cao et al., 2017).

H3K9 Methylation

H3K9 methylation is a histone modification that is usually an indicator of silenced gene transcription and heterochromatin structure (Barski et al., 2007); in general, H3K9me2 and H3K9me3 are higher in and spread evenly across silent genes, whereas H3K9me1 can be found near promoters of active genes (Barski et al., 2007; Kooistra and Helin, 2012). Although ChIP-Seq of human CD4$^+$ T cells indicates that only H3K9me2 is significant at silent genes and that H3K9me1 is also spread throughout the body of active genes and at enhancer regions (Barski et al., 2007; Wang et al., 2008; Rosenfeld et al., 2009), both di- and trimethylation of H3K9 are significantly enriched at pericentromeres and gene deserts, contributing to silent transcription (Rosenfeld et al., 2009). This is expected because H3K9me3 is a mark for constitutive heterochromatin, but H3K9me3 is one of the least significant marks for telomeres, indicating that telomeres are different from other noncoding regions (Rosenfeld et al., 2009). It has also been shown that a local increase of H3K9me2 and H3K9me3 leads to the inclusion of an alternative exon and demethylation resulting in exon skipping (Bieberstein et al., 2016). Furthermore, using human recombinant proteins and mouse embryonic cells,

SUV39H1-dependent H3K9me2 and H3K9me3 create a binding site for heterochromatin protein 1 (HP1) proteins, which are a family of heterochromatic adaptor molecules implicated in gene transcriptional silencing and supranucleosomal chromatin structure (Lachner et al., 2001). This enhances the notion of H3K9 methylation as a transcriptional repressive mark. Interestingly, in human and mouse embryonic cell lines, HP1 also interacts with DNA methyltransferase 3A/B (DNMT3A/B) at H3K9me3 heterochromatin sites, reinforcing the condensed chromatin structure (Fuks et al., 2003; Lehnertz et al., 2003); in fact, H3K9 methylation is necessary for DNA methylation in the mold *Neurospora crassa* and the weed *Arabidopsis thaliana* (Tamaru and Selker, 2001; Gendrel et al., 2002). In human cells, H3K9 methylation during DNA replication colocalizes ubiquitin-like with PHD and ring finger domains 1 (UHRF1) and DNMT1 for the maintenance of DNA methylation (Rothbart et al., 2012). In mice, the H3K9me1 and H3K9me2 producing methyltransferases Ehmt1 and Ehmt2 form a silencing complex with Dnmt3a through the bridging enzyme M-phase phosphoprotein 8 (MPP8), which may explain, in part, the co-occurrence of DNA methylation and H3K9 methylation (Chang et al., 2011; Rose and Klose, 2014). Unexpectedly, H3K9me2 and H3K9me3 interacting with HIPγ were found to increase during the activation of transcription and then removed upon gene repression, indicating a role for heterochromatin components with the transcriptional elongation of RNA polymerase II (Vakoc et al., 2005). Furthermore, H3K9me3 was found to be positively associated with active genes in multiple human cancer cell lines with marks found in promoter regions and downstream of TSSs (Wiencke et al., 2008). This also extended to an inverse relationship between promoter region DNA hypermethylation and H3K9me3 such that nonmethylated genes were overexpressed compared with methylated genes (Wiencke et al., 2008).

Regarding H3K9 methyltransferases, euchromatic histone lysine methyltransferase 1 (EHMT1/GLP) and 2 (EHMT2/G9A) catalyze mono- and dimethylation of H3K9 (Collins et al., 2008; Table 1). PRDM2, MECOM/PRDM3, and PRDM16 are H3K9me1 methyltransferases required for mammalian heterochromatin integrity (Wu et al., 2010a; Pinheiro et al., 2012). PRDM12 catalyzes H3K9 mono- and dimethylation (Yang and Shinkai, 2013). PRDM5 and PRDM8 are H3K9 dimethyltransferases (Eom et al., 2009; Porter et al., 2015). PRDM13 is H3K9 trimethyltransferase (Hanotel et al., 2014), as well as SETDB1 (Schultz et al., 2002), SETDB2 (Falandry et al., 2010), and SUV39H2 (Frontelo et al., 2004). SUV39H1 catalyzes both di- and trimethylation of H3K9 (Rea et al., 2000; Wang et al., 2012). In mouse embryonic stem cells, Suv39h-dependent H3K9me3 marks intact retrotransposons and silences long interspersed nuclear elements, providing a heterochromatin environment to restrict aberrant expression of retrotransposons (Bulut-Karslioglu et al., 2014).

Regarding H3K9 demethylases, hairless (HE) protein contains a jumonji C (JmjC) domain (Liu et al., 2014), JMJD1C (Kim et al., 2010); KDM3A and KDM3B (Yamane et al., 2006) can demethylate H3K9me1/2 (Table 2). KDM1A, also known as lysine-specific demethylase 1 (LSD1), is an H3K9me2 demethylase (Laurent et al., 2015). KDM7A/JHDM1D, PHF2, and PHF8 are also H3K9me2 demethylases (Horton et al., 2010; Baba et al., 2011). KDM4E is responsible for the demethylation of H3K9me2/3 (Hillringhaus et al., 2011). KDM4A, KDM4B, KDM4C, KDM4D, and RIOX2 are H3K9me3 demethylases (Whetstine et al., 2006; Lu et al., 2009).

Recently, it has been demonstrated that H3K9me mark can interact with the heterochromatin protein 1 (HP1; Bryan et al., 2017; Larson et al., 2017). The HP1 proteins can subsequently form phase-separated droplets that contain DNA elements that are part of heterochromatin components. These observations suggest that H3K9-methylation-mediated gene silencing may occur at least in part through the sequestration of compact chromatin in phase-separated HP1 protein droplets in a nuclear context-specific manner.

H3K27 Methylation

The methylation at H3K27 occurs in three forms, namely, mono- (H3K27me1), di- (H3K27me2), and trimethylation (H3K27me3). H3K27me1 is associated with active gene transcription, whereas H3K27me2 and H2K27me3 are linked to transcriptional repression (Barski et al., 2007; Cui et al., 2009a). A genome-wide characterization in mouse embryonic stem cells has shown that the three forms of H3K27 methylation are mutually exclusive and form spatially defined genomic domains. H3K27me1 accumulates specifically in intragenic regions of actively transcribed genes (the 5′ region downstream of the promoter) and correlates with the deposition of H3K36Me3, which promotes transcriptional elongation (to be addressed in the next section); H3K27me2 forms large intergenic and intragenic domains (Ferrari et al., 2014), whereas H3K27me3 is found at promoter regions, and in embryonic stem cells, H3K27me3 is found to be colocalized with the transcriptional activation mark H3K4me3 at the promoters of "bivalent genes" that are poised for either activation or repression (Bernstein et al., 2006). H3K27me3 is recognized by the polycomb repressive complex 1 (PRC1), which represses transcription by ubiquitination of histone H2A on lysine K119 and chromatin compaction (Margueron and Reinberg, 2011). H3K27me3 is a feature of facultative heterochromatin in many eukaryotes and maintains transcriptional repression established during early development, X-chromosome inactivation, genomic imprinting, and aberrant inactivation of genes in cancer (Wiles and Selker, 2017).

EZH1 and EZH2 are major histone methyltransferases for all three methylation forms of H3K27, and EHMT2/G9a can mediate the mono- and dimethylation of H3K27 (Margueron et al., 2008; Wu et al., 2011; Table 1). H3K27me3 is generated by the polycomb repressive complex 2 (PRC2), composed of the SET domain-containing histone methyltransferase EZH2 (enhancer of zeste homologue 2) or its functional homologue EZH1 and core accessory proteins (EED, SUZ12, and RbAp48; Cao et al., 2002; Margueron et al., 2008; Ezponda and Licht, 2014; Wiles and Selker, 2017). In *Drosophila*, polycomb protein binding is completed, locus specifically by both H3K9me3 (see the previous section) and H3K27me3, but there is no detectable colocalization between polycomb and H3K27me2 (Ringrose et al., 2004). In mouse embryonic stem cells, genome-wide intragenic deposition of H3K27me1 is dependent on PRC2 activity, H3K27me1 loss correlates with impaired transcription, and there is a preferential accumulation of H3K27me1 over H3K27me2 that is dependent on PRC2 activity (Ferrari et al., 2014). PRC2 also methylates the RNA polymerase II transcription elongation factor, namely, elongin A (EloA), and this is necessary in the repression of certain PRC2-targeted genes during embryonic stem cell differentiation (Ardehali et al., 2017).

As shown in Table 1, NSD2/REIIBP is another histone methyltransferase for H3K27 with transcriptional repression activity (Kim et al., 2008). NSD3/WHISTLE is a SET domain containing a protein with specific H3K4 and H3K27 histone methyltransferase activity (Kim et al., 2006). Prdm13 is found to cause H3K27 trimethylation that induces repressive chromatin (Hanotel et al., 2014).

KDM7A/JHDM1D is the major histone demethylase that demethylates H3K27me1/Me2 (Tsukada et al., 2010; Table 2). Removal of di- and trimethyl groups from H3K27 is performed by the histone demethylases KMD6A/UTX and KDM6B/JMJD3 (Hong et al., 2007; Lan et al., 2007a). KMD6A/UTX is part of a transcriptional activator complex including the H3K4 methyltransferase mixed-lineage leukemia 2/3 (MLL2/MLL3), suggesting a concerted mechanism in which the repressive H3K27 methyl marks are removed and H3K4 is methylated to activate transcription (Agger et al., 2007). UTY, also known as KDM6C, is found to catalyze the demethylation of H3K27me3 in vitro (Walport et al., 2014). Moreover, PHF8 is a histone demethylase for multiple substrates including H3K27me2 (Liu et al., 2010).

H3K36 Methylation

H3K36 methylation is considered a cotranscriptional event necessary to suppress histone exchange on transcribed genes. Specifically, in yeast, methylation of H3K36 leads to the activation of a histone deacetylase complex to maintain the hypoacetylated state of the coding region and suppress spurious cryptic transcripts from initiating within open reading frames so as to ensure the accuracy of transcription by RNA polymerase II (Venkatesh et al., 2012). Mono-, di-, and trimethylation of H3K36 are enriched at actively transcribed genes. In contrast to the promoter-proximal region-enriched H3K4 methylation, H3K36me3 peaks at the 3′ end and is the only covalent histone modification reported to be enriched in the 3′ end of active genes in higher eukaryotes (Bannister et al., 2005; Barski et al., 2007). H3K36me2 shows an intermediate distribution adjacent to promoters, and H3K36Me1 is enriched exclusively in the 5′ region downstream of the promoter (Bell et al., 2007). In fungi, H3K36me2 and H3K36me3, but not H3K36me1, correlate with transcriptional activation (Xu et al., 2008). Although H3K36 methylation is most commonly associated with the transcription of active euchromatin, it has also been implicated in diverse processes, including alternative splicing, dosage compensation, transcriptional repression, and DNA repair and recombination (de Almeida and Carmo-Fonseca, 2012; Wagner and Carpenter, 2012; Sorenson et al., 2016; Ardehali et al., 2017; Larson et al., 2017). Disrupted placement of methylated H3K36 within the chromatin landscape can lead to a range of human diseases.

Regarding methyltransferases responsible for H3K36 methylation, NSD1, NSD2, and NSD3 are mono- and dimethyltransferases for H3K36 (Li et al., 2009b; Table 1). In yeast, Set2 (ortholog of human SETD2) performs all three methylation reactions at H3K36 and couples H3K36Me3 with transcriptional elongation through an interaction with RNA polymerase II (Kizer et al., 2005). In contrast, the human SETD2 can only catalyze H3K36me3 in vivo and has been identified as a Huntingtin-interacting protein (Yuan et al., 2009). SETD3 can catalyze the formation of H3K36me1 and H3K36me2 (Eom et al., 2011). In *Drosophila*, two histone methyltransferases dMes4 (ortholog of human NSD family) and dHypb (ortholog of human SETD2/HYPB) are responsible for forming di- and trimethylation of H3K36, respectively (Bell et al., 2007). In vitro evidence has shown that PRDM9 is a highly active histone methyltransferase that can catalyze mono-, di-, and trimethylation of H3K36, although its H3K36me3 activity was validated mechanistically by overexpression of PRDM9 in HEK293 cells (Eram et al., 2014). SMYD2 and SETMAR are dimethyltransferases for H3K36 (Lee et al., 2005b; Abu-Farha et al., 2008; Fnu et al., 2011). ASH1-like histone lysine methyltransferase (ASH1L) mediates the di- and trimethylation of H3K36 (Gregory et al., 2007; An et al., 2011).

Regarding demethylases for methylated H3K36, mammalian JHDM1 subfamily members KDM2A/JHDM1A and KDM2B/JHDM1B reverse mono- and dimethylated H3K36

(Tsukada et al., 2006; Frescas et al., 2007; Blackledge et al., 2010), and the JMJD2 subfamily members KDM4A/JMJD2A and KDM4C/JMJD2C are trimethyl-specific demethylases for H3K36 (Whetstine et al., 2006; Table 2). KDM8/JMD5 and PHF8 are responsible for the demethylation of H3K36Me2 (Hsia et al., 2010; Loenarz et al., 2010), and histone demethylase RIOX1/NO66 removes the trimethylation at H3K36 (Brien et al., 2012).

H3K79 Methylation

Although both H3K36Me3 and H3K79 methylation in general both target the transcribed regions of active genes in the human genome, H3K36me3 peaks near the 3′ ends, whereas H3K79 methylation peaks at the 5′ ends of genes (Wang et al., 2008). The association of H3K79 methylation with either gene activation or repression is species-dependent. In yeast, H3K79me3 does not correlate with either active or silent genes (Pokholok et al., 2005), and H3K79 methylation is associated with telomere silencing, meiotic checkpoint control, and DNA damage response (Jones et al., 2008). In *Drosophila* (Schubeler et al., 2004) and in humans (Okada et al., 2005), H3K79me2 is linked to active transcription. H3K79me1 is moderately associated with transcriptional activation, whereas H3K79me3 is associated with transcriptional repression in human cells. H3K79Me2 does not specifically localize to either active or silent genes (Barski et al., 2007).

H3K79 methylation is unique in that it is the only residue to be methylated by a non-SET domain histone methyltransferase, which is DOT1-like histone lysine methyltransferase (DOT1L; Shi and Tsukada, 2013; Table 1). In humans, DOT1L is the only known histone H3 lysine 79 methyltransferase. DOT1L catalyzes the mono-, di-, and trimethylation of H3K79 (Feng et al., 2002; Kim et al., 2014b), and it binds to actively transcribing RNA polymerase II to regulate gene expression (Kim et al., 2012). Mouse embryos deficient in the *Dot1l* gene show multiple developmental abnormalities and die during midgestation, and this is correlated with the global loss of heterochromatic marks such as H3K9me2 and H4K20me3 (Jones et al., 2008; Ooga et al., 2013). This further illustrates the importance of Dot1l in heterochromatin formation and embryonic development.

To date, very little is known about H3K79 demethylases. KDM2B/FBXL10, which is known for its H3K36 demethylase activity, is recently found to be a di- and trimethyl H3K79 demethylase (Seo et al., 2017; Table 2).

H4K20 Methylation

H4K20 mono-, di-, and trimethylation have distinct genome distributions and functions. H4K20me1 is enriched in promoters or coding regions of actively transcribed genes (Talasz et al., 2005; Vakoc et al., 2006), whereas H4K20me2 and H4K20me3 are associated with heterochromatin formation and gene silencing (Schotta et al., 2004). H4K20me1 is a mark for transcriptional activation and is also important for cell-cycle regulation (Wang et al., 2008). H4K20me2 is important for cell-cycle control, particularly for marking points of origin for DNA replication (Kuo et al., 2012), and it is also important in DNA damage response (Botuyan et al., 2006). H4K20me3 is associated with the repression of transcription when present at promoters (Wang et al., 2008), and it appears to be important for the silencing of repetitive DNA and transposons (Schotta et al., 2004). The loss of H4K20me3 has been identified as a hallmark of cancer (Fraga et al., 2005).

Regarding the methyltransferases responsible for H4K20 methylation, KMT5A/PR-Set7 is responsible for dimethylation of H4K20 (Nishioka et al., 2002), whereas KMT5B and KMT5C are trimethyltransferases for H4K20 (Black et al., 2012; Greer and Shi, 2012; Table 1). NSD1 has been reported to dimethylate H4K20 (Li et al., 2009b). There are controversies in the role of NSD2 in H4K20 methylation, in that although NSD2/MMSET has been reported to trimethylate H4K20 (Marango et al., 2008), Gozani's group has demonstrated using NSD2-methylated histones that NSD2 has no activity toward H3K20 (Li et al., 2009b; Kuo et al., 2011). Last but not least, PRDM6 has been shown to be responsible for monomethylation of H4K20 (Mozzetta et al., 2015).

Regarding the demethylases responsible for methylated H4K20, PHF8 mediates H4K20me1 demethylation events involved in cell-cycle progression (Liu et al., 2010; Table 2). KDM4A/JMJD2A can catalyze the demethylation of H4K20me3 (Lee et al., 2008).

H4K5 Methylation

In mammalian genome, H4K5 methylation is a more recent methylation pattern discovered and has been suggested to regulate neoplastic disease (Van Aller et al., 2012), and its role in transcriptional regulation is elusive. The methyltransferase SMYD3 can catalyze the mono-, di-, and trimethylation of H4K5 (Van Aller et al., 2012; Peserico et al., 2015; Table 1). In budding yeast, monomethylation of H4 lysines 5, 8, and 12 by Set5, which is a histone H4 methyltransferase, is important for cell growth and stress responses (Green et al., 2012).

Arginine Methylation

There are three main forms of methylated arginine in mammalian cells: monomethylarginines, asymmetrical dimethylarginines, and symmetrical dimethylarginines.

There are two major classes of methyltransferases for histone arginine methylation: Type I arginine methyltransferases, which include CARM1/PRMT4, METTL23, protein arginine methyltransferase 1 (PRMT1), PRMT2, PRMT3, PRMT6, and PRMT8, form a monomethylarginine intermediate followed by an asymmetrical dimethylarginine product; type II arginine methyltransferases, which include PRMT5 and PRDM4, produce symmetrical dimethylarginine products. PRMT7 produces monomethylated arginine in a type III enzymatic activity (Di Lorenzo and Bedford, 2011). PRDM6 catalyzes H3R2me1, H3R2me2, and H4K20me2, whereas PRMT9 catalyzes H4R3Me2 (Table 1).

Regarding demethylases, JMJD6 can demethylate H3R4me2 and H4R3me2 (Chang et al., 2007), although it is not specific for arginine demethylation (Webby et al., 2009; Mantri et al., 2010). In addition, a subset of lysine demethylases that contain a jumonji C domain, namely, KDM3A, KDM4E, KDM5C, and KDM6B, also have site-specific arginine demethylase activity (Walport et al., 2016; Table 2).

H2AR3 and H4R3 asymmetrical dimethylation (H2AR3me2a and H4R3me2a) is associated with active gene transcription. The two marks are grouped together because the first five residues of H2A and H4 are identical; they are referred to as the "R3 motif" and provide an efficient means for signal amplification (Di Lorenzo and Bedford, 2011). The asymmetrical dimethylation of these two residues is catalyzed by PRMT1, PRMT6, and PRMT8 (Wang et al., 2001b; Hyllus et al., 2007; Iberg et al., 2008). H2AR3me2a and H4R3me2a have been detected at the promoters of genes encoding the breast cancer estrogen-inducible protein trefoil factor 1 (*TFF1/pS2*; Wang et al., 2004; Denis et al., 2009); the drug-metabolizing enzyme cytochrome P450 family 3 subfamily A member 4 (*CYP3A4*; Xie et al., 2009); and a competitive inhibitor of hypoxia-inducible factor 1-alpha (HIF1A) protein Cbp/p300-interacting transactivator with Glu/Asp-rich carboxy-terminal domain 2 (*CITED2*; Huang et al., 2005). A top-down mass spectrometry approach has revealed that less than 2 % of H4R3 is methylated, although asymmetrical and symmetrical methylations are not distinguishable (Pesavento et al., 2008). Interestingly, H4R3 methylation is often accompanied by lysine acetylation (Pesavento et al., 2008; Phanstiel et al., 2008), and this likely contributes to active gene transcription at R3 motif methylation sites (Phanstiel et al., 2008).

H3R2me2a is generally associated with transcriptional repression of gene promoters (Guccione et al., 2007; Iberg et al., 2008). In mammalian cells, PRMT6 is the primary enzyme for producing H3R2me2a, but it can also strongly methylate H3K4me1 and H3K4me2 and weakly methylate H3K4me3 (Iberg et al., 2008). H3R2Me2a inhibits transcriptional activators that bind to methylated H3K4 at transcriptional start sites, effectively making PRMT6 a transcriptional repressor (Di Lorenzo and Bedford, 2011). However, recent ChIP-Seq studies have found that H3R2me2a is associated with highly expressed genes; this may be more related to PRMT6 functioning as a coactivator with nuclear receptors (Harrison et al., 2010). More studies are needed to elucidate the likely site-specific effects of H3R2Me2a and PRMT6.

Glutamine Methylation

Glutamine methylation is a newly discovered methylation pattern observed on histone H2A at position 105 (Q105) in yeast and at Q104 in humans (Tessarz et al., 2014). This methylation mark is exclusively enriched at the 35S ribosomal DNA transcriptional unit in yeast, and it disrupts the interaction between the histone chaperone facilitator of chromatin transcription (FACT) complex and nucleosomes. Nop1 (yeast) and fibrillarin (humans) are the methyltransferases responsible for this methylation mark (Tessarz et al., 2014).

A summary of major histone methylation marks and associated enzymes for their methylation/demethylation is shown in Table 4.

REGULATION OF HISTONE METHYLATION

Because histone methylation plays intricate roles in transcription, replication, and cell division, precise regulation of methylation activity and localization is essential. There are several ways to regulate histone methylation, including posttranscriptional modifications and noncoding RNAs (ncRNAs).

Posttranscriptional Modifications

Ubiquitination of proteins leads to proteasomal degradation and can regulate lysine methyltransferases and demethylases. For example, the H4K20 methyltransferase KMT5A (also known as SETD8) is regulated by three different E3 ubiquitin ligase complexes during various phases of the cell cycle to ensure the normal progression of cell cycle. In the G_1- to S-phase transition, KMT5A is ubiquitinated by the Skp, Cullin, F-box containing complex (SCF complex) and degraded to initiate S phase (Yin et al., 2008). During S phase, KMT5A is degraded by the proliferating cell nuclear antigen (PCNA)-dependent E3 ubiquitin ligase CRL4[Cdt2] through a direct interaction with the DNA polymerase delta cofactor PCNA (Beck et al., 2012). Finally, during anaphase, KMT5A is removed from mitotic chromosomes and ubiquitinated by the anaphase-promoting complex/cadherin 1 (APC[CDH1]; Wu et al., 2010b).

Phosphorylation of methyltransferases can also alter catalytic activity and impact health. For example, phosphorylation of EZH2 by AKT at serine 21 impedes binding to H3 and reduces H3K27me3 (Cha et al., 2005). This results in the activation of repressed genes and enhanced cell growth, indicating that derepression of these genes may lead to oncogenesis (Cha et al., 2005; Black et al., 2012).

Noncoding RNAs

ncRNAs are functional RNA transcripts that are not translated into proteins. A few studies have looked at the importance of microRNAs (miRNAs), which are a subcategory of small ncRNAs, in regulating lysine methyltransferases. Demonstrated in mouse primary myoblasts during skeletal muscle differentiation, the H3K27 methyltransferase PRC2, which is a repressive mark, is disengaged from the transcription site of miR-214, leading to active transcription of miR-214, and miR-214 in turn targets the 3′-UTR of the catalytic subunit Ezh2 to accelerate differentiation as a negative feedback mechanism (Juan et al., 2009). Additionally, in mouse embryonic fibroblasts, the H3K27 demethylase Kdm2b is activated by fibroblast growth factor 2 (Fgf2) and interacts with Ezh2 synergistically to repress the Ezh2 inhibitor miR-101 (Kottakis et al., 2011).

lncRNAs have been shown to have roles associated with chromatin-remodeling complexes, and a few studies have started to reveal the molecular mechanisms by which lncRNAs regulate histone methylation (Khalil et al., 2009). For example, in the mouse placenta, the lncRNA Air is required for the recruitment of H3K9 methyltransferase Ehmt2 for paternal allele silencing of solute carrier family 22, member 3 (Slc22a3/Oct3), Slc22a2/Oct2, and insulin-like growth factor 2 receptor (Igf2r; Nagano et al., 2008). The lncRNA HOTAIR acts in trans to silence the 40 kb homeobox D (*HOXD*) locus by interacting with the H3K27 methylase complex PRC2 (Rinn et al., 2007). In human foreskin and lung fibroblasts, the 5′ domain of HOTAIR binds to PRC2, and the 3′ domain binds the H3K4 demethylase KDM1A (Tsai et al., 2010). These studies have demonstrated that lncRNAs can function as protein scaffolds to localize histone-modifying complexes to specific genomic regions. Furthermore, the lncRNA HOTTIP has been shown in primary human fibroblasts to bind WDR5 in the MLL complex and direct H3K4me3 to the 5′ HOXA gene cluster for active gene transcription (Wang et al., 2011). Interestingly, only endogenous HOTTIP transcribed upstream from HOXA can modulate levels of H3K4me3 (Wang et al., 2011). Thus, lncRNAs can localize methylation machinery for active or repressive gene transcription. Other lncRNAs have also been suggested to be important for three-dimensional architecture of enhancers and histone marks (Rinn et al., 2007; Nagano et al., 2008; De Santa et al., 2010; Orom et al., 2010; Tsai et al., 2010; Wang et al., 2011).

REGULATION OF XENOBIOTIC BIOTRANSFORMATION-RELATED GENES BY HISTONE METHYLATION

Drug-metabolizing enzymes and transporters (together called "drug-processing genes") are regulated by epigenetic modifications including histone methylation (Ingelman-Sundberg et al., 2013; Kim et al., 2014a; Tang and Chen, 2015). Histone methylation marks interact with distinct transcription factors at the target regulatory regions of drug-processing genes to modulate transcriptional output. The rifampicin-mediated transcriptional upregulation in the major phase I drug-metabolizing enzyme CYP3A4 depends on the recruitment of pregnane X receptor (PXR) to the *CYP3A4* enhancer region by PRMT1, which catalyzes histone arginine methylation (Xie et al., 2009). Human prostate adenocarcinoma cell line LNCaP lacks active transcription mark H3K4me2 at the *CYP1A1* gene regulatory region, and this at least partially explains the suppression of dioxin-mediated transactivation of CYP1A1 (Okino et al., 2006). During liver maturation, there is a molecular switch from expressing the perinatal isoforms CYP3A7 (humans) and Cyp3a16 (mice) to expressing the adult isoforms CYP3A4 (humans) and Cyp3a11 (mice). In mice, this developmental switch in the *Cyp3a* gene locus coincides with high enrichment of H3K4me2 around *Cyp3a16* gene in the neonatal liver and around *Cyp3a11* gene in the adult liver and low enrichment of H3K4me2 but high enrichment of H3K27me3, a suppressive epigenetic mark, around *Cyp3a16* gene in the adult livers (Li et al., 2009a). The developmental increase in the mRNA expression of the phase II UDP glucuronosyltransferase (Ugt) 2b and Ugt3a families, glutathione *S*-transferase z1 (Cui et al., 2010), and xenobiotic-sensing transcription factor aryl hydrocarbon receptor (AhR) also positively correlates with increased H3K4me2 around these gene loci in the mouse liver, whereas H3K27me3 and DNA methylation are very minimally present in these gene loci (Cui et al., 2009b; Choudhuri et al., 2010).

Following early-life exposure to xenobiotics, distinct drug-processing genes may be persistently regulated in adulthood accompanied by persistent changes in histone methylation patterns. For example, in mice, neonatal activation of the xenobiotic-sensing nuclear receptor constitutive androstane receptor (CAR) by TCPOBOP (a direct mouse CAR ligand) leads to persistently upregulated expression of the CAR target genes Cyp2b10 and Cyp2c37 when the pups reach adulthood. This coincides with a permanent increase of the permissive histone methylation marks H3K4me1, H3K4me2, and H3K4me3 but a permanent decrease of the suppressive histone methylation mark H3K9me3 within the *Cyp2b10* gene locus (Chen et al., 2012). Transcriptional coactivator activating signal cointegrator-2 and histone demethylase (KDM4D/JMJD2D) participate in this CAR-mediated epigenetic switch. Similarly, phenobarbital, which is an indirect CAR activator, also leads to persistent increase in the expression and enzyme activities of Cyp2b10 in the mouse liver (Tien et al., 2015).

In summary, understanding the interactions between distinct histone methylation marks and xenobiotic-sensing transcription factors is important for more accurate predictions in the biotransformation of chemicals and the associated toxicological outcomes.

HISTONE METHYLATION AND HUMAN DISEASES

Histone Methylation and Cancer

Changes in histone methylation have been linked to carcinogenesis in humans and experimental models. Among various histone methylation marks, lysine methylation/demethylation is the most extensively characterized target in cancer cells. Therefore, this epigenetic mark is emphasized in this section.

Regarding histone methylation, the suppressive epigenetic mark H3K27me3 is enriched in the promoters of six prostate-cancer-related genes, namely, RARβ2, ERα, PGR, and RGMA, in human prostate cancer tissues, leading to decreased expression of these genes and aggressive tumor features (Ngollo et al., 2014). The histone methyltransferase EHMT2 (G9a), which is responsible for H3K9me1 and H3K9me2, is upregulated in various types of cancers, and its upregulation is associated with poor prognosis (Casciello et al., 2015). Activation of c-Jun N-terminal kinase (JNK), which has been linked to hepatocellular carcinoma (HCC) in mouse models, positively associates with H3K4me3 and H3K9me3 in HCC tissues and increased expression of genes involved in cell growth but decreased expression of genes involved in cell differentiation and drug metabolism (Chang et al., 2009). Data from 66 specimens have shown that high expression of H3K9me3 is a strong predictor of rapid cell proliferation and distant metastasis in patients with salivary adenoid cystic carcinoma, leading to poor survival (Xia et al., 2013). Elevation of H3R2me2a and PRMT6 activities, which subsequently inhibit the interaction between UHRF1 (an accessory factor of DNA methyltransferase 1) and histone H3, has been suggested to contribute to global DNA hypomethylation in cancer (Veland et al., 2017).

Regarding histone demethylation, the histone demethylase LSD1 (KDM1A), which regulates stem cell maintenance and differentiation, cell motility, and metabolic reprogramming, is frequently overexpressed in various types of solid tumors and hematopoietic neoplasms (Hino et al., 2016). Several other histone demethylase isoforms in the KDM4/JMJD2 family, which target H3K9, H3K6, and H1.4K26, are overexpressed in breast, colon, lung, prostate, renal cell carcinoma, and other tumors, and their interactions with steroid hormone receptors have been suggested to contribute to this process (Berry and Janknecht, 2013). Glutamine deficiency decreases α-ketoglutarate, which is a key cofactor of JMJD histone demethylases, leading to histone hypermethylation and cancer cell dedifferentiation (Pan et al., 2016).

Because of the importance of histone methylation/demethylation in carcinogenesis, low-molecular-weight inhibitors targeting histone methyltransferases/demethylases are currently being developed as epigenetic drugs to facilitate cancer therapy (Kim and Huang, 2003; Simo-Riudalbas and Esteller, 2015; Gelato et al., 2016; Hamamoto and Nakamura, 2016; Song et al., 2016). Specific examples include KDM4 inhibitors (Berry and Janknecht, 2013; Guo and Zhang, 2017) and G9a inhibitors (Casciello et al., 2015). These chemicals have shown promising results in preclinical and clinical trials to improve cancer chemotherapy.

Histone Methylation and Neurological Disorders

In addition to cancer, histone methylation is also implicated in a wide spectrum of other human diseases (Polz and Jablonski, 1989; Greer and Shi, 2012). Single nucleotide polymorphisms in histone lysine methyltransferases, histone lysine demethylases, and histones have been shown to cause various genetic diseases and cancer (Van Rechem and Whetstine, 2014). Specifically, aberrant histone lysine methylation patterns correlate with several types of neurodevelopmental disorders (Kim et al., 2017). For H3K4 methylation, genetic mutations in the methyltransferases KMT2A, KMT2C, KMT2D, and MKT2F and genetic mutations in KDM1A, KDM5A, KDM5B, and KDM5C have been associated with neurodevelopmental or neuropsychiatric diseases (Ricq et al., 2016). Epigenetic drugs are under development for central nervous system disorders through modulating neuroplasticity via H3K4 methylation (Ricq et al., 2016).

Histone Methylation and Other Diseases

Histone methylation is implicated in many other diseases. For example, in a rat model of pathological cardiac hypertrophy, there is a pervasive loss of the suppressive mark H3K9me2 associated with reexpression of fetal genes, and this correlates with miRNA-217-mediated decrease in the expression of the histone methyltransferases EHMT1 and EHMT2, which are responsible for H3K9me2. Therefore, it has been suggested that targeting miR-217 and EHMT1/2 to prevent the loss of H3K9 methylation is a viable therapeutic approach for the treatment of heart disease (Thienpont et al., 2017).

Histone methylation is also involved in impaired intermediary metabolism. In inflammation and diabetes, the H3K4 methyltransferase SET7/9 is a novel coactivator of NFκB. In monocytes, SET7/9 is necessary in promoting TNFα-mediated upregulation of inflammatory genes and H3K4 methylation on these gene promoters, as well as monocyte adhesion to endothelial or smooth muscle cells. Increased inflammatory gene expression and SET7/9 recruitment have also been observed in macrophages of diabetic mice (Li et al., 2008). In a rat model of diet-induced hyperhomocysteinemia, which leads to neurological and vascular complications associated with increased homocysteine, global histone arginine methylation patterns are affected in a tissue-specific manner and affect histone arginine methylation in the brain (Esse et al., 2013). In diabetic rats, the DNA damage-response-related histone methylation marks H4K20me1 and H4K20me2 are upregulated in the retina, and this is reversed upon minocycline treatment corresponding to changes in DNA damage response biomarkers. Therefore, it has been suggested that the beneficial effect of minocycline on the retina of diabetic rodents is partially through its ability to normalize histone methylation levels (Wang et al., 2017). In brown adipocytes, which are important for thermogenesis and energy expenditure, the histone demethylase ubiquitously transcribed tetratricopeptide repeat on X-chromosome (UTX, also known as KDM6A; Table 2) promotes thermogenesis via coordinated regulation of H3K27me3 and histone acetylation. Specifically, UTX is recruited to the gene promoters of UCP1 and PCG1α following the activation of β-adrenergic signaling, leading to demethylation of H3K27me3 and subsequent recruitment of histone acetyltransferase CBP to promote H3K27 acetylation and increased transcription of these target genes (Zha et al., 2015).

Histone Methylation and the Toxicity of Chemicals

The therapeutic and environmental factors that modulate histone methylation status are detailed in Table 5.

TABLE 5   Therapeutic and Environmental Factors That Modulate Histone Methylation Status

Gene Target	Regulator	Methylation Site	Change	Model	References
Lysine methyltransferases					
DOT1L	EPZ0047777	H3K79Me2	↓	MLL-rearranged acute myelogenous leukemia cells	Daigle et al. (2011)
	EPZ-5676	H3K79Me1/2	↓	Human leukemia cell lines	Daigle et al. (2013)
	SYC-522	H3K79Me2	↓	MLL-rearranged leukemia cell lines	Liu et al. (2014)
EHMT1/2	UNC0638	H3K9Me2	↓	MDA-MB-231 breast adenocarcinoma cells	Vedadi et al. (2011)
	Cardiac hypertrophy, *miR217*	H3K9Me2	↓	Male Sprague-Dawley rats	Thienpont et al. (2017)
EHMT2	BIX-01294	H3K9Me2	↓	LA1-55n neuroblastoma cells	Lu et al. (2013)
	UNC0321				Liu et al. (2010)
	BRD4770	H3K9Me2/3	↓	PANC-1 pancreatic cancer cell line	Yuan et al. (2012)
	Hypoxia	H3K9Me2	↑	Mammalian cells	Chen et al. (2006)
	BIX01294	H3K9Me1	↓	Male C57/B6 mice	Maeda et al. (2017)
	Arsenic	H3K9Me2/3	↑	C57BL/6 mice and human hepatocellular carcinoma cells	Zhang et al. (2016)
	Chromate	H3K9Me2/3, H3K4Me2/3, H3K27Me3, H3R2Me2	↑ ↑ ↓ ↓	Human lung A549 adenocarcinoma cells	Sun et al. (2009) and Zhou et al. (2009)
EZH1/2	UNC1999	H3K27Me3	↓	DB diffuse large B-cell lymphoma cells	Konze et al. (2013)
EZH2	BPA	H3K27Me3	↑	Human breast cancer cell line	Singh and Li (2012)

TABLE 5 Therapeutic and Environmental Factors That Modulate Histone Methylation Status—cont'd

Gene Target	Regulator	Methylation Site	Change	Model	References
	DES	H3K27Me3	↑	Mouse mammary gland	Doherty et al. (2010)
	GSK126	H3K27Me3	↓	B-cell lymphoma cell lines	McCabe et al. (2012)
	GSK343	H3K27Me3	↓	Epithelial ovarian cancer cells	Amatangelo et al. (2013)
	EPZ005687	H3K27Me2/3	↓	Lymphoma cells	Knutson et al. (2012)
	EI1	H3K27Me2/3	↓	B-cell lymphoma cell lines	Qi et al. (2012)
	EPZ-6438	H3K27Me3	↓	SCID mice with G401 (rhabdoid tumor) xenografts	Knutson et al. (2013)
	1,3-Butadiene	H3K9Me3, H3K27Me3, H4K20Me3	↓ ↓ ↓	Mouse liver	Koturbash et al. (2011a,b)
	Genistein	H3K27Me3	↓	Eker rats	Greathouse et al. (2012)
KDM4D	Development and neonatal exposure to TCPOBOP	H3K4Me1, H3K4Me2, H3K4Me3	↑ ↑ ↑	Mouse liver	Chen et al. (2012)
KMT2A	Clozapine	H3K4Me3	↑	Humans and mice	Ricq et al. (2016)
	MM-401	H3K4Me2/3	↓	MLL-AF9 mouse leukemia cells	Cao et al. (2014)
Kmt2a and Setdla	LPS	H3K4Me3	↑	RAW 264.7, murine macrophage cell line	Ara et al. (2008)
PCR2 complex	3-Deazaneplanocin A	H3K27Me3, H4K20Me3	↓ ↓	Breast cancer MCF-7 and colorectal cancer HCT116 cells	Tan et al. (2007) and Miranda et al. (2009)
SETDB1	Mithramycin			Human lung cancer cell lines	Rodriguez-Paredes et al. (2014)
SMYD2	AZ505				Ferguson et al. (2011)

Continued

1. HISTONE MODIFICATIONS AND CHROMATIN STRUCTURE

TABLE 5 Therapeutic and Environmental Factors That Modulate Histone Methylation Status—cont'd

Gene Target	Regulator	Methylation Site	Change	Model	References
	LLY-507	H3K, H4K	NCG[a]	Human embryonic kidney (HEK) 293 cells	Nguyen et al. (2015)
SUV39H1	Chaetocin	H3K9Me3	↓	U937 and K562 leukemia cell lines	Chaib et al. (2012)
Not specified	4-Aminobiphenyl	H3K4Me1	↓	Human mammary epithelial cells	Chappell et al. (2016)
	Aflatoxin	H3K4Me2, H3K9Me3, H3K27Me3, H4K20Me3	↓ ↑ ↓ ↑	Mouse oocytes	Zhu et al. (2014) and Liu et al. (2015)
	Arsenic	H3K9Me3	↓	Human CD4[+] blood cells	Pournara et al. (2016)
	Arsenic	H3K4Me3	↑	Mouse brain	Tyler et al. (2015)
	Arsenite	H3K4Me2, H3K4Me2	↑ ↑	Human lung carcinoma A549 cells	Zhou et al. (2009)
	Benzyl butyl phthalate	H3K9Me3	↓	Mesenchymal stem cells	Sonkar et al. (2016)
	BPA	H3K9Me3, H3K27Me3, H3K4Me3	↑ ↓ ↓	Human prostaspheres	Ho et al. (2015)
	BPA	H3K9Me3, H3K4Me3	↑ ↓	Human neuroblastoma SH-SY5Y cells	Senyildiz et al. (2017)
	BPA	H3K4Me3	↑	Wistar-Furth rats	Dhimolea et al. (2014)
	BPA	H3K9Me3	↓	Mouse oocytes	Trapphoff et al. (2013)
	Benzo[a]pyrene	H3K4Me2	↓	MCF7 breast cancer cells	Khanal et al. (2015)
	Benzo[a]pyrene	H3K4Me3	↑	Mouse Hepa-1 cells	Schnekenburger et al. (2007) and Ovesen et al. (2011)
	Cadmium	H3K4Me3, H3K9Me2	↑	Human bronchial epithelial cells	Xiao et al. (2015)

1. HISTONE MODIFICATIONS AND CHROMATIN STRUCTURE

TABLE 5 Therapeutic and Environmental Factors That Modulate Histone Methylation Status—cont'd

Gene Target	Regulator	Methylation Site	Change	Model	References
	Development	H3K4Me2, H3K27Me3	↑ ↓	Mouse liver (Ugt2b, Ugt3a, and Gstzl loci)	Cui et al. (2010) and Choudhuri et al. (2010)
	Streptozotocin-induced diabetes	H4K20Me1, H4K20Me2	↑ ↑	Diabetic rats	Wang et al. (2017)
	Nickel	H3K9Me2	↑	BEAS-2B human bronchial epithelium cells	Li et al. (2017)
	Nickel	H3K9Me2, H3K9Me1	↑	G12 Chinese hamster cells	Cholewinski and Wilkin (1988) and Chen et al. (2006)
	Nickel	H3K4Me3, H3K9Me2	↑	Human lung carcinoma A549 cells	Zhou et al. (2009)
	Nickel	H3K4Me3, H3K9Me2	↑ ↓	Human peripheral blood mononuclear cells	Arita et al. (2012a, b)
	Valproic acid	H3K4, H3K9	↑ ↓	Pregnant mice	Tung and Winn (2010)
	WY-14643	H3K9Me3, H4K20Me3	↓	Mouse liver	Pogribny et al. (2007)
Arginine methyltransferases					
CARM1	Ellagic acid (TBBD)	H3R17Me2	↓	HeLa cells	Selvi et al. (2010)
PRMT3	SGC707	H4R3Me2a	↓	Human lung A549 adenocarcinoma cells and HEK cells	Kaniskan et al. (2015)
PRMT5	EPZ015666	H3R8Me2s, H4R3Me2s	NCG	B-cell non-Hodgkin's lymphoma Z-138 cells	Chan-Penebre et al. (2015)
PRMT6	EPZ020411	H3R2Me2a	↓	A375 melanoma cells	Mitchell et al. (2015)
Type 1 PRMTs (PRMT1, PRMT3, PRMT4, PRMT6, and PRMT8)	MS023	H3R2Me2a, RMe1, RMe2s	↓ ↑ ↑	HEK293 cells	Eram et al. (2016)

Continued

1. HISTONE MODIFICATIONS AND CHROMATIN STRUCTURE

TABLE 5   Therapeutic and Environmental Factors That Modulate Histone Methylation Status—cont'd

Gene Target	Regulator	Methylation Site	Change	Model	References
Not specified	Methionine-deficient diet and vitamin B deficiency	H3R8Me2a	↓	Female Wistar rats	Esse et al. (2013)
Lysine demethylases					
JMJC domain (KDM4 family)	8-Hydroxyquinolines	H3K9Me3	↑	HeLa cells	King et al. (2010)
	Pyridine hydrazones				Chang et al. (2011)
KDM1A	Tranylcypromine	H3K4Me2	↑	Acute myelogenous leukemia engrafted mice	Schenk et al. (2012)
	ORY-1001				EU Clinical Trials Register [Internet] (2013)
	Pargyline	H3K4Me1/2	↑	Sorafenib-resistant hepatoma PLC cells	Huang et al. (2017)
	GSK2879552	H3K4Me1/2	↑	Sorafenib-resistant hepatoma PLC cells	Huang et al. (2017)
KDM3A	Hypoxia, nickel	H3K9Me2	↑	Human bronchial epithelial BEAS-2B cells	Chen et al. (2010) and Chervona and Costa (2012)
KDM4A	Hypoxia	H3K9Me1/2	↑	Human bronchial epithelial BEAS-2B cells	Chervona and Costa (2012)
	R-2-Hydroxyglutarate	H3K9Me3	↑	HeLa cells	Chowdhury et al. (2011) and Kaelin and McKnight (2013)
	Fumarate, succinate	H3K4Me1/3, H3K27Me2, H3K79Me2	↑ ↑ ↑		Xiao et al. (2012) and Kaelin and McKnight (2013)
KDM4D (Jmjc domain)	Succinate	H3K36Me2, H3K9Me3	↑ ↑	Yeast, HEK 293T cells	Smith et al. (2007) and Kaelin and McKnight (2013)

1. HISTONE MODIFICATIONS AND CHROMATIN STRUCTURE

TABLE 5 Therapeutic and Environmental Factors That Modulate Histone Methylation Status—cont'd

Gene Target	Regulator	Methylation Site	Change	Model	References
KDM5A	Hypoxia	H3K4Me3	↑	Human bronchial epithelial BEAS-2B cells	Zhou et al. (2010)
KDM6A	Cold exposure	H3K27Me3	↓	Mouse brown adipocytes	Zha et al. (2015)
KDM6 family	GSK-J4	H3K27Me3	↑	HeLa cells	Kruidenier et al. (2012)

[a] *No change globally (NCG).*

Heavy Metals

Heavy metals as a classic group of environmental toxicants adversely impact human health such as causing various types of cancers, neurological disorders, metabolic disorders, and toxicities of other organs. Among all heavy metals, arsenic is a well-studied epigenetic modifier leading to alterations in DNA methylation, histone modifications, and ncRNAs. Regarding histone methylation, in the mouse brain, developmental exposure to arsenic induces differential H3K4me3 enrichment on genes and pathways associated with cellular development and growth, cell death and survival, neurological disorders, and cancer in adult males (Tyler et al., 2015). In human lung carcinoma A549 cell lines, 7 days of exposure to low levels of arsenite (0.1, 0.5, and 1 μM) leads to an acute increase in H3K4me3 and H3K9me2 in different genomic regions and a persistent increase in H3K4me3 7 days after the exposure (Zhou et al., 2009). H3K9 methylation is also disrupted by arsenic exposure. For example, a global increase in H3K9me2 and a decrease in H3K9ac have been found in human subjects exposed to arsenic (Arita et al., 2012b). In mice, long-term exposure to arsenic produces liver cancer, and this coincides with a decrease in the expression of the tumor suppressor gene p16, corresponding to an increase in the suppressive histone methylation mark H3K9me2 and recruitment of H3K9 histone methyltransferase G9a around the *p16* gene promoter region (Suzuki and Nohara, 2013). Arsenic silences the mitochondrial protein pyruvate dehydrogenase kinase 4 and subsequently disrupts the TCA cycle, through the activation of histone H3K9 methyltransferases G9a and Suv39H in human liver cancer-derived cell lines (Zhang et al., 2016). In sorted blood cells from humans exposed to arsenic through drinking water, there is a decrease in global H3K9me3 in CD4$^+$ T cells in a dose-response manner (Pournara et al., 2016). Histone methylation marks also act in concert with other types of epigenetic marks to regulate the fate of gene transcription. For example, in arsenite-transformed prostate epithelial cells, the majority of DNA hypermethylation events colocalize with the preexisting suppressive histone methylation mark H3K27me3, leading to the reinforcement of gene silencing. DNA methylation also couples with another suppressive histone methylation mark H3K9me3 around the loci of C2H2-zinc-finger (ZNF) genes, which encode cancer-related transcription factors, resulting in a decrease in C2H2-ZNF gene expression in the arsenic-transformed cells (Severson et al., 2013).

Chromium in its hexavalent state is a human carcinogen leading to lung cancer and other toxicities via occupational or environmental exposure at high concentrations. In human carcinoma A549 cell lines, hexavalent chromium increases global H3K9me2, H3K9me3, H3K4me2, and H3K4me3 but decreases H3K27me3 and H3R2Me2. Increased H3K9 methylation correlates with decreased expression of the tumor suppressor gene MLH1, and this may contribute to chromium-induced carcinogenesis. The chromium-mediated increase in H3K9me2 appears to be due to an increase in its corresponding histone methyltransferase G9a, whereas ascorbate supplementation that reduces chromium reverses the H3K9me2 pattern (Sun et al., 2009). The acute and persistent increase in H3K4me3 and H3K9me2 following chromate exposure at 10 μM has also been shown in A549 cells in another study (Zhou et al., 2009).

Nickel is another well-established human carcinogen and skin irritant. In nickel-exposed human subjects, there is an increase in H3K4me3 and a decrease in H3K9me2 in peripheral blood mononuclear cells (Arita et al., 2012a,b). In human lung bronchial epithelial cell line (BEAS-2B), exposure to nickel chloride increases the suppressive mark H3K9me2 and decreases the H3K9me2 target genes (MAP2K3 and DKK1). Repression of nicotinamide N-methyltransferase (NNMT) inhibits nickel-induced H3K9me2 by altering the cellular SAM/SAH ratio (Li et al., 2017). In human carcinoma A549 cell lines, nickel exposure increases H3K4me3 and H3K9me2 acutely and persistently in a dose-dependent manner (Zhou et al., 2009). In G12 Chinese hamster cell lines, nickel ion exposure increases global H3K9me1 and H3K9me2, both of which colocalize with DNA methylation and lead to long-term gene silencing (Cholewinski and Wilkin, 1988; Chen et al., 2006). In addition, exposure to nickel ions increases H3K9me2 at the glutamic-pyruvic transaminase (GPT) transgene locus, leading to repression of this gene, and this is mediated by the inhibition of the histone H3K9 demethylation (Chen et al., 2006).

Other transition metals have also been shown to alter histone methylation patterns. For example, in normal human bronchial epithelial cells, cadmium exposure within a dose range of 0.625–5.0 μM leads to both an acute and persistent increase in global H3K4me3 and H3K9me2 patterns through inhibiting histone demethylases (Xiao et al., 2015). To note, cadmium is a well-established toxicant that causes lung cancer, kidney injury, and bone toxicities in humans. In mouse embryonic stem cells, acute exposure to cadmium, mercury, arsenic, and nickel decreases H3K27me1 and decreases cell proliferation; arsenic and mercury exposure also decreases total histone protein production (Gadhia et al., 2012).

In summary, heavy metals disrupt various histone methylation patterns in a metal- and cell-type-specific manner, and such alterations are mediated by aberrant regulation of histone methyltransferases and demethylases, leading to altered expression of target genes locally and/or globally and the adverse toxic outcomes such as carcinogenesis.

Endocrine Disruptors (EDCs)

Various EDCs such as bisphenol A (BPA), diethylstilbestrol (DES), genistein, organochlorines, polybrominated flame retardants, perfluorinated compounds, alkyphenols, phthalates, pesticides, and polycyclic aromatic hydrocarbons have been linked to adverse health effects such as cancer, metabolic syndrome, neurotoxicity, and reproductive disorders (De Coster and van Larebeke, 2012; Kajta and Wojtowicz, 2013; Costa et al., 2014; Kirkley and Sargis, 2014). Some of these EDCs have been shown to lead to aberrant DNA methylation and histone modification patterns, as well as genomic instability (De Coster and van Larebeke, 2012; Maqbool et al., 2016). Importantly, early-life exposure to certain EDCs leads to not only acute but also persistent adverse health effects in adulthood through epigenetic reprogramming (Xin et al., 2015; Alavian-Ghavanini and Ruegg, 2017).

BPA is among the most investigated environmental chemicals that has been shown to disrupt histone methylation. For example, in a human breast cancer cell line and mice, BPA promotes H3K27me3 through EZH2 (Singh and Li, 2012). In the presence of BPA in breast cancer cells, the histone methyltransferase KMT2C/MLL3 binds to the promoter of HOXB9 gene, leading to the transactivation of HOXB9, which plays a key role in mammary gland development and breast cancer (Deb et al., 2016). In rats, prenatal exposure to BPA leads to higher levels of the permissive histone methylation mark H3K4me3 at the transcription start site of α-lactalbumin gene and increased mRNA expression at postnatal day 4, and this may contribute to preneoplastic and neoplastic lesions in adult mammary gland (Dhimolea et al., 2014). In mouse oocytes, BPA exposure leads to decreased H3K9me3 and accelerated follicle development, as well as increased allele methylation errors in differentially methylated regions on maternally imprinted genes, and this may at least partially explain the adverse health outcomes of the offspring (Trapphoff et al., 2013). In human prostaspheres enriched in epithelial stemlike/progenitor cells (which is a model for prostate cancer), BPA-mediated gene silencing is associated with altered recruitment of H3K9me3, H3K27me3, and H3K4me3 to the 5′ regulatory/exonic sequences of five small nucleolar RNAs with C/D motif, which encode ncRNAs that regulate ribosomal RNA assembly and function (Ho et al., 2015). In human neuroblastoma SH-SY5Y cells, BPA increases H3K9me3 but decreases H3K4me3 in a dose-dependent manner, and this corresponds to changes in the expression of various enzymes involved in histone methylation such as G9a, EZH2, SUV39H1, SETD1A, and SETD8 (Senyildiz et al., 2017).

Besides BPA, other EDCs also have been reported to alter histone methylation patterns. For example, in mouse mammary gland of the F1 generation, perinatal exposure to DES increases the mRNA and protein expression of the histone methyltransferase Ezh2 expression and the corresponding suppressive mark H3K37Me3 (Doherty et al., 2010). In the rat uterus of the F1 generation, postnatal exposure to DES reduces H3K27me3 with the involvement of estrogen receptor and PI3K/Akt pathway (Bredfeldt et al., 2010). In developing rat uterus, the soy phytoestrogen genistein decreases the EZH2, which is responsible for generating the suppressive mark H3K27me3, leading to a decrease in H3K27me3 in chromatin, and this correlates with the effect of genistein on tumorigenesis (Greathouse et al., 2012). In mesenchymal stem cells, benzyl butyl phthalate exposure decreases H3K9me3 and downregulates histone methyltransferases SETDB1 and G9 through PPARγ, and this is accompanied with increased lipid accumulation and adipogenesis (Sonkar et al., 2016). In the testis, exposure to vinclozolin and dibutyl phthalate leads to coexpression patterns of histone methylation modifiers such as Kdm1, and this may contribute to the endocrine disruptor mode of action (Anderson et al., 2012).

In summary, EDCs disrupt histone methylation marks in multiple cell types including those related to mammary, reproductive, and neuronal tissues and in stem cells, and this may contribute to the toxicities of this chemical in the target organs. It remains a controversial issue that during toxic exposure, whether the changes in histone posttranscriptional modifications are a cause or a consequence, it is often assumed that changes in posttranscriptional modifications are key drivers of the epigenomic and transcriptomic changes. Further and more careful studies are needed to precisely clarify this issue.

Human Genotoxic Chemical Carcinogens From Occupational and Environmental Exposure

The effect of human genotoxic carcinogens on histone methylation and other epigenetic marks has been summarized in a comprehensive review article by Dr. Ivan Rusyn's group (Chappell et al., 2016).

For benzo[a]pyrene (BaP), MCF7 breast cancer cells and the livers of mice exposed to BaP have decreased H3K4me2 in the promoter of estrogen receptor, and this is associated with downregulation of estrogen receptor and increased BaP-induced oxidative injury (Khanal et al., 2015). In contrast, in mouse hepatoma Hepa-1 cells, BaP increases H3K4me3 at the *Cyp1a1* gene promoter, accompanied by AhR recruitment and increased Cyp1a1 gene expression (Schnekenburger et al., 2007; Ovesen et al., 2011).

For aflatoxin, in oocytes from aflatoxin B1-fed mice, there is an increase in the transcriptional suppression mark H3K9me3 and the transcriptional activation mark H4K20me3 but a decrease in the transcriptional silencing mark H3K27me3 and the transcriptional activation marks H4K20me3 and H3K4me2 (Zhu et al., 2014; Liu et al., 2015). These aberrant alterations in histone methylation marks are accompanied with DNA hypermethylation and have been suggested to contribute to decreased developmental competence of oocytes in aflatoxin B1-exposed mice (Zhu et al., 2014). For benzene, patients with benzene exposure have reduced H3K4 methylation but increased H3K9 methylation in the gene promoter region of topoisomerase II2a (Topo IIa), and this is accompanied with decreased Topo IIa expression and activity and has been linked to benzene-induced hematotoxicity (Yu et al., 2011). For coke, which is produced from coal carbonization and metal smelting process, there is a decrease in H3K9me3, H3K27me3, and H4K20me3 in the livers of mice exposed to 1,3-butadiene in a dose-dependent manner (Koturbash et al., 2011a,b). These histone marks are important for the maintenance of proper chromatin structure and cell cycle. For 4-aminobiphenyl, which is used to manufacture azo dyes, normal human mammary epithelial cells exposed to 4-aminophenyl have decreased H3K4me1. For TCDD, although its carcinogenic role in animals is better validated as compared with humans, studies have shown that TCDD increases the association of H3K9me3 together with AhR and other epigenetic marks around the gene promoter of breast cancer 1 (*BRCA-1*), and such association is antagonized at physiologically relevant doses by the phytoalexin resveratrol (Papoutsis et al., 2012). In summary, histone methylation is disrupted by multiple known human carcinogens from occupational and environmental exposure and thus may serve as a potential drug target for cancer prevention and treatment.

Other Xenobiotics

Histone methylation is also a target for other environmental chemicals, drugs, and experimental model compounds. For example, in primary human bronchial epithelial cells, ozone-induced expression of genes involved in inflammatory response such as IL-6, IL-8, HMOX1, and COX2 is regulated by H3K4me3 and H3K27me2/3, and the baseline levels of these histone methylation marks are correlated with interindividual variability in ozone-mediated induction of the target genes HMOX1 and COX2 (McCullough et al., 2016). In the livers of mice, dietary exposure to the prototypical PPARα ligand WY-14643 leads to decreased H4K20me3 and H3K9me3 in a PPARα-dependent manner, suggesting that these histone marks may be involved in PPAR-mediated liver tumor formation in mice (Pogribny et al., 2007). In mouse neurons, the drug of abuse cocaine leads to dynamic alterations in H3K9 methylation patterns between dopamine 1 (D1) and D2 nuclei, and these alterations are exclusively persistent in the D1 nuclei (Jordi et al., 2013). Exposing pregnant mice to a teratogenic dose of the anticonvulsant drug valproic acid results in an increase in H3K4 methylation in embryonic neuroepithelium but a decrease in H3K9 methylation in embryonic neuroepithelium and somite (Tung and Winn, 2010). In mice, dietary folate deficiency and γ-irradiation together upregulate H3K4 histone methyltransferase but downregulate H3K27 histone methyltransferase in a dose- and time-dependent manner, and it has been suggested that the maintenance of histone methylation pattern under γ-irradiation stress may be a dynamic process that can be modulated by folate deficiency leading to the formation of epigenetically reprogrammed cells (Batra and Devasagayam, 2012). In RAW cells derived from mouse tumors, the bacterial product lipopolysaccharide (LPS) increases the binding of H3K4me3 to the TNFα gene promoter, and this is blocked by SAMe or methylthioadenosine treatment (Ara et al., 2008). The effect of various xenobiotics on histone methylation and other epigenetic marks has been extensively reviewed in the literature (Bollati and Baccarelli, 2010; Choudhuri et al., 2010; Hou et al., 2012; Pulliero et al., 2015; Chappell et al., 2016), and the readers are encouraged to read these reviews for additional information.

The regulation of histone methylation status by therapeutic and environmental factors is summarized in Table 5.

CONCLUSION

Taken together, this book chapter has systematically reviewed various histone methylation marks and their key regulators implicated in toxicology. Methylation at different amino acid residues on histone proteins leads to different transcriptional outcomes of target genes. Global and local changes in histone methylation patterns can be used as biomarkers for diseases and toxic exposures, and many of the histone methylation marks also colocalize or interfere with other epigenetic marks such as DNA methylation and histone acetylation and act in concert with transcription factors to determine the fate of chromatin status and gene transcription. Among various human diseases and chemical-induced toxicities, cancer has been the most studied toxicological end point associated with histone methylation; however, emerging evidence in the literature has also demonstrated that this epigenetic mark is crucial for development, susceptibility to chemicals, metabolic syndrome, neurological disorders, and many other diseases and chemical-induced toxicities. Understanding the molecular mechanisms underlying the histone-methylation-mediated changes in the pathogenesis and progression of human diseases will pave the path for designing epigenetic drugs targeting this epigenetic mark for disease prevention and treatment.

References

Abu-Farha, M., Lambert, J.P., Al-Madhoun, A.S., Elisma, F., Skerjanc, I.S., Figeys, D., 2008. The tale of two domains: proteomics and genomics analysis of SMYD2, a new histone methyltransferase. Mol. Cell. Proteomics 7, 560–572.

Agger, K., Cloos, P.A., Christensen, J., Pasini, D., Rose, S., Rappsilber, J., Issaeva, I., Canaani, E., Salcini, A.E., Helin, K., 2007. UTX and JMJD3 are histone H3K27 demethylases involved in HOX gene regulation and development. Nature 449, 731–734.

Alavian-Ghavanini, A., Ruegg, J., 2017. Understanding epigenetic effects of endocrine disrupting chemicals: from mechanisms to novel test methods. Basic Clin. Pharmacol. Toxicol. 122, 38–45.

Allfrey, V.G., Mirsky, A.E., 1964. Structural modifications of histones and their possible role in the regulation of RNA synthesis. Science 144, 559.

Amatangelo, M.D., Garipov, A., Li, H., Conejo-Garcia, J.R., Speicher, D.W., Zhang, R., 2013. Three-dimensional culture sensitizes epithelial ovarian cancer cells to EZH2 methyltransferase inhibition. Cell Cycle 12, 2113–2119.

An, S., Yeo, K.J., Jeon, Y.H., Song, J.J., 2011. Crystal structure of the human histone methyltransferase ASH1L catalytic domain and its implications for the regulatory mechanism. J. Biol. Chem. 286, 8369–8374.

Anderson, A.M., Carter, K.W., Anderson, D., Wise, M.J., 2012. Coexpression of nuclear receptors and histone methylation modifying genes in the testis: implications for endocrine disruptor modes of action. PLoS One 7, e34158.

Ara, A.I., Xia, M., Ramani, K., Mato, J.M., Lu, S.C., 2008. S-adenosylmethionine inhibits lipopolysaccharide-induced gene expression via modulation of histone methylation. Hepatology 47, 1655–1666.

Ardehali, M.B., Anselmo, A., Cochrane, J.C., Kundu, S., Sadreyev, R.I., Kingston, R.E., 2017. Polycomb repressive complex 2 methylates elongin A to regulate transcription. Mol. Cell 68, 872–884. e876.

Arita, A., Niu, J., Qu, Q., Zhao, N., Ruan, Y., Nadas, A., Chervona, Y., Wu, F., Sun, H., Hayes, R.B., Costa, M., 2012a. Global levels of histone modifications in peripheral blood mononuclear cells of subjects with exposure to nickel. Environ. Health Perspect. 120, 198–203.

Arita, A., Shamy, M.Y., Chervona, Y., Clancy, H.A., Sun, H., Hall, M.N., Qu, Q., Gamble, M.V., Costa, M., 2012b. The effect of exposure to carcinogenic metals on histone tail modifications and gene expression in human subjects. J. Trace Elem. Med. Biol. 26, 174–178.

Baba, A., Ohtake, F., Okuno, Y., Yokota, K., Okada, M., Imai, Y., Ni, M., Meyer, C.A., Igarashi, K., Kanno, J., Brown, M., Kato, S., 2011. PKA-dependent regulation of the histone lysine demethylase complex PHF2-ARID5B. Nat. Cell Biol. 13, 668–675.

Bannister, A.J., Schneider, R., Myers, F.A., Thorne, A.W., Crane-Robinson, C., Kouzarides, T., 2005. Spatial distribution of di- and tri-methyl lysine 36 of histone H3 at active genes. J. Biol. Chem. 280, 17732–17736.

Barrera, L.O., Li, Z., Smith, A.D., Arden, K.C., Cavenee, W.K., Zhang, M.Q., Green, R.D., Ren, B., 2008. Genome-wide mapping and analysis of active promoters in mouse embryonic stem cells and adult organs. Genome Res. 18, 46–59.

Barski, A., Cuddapah, S., Cui, K., Roh, T.Y., Schones, D.E., Wang, Z., Wei, G., Chepelev, I., Zhao, K., 2007. High-resolution profiling of histone methylations in the human genome. Cell 129, 823–837.

Batra, V., Devasagayam, T.P., 2012. Interaction between gamma-radiation and dietary folate starvation metabolically reprograms global hepatic histone H3 methylation at lysine 4 and lysine 27 residues. Food Chem. Toxicol. 50, 464–472.

Beck, D.B., Oda, H., Shen, S.S., Reinberg, D., 2012. PR-Set7 and H4K20me1: at the crossroads of genome integrity, cell cycle, chromosome condensation, and transcription. Genes Dev. 26, 325–337.

Bedford, M.T., 2007. Arginine methylation at a glance. J. Cell Sci. 120, 4243–4246.

Bell, O., Wirbelauer, C., Hild, M., Scharf, A.N., Schwaiger, M., MacAlpine, D.M., Zilbermann, F., van Leeuwen, F., Bell, S.P., Imhof, A., Garza, D., Peters, A.H., Schubeler, D., 2007. Localized H3K36 methylation states define histone H4K16 acetylation during transcriptional elongation in Drosophila. EMBO J. 26, 4974–4984.

Bernstein, B.E., Mikkelsen, T.S., Xie, X., Kamal, M., Huebert, D.J., Cuff, J., Fry, B., Meissner, A., Wernig, M., Plath, K., Jaenisch, R., Wagschal, A., Feil, R., Schreiber, S.L., Lander, E.S., 2006. A bivalent chromatin structure marks key developmental genes in embryonic stem cells. Cell 125, 315–326.

Berry, W.L., Janknecht, R., 2013. KDM4/JMJD2 histone demethylases: epigenetic regulators in cancer cells. Cancer Res. 73, 2936–2942.

Bieberstein, N.I., Kozakova, E., Huranova, M., Thakur, P.K., Krchnakova, Z., Krausova, M., Carrillo Oesterreich, F., Stanek, D., 2016. TALE-directed local modulation of H3K9 methylation shapes exon recognition. Sci. Rep. 6, 29961.

Black, J.C., Van Rechem, C., Whetstine, J.R., 2012. Histone lysine methylation dynamics: establishment, regulation, and biological impact. Mol. Cell 48, 491–507.

Blackledge, N.P., Zhou, J.C., Tolstorukov, M.Y., Farcas, A.M., Park, P.J., Klose, R.J., 2010. CpG islands recruit a histone H3 lysine 36 demethylase. Mol. Cell 38, 179–190.

Blazer, L.L., Lima-Fernandes, E., Gibson, E., Eram, M.S., Loppnau, P., Arrowsmith, C.H., Schapira, M., Vedadi, M., 2016. PR domain-containing protein 7 (PRDM7) is a histone 3 lysine 4 trimethyltransferase. J. Biol. Chem. 291, 13509–13519.

Bollati, V., Baccarelli, A., 2010. Environmental epigenetics. Heredity (Edinb.) 105, 105–112.

Borun, T.W., Pearson, D., WK, P., 1972. Studies of histone methylation during the HeLa S-3 cell cycle. J. Biol. Chem. 247, 4288–4298.

Botuyan, M.V., Lee, J., Ward, I.M., Kim, J.E., Thompson, J.R., Chen, J., Mer, G., 2006. Structural basis for the methylation state-specific recognition of histone H4-K20 by 53BP1 and Crb2 in DNA repair. Cell 127, 1361–1373.

Bredfeldt, T.G., Greathouse, K.L., Safe, S.H., Hung, M.C., Bedford, M.T., Walker, C.L., 2010. Xenoestrogen-induced regulation of EZH2 and histone methylation via estrogen receptor signaling to PI3K/AKT. Mol. Endocrinol. 24, 993–1006.

Brien, G.L., Gambero, G., O'Connell, D.J., Jerman, E., Turner, S.A., Egan, C.M., Dunne, E.J., Jurgens, M.C., Wynne, K., Piao, L., Lohan, A.J., Ferguson, N., Shi, X., Sinha, K.M., Loftus, B.J., Cagney, G., Bracken, A.P., 2012. Polycomb PHF19 binds H3K36me3 and recruits PRC2 and demethylase NO66 to embryonic stem cell genes during differentiation. Nat. Struct. Mol. Biol. 19, 1273–1281.

Briggs, S.D., Bryk, M., Strahl, B.D., Cheung, W.L., Davie, J.K., Dent, S.Y., Winston, F., Allis, C.D., 2001. Histone H3 lysine 4 methylation is mediated by Set1 and required for cell growth and rDNA silencing in Saccharomyces cerevisiae. Genes Dev. 15, 3286–3295.

Bryan, L.C., Weilandt, D.R., Bachmann, A.L., Kilic, S., Lechner, C.C., Odermatt, P.D., Fantner, G.E., Georgeon, S., Hantschel, O., Hatzimanikatis, V., Fierz, B., 2017. Single-molecule kinetic analysis of HP1-chromatin binding reveals a dynamic network of histone modification and DNA interactions. Nucleic Acids Res. 45, 10504–10517.

Bulut-Karslioglu, A., De La Rosa-Velazquez, I.A., Ramirez, F., Barenboim, M., Onishi-Seebacher, M., Arand, J., Galan, C., Winter, G.E., Engist, B., Gerle, B., O'Sullivan, R.J., Martens, J.H., Walter, J., Manke, T., Lachner, M., Jenuwein, T., 2014. Suv39h-dependent H3K9me3 marks intact retrotransposons and silences LINE elements in mouse embryonic stem cells. Mol. Cell. 55, 277–290.

Byrd, K.N., Shearn, A., 2003. ASH1, a Drosophila trithorax group protein, is required for methylation of lysine 4 residues on histone H3. Proc. Natl. Acad. Sci. USA 100, 11535–11540.

Byvoet, P., Shepherd, G.R., Hardin, J.M., Noland, B.J., 1972. The distribution and turnover of labeled methyl groups in histone fractions of cultured mammalian cells. Arch. Biochem. Biophys. 148, 558–567.

Cao, R., Wang, L., Wang, H., Xia, L., Erdjument-Bromage, H., Tempst, P., Jones, R.S., Zhang, Y., 2002. Role of histone H3 lysine 27 methylation in polycomb-group silencing. Science 298, 1039–1043.

Cao, F., Townsend, E.C., Karatas, H., Xu, J., Li, L., Lee, S., Liu, L., Chen, Y., Ouillette, P., Zhu, J., Hess, J.L., Atadja, P., Lei, M., Qin, Z.S., Malek, S., Wang, S., Dou, Y., 2014. Targeting MLL1 H3K4 methyltransferase activity in mixed-lineage leukemia. Mol. Cell 53, 247–261.

Cao, F., Fang, Y., Tan, H.K., Goh, Y., Choy, J.Y.H., Koh, B.T.H., Hao Tan, J., Bertin, N., Ramadass, A., Hunter, E., Green, J., Salter, M., Akoulitchev, A., Wang, W., Chng, W.J., Tenen, D.G., Fullwood, M.J., 2017. Super-enhancers and broad H3K4me3 domains form complex gene regulatory circuits involving chromatin interactions. Sci. Rep. 7, 2186.

Casciello, F., Windloch, K., Gannon, F., Lee, J.S., 2015. Functional role of g9a histone methyltransferase in cancer. Front. Immunol. 6, 487.

Cha, T.L., Zhou, B.P., Xia, W., Wu, Y., Yang, C.C., Chen, C.T., Ping, B., Otte, A.P., Hung, M.C., 2005. Akt-mediated phosphorylation of EZH2 suppresses methylation of lysine 27 in histone H3. Science 310, 306–310.

Chaib, H., Nebbioso, A., Prebet, T., Castellano, R., Garbit, S., Restouin, A., Vey, N., Altucci, L., Collette, Y., 2012. Anti-leukemia activity of chaetocin via death receptor-dependent apoptosis and dual modulation of the histone methyl-transferase SUV39H1. Leukemia 26, 662–674.

Chang, B., Chen, Y., Zhao, Y., Bruick, R.K., 2007. JMJD6 is a histone arginine demethylase. Science 318, 444–447.

Chang, Q., Zhang, Y., Beezhold, K.J., Bhatia, D., Zhao, H., Chen, J., Castranova, V., Shi, X., Chen, F., 2009. Sustained JNK1 activation is associated with altered histone H3 methylations in human liver cancer. J. Hepatol. 50, 323–333.

Chang, Y., Sun, L., Kokura, K., Horton, J.R., Fukuda, M., Espejo, A., Izumi, V., Koomen, J.M., Bedford, M.T., Zhang, X., Shinkai, Y., Fang, J., Cheng, X., 2011. MPP8 mediates the interactions between DNA methyltransferase Dnmt3a and H3K9 methyltransferase GLP/G9a. Nat. Commun. 2, 533.

Chan-Penebre, E., Kuplast, K.G., Majer, C.R., Boriack-Sjodin, P.A., Wigle, T.J., Johnston, L.D., Rioux, N., Munchhof, M.J., Jin, L., Jacques, S.L., West, K.A., Lingaraj, T., Stickland, K., Ribich, S.A., Raimondi, A., Scott, M.P., Waters, N.J., Pollock, R.M., Smith, J.J., Barbash, O., Pappalardi, M., Ho, T.F., Nurse, K., Oza, K.P., Gallagher, K.T., Kruger, R., Moyer, M.P., Copeland, R.A., Chesworth, R., Duncan, K.W., 2015. A selective inhibitor of PRMT5 with in vivo and in vitro potency in MCL models. Nat. Chem. Biol. 11, 432–437.

Chappell, G., Pogribny, I.P., Guyton, K.Z., Rusyn, I., 2016. Epigenetic alterations induced by genotoxic occupational and environmental human chemical carcinogens: a systematic literature review. Mutat. Res. Rev. Mutat. Res. 768, 27–45.

Chen, D., Ma, H., Hong, H., Koh, S.S., Huang, S.M., Schurter, B.T., Aswad, D.W., Stallcup, M.R., 1999. Regulation of transcription by a protein methyltransferase. Science 284, 2174–2177.

Chen, H., Ke, Q., Kluz, T., Yan, Y., Costa, M., 2006. Nickel ions increase histone H3 lysine 9 dimethylation and induce transgene silencing. Mol. Cell. Biol. 26, 3728–3737.

Chen, L., Li, Z., Zwolinska, A.K., Smith, M.A., Cross, B., Koomen, J., Yuan, Z.M., Jenuwein, T., Marine, J.C., Wright, K.L., Chen, J., 2010. MDM2 recruitment of lysine methyltransferases regulates p53 transcriptional output. EMBO J. 29, 2538–2552.

Chen, W.D., Fu, X., Dong, B., Wang, Y.D., Shiah, S., Moore, D.D., Huang, W., 2012. Neonatal activation of the nuclear receptor CAR results in epigenetic memory and permanent change of drug metabolism in mouse liver. Hepatology 56, 1499–1509.

Chen, K., Chen, Z., Wu, D., Zhang, L., Lin, X., Su, J., Rodriguez, B., Xi, Y., Xia, Z., Chen, X., Shi, X., Wang, Q., Li, W., 2015. Broad H3K4me3 is associated with increased transcription elongation and enhancer activity at tumor-suppressor genes. Nat. Genet. 47, 1149–1157.

Chen, Y.J., Tsai, C.H., Wang, P.Y., Teng, S.C., 2017. SMYD3 promotes homologous recombination via regulation of H3K4-mediated gene expression. Sci. Rep. 7, 3842.

Chervona, Y., Costa, M., 2012. The control of histone methylation and gene expression by oxidative stress, hypoxia, and metals. Free Radic. Biol. Med. 53, 1041–1047.

Chittka, A., Nitarska, J., Grazini, U., Richardson, W.D., 2012. Transcription factor positive regulatory domain 4 (PRDM4) recruits protein arginine methyltransferase 5 (PRMT5) to mediate histone arginine methylation and control neural stem cell proliferation and differentiation. J. Biol. Chem. 287, 42995–43006.

Cholewinski, A.J., Wilkin, G.P., 1988. Astrocytes from forebrain, cerebellum, and spinal cord differ in their responses to vasoactive intestinal peptide. J. Neurochem. 51, 1626–1633.

Choudhuri, S., Cui, Y., Klaassen, C.D., 2010. Molecular targets of epigenetic regulation and effectors of environmental influences. Toxicol. Appl. Pharmacol. 245, 378–393.

Chowdhury, R., Yeoh, K.K., Tian, Y.M., Hillringhaus, L., Bagg, E.A., Rose, N.R., Leung, I.K., Li, X.S., Woon, E.C., Yang, M., McDonough, M.A., King, O.N., Clifton, I.J., Klose, R.J., Claridge, T.D., Ratcliffe, P.J., Schofield, C.J., Kawamura, A., 2011. The oncometabolite 2-hydroxyglutarate inhibits histone lysine demethylases. EMBO Rep. 12, 463–469.

Christensen, J., Agger, K., Cloos, P.A., Pasini, D., Rose, S., Sennels, L., Rappsilber, J., Hansen, K.H., Salcini, A.E., Helin, K., 2007. RBP2 belongs to a family of demethylases, specific for tri-and dimethylated lysine 4 on histone 3. Cell 128, 1063–1076.

Chuikov, S., Kurash, J.K., Wilson, J.R., Xiao, B., Justin, N., Ivanov, G.S., McKinney, K., Tempst, P., Prives, C., Gamblin, S.J., Barlev, N.A., Reinberg, D., 2004. Regulation of p53 activity through lysine methylation. Nature 432, 353–360.

Ciccone, D.N., Su, H., Hevi, S., Gay, F., Lei, H., Bajko, J., Xu, G., Li, E., Chen, T., 2009. KDM1B is a histone H3K4 demethylase required to establish maternal genomic imprints. Nature 461, 415–418.

Collins, R.E., Northrop, J.P., Horton, J.R., Lee, D.Y., Zhang, X., Stallcup, M.R., Cheng, X., 2008. The ankyrin repeats of G9a and GLP histone methyltransferases are mono- and dimethyllysine binding modules. Nat. Struct. Mol. Biol. 15, 245–250.

Costa, E.M., Spritzer, P.M., Hohl, A., Bachega, T.A., 2014. Effects of endocrine disruptors in the development of the female reproductive tract. Arq. Bras. Endocrinol. Metabol. 58, 153–161.

Cui, K., Zang, C., Roh, T.Y., Schones, D.E., Childs, R.W., Peng, W., Zhao, K., 2009a. Chromatin signatures in multipotent human hematopoietic stem cells indicate the fate of bivalent genes during differentiation. Cell Stem Cell 4, 80–93.

Cui, Y.J., Yeager, R.L., Zhong, X.B., Klaassen, C.D., 2009b. Ontogenic expression of hepatic Ahr mRNA is associated with histone H3K4 di-methylation during mouse liver development. Toxicol. Lett. 189, 184–190.

Cui, J.Y., Choudhuri, S., Knight, T.R., Klaassen, C.D., 2010. Genetic and epigenetic regulation and expression signatures of glutathione S-transferases in developing mouse liver. Toxicol. Sci. 116, 32–43.

Daigle, S.R., Olhava, E.J., Therkelsen, C.A., Majer, C.R., Sneeringer, C.J., Song, J., Johnston, L.D., Scott, M.P., Smith, J.J., Xiao, Y., Jin, L., Kuntz, K.W., Chesworth, R., Moyer, M.P., Bernt, K.M., Tseng, J.C., Kung, A.L., Armstrong, S.A., Copeland, R.A., Richon, V.M., Pollock, R.M., 2011. Selective killing of mixed lineage leukemia cells by a potent small-molecule DOT1L inhibitor. Cancer Cell 20, 53–65.

Daigle, S.R., Olhava, E.J., Therkelsen, C.A., Basavapathruni, A., Jin, L., Boriack-Sjodin, P.A., Allain, C.J., Klaus, C.R., Raimondi, A., Scott, M.P., Waters, N.J., Chesworth, R., Moyer, M.P., Copeland, R.A., Richon, V.M., Pollock, R.M., 2013. Potent inhibition of DOT1L as treatment of MLL-fusion leukemia. Blood 122, 1017–1025.

Davis, C.A., Haberland, M., Arnold, M.A., Sutherland, L.B., McDonald, O.G., Richardson, J.A., Childs, G., Harris, S., Owens, G.K., Olson, E.N., 2006. PRISM/PRDM6, a transcriptional repressor that promotes the proliferative gene program in smooth muscle cells. Mol. Cell Biol. 26, 2626–2636.

de Almeida, S.F., Carmo-Fonseca, M., 2012. Design principles of interconnections between chromatin and pre-mRNA splicing. Trends Biochem. Sci. 37, 248–253.

De Coster, S., van Larebeke, N., 2012. Endocrine-disrupting chemicals: associated disorders and mechanisms of action. J. Environ. Public Health 2012, 713696.

De Santa, F., Barozzi, I., Mietton, F., Ghisletti, S., Polletti, S., Tusi, B.K., Muller, H., Ragoussis, J., Wei, C.L., Natoli, G., 2010. A large fraction of extragenic RNA pol II transcription sites overlap enhancers. PLoS Biol. 8, e1000384.

Deb, P., Bhan, A., Hussain, I., Ansari, K.I., Bobzean, S.A., Pandita, T.K., Perrotti, L.I., Mandal, S.S., 2016. Endocrine disrupting chemical, bisphenol-A, induces breast cancer associated gene HOXB9 expression in vitro and in vivo. Gene 590, 234–243.

Dehe, P.M., Geli, V., 2006. The multiple faces of Set1. Biochem. Cell Biol. 84, 536–548.

Demers, C., Chaturvedi, C.P., Ranish, J.A., Juban, G., Lai, P., Morle, F., Aebersold, R., Dilworth, F.J., Groudine, M., Brand, M., 2007. Activator-mediated recruitment of the MLL2 methyltransferase complex to the beta-globin locus. Mol. Cell 27, 573–584.

Denis, H., Deplus, R., Putmans, P., Yamada, M., Metivier, R., Fuks, F., 2009. Functional connection between deimination and deacetylation of histones. Mol. Cell. Biol. 29, 4982–4993.

Dhami, P., Saffrey, P., Bruce, A.W., Dillon, S.C., Chiang, K., Bonhoure, N., Koch, C.M., Bye, J., James, K., Foad, N.S., Ellis, P., Watkins, N.A., Ouwehand, W.H., Langford, C., Andrews, R.M., Dunham, I., Vetrie, D., 2010. Complex exon-intron marking by histone modifications is not determined solely by nucleosome distribution. PLoS One 5, e12339.

Dhimolea, E., Wadia, P.R., Murray, T.J., Settles, M.L., Treitman, J.D., Sonnenschein, C., Shioda, T., Soto, A.M., 2014. Prenatal exposure to BPA alters the epigenome of the rat mammary gland and increases the propensity to neoplastic development. PLoS One 9, e99800.

Di Lorenzo, A., Bedford, M.T., 2011. Histone arginine methylation. FEBS Lett. 585, 2024–2031.

Doherty, L.F., Bromer, J.G., Zhou, Y., Aldad, T.S., Taylor, H.S., 2010. In utero exposure to diethylstilbestrol (DES) or bisphenol-A (BPA) increases EZH2 expression in the mammary gland: an epigenetic mechanism linking endocrine disruptors to breast cancer. Horm. Cancer 1, 146–155.

El Messaoudi, S., Fabbrizio, E., Rodriguez, C., Chuchana, P., Fauquier, L., Cheng, D., Theillet, C., Vandel, L., Bedford, M.T., Sardet, C., 2006. Coactivator-associated arginine methyltransferase 1 (CARM1) is a positive regulator of the Cyclin E1 gene. Proc. Natl. Acad. Sci. U. S. A. 103, 13351–13356.

Eom, G.H., Kim, K., Kim, S.M., Kee, H.J., Kim, J.Y., Jin, H.M., Kim, J.R., Kim, J.H., Choe, N., Kim, K.B., Lee, J., Kook, H., Kim, N., Seo, S.B., 2009. Histone methyltransferase PRDM8 regulates mouse testis steroidogenesis. Biochem. Biophys. Res. Commun. 388, 131–136.

Eom, G.H., Kim, K.B., Kim, J.H., Kim, J.Y., Kim, J.R., Kee, H.J., Kim, D.W., Choe, N., Park, H.J., Son, H.J., Choi, S.Y., Kook, H., Seo, S.B., 2011. Histone methyltransferase SETD3 regulates muscle differentiation. J. Biol. Chem. 286, 34733–34742.

Eram, M.S., Bustos, S.P., Lima-Fernandes, E., Siarheyeva, A., Senisterra, G., Hajian, T., Chau, I., Duan, S., Wu, H., Dombrovski, L., Schapira, M., Arrowsmith, C.H., Vedadi, M., 2014. Trimethylation of histone H3 lysine 36 by human methyltransferase PRDM9 protein. J. Biol. Chem. 289, 12177–12188.

Eram, M.S., Shen, Y., Szewczyk, M., Wu, H., Senisterra, G., Li, F., Butler, K.V., Kaniskan, H.U., Speed, B.A., Dela Sena, C., Dong, A., Zeng, H., Schapira, M., Brown, P.J., Arrowsmith, C.H., Barsyte-Lovejoy, D., Liu, J., Vedadi, M., Jin, J., 2016. A potent, selective, and cell-active inhibitor of human type I protein arginine methyltransferases. ACS Chem. Biol. 11, 772–781.

Esse, R., Florindo, C., Imbard, A., Rocha, M.S., de Vriese, A.S., Smulders, Y.M., Teerlink, T., Tavares de Almeida, I., Castro, R., Blom, H.J., 2013. Global protein and histone arginine methylation are affected in a tissue-specific manner in a rat model of diet-induced hyperhomocysteinemia. Biochim. Biophys. Acta 1832, 1708–1714.

Ezponda, T., Licht, J.D., 2014. Molecular pathways: deregulation of histone h3 lysine 27 methylation in cancer-different paths, same destination. Clin. Cancer Res. 20, 5001–5008.

EU Clinical Trials Register [Internet], 2013. A phase I study of Human Pharmacokinetics and Safety of ORY-1001, and LSD1 inhibitor, in relapsed or refractory acute leukaemia (AL). European Medicines Agency (European Union), London. Identifier 2013-002447-29. Available from: https://www.clinicaltrialsregister.eu/ctr-search/trial/2013-002447-29/ES [cited 8 August 2018].

Falandry, C., Fourel, G., Galy, V., Ristriani, T., Horard, B., Bensimon, E., Salles, G., Gilson, E., Magdinier, F., 2010. CLLD8/KMT1F is a lysine methyltransferase that is important for chromosome segregation. J. Biol. Chem. 285, 20234–20241.

Feng, Q., Wang, H., Ng, H.H., Erdjument-Bromage, H., Tempst, P., Struhl, K., Zhang, Y., 2002. Methylation of H3-lysine 79 is mediated by a new family of HMTases without a SET domain. Curr. Biol. 12, 1052–1058.

Ferguson, A.D., Larsen, N.A., Howard, T., Pollard, H., Green, I., Grande, C., Cheung, T., Garcia-Arenas, R., Cowen, S., Wu, J., Godin, R., Chen, H., Keen, N., 2011. Structural basis of substrate methylation and inhibition of SMYD2. Structure 19, 1262–1273.

Ferrari, K.J., Scelfo, A., Jammula, S., Cuomo, A., Barozzi, I., Stutzer, A., Fischle, W., Bonaldi, T., Pasini, D., 2014. Polycomb-dependent H3K27me1 and H3K27me2 regulate active transcription and enhancer fidelity. Mol. Cell 53, 49–62.

Fnu, S., Williamson, E.A., De Haro, L.P., Brenneman, M., Wray, J., Shaheen, M., Radhakrishnan, K., Lee, S.H., Nickoloff, J.A., Hromas, R., 2011. Methylation of histone H3 lysine 36 enhances DNA repair by nonhomologous end-joining. Proc. Natl. Acad. Sci. USA 108, 540–545.

Forneris, F., Binda, C., Vanoni, M.A., Battaglioli, E., Mattevi, A., 2005. Human histone demethylase LSD1 reads the histone code. J. Biol. Chem. 280, 41360–41365.

Fraga, M.F., Ballestar, E., Villar-Garea, A., Boix-Chornet, M., Espada, J., Schotta, G., Bonaldi, T., Haydon, C., Ropero, S., Petrie, K., Iyer, N.G., Perez-Rosado, A., Calvo, E., Lopez, J.A., Cano, A., Calasanz, M.J., Colomer, D., Piris, M.A., Ahn, N., Imhof, A., Caldas, C., Jenuwein, T., Esteller, M., 2005. Loss of acetylation at Lys16 and trimethylation at Lys20 of histone H4 is a common hallmark of human cancer. Nat. Genet. 37, 391–400.

Frescas, D., Guardavaccaro, D., Bassermann, F., Koyama-Nasu, R., Pagano, M., 2007. JHDM1B/FBXL10 is a nucleolar protein that represses transcription of ribosomal RNA genes. Nature 450, 309–313.

Frontelo, P., Leader, J.E., Yoo, N., Potocki, A.C., Crawford, M., Kulik, M., Lechleider, R.J., 2004. Suv39h histone methyltransferases interact with Smads and cooperate in BMP-induced repression. Oncogene 23, 5242–5251.

Fuks, F., Hurd, P.J., Deplus, R., Kouzarides, T., 2003. The DNA methyltransferases associate with HP1 and the SUV39H1 histone methyltransferase. Nucleic Acids Res. 31, 2305–2312.

Fumasoni, I., Meani, N., Rambaldi, D., Scafetta, G., Alcalay, M., Ciccarelli, F.D., 2007. Family expansion and gene rearrangements contributed to the functional specialization of PRDM genes in vertebrates. BMC Evol. Biol. 7, 187.

Gadhia, S.R., Calabro, A.R., Barile, F.A., 2012. Trace metals alter DNA repair and histone modification pathways concurrently in mouse embryonic stem cells. Toxicol. Lett. 212, 169–179.

Garcia-Gimenez, J.L., Olaso, G., Hake, S.B., Bonisch, C., Wiedemann, S.M., Markovic, J., Dasi, F., Gimeno, A., Perez-Quilis, C., Palacios, O., Capdevila, M., Vina, J., Pallardo, F.V., 2013. Histone h3 glutathionylation in proliferating mammalian cells destabilizes nucleosomal structure. Antioxid. Redox Signal. 19, 1305–1320.

Gelato, K.A., Shaikhibrahim, Z., Ocker, M., Haendler, B., 2016. Targeting epigenetic regulators for cancer therapy: modulation of bromodomain proteins, methyltransferases, demethylases, and microRNAs. Expert Opin. Ther. Targets 20, 783–799.

Gendrel, A.V., Lippman, Z., Yordan, C., Colot, V., Martienssen, R.A., 2002. Dependence of heterochromatic histone H3 methylation patterns on the Arabidopsis gene DDM1. Science 297, 1871–1873.

Gershey, E.L., Haslett, G.W., Vidali, G., Allfrey, V.G., 1969. Chemical studies of histone methylation. Evidence for the occurrence of 3-methylhistidine in avian erythrocyte histone fractions. J. Biol. Chem. 244, 4871–4877.

Girardot, M., Hirasawa, R., Kacem, S., Fritsch, L., Pontis, J., Kota, S.K., Filipponi, D., Fabbrizio, E., Sardet, C., Lohmann, F., Kadam, S., Ait-Si-Ali, S., Feil, R., 2014. PRMT5-mediated histone H4 arginine-3 symmetrical dimethylation marks chromatin at G + C-rich regions of the mouse genome. Nucleic Acids Res. 42, 235–248.

Greathouse, K.L., Bredfeldt, T., Everitt, J.I., Lin, K., Berry, T., Kannan, K., Mittelstadt, M.L., Ho, S.M., Walker, C.L., 2012. Environmental estrogens differentially engage the histone methyltransferase EZH2 to increase risk of uterine tumorigenesis. Mol. Cancer Res. 10, 546–557.

Green, E.M., Mas, G., Young, N.L., Garcia, B.A., Gozani, O., 2012. Methylation of H4 lysines 5, 8 and 12 by yeast Set5 calibrates chromatin stress responses. Nat. Struct. Mol. Biol. 19, 361–363.

Greer, E.L., Shi, Y., 2012. Histone methylation: a dynamic mark in health, disease and inheritance. Nat. Rev. Genet. 13, 343–357.

Gregory, G.D., Vakoc, C.R., Rozovskaia, T., Zheng, X., Patel, S., Nakamura, T., Canaani, E., Blobel, G.A., 2007. Mammalian ASH1L is a histone methyltransferase that occupies the transcribed region of active genes. Mol. Cell. Biol. 27, 8466–8479.

Guccione, E., Bassi, C., Casadio, F., Martinato, F., Cesaroni, M., Schuchlautz, H., Luscher, B., Amati, B., 2007. Methylation of histone H3R2 by PRMT6 and H3K4 by an MLL complex are mutually exclusive. Nature 449, 933–937.

Guo, X., Zhang, Q., 2017. The emerging role of histone demethylases in renal cell carcinoma. J. Kidney Cancer VHL 4, 1–5.

Hahn, P., Wegener, I., Burrells, A., Bose, J., Wolf, A., Erck, C., Butler, D., Schofield, C.J., Bottger, A., Lengeling, A., 2010. Analysis of Jmjd6 cellular localization and testing for its involvement in histone demethylation. PLoS One 5, e13769.

Hamamoto, R., Nakamura, Y., 2016. Dysregulation of protein methyltransferases in human cancer: an emerging target class for anticancer therapy. Cancer Sci. 107, 377–384.

Hamamoto, R., Furukawa, Y., Morita, M., Iimura, Y., Silva, F.P., Li, M., Yagyu, R., Nakamura, Y., 2004. SMYD3 encodes a histone methyltransferase involved in the proliferation of cancer cells. Nat. Cell Biol. 6, 731–740.

Hanotel, J., Bessodes, N., Thelie, A., Hedderich, M., Parain, K., Van Driessche, B., Brandao Kde, O., Kricha, S., Jorgensen, M.C., Grapin-Botton, A., Serup, P., Van Lint, C., Perron, M., Pieler, T., Henningfeld, K.A., Bellefroid, E.J., 2014. The Prdm13 histone methyltransferase encoding gene is a Ptf1a-Rbpj downstream target that suppresses glutamatergic and promotes GABAergic neuronal fate in the dorsal neural tube. Dev. Biol. 386, 340–357.

Harrison, M.J., Tang, Y.H., Dowhan, D.H., 2010. Protein arginine methyltransferase 6 regulates multiple aspects of gene expression. Nucleic Acids Res. 38, 2201–2216.

He, Y., Gorkin, D.U., Dickel, D.E., Nery, J.R., Castanon, R.G., Lee, A.Y., Shen, Y., Visel, A., Pennacchio, L.A., Ren, B., Ecker, J.R., 2017. Improved regulatory element prediction based on tissue-specific local epigenomic signatures. Proc. Natl. Acad. Sci. USA 114, E1633–E1640.

Heintzman, N.D., Stuart, R.K., Hon, G., Fu, Y., Ching, C.W., Hawkins, R.D., Barrera, L.O., Van Calcar, S., Qu, C., Ching, K.A., Wang, W., Weng, Z., Green, R.D., Crawford, G.E., Ren, B., 2007. Distinct and predictive chromatin signatures of transcriptional promoters and enhancers in the human genome. Nat. Genet. 39, 311–318.

Hillringhaus, L., Yue, W.W., Rose, N.R., Ng, S.S., Gileadi, C., Loenarz, C., Bello, S.H., Bray, J.E., Schofield, C.J., Oppermann, U., 2011. Structural and evolutionary basis for the dual substrate selectivity of human KDM4 histone demethylase family. J. Biol. Chem. 286, 41616–41625.

Hino, S., Kohrogi, K., Nakao, M., 2016. Histone demethylase LSD1 controls the phenotypic plasticity of cancer cells. Cancer Sci. 107, 1187–1192.

Ho, S.M., Cheong, A., Lam, H.M., Hu, W.Y., Shi, G.B., Zhu, X., Chen, J., Zhang, X., Medvedovic, M., Leung, Y.K., Prins, G.S., 2015. Exposure of human prostaspheres to bisphenol A epigenetically regulates SNORD family noncoding RNAs via histone modification. Endocrinology 156, 3984–3995.

Hon, G., Wang, W., Ren, B., 2009. Discovery and annotation of functional chromatin signatures in the human genome. PLoS Comput. Biol. 5, e1000566.

Hong, S., Cho, Y.W., Yu, L.R., Yu, H., Veenstra, T.D., Ge, K., 2007. Identification of JmjC domain-containing UTX and JMJD3 as histone H3 lysine 27 demethylases. Proc. Natl. Acad. Sci. USA 104, 18439–18444.

Horton, J.R., Upadhyay, A.K., Qi, H.H., Zhang, X., Shi, Y., Cheng, X., 2010. Enzymatic and structural insights for substrate specificity of a family of jumonji histone lysine demethylases. Nat. Struct. Mol. Biol. 17, 38–43.

Hou, L., Zhang, X., Wang, D., Baccarelli, A., 2012. Environmental chemical exposures and human epigenetics. Int. J. Epidemiol. 41, 79–105.

Hsia, D.A., Tepper, C.G., Pochampalli, M.R., Hsia, E.Y., Izumiya, C., Huerta, S.B., Wright, M.E., Chen, H.W., Kung, H.J., Izumiya, Y., 2010. KDM8, a H3K36me2 histone demethylase that acts in the cyclin A1 coding region to regulate cancer cell proliferation. Proc. Natl. Acad. Sci. USA 107, 9671–9676.

Huang, S., Litt, M., Felsenfeld, G., 2005. Methylation of histone H4 by arginine methyltransferase PRMT1 is essential in vivo for many subsequent histone modifications. Genes Dev. 19, 1885–1893.

Huang, J., Perez-Burgos, L., Placek, B.J., Sengupta, R., Richter, M., Dorsey, J.A., Kubicek, S., Opravil, S., Jenuwein, T., Berger, S.L., 2006. Repression of p53 activity by Smyd2-mediated methylation. Nature 444, 629–632.

Huang, J., Dorsey, J., Chuikov, S., Perez-Burgos, L., Zhang, X., Jenuwein, T., Reinberg, D., Berger, S.L., 2010. G9a and Glp methylate lysine 373 in the tumor suppressor p53. J. Biol. Chem. 285, 9636–9641.

Huang, M., Chen, C., Geng, J., Han, D., Wang, T., Xie, T., Wang, L., Wang, Y., Wang, C., Lei, Z., Chu, X., 2017. Targeting KDM1A attenuates Wnt/beta-catenin signaling pathway to eliminate sorafenib-resistant stem-like cells in hepatocellular carcinoma. Cancer Lett. 398, 12–21.

Hyllus, D., Stein, C., Schnabel, K., Schiltz, E., Imhof, A., Dou, Y., Hsieh, J., Bauer, U.M., 2007. PRMT6-mediated methylation of R2 in histone H3 antagonizes H3 K4 trimethylation. Genes Dev. 21, 3369–3380.

Hyun, K., Jeon, J., Park, K., Kim, J., 2017. Writing, erasing and reading histone lysine methylations. Exp. Mol. Med. 49, e324.

Iberg, A.N., Espejo, A., Cheng, D., Kim, D., Michaud-Levesque, J., Richard, S., Bedford, M.T., 2008. Arginine methylation of the histone H3 tail impedes effector binding. J. Biol. Chem. 283, 3006–3010.

Ingelman-Sundberg, M., Zhong, X.B., Hankinson, O., Beedanagari, S., Yu, A.M., Peng, L., Osawa, Y., 2013. Potential role of epigenetic mechanisms in the regulation of drug metabolism and transport. Drug Metab. Dispos. 41, 1725–1731.

Iwase, S., Lan, F., Bayliss, P., de la Torre-Ubieta, L., Huarte, M., Qi, H.H., Whetstine, J.R., Bonni, A., Roberts, T.M., Shi, Y., 2007. The X-linked mental retardation gene SMCX/JARID1C defines a family of histone H3 lysine 4 demethylases. Cell 128, 1077–1088.

Jenuwein, T., 2006. The epigenetic magic of histone lysine methylation. FEBS J. 273, 3121–3135.

Jones, B., Su, H., Bhat, A., Lei, H., Bajko, J., Hevi, S., Baltus, G.A., Kadam, S., Zhai, H., Valdez, R., Gonzalo, S., Zhang, Y., Li, E., Chen, T., 2008. The histone H3K79 methyltransferase Dot1L is essential for mammalian development and heterochromatin structure. PLoS Genet. 4, e1000190.

Jordi, E., Heiman, M., Marion-Poll, L., Guermonprez, P., Cheng, S.K., Nairn, A.C., Greengard, P., Girault, J.A., 2013. Differential effects of cocaine on histone posttranslational modifications in identified populations of striatal neurons. Proc. Natl. Acad. Sci. USA 110, 9511–9516.

Juan, A.H., Kumar, R.M., Marx, J.G., Young, R.A., Sartorelli, V., 2009. Mir-214-dependent regulation of the polycomb protein Ezh2 in skeletal muscle and embryonic stem cells. Mol. Cell 36, 61–74.

Jung, H.R., Pasini, D., Helin, K., Jensen, O.N., 2010. Quantitative mass spectrometry of histones H3.2 and H3.3 in Suz12-deficient mouse embryonic stem cells reveals distinct, dynamic post-translational modifications at Lys-27 and Lys-36. Mol. Cell. Proteomics 9, 838–850.

Kaelin Jr., W.G., McKnight, S.L., 2013. Influence of metabolism on epigenetics and disease. Cell 153, 56–69.

Kajta, M., Wojtowicz, A.K., 2013. Impact of endocrine-disrupting chemicals on neural development and the onset of neurological disorders. Pharmacol. Rep. 65, 1632–1639.

Kaniskan, H.U., Szewczyk, M.M., Yu, Z., Eram, M.S., Yang, X., Schmidt, K., Luo, X., Dai, M., He, F., Zang, I., Lin, Y., Kennedy, S., Li, F., Dobrovetsky, E., Dong, A., Smil, D., Min, S.J., Landon, M., Lin-Jones, J., Huang, X.P., Roth, B.L., Schapira, M., Atadja, P., Barsyte-Lovejoy, D., Arrowsmith, C.H., Brown, P.J., Zhao, K., Jin, J., Vedadi, M., 2015. A potent, selective and cell-active allosteric inhibitor of protein arginine methyltransferase 3 (PRMT3). Angew. Chem. Int. Ed. Engl. 54, 5166–5170.

Khalil, A.M., Guttman, M., Huarte, M., Garber, M., Raj, A., Rivea Morales, D., Thomas, K., Presser, A., Bernstein, B.E., van Oudenaarden, A., Regev, A., Lander, E.S., Rinn, J.L., 2009. Many human large intergenic noncoding RNAs associate with chromatin-modifying complexes and affect gene expression. Proc. Natl. Acad. Sci. USA 106, 11667–11672.

Khanal, T., Kim, D., Johnson, A., Choubey, D., Kim, K., 2015. Deregulation of NR2E3, an orphan nuclear receptor, by benzo(a)pyrene-induced oxidative stress is associated with histone modification status change of the estrogen receptor gene promoter. Toxicol. Lett. 237, 228–236.

Kim, K.C., Huang, S., 2003. Histone methyltransferases in tumor suppression. Cancer Biol Ther 2, 491–499.

Kim, S.M., Kee, H.J., Eom, G.H., Choe, N.W., Kim, J.Y., Kim, Y.S., Kim, S.K., Kook, H., Kook, H., Seo, S.B., 2006. Characterization of a novel WHSC1-associated SET domain protein with H3K4 and H3K27 methyltransferase activity. Biochem. Biophys. Res. Commun. 345, 318–323.

1. HISTONE MODIFICATIONS AND CHROMATIN STRUCTURE

Kim, J.Y., Kee, H.J., Choe, N.W., Kim, S.M., Eom, G.H., Baek, H.J., Kook, H., Kook, H., Seo, S.B., 2008. Multiple-myeloma-related WHSC1/MMSET isoform RE-IIBP is a histone methyltransferase with transcriptional repression activity. Mol. Cell. Biol. 28, 2023–2034.

Kim, S.M., Kim, J.Y., Choe, N.W., Cho, I.H., Kim, J.R., Kim, D.W., Seol, J.E., Lee, S.E., Kook, H., Nam, K.I., Kook, H., Bhak, Y.Y., Seo, S.B., 2010. Regulation of mouse steroidogenesis by WHISTLE and JMJD1C through histone methylation balance. Nucleic Acids Res. 38, 6389–6403.

Kim, S.K., Jung, I., Lee, H., Kang, K., Kim, M., Jeong, K., Kwon, C.S., Han, Y.M., Kim, Y.S., Kim, D., Lee, D., 2012. Human histone H3K79 methyltransferase DOT1L protein [corrected] binds actively transcribing RNA polymerase II to regulate gene expression. J. Biol. Chem. 287, 39698–39709.

Kim, I.W., Han, N., Burckart, G.J., Oh, J.M., 2014a. Epigenetic changes in gene expression for drug-metabolizing enzymes and transporters. Pharmacotherapy 34, 140–150.

Kim, W., Choi, M., Kim, J.E., 2014b. The histone methyltransferase Dot1/DOT1L as a critical regulator of the cell cycle. Cell Cycle 13, 726–738.

Kim, J.H., Lee, J.H., Lee, I.S., Lee, S.B., Cho, K.S., 2017. Histone lysine methylation and neurodevelopmental disorders. Int. J. Mol. Sci. 18, E1404. https://dx.doi.org/10.3390/ijms18071404.

King, O.N., Li, X.S., Sakurai, M., Kawamura, A., Rose, N.R., Ng, S.S., Quinn, A.M., Rai, G., Mott, B.T., Beswick, P., Klose, R.J., Oppermann, U., Jadhav, A., Heightman, T.D., Maloney, D.J., Schofield, C.J., Simeonov, A., 2010. Quantitative high-throughput screening identifies 8-hydroxyquinolines as cell-active histone demethylase inhibitors. PLoS ONE 5, e15535.

Kirkley, A.G., Sargis, R.M., 2014. Environmental endocrine disruption of energy metabolism and cardiovascular risk. Curr. Diab. Rep. 14, 494.

Kizer, K.O., Phatnani, H.P., Shibata, Y., Hall, H., Greenleaf, A.L., Strahl, B.D., 2005. A novel domain in Set2 mediates RNA polymerase II interaction and couples histone H3 K36 methylation with transcript elongation. Mol. Cell. Biol. 25, 3305–3316.

Klose, R.J., Yan, Q., Tothova, Z., Yamane, K., Erdjument-Bromage, H., Tempst, P., Gilliland, D.G., Zhang, Y., Kaelin Jr., W.G., 2007. The retinoblastoma binding protein RBP2 is an H3K4 demethylase. Cell 128, 889–900.

Knutson, S.K., Wigle, T.J., Warholic, N.M., Sneeringer, C.J., Allain, C.J., Klaus, C.R., Sacks, J.D., Raimondi, A., Majer, C.R., Song, J., Scott, M.P., Jin, L., Smith, J.J., Olhava, E.J., Chesworth, R., Moyer, M.P., Richon, V.M., Copeland, R.A., Keilhack, H., Pollock, R.M., Kuntz, K.W., 2012. A selective inhibitor of EZH2 blocks H3K27 methylation and kills mutant lymphoma cells. Nat. Chem. Biol. 8, 890–896.

Knutson, S.K., Warholic, N.M., Wigle, T.J., Klaus, C.R., Allain, C.J., Raimondi, A., Porter Scott, M., Chesworth, R., Moyer, M.P., Copeland, R.A., Richon, V.M., Pollock, R.M., Kuntz, K.W., Keilhack, H., 2013. Durable tumor regression in genetically altered malignant rhabdoid tumors by inhibition of methyltransferase EZH2. Proc. Natl. Acad. Sci. U. S. A. 110, 7922–7927.

Konze, K.D., Ma, A., Li, F., Barsyte-Lovejoy, D., Parton, T., Macnevin, C.J., Liu, F., Gao, C., Huang, X.P., Kuznetsova, E., Rougie, M., Jiang, A., Pattenden, S.G., Norris, J.L., James, L.I., Roth, B.L., Brown, P.J., Frye, S.V., Arrowsmith, C.H., Hahn, K.M., Wang, G.G., Vedadi, M., Jin, J., 2013. An orally bioavailable chemical probe of the lysine methyltransferases EZH2 and EZH1. ACS Chem. Biol. 8, 1324–1334.

Kooistra, S.M., Helin, K., 2012. Molecular mechanisms and potential functions of histone demethylases. Nat. Rev. Mol. Cell Biol. 13, 297–311.

Kottakis, F., Polytarchou, C., Foltopoulou, P., Sanidas, I., Kampranis, S.C., Tsichlis, P.N., 2011. FGF-2 regulates cell proliferation, migration, and angiogenesis through an NDY1/KDM2B-miR-101-EZH2 pathway. Mol. Cell 43, 285–298.

Koturbash, I., Scherhag, A., Sorrentino, J., Sexton, K., Bodnar, W., Swenberg, J.A., Beland, F.A., Pardo-Manuel Devillena, F., Rusyn, I., Pogribny, I.P., 2011a. Epigenetic mechanisms of mouse interstrain variability in genotoxicity of the environmental toxicant 1,3-butadiene. Toxicol. Sci. 122, 448–456.

Koturbash, I., Scherhag, A., Sorrentino, J., Sexton, K., Bodnar, W., Tryndyak, V., Latendresse, J.R., Swenberg, J.A., Beland, F.A., Pogribny, I.P., Rusyn, I., 2011b. Epigenetic alterations in liver of C57BL/6J mice after short-term inhalational exposure to 1,3-butadiene. Environ. Health Perspect. 119, 635–640.

Kouzarides, T., 2007. Chromatin modifications and their function. Cell 128, 693–705.

Kruidenier, L., Chung, C.W., Cheng, Z., Liddle, J., Che, K., Joberty, G., Bantscheff, M., Bountra, C., Bridges, A., Diallo, H., Eberhard, D., Hutchinson, S., Jones, E., Katso, R., Leveridge, M., Mander, P.K., Mosley, J., Ramirez-Molina, C., Rowland, P., Schofield, C.J., Sheppard, R.J., Smith, J.E., Swales, C., Tanner, R., Thomas, P., Tumber, A., Drewes, G., Oppermann, U., Patel, D.J., Lee, K., Wilson, D.M., 2012. A selective jumonji H3K27 demethylase inhibitor modulates the proinflammatory macrophage response. Nature 488, 404–408.

Kuo, A.J., Cheung, P., Chen, K., Zee, B.M., Kioi, M., Lauring, J., Xi, Y., Park, B.H., Shi, X., Garcia, B.A., Li, W., Gozani, O., 2011. NSD2 links dimethylation of histone H3 at lysine 36 to oncogenic programming. Mol. Cell 44, 609–620.

Kuo, A.J., Song, J., Cheung, P., Ishibe-Murakami, S., Yamazoe, S., Chen, J.K., Patel, D.J., Gozani, O., 2012. The BAH domain of ORC1 links H4K20me2 to DNA replication licensing and Meier-Gorlin syndrome. Nature 484, 115–119.

Lachner, M., O'Carroll, D., Rea, S., Mechtler, K., Jenuwein, T., 2001. Methylation of histone H3 lysine 9 creates a binding site for HP1 proteins. Nature 410, 116–120.

Lacroix, M., El Messaoudi, S., Rodier, G., Le Cam, A., Sardet, C., Fabbrizio, E., 2008. The histone-binding protein COPR5 is required for nuclear functions of the protein arginine methyltransferase PRMT5. EMBO Rep. 9, 452–458.

Lakowski, T.M., Frankel, A., 2009. Kinetic analysis of human protein arginine N-methyltransferase 2: formation of monomethyl- and asymmetric dimethyl-arginine residues on histone H4. Biochem. J. 421, 253–261.

Lan, F., Bayliss, P.E., Rinn, J.L., Whetstine, J.R., Wang, J.K., Chen, S., Iwase, S., Alpatov, R., Issaeva, I., Canaani, E., Roberts, T.M., Chang, H.Y., Shi, Y., 2007a. A histone H3 lysine 27 demethylase regulates animal posterior development. Nature 449, 689–694.

Lan, F., Collins, R.E., De Cegli, R., Alpatov, R., Horton, J.R., Shi, X., Gozani, O., Cheng, X., Shi, Y., 2007b. Recognition of unmethylated histone H3 lysine 4 links BHC80 to LSD1-mediated gene repression. Nature 448, 718–722.

Larson, A.G., Elnatan, D., Keenen, M.M., Trnka, M.J., Johnston, J.B., Burlingame, A.L., Agard, D.A., Redding, S., Narlikar, G.J., 2017. Liquid droplet formation by HP1alpha suggests a role for phase separation in heterochromatin. Nature 547, 236–240.

Laurent, B., Ruitu, L., Murn, J., Hempel, K., Ferrao, R., Xiang, Y., Liu, S., Garcia, B.A., Wu, H., Wu, F., Steen, H., Shi, Y., 2015. A specific LSD1/KDM1A isoform regulates neuronal differentiation through H3K9 demethylation. Mol. Cell 57, 957–970.

Lee, M.G., Wynder, C., Cooch, N., Shiekhattar, R., 2005a. An essential role for CoREST in nucleosomal histone 3 lysine 4 demethylation. Nature 437, 432–435.

Lee, S.H., Oshige, M., Durant, S.T., Rasila, K.K., Williamson, E.A., Ramsey, H., Kwan, L., Nickoloff, J.A., Hromas, R., 2005b. The SET domain protein Metnase mediates foreign DNA integration and links integration to nonhomologous end-joining repair. Proc. Natl. Acad. Sci. USA 102, 18075–18080.

Lee, S.H., Oshige, M., Durant, S.T., Rasila, K.K., Williamson, E.A., Ramsey, H., Kwan, L., Nickoloff, J.A., Hromas, R., 2005c. The SET domain protein Metnase mediates foreign DNA integration and links integration to nonhomologous end-joining repair. Proc. Natl. Acad. Sci. U. S. A. 102, 18075–18080.

Lee, J.H., Tate, C.M., You, J.S., Skalnik, D.G., 2007. Identification and characterization of the human Set1B histone H3-Lys4 methyltransferase complex. J. Biol. Chem. 282, 13419–13428.

Lee, J., Thompson, J.R., Botuyan, M.V., Mer, G., 2008. Distinct binding modes specify the recognition of methylated histones H3K4 and H4K20 by JMJD2A-tudor. Nat. Struct. Mol. Biol. 15, 109–111.

Lehnertz, B., Ueda, Y., Derijck, A.A., Braunschweig, U., Perez-Burgos, L., Kubicek, S., Chen, T., Li, E., Jenuwein, T., Peters, A.H., 2003. Suv39h-mediated histone H3 lysine 9 methylation directs DNA methylation to major satellite repeats at pericentric heterochromatin. Curr. Biol. 13, 1192–1200.

Levy, D., Kuo, A.J., Chang, Y., Schaefer, U., Kitson, C., Cheung, P., Espejo, A., Zee, B.M., Liu, C.L., Tangsombatvisit, S., Tennen, R.I., Kuo, A.Y., Tanjing, S., Cheung, R., Chua, K.F., Utz, P.J., Shi, X., Prinjha, R.K., Lee, K., Garcia, B.A., Bedford, M.T., Tarakhovsky, A., Cheng, X., Gozani, O., 2011. Lysine methylation of the NF-kappaB subunit RelA by SETD6 couples activity of the histone methyltransferase GLP at chromatin to tonic repression of NF-kappaB signaling. Nat. Immunol. 12, 29–36.

Li, H., Park, S., Kilburn, B., Jelinek, M.A., Henschen-Edman, A., Aswad, D.W., Stallcup, M.R., Laird-Offringa, I.A., 2002. Lipopolysaccharide-induced methylation of HuR, an mRNA-stabilizing protein, by CARM1. Coactivator-associated arginine methyltransferase. J. Biol. Chem. 277, 44623–44630.

Li, Y., Reddy, M.A., Miao, F., Shanmugam, N., Yee, J.K., Hawkins, D., Ren, B., Natarajan, R., 2008. Role of the histone H3 lysine 4 methyltransferase, SET7/9, in the regulation of NF-kappaB-dependent inflammatory genes. Relevance to diabetes and inflammation. J. Biol. Chem. 283, 26771–26781.

Li, Y., Cui, Y., Hart, S.N., Klaassen, C.D., Zhong, X.B., 2009a. Dynamic patterns of histone methylation are associated with ontogenic expression of the Cyp3a genes during mouse liver maturation. Mol. Pharmacol. 75, 1171–1179.

Li, Y., Trojer, P., Xu, C.F., Cheung, P., Kuo, A., Drury 3rd, W.J., Qiao, Q., Neubert, T.A., Xu, R.M., Gozani, O., Reinberg, D., 2009b. The target of the NSD family of histone lysine methyltransferases depends on the nature of the substrate. J. Biol. Chem. 284, 34283–34295.

Li, Q., He, M.D., Mao, L., Wang, X., Jiang, Y.L., Li, M., Lu, Y.H., Yu, Z.P., Zhou, Z., 2017. Nicotinamide N-methyltransferase suppression participates in nickel-induced histone H3 lysine9 dimethylation in BEAS-2B cells. Cell. Physiol. Biochem. 41, 2016–2026.

Linghu, C., Zheng, H., Zhang, L., Zhang, J., 2013. Discovering common combinatorial histone modification patterns in the human genome. Gene 518, 171–178.

Liu, W., Tanasa, B., Tyurina, O.V., Zhou, T.Y., Gassmann, R., Liu, W.T., Ohgi, K.A., Benner, C., Garcia-Bassets, I., Aggarwal, A.K., Desai, A., Dorrestein, P.C., Glass, C.K., Rosenfeld, M.G., 2010. PHF8 mediates histone H4 lysine 20 demethylation events involved in cell cycle progression. Nature 466, 508–512.

Liu, L., Kim, H., Casta, A., Kobayashi, Y., Shapiro, L.S., Christiano, A.M., 2014. Hairless is a histone H3K9 demethylase. FASEB J. 28, 1534–1542.

Liu, J., Wang, Q.C., Han, J., Xiong, B., Sun, S.C., 2015. Aflatoxin B1 is toxic to porcine oocyte maturation. Mutagenesis 30, 527–535.

Loenarz, C., Ge, W., Coleman, M.L., Rose, N.R., Cooper, C.D., Klose, R.J., Ratcliffe, P.J., Schofield, C.J., 2010. PHF8, a gene associated with cleft lip/palate and mental retardation, encodes for an Nepsilon-dimethyl lysine demethylase. Hum. Mol. Genet. 19, 217–222.

Lu, Y., Chang, Q., Zhang, Y., Beezhold, K., Rojanasakul, Y., Zhao, H., Castranova, V., Shi, X., Chen, F., 2009. Lung cancer-associated JmjC domain protein mdig suppresses formation of tri-methyl lysine 9 of histone H3. Cell Cycle 8, 2101–2109.

Lu, Z., Tian, Y., Salwen, H.R., Chlenski, A., Godley, L.A., Raj, J.U., Yang, Q., 2013. Histone-lysine methyltransferase EHMT2 is involved in proliferation, apoptosis, cell invasion, and DNA methylation of human neuroblastoma cells. Anticancer Drugs 24, 484–493.

Maeda, K., Doi, S., Nakashima, A., Nagai, T., Irifuku, T., Ueno, T., Masaki, T., 2017. Inhibition of H3K9 methyltransferase G9a ameliorates methylglyoxal-induced peritoneal fibrosis. PLOS ONE 12, e0173706.

Majumder, S., Alinari, L., Roy, S., Miller, T., Datta, J., Sif, S., Baiocchi, R., Jacob, S.T., 2010. Methylation of histone H3 and H4 by PRMT5 regulates ribosomal RNA gene transcription. J. Cell Biochem. 109, 553–563.

Mantri, M., Krojer, T., Bagg, E.A., Webby, C.A., Butler, D.S., Kochan, G., Kavanagh, K.L., Oppermann, U., McDonough, M.A., Schofield, C.J., 2010. Crystal structure of the 2-oxoglutarate- and Fe(II)-dependent lysyl hydroxylase JMJD6. J. Mol. Biol 401, 211–222.

Maqbool, F., Mostafalou, S., Bahadar, H., Abdollahi, M., 2016. Review of endocrine disorders associated with environmental toxicants and possible involved mechanisms. Life Sci. 145, 265–273.

Marango, J., Shimoyama, M., Nishio, H., Meyer, J.A., Min, D.J., Sirulnik, A., Martinez-Martinez, Y., Chesi, M., Bergsagel, P.L., Zhou, M.M., Waxman, S., Leibovitch, B.A., Walsh, M.J., Licht, J.D., 2008. The MMSET protein is a histone methyltransferase with characteristics of a transcriptional corepressor. Blood 111, 3145–3154.

Margueron, R., Reinberg, D., 2011. The polycomb complex PRC2 and its mark in life. Nature 469, 343–349.

Margueron, R., Li, G., Sarma, K., Blais, A., Zavadil, J., Woodcock, C.L., Dynlacht, B.D., Reinberg, D., 2008. Ezh1 and Ezh2 maintain repressive chromatin through different mechanisms. Mol. Cell 32, 503–518.

McCabe, M.T., Ott, H.M., Ganji, G., Korenchuk, S., Thompson, C., Van Aller, G.S., Liu, Y., Graves, A.P., Della Pietra 3rd, A., Diaz, E., LaFrance, L.V., Mellinger, M., Duquenne, C., Tian, X., Kruger, R.G., McHugh, C.F., Brandt, M., Miller, W.H., Dhanak, D., Verma, S.K., Tummino, P.J., Creasy, C.L., 2012. EZH2 inhibition as a therapeutic strategy for lymphoma with EZH2-activating mutations. Nature 492, 108–112.

McCullough, S.D., Bowers, E.C., On, D.M., Morgan, D.S., Dailey, L.A., Hines, R.N., Devlin, R.B., Diaz-Sanchez, D., 2016. Baseline chromatin modification levels may predict interindividual variability in ozone-induced gene expression. Toxicol. Sci. 150, 216–224.

Miao, F., Li, S., Chavez, V., Lanting, L., Natarajan, R., 2006. Coactivator-associated arginine methyltransferase-1 enhances nuclear factor-kappaB-mediated gene transcription through methylation of histone H3 at arginine 17. Mol. Endocrinol. 20, 1562–1573.

Migliori, V., Phalke, S., Bezzi, M., Guccione, E., 2010. Arginine/lysine-methyl/methyl switches: biochemical role of histone arginine methylation in transcriptional regulation. Epigenomics 2, 119–137.

Mikheyeva, I.V., Grady, P.J., Tamburini, F.B., Lorenz, D.R., Cam, H.P., 2014. Multifaceted genome control by Set1 dependent and independent of H3K4 methylation and the Set1C/COMPASS complex. PLoS Genet. 10, e1004740.

Mikkelsen, T.S., Ku, M., Jaffe, D.B., Issac, B., Lieberman, E., Giannoukos, G., Alvarez, P., Brockman, W., Kim, T.K., Koche, R.P., Lee, W., Mendenhall, E., O'Donovan, A., Presser, A., Russ, C., Xie, X., Meissner, A., Wernig, M., Jaenisch, R., Nusbaum, C., Lander, E.S., Bernstein, B.E., 2007. Genome-wide maps of chromatin state in pluripotent and lineage-committed cells. Nature 448, 553–560.

Miranda, T.B., Cortez, C.C., Yoo, C.B., Liang, G., Abe, M., Kelly, T.K., Marquez, V.E., Jones, P.A., 2009. DZNep is a global histone methylation inhibitor that reactivates developmental genes not silenced by DNA methylation. Mol. Cancer Ther. 8, 1579–1588.

Mitchell, L.H., Drew, A.E., Ribich, S.A., Rioux, N., Swinger, K.K., Jacques, S.L., Lingaraj, T., Boriack-Sjodin, P.A., Waters, N.J., Wigle, T.J., Moradei, O., Jin, L., Riera, T., Porter-Scott, M., Moyer, M.P., Smith, J.J., Chesworth, R., Copeland, R.A., 2015. Aryl pyrazoles as potent inhibitors of arginine methyltransferases: identification of the first PRMT6 tool compound. ACS Med. Chem. Lett. 6, 655–659.

Mozzetta, C., Boyarchuk, E., Pontis, J., Ait-Si-Ali, S., 2015. Sound of silence: the properties and functions of repressive Lys methyltransferases. Nat. Rev. Mol. Cell Biol. 16, 499–513.

Murray, K., 1964. The occurrence of epsilon-N-methyl lysine in histones. Biochemistry 3, 10–15.

Nagandla, H., Lopez, S., Yu, W., Rasmussen, T.L., Tucker, H.O., Schwartz, R.J., Stewart, M.D., 2016. Defective myogenesis in the absence of the muscle-specific lysine methyltransferase SMYD1. Dev. Biol. 410, 86–97.

Nagano, T., Mitchell, J.A., Sanz, L.A., Pauler, F.M., Ferguson-Smith, A.C., Feil, R., Fraser, P., 2008. The Air noncoding RNA epigenetically silences transcription by targeting G9a to chromatin. Science 322, 1717–1720.

Ng, S.S., Yue, W.W., Oppermann, U., Klose, R.J., 2009. Dynamic protein methylation in chromatin biology. Cell Mol. Life Sci. 66, 407–422.

Ngollo, M., Lebert, A., Dagdemir, A., Judes, G., Karsli-Ceppioglu, S., Daures, M., Kemeny, J.L., Penault-Llorca, F., Boiteux, J.P., Bignon, Y.J., Guy, L., Bernard-Gallon, D., 2014. The association between histone 3 lysine 27 trimethylation (H3K27me3) and prostate cancer: relationship with clinicopathological parameters. BMC Cancer 14, 994.

Nguyen, H., Allali-Hassani, A., Antonysamy, S., Chang, S., Chen, L.H., Curtis, C., Emtage, S., Fan, L., Gheyi, T., Li, F., Liu, S., Martin, J.R., Mendel, D., Olsen, J.B., Pelletier, L., Shatseva, T., Wu, S., Zhang, F.F., Arrowsmith, C.H., Brown, P.J., Campbell, R.M., Garcia, B.A., Barsyte-Lovejoy, D., Mader, M., Vedadi, M., 2015. LLY-507, a cell-active, potent, and selective inhibitor of protein-lysine methyltransferase SMYD2. J. Biol. Chem. 290, 13641–13653.

Nishikata, I., Sasaki, H., Iga, M., Tateno, Y., Imayoshi, S., Asou, N., Nakamura, T., Morishita, K., 2003. A novel EVI1 gene family, MEL1, lacking a PR domain (MEL1S) is expressed mainly in t(1;3)(p36;q21)-positive AML and blocks G-CSF-induced myeloid differentiation. Blood 102, 3323–3332.

Nishioka, K., Rice, J.C., Sarma, K., Erdjument-Bromage, H., Werner, J., Wang, Y., Chuikov, S., Valenzuela, P., Tempst, P., Steward, R., Lis, J.T., Allis, C.D., Reinberg, D., 2002. PR-Set7 is a nucleosome-specific methyltransferase that modifies lysine 20 of histone H4 and is associated with silent chromatin. Mol. Cell 9, 1201–1213.

Okada, Y., Feng, Q., Lin, Y., Jiang, Q., Li, Y., Coffield, V.M., Su, L., Xu, G., Zhang, Y., 2005. hDOT1L links histone methylation to leukemogenesis. Cell 121, 167–178.

Okino, S.T., Pookot, D., Li, L.C., Zhao, H., Urakami, S., Shiina, H., Igawa, M., Dahiya, R., 2006. Epigenetic inactivation of the dioxin-responsive cytochrome P4501A1 gene in human prostate cancer. Cancer Res. 66, 7420–7428.

Ooga, M., Suzuki, M.G., Aoki, F., 2013. Involvement of DOT1L in the remodeling of heterochromatin configuration during early preimplantation development in mice. Biol. Reprod. 89, 145.

Orom, U.A., Derrien, T., Beringer, M., Gumireddy, K., Gardini, A., Bussotti, G., Lai, F., Zytnicki, M., Notredame, C., Huang, Q., Guigo, R., Shiekhattar, R., 2010. Long noncoding RNAs with enhancer-like function in human cells. Cell 143, 46–58.

Ovesen, J.L., Schnekenburger, M., Puga, A., 2011. Aryl hydrocarbon receptor ligands of widely different toxic equivalency factors induce similar histone marks in target gene chromatin. Toxicol. Sci. 121, 123–131.

Pan, M., Reid, M.A., Lowman, X.H., Kulkarni, R.P., Tran, T.Q., Liu, X., Yang, Y., Hernandez-Davies, J.E., Rosales, K.K., Li, H., Hugo, W., Song, C., Xu, X., Schones, D.E., Ann, D.K., Gradinaru, V., Lo, R.S., Locasale, J.W., Kong, M., 2016. Regional glutamine deficiency in tumours promotes dedifferentiation through inhibition of histone demethylation. Nat. Cell Biol. 18, 1090–1101.

1. HISTONE MODIFICATIONS AND CHROMATIN STRUCTURE

Papoutsis, A.J., Borg, J.L., Selmin, O.I., Romagnolo, D.F., 2012. BRCA-1 promoter hypermethylation and silencing induced by the aromatic hydrocarbon receptor-ligand TCDD are prevented by resveratrol in MCF-7 cells. J. Nutr. Biochem. 23, 1324–1332.

Pekowska, A., Benoukraf, T., Ferrier, P., Spicuglia, S., 2010. A unique H3K4me2 profile marks tissue-specific gene regulation. Genome Res. 20, 1493–1502.

Pesavento, J.J., Bullock, C.R., LeDuc, R.D., Mizzen, C.A., Kelleher, N.L., 2008. Combinatorial modification of human histone H4 quantitated by two-dimensional liquid chromatography coupled with top down mass spectrometry. J. Biol. Chem. 283, 14927–14937.

Peserico, A., Germani, A., Sanese, P., Barbosa, A.J., Di Virgilio, V., Fittipaldi, R., Fabini, E., Bertucci, C., Varchi, G., Moyer, M.P., Caretti, G., Del Rio, A., Simone, C., 2015. A SMYD3 small-molecule inhibitor impairing cancer cell growth. J. Cell. Physiol. 230, 2447–2460.

Phanstiel, D., Brumbaugh, J., Berggren, W.T., Conard, K., Feng, X., Levenstein, M.E., McAlister, G.C., Thomson, J.A., Coon, J.J., 2008. Mass spectrometry identifies and quantifies 74 unique histone H4 isoforms in differentiating human embryonic stem cells. Proc. Natl. Acad. Sci. USA 105, 4093–4098.

Pinheiro, I., Margueron, R., Shukeir, N., Eisold, M., Fritzsch, C., Richter, F.M., Mittler, G., Genoud, C., Goyama, S., Kurokawa, M., Son, J., Reinberg, D., Lachner, M., Jenuwein, T., 2012. Prdm3 and Prdm16 are H3K9me1 methyltransferases required for mammalian heterochromatin integrity. Cell 150, 948–960.

Pogribny, I.P., Tryndyak, V.P., Woods, C.G., Witt, S.E., Rusyn, I., 2007. Epigenetic effects of the continuous exposure to peroxisome proliferator WY-14,643 in mouse liver are dependent upon peroxisome proliferator activated receptor alpha. Mutat. Res. 625, 62–71.

Pokholok, D.K., Harbison, C.T., Levine, S., Cole, M., Hannett, N.M., Lee, T.I., Bell, G.W., Walker, K., Rolfe, P.A., Herbolsheimer, E., Zeitlinger, J., Lewitter, F., Gifford, D.K., Young, R.A., 2005. Genome-wide map of nucleosome acetylation and methylation in yeast. Cell 122, 517–527.

Polz, M., Jablonski, L., 1989. Viral hepatitis among the hospital staff. J. Hyg. Epidemiol. Microbiol. Immunol. 33, 157–161.

Porter, L.F., Galli, G.G., Williamson, S., Selley, J., Knight, D., Elcioglu, N., Aydin, A., Elcioglu, M., Venselaar, H., Lund, A.H., Bonshek, R., Black, G.C., Manson, F.D., 2015. A role for repressive complexes and H3K9 di-methylation in PRDM5-associated brittle cornea syndrome. Hum. Mol. Genet. 24, 6565–6579.

Pournara, A., Kippler, M., Holmlund, T., Ceder, R., Grafstrom, R., Vahter, M., Broberg, K., Wallberg, A.E., 2016. Arsenic alters global histone modifications in lymphocytes in vitro and in vivo. Cell Biol. Toxicol. 32, 275–284.

Pradhan, S., Chin, H.G., Esteve, P.O., Jacobsen, S.E., 2009. SET7/9 mediated methylation of non-histone proteins in mammalian cells. Epigenetics 4, 383–387.

Pulliero, A., Cao, J., Vasques Ldos, R., Pacchierotti, F., 2015. Genetic and epigenetic effects of environmental mutagens and carcinogens. Biomed. Res. Int. 2015, 608054.

Qi, W., Chan, H., Teng, L., Li, L., Chuai, S., Zhang, R., Zeng, J., Li, M., Fan, H., Lin, Y., Gu, J., Ardayfio, O., Zhang, J.H., Yan, X., Fang, J., Mi, Y., Zhang, M., Zhou, T., Feng, G., Chen, Z., Li, G., Yang, T., Zhao, K., Liu, X., Yu, Z., Lu, C.X., Atadja, P., Li, E., 2012. Selective inhibition of Ezh2 by a small molecule inhibitor blocks tumor cells proliferation. Proc. Natl. Acad. Sci. U. S. A. 109, 21360–21365.

Qiao, Q., Li, Y., Chen, Z., Wang, M., Reinberg, D., Xu, R.M., 2011. The structure of NSD1 reveals an autoregulatory mechanism underlying histone H3K36 methylation. J. Biol. Chem. 286, 8361–8368.

Racine, A., Page, V., Nagy, S., Grabowski, D., Tanny, J.C., 2012. Histone H2B ubiquitylation promotes activity of the intact Set1 histone methyltransferase complex in fission yeast. J. Biol. Chem. 287, 19040–19047.

Rada-Iglesias, A., Bajpai, R., Swigut, T., Brugmann, S.A., Flynn, R.A., Wysocka, J., 2011. A unique chromatin signature uncovers early developmental enhancers in humans. Nature 470, 279–283.

Rahman, S., Sowa, M.E., Ottinger, M., Smith, J.A., Shi, Y., Harper, J.W., Howley, P.M., 2011. The Brd4 extraterminal domain confers transcription activation independent of pTEFb by recruiting multiple proteins, including NSD3. Mol. Cell Biol. 31, 2641–2652.

Rea, S., Eisenhaber, F., O'Carroll, D., Strahl, B.D., Sun, Z.W., Schmid, M., Opravil, S., Mechtler, K., Ponting, C.P., Allis, C.D., Jenuwein, T., 2000. Regulation of chromatin structure by site-specific histone H3 methyltransferases. Nature 406, 593–599.

Ricq, E.L., Hooker, J.M., Haggarty, S.J., 2016. Toward development of epigenetic drugs for central nervous system disorders: modulating neuroplasticity via H3K4 methylation. Psychiatry Clin. Neurosci. 70, 536–550.

Ringrose, L., Ehret, H., Paro, R., 2004. Distinct contributions of histone H3 lysine 9 and 27 methylation to locus-specific stability of polycomb complexes. Mol. Cell 16, 641–653.

Rinn, J.L., Kertesz, M., Wang, J.K., Squazzo, S.L., Xu, X., Brugmann, S.A., Goodnough, L.H., Helms, J.A., Farnham, P.J., Segal, E., Chang, H.Y., 2007. Functional demarcation of active and silent chromatin domains in human HOX loci by noncoding RNAs. Cell 129, 1311–1323.

Roadmap Epigenomics, C., Kundaje, A., Meuleman, W., Ernst, J., Bilenky, M., Yen, A., Heravi-Moussavi, A., Kheradpour, P., Zhang, Z., Wang, J., Ziller, M.J., Amin, V., Whitaker, J.W., Schultz, M.D., Ward, L.D., Sarkar, A., Quon, G., Sandstrom, R.S., Eaton, M.L., Wu, Y.C., Pfenning, A.R., Wang, X., Claussnitzer, M., Liu, Y., Coarfa, C., Harris, R.A., Shoresh, N., Epstein, C.B., Gjoneska, E., Leung, D., Xie, W., Hawkins, R.D., Lister, R., Hong, C., Gascard, P., Mungall, A.J., Moore, R., Chuah, E., Tam, A., Canfield, T.K., Hansen, R.S., Kaul, R., Sabo, P.J., Bansal, M.S., Carles, A., Dixon, J.R., Farh, K.H., Feizi, S., Karlic, R., Kim, A.R., Kulkarni, A., Li, D., Lowdon, R., Elliott, G., Mercer, T.R., Neph, S.J., Onuchic, V., Polak, P., Rajagopal, N., Ray, P., Sallari, R.C., Siebenthall, K.T., Sinnott-Armstrong, N.A., Stevens, M., Thurman, R.E., Wu, J., Zhang, B., Zhou, X., Beaudet, A.E., Boyer, L.A., De Jager, P.L., Farnham, P.J., Fisher, S.J., Haussler, D., Jones, S.J., Li, W., Marra, M.A., McManus, M.T., Sunyaev, S., Thomson, J.A., Tlsty, T.D., Tsai, L.H., Wang, W., Waterland, R.A., Zhang, M.Q., Chadwick, L.H., Bernstein, B.E., Costello, J.F., Ecker, J.R., Hirst, M., Meissner, A., Milosavljevic, A., Ren, B., Stamatoyannopoulos, J.A., Wang, T., Kellis, M., 2015. Integrative analysis of 111 reference human epigenomes. Nature 518, 317–330.

Rodriguez-Paredes, M., Martinez de Paz, A., Simo-Riudalbas, L., Sayols, S., Moutinho, C., Moran, S., Villanueva, A., Vazquez-Cedeira, M., Lazo, P.A., Carneiro, F., Moura, C.S., Vieira, J., Teixeira, M.R., Esteller, M., 2014. Gene amplification of the histone methyltransferase SETDB1 contributes to human lung tumorigenesis. Oncogene 33, 2807–2813.

Roguev, A., Schaft, D., Shevchenko, A., Aasland, R., Shevchenko, A., Stewart, A.F., 2003. High conservation of the Set1/Rad6 axis of histone 3 lysine 4 methylation in budding and fission yeasts. J. Biol. Chem. 278, 8487–8493.

Rose, N.R., Klose, R.J., 2014. Understanding the relationship between DNA methylation and histone lysine methylation. Biochim. Biophys. Acta 1839, 1362–1372.

Rosenfeld, J.A., Wang, Z., Schones, D.E., Zhao, K., DeSalle, R., Zhang, M.Q., 2009. Determination of enriched histone modifications in non-genic portions of the human genome. BMC Genomics 10, 143.

Rothbart, S.B., Krajewski, K., Nady, N., Tempel, W., Xue, S., Badeaux, A.I., Barsyte-Lovejoy, D., Martinez, J.Y., Bedford, M.T., Fuchs, S.M., Arrowsmith, C.H., Strahl, B.D., 2012. Association of UHRF1 with methylated H3K9 directs the maintenance of DNA methylation. Nat. Struct. Mol. Biol. 19, 1155–1160.

Santos-Rosa, H., Schneider, R., Bannister, A.J., Sherriff, J., Bernstein, B.E., Emre, N.C., Schreiber, S.L., Mellor, J., Kouzarides, T., 2002. Active genes are tri-methylated at K4 of histone H3. Nature 419, 407–411.

Schenk, T., Chen, W.C., Gollner, S., Howell, L., Jin, L., Hebestreit, K., Klein, H.U., Popescu, A.C., Burnett, A., Mills, K., Casero Jr., R.A., Marton, L., Woster, P., Minden, M.D., Dugas, M., Wang, J.C., Dick, J.E., Muller-Tidow, C., Petrie, K., Zelent, A., 2012. Inhibition of the LSD1 (KDM1A) demethylase reactivates the all-trans-retinoic acid differentiation pathway in acute myeloid leukemia. Nat. Med. 18, 605–611.

Schnekenburger, M., Peng, L., Puga, A., 2007. HDAC1 bound to the Cyp1a1 promoter blocks histone acetylation associated with Ah receptor-mediated trans-activation. Biochim. Biophys. Acta 1769, 569–578.

Schotta, G., Lachner, M., Sarma, K., Ebert, A., Sengupta, R., Reuter, G., Reinberg, D., Jenuwein, T., 2004. A silencing pathway to induce H3-K9 and H4-K20 trimethylation at constitutive heterochromatin. Genes Dev. 18, 1251–1262.

Schotta, G., Sengupta, R., Kubicek, S., Malin, S., Kauer, M., Callen, E., Celeste, A., Pagani, M., Opravil, S., De La Rosa-Velazquez, I.A., Espejo, A., Bedford, M.T., Nussenzweig, A., Busslinger, M., Jenuwein, T., 2008. A chromatin-wide transition to H4K20 monomethylation impairs genome integrity and programmed DNA rearrangements in the mouse. Genes Dev. 22, 2048–2061.

Schubeler, D., MacAlpine, D.M., Scalzo, D., Wirbelauer, C., Kooperberg, C., van Leeuwen, F., Gottschling, D.E., O'Neill, L.P., Turner, B.M., Delrow, J., Bell, S.P., Groudine, M., 2004. The histone modification pattern of active genes revealed through genome-wide chromatin analysis of a higher eukaryote. Genes Dev. 18, 1263–1271.

Schultz, D.C., Ayyanathan, K., Negorev, D., Maul, G.G., Rauscher 3rd, F.J., 2002. SETDB1: a novel KAP-1-associated histone H3, lysine 9-specific methyltransferase that contributes to HP1-mediated silencing of euchromatic genes by KRAB zinc-finger proteins. Genes Dev. 16, 919–932.

Selvi, B.R., Batta, K., Kishore, A.H., Mantelingu, K., Varier, R.A., Balasubramanyam, K., Pradhan, S.K., Dasgupta, D., Sriram, S., Agrawal, S., Kundu, T.K., 2010. Identification of a novel inhibitor of coactivator-associated arginine methyltransferase 1 (CARM1)-mediated methylation of histone H3 Arg-17. J. Biol. Chem. 285, 7143–7152.

Senyildiz, M., Karaman, E.F., Bas, S.S., Pirincci, P.A., Ozden, S., 2017. Effects of BPA on global DNA methylation and global histone 3 lysine modifications in SH-SY5Y cells: an epigenetic mechanism linking the regulation of chromatin modifiying genes. Toxicol. in Vitro 44, 313–321.

1. HISTONE MODIFICATIONS AND CHROMATIN STRUCTURE

Seo, S.B., Kang, J.Y., Kim, J.Y., Kim, K.B., Park, J.W., Cho, H., Hahm, J.Y., Chae, Y.C., Kim, D., Kook, H., Rhee, S., NC, H., 2017. KDM2B is a histone H3K79 demethylase and induces transcriptional repression via SIRT1-mediated chromatin silencing. bioRxiv.

Severson, P.L., Tokar, E.J., Vrba, L., Waalkes, M.P., Futscher, B.W., 2013. Coordinate H3K9 and DNA methylation silencing of ZNFs in toxicant-induced malignant transformation. Epigenetics 8, 1080–1088.

Shi, Y.G., Tsukada, Y., 2013. The discovery of histone demethylases. Cold Spring Harb. Perspect. Biol. 5.

Shi, Y., Lan, F., Matson, C., Mulligan, P., Whetstine, J.R., Cole, P.A., Casero, R.A., Shi, Y., 2004. Histone demethylation mediated by the nuclear amine oxidase homolog LSD1. Cell 119, 941–953.

Shi, X., Kachirskaia, I., Yamaguchi, H., West, L.E., Wen, H., Wang, E.W., Dutta, S., Appella, E., Gozani, O., 2007. Modulation of p53 function by SET8-mediated methylation at lysine 382. Mol. Cell 27, 636–646.

Shilatifard, A., 2012. The COMPASS family of histone H3K4 methylases: mechanisms of regulation in development and disease pathogenesis. Annu. Rev. Biochem. 81, 65–95.

Shinsky, S.A., Monteith, K.E., Viggiano, S., Cosgrove, M.S., 2015. Biochemical reconstitution and phylogenetic comparison of human SET1 family core complexes involved in histone methylation. J. Biol. Chem. 290, 6361–6375.

Simo-Riudalbas, L., Esteller, M., 2015. Targeting the histone orthography of cancer: drugs for writers, erasers and readers. Br. J. Pharmacol. 172, 2716–2732.

Singh, S., Li, S.S., 2012. Epigenetic effects of environmental chemicals bisphenol A and phthalates. Int. J. Mol. Sci. 13, 10143–10153.

Sinha, K.M., Yasuda, H., Coombes, M.M., Dent, S.Y., de Crombrugghe, B., 2010. Regulation of the osteoblast-specific transcription factor Osterix by NO66, a Jumonji family histone demethylase. EMBO J. 29, 68–79.

Smith, E.H., Janknecht, R., Maher 3rd, L.J., 2007. Succinate inhibition of alpha-ketoglutarate-dependent enzymes in a yeast model of paraganglioma. Hum. Mol. Genet. 16, 3136–3148.

Smith, A.E., Chronis, C., Christodoulakis, M., Orr, S.J., Lea, N.C., Twine, N.A., Bhinge, A., Mufti, G.J., Thomas, N.S., 2009. Epigenetics of human T cells during the G0–>G1 transition. Genome Res. 19, 1325–1337.

Soares, L.M., He, P.C., Chun, Y., Suh, H., Kim, T., Buratowski, S., 2017. Determinants of histone H3K4 methylation patterns. Mol. Cell 68, 773–785. e776.

Song, Y., Wu, F., Wu, J., 2016. Targeting histone methylation for cancer therapy: enzymes, inhibitors, biological activity and perspectives. J. Hematol. Oncol. 9, 49.

Sonkar, R., Powell, C.A., Choudhury, M., 2016. Benzyl butyl phthalate induces epigenetic stress to enhance adipogenesis in mesenchymal stem cells. Mol. Cell. Endocrinol. 431, 109–122.

Sorenson, M.R., Jha, D.K., Ucles, S.A., Flood, D.M., Strahl, B.D., Stevens, S.W., Kress, T.L., 2016. Histone H3K36 methylation regulates pre-mRNA splicing in Saccharomyces cerevisiae. RNA Biol. 13, 412–426.

Southall, S.M., Wong, P.S., Odho, Z., Roe, S.M., Wilson, J.R., 2009. Structural basis for the requirement of additional factors for MLL1 SET domain activity and recognition of epigenetic marks. Mol. Cell 33, 181–191.

Spies, N., Nielsen, C.B., Padgett, R.A., Burge, C.B., 2009. Biased chromatin signatures around polyadenylation sites and exons. Mol. Cell 36, 245–254.

Steinmeyer, J., Kalbhen, D.A., 1991. Influence of some natural and semisynthetic agents on elastase and cathepsin G from polymorphonuclear granulocytes. Arzneimittelforschung 41, 77–80.

Strahl, B.D., Allis, C.D., 2000. The language of covalent histone modifications. Nature 403, 41–45.

Stunnenberg, H.G., International Human Epigenome Consortium, Hirst, M., 2016. The International Human Epigenome Consortium: a blueprint for scientific collaboration and discovery. Cell 167, 1145–1149.

Sun, X.J., Wei, J., Wu, X.Y., Hu, M., Wang, L., Wang, H.H., Zhang, Q.H., Chen, S.J., Huang, Q.H., Chen, Z., 2005. Identification and characterization of a novel human histone H3 lysine 36-specific methyltransferase. J. Biol. Chem. 280, 35261–35271.

Sun, H., Zhou, X., Chen, H., Li, Q., Costa, M., 2009. Modulation of histone methylation and MLH1 gene silencing by hexavalent chromium. Toxicol. Appl. Pharmacol. 237, 258–266.

Suzuki, T., Nohara, K., 2013. Long-term arsenic exposure induces histone H3 Lys9 dimethylation without altering DNA methylation in the promoter region of p16(INK4a) and down-regulates its expression in the liver of mice. J. Appl. Toxicol. 33, 951–958.

Talasz, H., Lindner, H.H., Sarg, B., Helliger, W., 2005. Histone H4-lysine 20 monomethylation is increased in promoter and coding regions of active genes and correlates with hyperacetylation. J. Biol. Chem. 280, 38814–38822.

Tamaru, H., Selker, E.U., 2001. A histone H3 methyltransferase controls DNA methylation in Neurospora crassa. Nature 414, 277–283.

1. HISTONE MODIFICATIONS AND CHROMATIN STRUCTURE

Tan, X., Rotllant, J., Li, H., De Deyne, P., Du, S.J., 2006. SmyD1, a histone methyltransferase, is required for myofibril organization and muscle contraction in zebrafish embryos. Proc. Natl. Acad. Sci. USA 103, 2713–2718.

Tan, J., Yang, X., Zhuang, L., Jiang, X., Chen, W., Lee, P.L., Karuturi, R.K., Tan, P.B., Liu, E.T., Yu, Q., 2007. Pharmacologic disruption of Polycomb-repressive complex 2-mediated gene repression selectively induces apoptosis in cancer cells. Genes Dev. 21, 1050–1063.

Tan, M., Luo, H., Lee, S., Jin, F., Yang, J.S., Montellier, E., Buchou, T., Cheng, Z., Rousseaux, S., Rajagopal, N., Lu, Z., Ye, Z., Zhu, Q., Wysocka, J., Ye, Y., Khochbin, S., Ren, B., Zhao, Y., 2011. Identification of 67 histone marks and histone lysine crotonylation as a new type of histone modification. Cell 146, 1016–1028.

Tanaka, Y., Katagiri, Z., Kawahashi, K., Kioussis, D., Kitajima, S., 2007. Trithorax-group protein ASH1 methylates histone H3 lysine 36. Gene 397, 161–168.

Tang, X., Chen, S., 2015. Epigenetic regulation of cytochrome P450 enzymes and clinical implication. Curr. Drug Metab. 16, 86–96.

Tang, Z., Chen, W.Y., Shimada, M., Nguyen, U.T., Kim, J., Sun, X.J., Sengoku, T., McGinty, R.K., Fernandez, J.P., Muir, T.W., Roeder, R.G., 2013. SET1 and p300 act synergistically, through coupled histone modifications, in transcriptional activation by p53. Cell 154, 297–310.

Tee, W.W., Pardo, M., Theunissen, T.W., Yu, L., Choudhary, J.S., Hajkova, P., Surani, M.A., 2010. Prmt5 is essential for early mouse development and acts in the cytoplasm to maintain ES cell pluripotency. Genes Dev. 24, 2772–2777.

Tessarz, P., Santos-Rosa, H., Robson, S.C., Sylvestersen, K.B., Nelson, C.J., Nielsen, M.L., Kouzarides, T., 2014. Glutamine methylation in histone H2A is an RNA-polymerase-I-dedicated modification. Nature 505, 564–568.

Thienpont, B., Aronsen, J.M., Robinson, E.L., Okkenhaug, H., Loche, E., Ferrini, A., Brien, P., Alkass, K., Tomasso, A., Agrawal, A., Bergmann, O., Sjaastad, I., Reik, W., Roderick, H.L., 2017. The H3K9 dimethyltransferases EHMT1/2 protect against pathological cardiac hypertrophy. J. Clin. Invest. 127, 335–348.

Tian, Y., Jia, Z., Wang, J., Huang, Z., Tang, J., Zheng, Y., Tang, Y., Wang, Q., Tian, Z., Yang, D., Zhang, Y., Fu, X., Song, J., Liu, S., van Velkinburgh, J.C., Wu, Y., Ni, B., 2011. Global mapping of H3K4me1 and H3K4me3 reveals the chromatin state-based cell type-specific gene regulation in human Treg cells. PLoS One 6, e27770.

Tien, Y.C., Liu, K., Pope, C., Wang, P., Ma, X., Zhong, X.B., 2015. Dose of phenobarbital and age of treatment at early life are two key factors for the persistent induction of cytochrome P450 enzymes in adult mouse liver. Drug Metab. Dispos. 43, 1938–1945.

Trapphoff, T., Heiligentag, M., El Hajj, N., Haaf, T., Eichenlaub-Ritter, U., 2013. Chronic exposure to a low concentration of bisphenol A during follicle culture affects the epigenetic status of germinal vesicles and metaphase II oocytes. Fertil. Steril. 100, 1758–1767. e1751.

Tsai, M.C., Manor, O., Wan, Y., Mosammaparast, N., Wang, J.K., Lan, F., Shi, Y., Segal, E., Chang, H.Y., 2010. Long noncoding RNA as modular scaffold of histone modification complexes. Science 329, 689–693.

Tsukada, Y., Fang, J., Erdjument-Bromage, H., Warren, M.E., Borchers, C.H., Tempst, P., Zhang, Y., 2006. Histone demethylation by a family of JmjC domain-containing proteins. Nature 439, 811–816.

Tsukada, Y., Ishitani, T., Nakayama, K.I., 2010. KDM7 is a dual demethylase for histone H3 Lys 9 and Lys 27 and functions in brain development. Genes Dev. 24, 432–437.

Tung, E.W., Winn, L.M., 2010. Epigenetic modifications in valproic acid-induced teratogenesis. Toxicol. Appl. Pharmacol. 248, 201–209.

Tyler, C.R., Weber, J.A., Labrecque, M., Hessinger, J.M., Edwards, J.S., AM, A., 2015. ChIP-Seq analysis of the adult male mouse brain after developmental exposure to arsenic. Data Brief 5, 248–254.

Vakoc, C.R., Mandat, S.A., Olenchock, B.A., Blobel, G.A., 2005. Histone H3 lysine 9 methylation and HP1gamma are associated with transcription elongation through mammalian chromatin. Mol. Cell 19, 381–391.

Vakoc, C.R., Sachdeva, M.M., Wang, H., Blobel, G.A., 2006. Profile of histone lysine methylation across transcribed mammalian chromatin. Mol. Cell. Biol. 26, 9185–9195.

Van Aller, G.S., Reynoird, N., Barbash, O., Huddleston, M., Liu, S., Zmoos, A.F., McDevitt, P., Sinnamon, R., Le, B., Mas, G., Annan, R., Sage, J., Garcia, B.A., Tummino, P.J., Gozani, O., Kruger, R.G., 2012. Smyd3 regulates cancer cell phenotypes and catalyzes histone H4 lysine 5 methylation. Epigenetics 7, 340–343.

Van Rechem, C., Whetstine, J.R., 2014. Examining the impact of gene variants on histone lysine methylation. Biochim. Biophys. Acta 1839, 1463–1476.

Vedadi, M., Barsyte-Lovejoy, D., Liu, F., Rival-Gervier, S., Allali-Hassani, A., Labrie, V., Wigle, T.J., Dimaggio, P.A., Wasney, G.A., Siarheyeva, A., Dong, A., Tempel, W., Wang, S.C., Chen, X., Chau, I., Mangano, T.J., Huang, X.P., Simpson, C.D., Pattenden, S.G., Norris, J.L., Kireev, D.B., Tripathy, A., Edwards, A., Roth, B.L., Janzen, W.P.,

Garcia, B.A., Petronis, A., Ellis, J., Brown, P.J., Frye, S.V., Arrowsmith, C.H., Jin, J., 2011. A chemical probe selectively inhibits G9a and GLP methyltransferase activity in cells. Nat. Chem. Biol. 7, 566–574.

Veland, N., Hardikar, S., Zhong, Y., Gayatri, S., Dan, J., Strahl, B.D., Rothbart, S.B., Bedford, M.T., Chen, T., 2017. The arginine methyltransferase PRMT6 regulates DNA methylation and contributes to global DNA hypomethylation in cancer. Cell Rep. 21, 3390–3397.

Venkatesh, S., Smolle, M., Li, H., Gogol, M.M., Saint, M., Kumar, S., Natarajan, K., Workman, J.L., 2012. Set2 methylation of histone H3 lysine 36 suppresses histone exchange on transcribed genes. Nature 489, 452–455.

Wagner, E.J., Carpenter, P.B., 2012. Understanding the language of Lys36 methylation at histone H3. Nat. Rev. Mol. Cell Biol. 13, 115–126.

Walport, L.J., Hopkinson, R.J., Vollmar, M., Madden, S.K., Gileadi, C., Oppermann, U., Schofield, C.J., Johansson, C., 2014. Human UTY(KDM6C) is a male-specific N-methyl lysyl demethylase. J. Biol. Chem. 289, 18302–18313.

Walport, L.J., Hopkinson, R.J., Chowdhury, R., Schiller, R., Ge, W., Kawamura, A., Schofield, C.J., 2016. Arginine demethylation is catalysed by a subset of JmjC histone lysine demethylases. Nat. Commun. 7, 11974.

Wang, H., Cao, R., Xia, L., Erdjument-Bromage, H., Borchers, C., Tempst, P., Zhang, Y., 2001a. Purification and functional characterization of a histone H3-lysine 4-specific methyltransferase. Mol. Cell 8, 1207–1217.

Wang, H., Huang, Z.Q., Xia, L., Feng, Q., Erdjument-Bromage, H., Strahl, B.D., Briggs, S.D., Allis, C.D., Wong, J., Tempst, P., Zhang, Y., 2001b. Methylation of histone H4 at arginine 3 facilitating transcriptional activation by nuclear hormone receptor. Science 293, 853–857.

Wang, Y., Wysocka, J., Sayegh, J., Lee, Y.H., Perlin, J.R., Leonelli, L., Sonbuchner, L.S., McDonald, C.H., Cook, R.G., Dou, Y., Roeder, R.G., Clarke, S., Stallcup, M.R., Allis, C.D., Coonrod, S.A., 2004. Human PAD4 regulates histone arginine methylation levels via demethylimination. Science 306, 279–283.

Wang, Z., Zang, C., Rosenfeld, J.A., Schones, D.E., Barski, A., Cuddapah, S., Cui, K., Roh, T.Y., Peng, W., Zhang, M.Q., Zhao, K., 2008. Combinatorial patterns of histone acetylations and methylations in the human genome. Nat. Genet. 40, 897–903.

Wang, K.C., Yang, Y.W., Liu, B., Sanyal, A., Corces-Zimmerman, R., Chen, Y., Lajoie, B.R., Protacio, A., Flynn, R.A., Gupta, R.A., Wysocka, J., Lei, M., Dekker, J., Helms, J.A., Chang, H.Y., 2011. A long noncoding RNA maintains active chromatin to coordinate homeotic gene expression. Nature 472, 120–124.

Wang, T., Xu, C., Liu, Y., Fan, K., Li, Z., Sun, X., Ouyang, H., Zhang, X., Zhang, J., Li, Y., Mackenzie, F., Min, J., Tu, X., 2012. Crystal structure of the human SUV39H1 chromodomain and its recognition of histone H3K9me2/3. PLoS One 7, e52977.

Wang, W., Sidoli, S., Zhang, W., Wang, Q., Wang, L., Jensen, O.N., Guo, L., Zhao, X., Zheng, L., 2017. Abnormal levels of histone methylation in the retinas of diabetic rats are reversed by minocycline treatment. Sci. Rep. 7, 45103.

Webby, C.J., Wolf, A., Gromak, N., Dreger, M., Kramer, H., Kessler, B., Nielsen, M.L., Schmitz, C., Butler, D.S., Yates 3rd, J.R., Delahunty, C.M., Hahn, P., Lengeling, A., Mann, M., Proudfoot, N.J., Schofield, C.J., Bottger, A., 2009. Jmjd6 catalyses lysyl-hydroxylation of U2AF65, a protein associated with RNA splicing. Science 325, 90–93.

Whetstine, J.R., Nottke, A., Lan, F., Huarte, M., Smolikov, S., Chen, Z., Spooner, E., Li, E., Zhang, G., Colaiacovo, M., Shi, Y., 2006. Reversal of histone lysine trimethylation by the JMJD2 family of histone demethylases. Cell 125, 467–481.

Wiencke, J.K., Zheng, S., Morrison, Z., Yeh, R.F., 2008. Differentially expressed genes are marked by histone 3 lysine 9 trimethylation in human cancer cells. Oncogene 27, 2412–2421.

Wiles, E.T., Selker, E.U., 2017. H3K27 methylation: a promiscuous repressive chromatin mark. Curr. Opin. Genet. Dev. 43, 31–37.

Wilson, J.R., Jing, C., Walker, P.A., Martin, S.R., Howell, S.A., Blackburn, G.M., Gamblin, S.J., Xiao, B., 2002. Crystal structure and functional analysis of the histone methyltransferase SET7/9. Cell 111, 105–115.

Wu, H., Min, J., Lunin, V.V., Antoshenko, T., Dombrovski, L., Zeng, H., Allali-Hassani, A., Campagna-Slater, V., Vedadi, M., Arrowsmith, C.H., Plotnikov, A.N., Schapira, M., 2010a. Structural biology of human H3K9 methyltransferases. PLoS One 5, e8570.

Wu, S., Wang, W., Kong, X., Congdon, L.M., Yokomori, K., Kirschner, M.W., Rice, J.C., 2010b. Dynamic regulation of the PR-Set7 histone methyltransferase is required for normal cell cycle progression. Genes Dev. 24, 2531–2542.

Wu, H., Chen, X., Xiong, J., Li, Y., Li, H., Ding, X., Liu, S., Chen, S., Gao, S., Zhu, B., 2011. Histone methyltransferase G9a contributes to H3K27 methylation in vivo. Cell Res. 21, 365–367.

Xia, R., Zhou, R., Tian, Z., Zhang, C., Wang, L., Hu, Y., Han, J., Li, J., 2013. High expression of H3K9me3 is a strong predictor of poor survival in patients with salivary adenoid cystic carcinoma. Arch. Pathol. Lab. Med. 137, 1761–1769.

Xiao, B., Jing, C., Wilson, J.R., Walker, P.A., Vasisht, N., Kelly, G., Howell, S., Taylor, I.A., Blackburn, G.M., Gamblin, S.J., 2003. Structure and catalytic mechanism of the human histone methyltransferase SET7/9. Nature 421, 652–656.

Xiao, M., Yang, H., Xu, W., Ma, S., Lin, H., Zhu, H., Liu, L., Liu, Y., Yang, C., Xu, Y., Zhao, S., Ye, D., Xiong, Y., Guan, K.L., 2012. Inhibition of alpha-KG-dependent histone and DNA demethylases by fumarate and succinate that are accumulated in mutations of FH and SDH tumor suppressors. Genes Dev. 26, 1326–1338.

Xiao, C., Liu, Y., Xie, C., Tu, W., Xia, Y., Costa, M., Zhou, X., 2015. Cadmium induces histone H3 lysine methylation by inhibiting histone demethylase activity. Toxicol. Sci. 145, 80–89.

Xie, Y., Ke, S., Ouyang, N., He, J., Xie, W., Bedford, M.T., Tian, Y., 2009. Epigenetic regulation of transcriptional activity of pregnane X receptor by protein arginine methyltransferase 1. J. Biol. Chem. 284, 9199–9205.

Xin, F., Susiarjo, M., Bartolomei, M.S., 2015. Multigenerational and transgenerational effects of endocrine disrupting chemicals: a role for altered epigenetic regulation? Semin. Cell Dev. Biol. 43, 66–75.

Xu, L., Zhao, Z., Dong, A., Soubigou-Taconnat, L., Renou, J.P., Steinmetz, A., Shen, W.H., 2008. Di- and tri- but not monomethylation on histone H3 lysine 36 marks active transcription of genes involved in flowering time regulation and other processes in Arabidopsis thaliana. Mol. Cell. Biol. 28, 1348–1360.

Yamane, K., Toumazou, C., Tsukada, Y., Erdjument-Bromage, H., Tempst, P., Wong, J., Zhang, Y., 2006. JHDM2A, a JmjC-containing H3K9 demethylase, facilitates transcription activation by androgen receptor. Cell 125, 483–495.

Yamane, K., Tateishi, K., Klose, R.J., Fang, J., Fabrizio, L.A., Erdjument-Bromage, H., Taylor-Papadimitriou, J., Tempst, P., Zhang, Y., 2007. PLU-1 is an H3K4 demethylase involved in transcriptional repression and breast cancer cell proliferation. Mol. Cell 25, 801–812.

Yang, C.M., Shinkai, Y., 2013. Prdm12 is induced by retinoic acid and exhibits anti-proliferative properties through the cell cycle modulation of P19 embryonic carcinoma cells. Cell Struct. Funct. 38, 197–206.

Yin, Y., Yu, V.C., Zhu, G., Chang, D.C., 2008. SET8 plays a role in controlling G1/S transition by blocking lysine acetylation in histone through binding to H4 N-terminal tail. Cell Cycle 7, 1423–1432.

Yu, K., Shi, Y.F., Yang, K.Y., Zhuang, Y., Zhu, R.H., Xu, X., Cai, G., 2011. Decreased topoisomerase IIalpha expression and altered histone and regulatory factors of topoisomerase IIalpha promoter in patients with chronic benzene poisoning. Toxicol. Lett. 203, 111–117.

Yuan, W., Xie, J., Long, C., Erdjument-Bromage, H., Ding, X., Zheng, Y., Tempst, P., Chen, S., Zhu, B., Reinberg, D., 2009. Heterogeneous nuclear ribonucleoprotein L Is a subunit of human KMT3a/Set2 complex required for H3 Lys-36 trimethylation activity in vivo. J. Biol. Chem. 284, 15701–15707.

Yuan, Y., Wang, Q., Paulk, J., Kubicek, S., Kemp, M.M., Adams, D.J., Shamji, A.F., Wagner, B.K., Schreiber, S.L., 2012. A small-molecule probe of the histone methyltransferase G9a induces cellular senescence in pancreatic adenocarcinoma. ACS Chem. Biol. 7, 1152–1157.

Zha, L., Li, F., Wu, R., Artinian, L., Rehder, V., Yu, L., Liang, H., Xue, B., Shi, H., 2015. The histone demethylase UTX promotes brown adipocyte thermogenic program via coordinated regulation of H3K27 demethylation and acetylation. J. Biol. Chem. 290, 25151–25163.

Zhang, J., Parvin, J., Huang, K., 2012a. Redistribution of H3K4me2 on neural tissue specific genes during mouse brain development. BMC Genomics 13 (Suppl. 8), S5.

Zhang, X., Wen, H., Shi, X., 2012b. Lysine methylation: beyond histones. Acta Biochim. Biophys. Sin. Shanghai 44, 14–27.

Zhang, X., Wu, J., Choiniere, J., Yang, Z., Huang, Y., Bennett, J., Wang, L., 2016. Arsenic silences hepatic PDK4 expression through activation of histone H3K9 methylatransferase G9a. Toxicol. Appl. Pharmacol. 304, 42–47.

Zhou, X., Li, Q., Arita, A., Sun, H., Costa, M., 2009. Effects of nickel, chromate, and arsenite on histone 3 lysine methylation. Toxicol. Appl. Pharmacol. 236, 78–84.

Zhou, X., Sun, H., Chen, H., Zavadil, J., Kluz, T., Arita, A., Costa, M., 2010. Hypoxia induces trimethylated H3 lysine 4 by inhibition of JARID1A demethylase. Cancer Res. 70, 4214–4221.

Zhu, C.C., Hou, Y.J., Han, J., Liu, H.L., Cui, X.S., Kim, N.H., Sun, S.C., 2014. Effect of mycotoxin-containing diets on epigenetic modifications of mouse oocytes by fluorescence microscopy analysis. Microsc. Microanal. 20, 1158–1166.

1. HISTONE MODIFICATIONS AND CHROMATIN STRUCTURE

Chromatin Accessibility as a Strategy to Detect Changes Associated With Development, Disease, and Exposure and Susceptibility to Chemical Toxins

Ian J. Davis[*,†], *Samantha G. Pattenden*[*,‡]

[*]Lineberger Comprehensive Cancer Center, The University of North Carolina, Chapel Hill, NC, United States [†]Departments of Genetics and Pediatrics, The University of North Carolina, Chapel Hill, NC, United States [‡]Division of Chemical Biology and Medicinal Chemistry, Center for Integrative Chemical Biology and Drug Discovery, Eshelman School of Pharmacy, The University of North Carolina, Chapel Hill, NC, United States

OUTLINE

Packaging of DNA into a cell nucleus requires multiple levels of organization. DNA is wrapped around histone octamers to form nucleosomes, the repeating unit of chromatin. By restricting access to DNA, nucleosomes present a barrier to DNA-templated processes such as transcription and replication. However, nucleosome positioning is dynamic. Multiple regulatory processes influence the precise placement by nucleosome sliding, eviction, or histone posttranslational modification. Several informative and widely used assays have been developed to identify changes to the epigenome including DNA methylation (e.g., bisulfite sequencing) or histone posttranslational modifications and the occupancy of chromatin-associated complexes [e.g., chromatin immunoprecipitation (ChIP)]. Because they require a priori knowledge of potential biological mechanisms, these assays reveal specific epigenetic information. In contrast, assays that interrogate patterns of chromatin accessibility are mechanism agnostic. In this chapter, we will discuss the role of chromatin accessibility in normal development and diseases, specific assays for chromatin architecture, and the potential use of disease-specific chromatin variation as a diagnostic or therapeutic discovery target. We will refer to cell-specific chromatin accessibility as a "chromatin signature," which is defined as the unique pattern of accessible and inaccessible regions that represents the convergence of multiple cellular processes such as transcriptional activation, DNA damage repair, replication, RNA processing, and nuclear organization (Fig. 1).

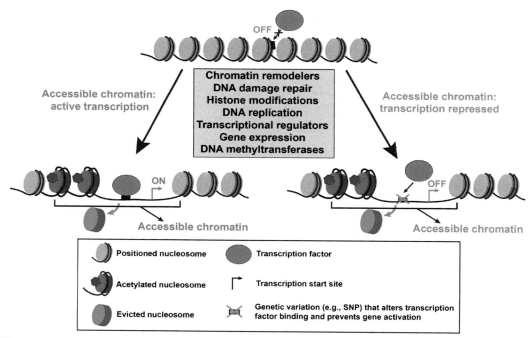

FIG. 1 Chromatin accessibility is influenced by both genetic variation and a variety of cellular processes and is not necessarily correlated with gene transcription. Positioned nucleosomes prevent a sequence-specific DNA-binding protein from accessing DNA. Many cellular processes independently or in combination influence chromatin accessibility. Once accessible, a sequence-specific DNA-binding protein such as a transcription factor can access DNA and activate gene transcription (left outcome). Genetic variation (e.g., SNP or disease-specific mutation) can result in a change in the DNA sequence that prevents the binding of the DNA-binding protein (right outcome). In this case, chromatin accessibility is present at a transcriptionally repressed gene.

1. HISTONE MODIFICATIONS AND CHROMATIN STRUCTURE

CHROMATIN ACCESSIBILITY AS A MARKER OF CELL LINEAGE

Early studies showed that alterations in chromatin accessibility could be either inducible (Wu et al., 1979b; Weintraub and Groudine, 1976) or stable and heritable (Groudine and Weintraub, 1982). DNase hypersensitivity assays, which identify sites of accessible chromatin through limited digestion with the endonuclease DNase I, were among the first enzymatic approaches to interrogate chromatin accessibility (Wu et al., 1979a,b; Groudine and Weintraub, 1982; Weintraub and Groudine, 1976). Weintraub and Groudine used DNase I hypersensitivity to demonstrate a correlation between the transcriptional activation and the formation of distinct sites of accessible chromatin at the beta-globin gene locus in chicken erythrocyte nuclei (Weintraub and Groudine, 1976). In *Drosophila* Schneider cell nuclei, sites of DNase I hypersensitivity at heat-shock genes were gained and lost following temperature shifts (Wu et al., 1979b). These studies linked changes in chromatin accessibility to the activation of gene transcription.

Groudine and Weintraub also demonstrated that chromatin accessibility could be heritable. Infection of chicken embryo fibroblast (CEF) cells with the Rous sarcoma virus (RSV) resulted in the activation of beta-globin genes and the formation of distinct sites of accessible chromatin (Groudine and Weintraub, 1975, 1980; Weintraub et al., 1981). The use of DNase I hypersensitivity assays demonstrated that virus-induced alterations in chromatin accessibility persisted in RSV-infected daughter cells after 20 cell doublings even though the beta-globin genes were no longer actively transcribed (Groudine and Weintraub, 1982). Thus, stable changes in chromatin architecture did not appear to be restricted only to active gene promoters (Stalder et al., 1980; Weintraub et al., 1981; Wu and Gilbert, 1981), suggesting that chromatin accessibility is not simply a function of gene expression.

Contemporary studies further support the heritability of chromatin architecture. In lymphoblastoid cells from two independent families, both individual-specific and allele-specific DHS were transmitted from parent to child (McDaniell et al., 2010). Comparisons of DHS patterns across different cell types have shown that DHS located away from transcription start sites (TSS) contain DNA-binding motifs associated with lineage-specific cell identity and can predict cell-type-specific functional behaviors (Boyle et al., 2011; Song et al., 2011; Thurman et al., 2012). This feature has been commonly exploited through the study of regions of chromatin accessibility to identify cis-regulatory elements without a priori knowledge of the identity of factors interacting with those elements (Rivera and Ren, 2013).

The heritability of chromatin variation raised the possibility that patterns of chromatin accessibility were cell- or tissue-specific. Certainly, cell fate decisions during normal development are significantly influenced by transcription-factor-dependent gene regulation. There is, however, growing evidence that throughout development, lineage-specific chromatin variation reflects regions that are distal to gene promoters. Sites of chromatin accessibility do not universally reflect gene expression. Indeed, recent studies suggest that chromatin signature enforces lineage-specific transcriptional constraints during development and is therefore a better early predictor of stem cell fate than gene expression patterns alone (Stergachis et al., 2013; Yu et al., 2016). The DHS landscape has now been examined in multiple cells at different stages of normal development (Stergachis et al., 2013; Thurman et al., 2012; ENCODE Project Consortium, 2004). A comparison of DHS in human embryonic stem cells with that of lineage-defined cells indicates that chromatin accessibility decreases as cells differentiate. This shift in chromatin signature is coordinated in part by the expression of lineage-specific transcription factors and is cell-type-specific (Stergachis et al., 2013). These data suggest that development is accompanied by restriction in epigenetic plasticity, a concept that is further supported by a study examining the cellular niche in hematopoietic cell development (Yu et al., 2016). Following transplant, single hematopoietic stem cell (HSC) clones exhibit myeloid, lymphoid, or megakaryocytic lineage bias (reviewed in Obier and Bonifer, 2016). To interrogate the mechanisms underlying this bias, HSCs were clonally tracked in mice after transplant or tissue injury. Incredibly, the chromatin signature of each clone was preserved regardless of niche environment or stress, suggesting that these cells have both a constrained epigenetic memory and limited epigenetic plasticity, thereby restricting response to external conditions (Yu et al., 2016). Therefore, chromatin accessibility is both heritable and specific and can serve as an important marker of cell identity during normal development.

METHODS FOR DETERMINING CHROMATIN ARCHITECTURE AND ACCESSIBILITY

Chromatin architecture was first explored using nucleases that preferentially cleave DNA in the linker region between nucleosomes (reviewed in Felsenfeld, 1978). These assays revealed that nucleosome positioning was critical for DNA-templated processes such as replication and transcription. In recent years, the repertoire of assays to map regions of accessible chromatin has significantly expanded (Tsompana and Buck, 2014; Rivera and Ren, 2013). Here, we will outline the current assays used to study chromatin accessibility (Fig. 2).

DNase I Hypersensitivity Assays

DNase I hypersensitivity assays map the location of accessible chromatin through the digestion of chromatin in intact nuclei with limiting concentrations of DNase I. DNase I generally nicks accessible DNA in the minor groove at 10 base-pair intervals. Nucleosome-free regions demonstrate preferential cleavage. Digested DNA appears as a smear when separated on a nondenaturing agarose gel (Felsenfeld, 1978; Noll, 1974). Prior to the advent of sequencing-based approaches, the chromatin status of specific regions was established by following DNase I treatment with restriction enzyme digestion and hybridization to a radioactively labeled locus-specific DNA probe (Southern, 1975; Wu et al., 1979a). With the application of high-throughput sequencing, DNase-digested DNA is now assayed directly (DNase-seq) (John et al., 2013; Song and Crawford, 2010). Prior to performing the assay, DNase I digestion must be carefully titrated for each cell type to ensure that accessible chromatin is selectively cleaved (John et al., 2013). Successful DNase-seq assays are capable of identifying regulatory elements across the genome and, to some extent, identifying sites of nucleosome occupancy and transcription factor binding (DNase I footprinting) (Boyle et al., 2011; Song and Crawford, 2010; Hesselberth et al., 2009). DNase I has been shown to have a sequence bias for DNA cleavage, which can result in false-positive identification of transcription factor footprinting (He et al., 2014; Tsompana and Buck, 2014). As a result, DNase-seq data from intact nuclei should be compared with data from purified DNase

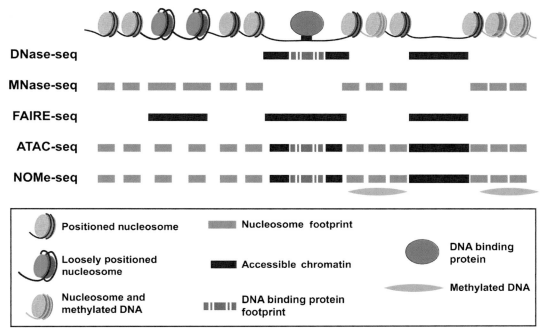

FIG. 2 Methods for determining chromatin architecture and accessibility. *Bars* represent the predicted regions identified by high-throughput sequencing. Both assays that identify chromatin architecture (MNase-seq and NOMe-seq) or accessibility (DNase-seq, FAIRE-seq, and ATAC-seq) are represented. A legend is located at the bottom of the figure.

I-digested DNA to eliminate cleavage bias during data analysis (Song and Crawford, 2010), and putative transcription factor binding sites should be confirmed with an assay such as ChIP-seq (Boyle et al., 2011; He et al., 2014).

Micrococcal Nuclease Assays

Micrococcal nuclease (MNase) assays are useful for defining nucleosome position and chromatin architecture (Rivera and Ren, 2013; Tsompana and Buck, 2014). This enzyme preferentially cleaves the linker region between nucleosomes and then digests the free DNA ends toward the core nucleosome. Since the nucleosome limits further digestion, the resulting DNA fragments reflect nucleosome placement (Felsenfeld, 1978; Keene and Elgin, 1981; Levy and Noll, 1981; Wu et al., 1979a; Dingwall et al., 1981; Fittler and Zachau, 1979; Horz and Altenburger, 1981). As with DNase I, MNase assays were originally performed by enzymatic digestion followed by Southern blot with a radioactive DNA probe to identify regions of interest. Since nucleosome positioning can change during chromatin preparation, some assays also include a formaldehyde cross-linking step prior to nuclear isolation and MNase digestion to arrest nucleosomes in an in vivo conformation (Byrum et al., 2011; Zhang and Pugh, 2011). MNase digestion followed by high-throughput sequencing (MNase-seq) is now commonly used to map nucleosome positioning in a variety of organisms (Cui and Zhao, 2012; Gaffney et al., 2012; Kaplan et al., 2009; Zhang and Pugh, 2011; Mueller et al., 2017). One such study revealed that the majority (~80%) of nucleosomes across the genome of human lymphoblastoid cells are consistently positioned (Gaffney et al., 2012). The degree of digestion necessary to isolate mononucleosomes is influenced by the underlying chromatin state. Using a range of MNase concentrations, in combination with a chromatin accessibility assay, demonstrated that the induction of rapid transcription in *Drosophila* resulted in changes to nucleosome accessibility that were not accompanied by changes in nucleosome position (Mueller et al., 2017). Indeed, MNase-seq reveals the position of nucleosomes. Chromatin accessibility, which commonly reflects the loss of nucleosomes, is only indirectly reflected by MNase-seq. Thus, caution should be used when interpreting MNase-seq as an indicator of chromatin accessibility.

Although MNase-seq is potentially a powerful technique to explore nucleosome position, sequencing data interpretation can be problematic. MNase has a bias for A-T-rich regions, with enzymatic cleavage estimated to be 30 times greater at the 5' side of an A or T compared with a G or C (Dingwall et al., 1981; Horz and Altenburger, 1981). In addition, MNase digestion between nucleosomes is not consistent, resulting in the need to estimate nucleosome position based on the center point of multiple DNA fragments from a single region (Rivera and Ren, 2013; Tsompana and Buck, 2014). Taken together, these factors can complicate the interpretation of MNase-seq data from different experiments. Thus, it is important to standardize MNase-seq protocols if comparing data from multiple conditions (Rizzo et al., 2012). Depending on the goal of the experiment, data should be collected at several MNase concentrations to provide information for nucleosomes in regions of the genome that are sensitive to cleavage at varying MNase concentrations (Chereji et al., 2017; Iwafuchi-Doi et al., 2016; Mieczkowski et al., 2016; Rizzo et al., 2012; Mueller et al., 2017).

Formaldehyde Assisted Isolation of Regulatory Elements (FAIRE)

FAIRE is a biochemical method that isolates accessible chromatin from formaldehyde cross-linked cells. Accessible chromatin, which is relatively protein-free, maintains the net negative charge of the DNA phosphate backbone. As a result, it can be biochemically separated from nucleosomal chromatin using phenol-chloroform to extract the hydrophilic, accessible chromatin in the aqueous phase from the protein-bound chromatin that partitions with the organic phase (Giresi et al., 2007; Giresi and Lieb, 2009; Simon et al., 2013). FAIRE-enriched DNA can be analyzed by quantitative polymerase chain reaction (qPCR) for specific regions or by high-throughput sequencing (FAIRE-seq). Recently, the FAIRE-qPCR assay was adapted for high-throughput screening of small-molecule compounds. To increase throughput, organic extraction was replaced with silica matrix columns in a 96-well format to isolate accessible chromatin (Pattenden et al., 2016). The FAIRE assay is unique from other enzyme-based assays such as DNase I or ATAC (see below) in that it also identifies regions containing destabilized nucleosomes, in addition to nucleosome-depleted regions of chromatin (Gomez et al., 2016; Fig. 2). Comparison of combined FAIRE-seq and DNase-seq analyses from multiple cell types showed a discrepancy in accessible chromatin signal, especially at distal regulatory elements where FAIRE isolated accessible chromatin that was not present in the

DNase assays (Song et al., 2011). When FAIRE-seq was performed on human embryonic stem cells, specific classes of repetitive elements demonstrated an increase in chromatin accessibility that was not evident in DNase-seq data. Further comparisons to MNase-seq data from these cells demonstrated the presence of nucleosomes across the repetitive elements (Gomez et al., 2016). This result is consistent with another study indicating that changes in nucleosome accessibility are not necessarily accompanied by changes in nucleosome occupancy at rapidly induced genes in *Drosophila* (Mueller et al., 2017). The discrepancies in chromatin accessibility detected by FAIRE-seq compared with DNase-seq may be explained by the possibility that nucleosomes that are loosely associated with chromatin (e.g., hyperacetylated histones) are lost during biochemical purification, whereas limited enzymatic digestion by DNase I depends on the presence of exposed DNA, thereby eliminating regions with well-positioned nucleosomes (Boulay et al., 2017; Gomez et al., 2016; Mueller et al., 2017).

Although a simple technique overcomes issues with bias and concentration in enzyme-based accessibility assays, FAIRE requires optimization at the fixation and fragmentation steps. The optimum fixation time for mammalian cells is 5 min (Simon et al., 2013); however, organisms with a cell wall such as fungi or plants or dissected tissue samples may require longer fixation times (Hogan et al., 2006; Omidbakhshfard et al., 2014; Davie et al., 2015). Prior to biochemical separation of accessible regions, the FAIRE assay requires that nuclei be subjected to acoustic sonication to fragment chromatin. Since accessible chromatin is more sensitive to acoustic sonication compared with inaccessible chromatin, sonication times must be carefully titrated to avoid the loss of the accessible chromatin signal (Teytelman et al., 2009). In addition, high-throughput sequencing-based analysis of FAIRE can be challenging because of the limited signal relative to background.

Assay for Transposase-Accessible Chromatin (ATAC)

ATAC uses a hyperactive mutant of the bacterial Tn5 transposase (Goryshin et al., 1998; Weinreich et al., 1994; Zhou et al., 1998; Zhou and Reznikoff, 1997; Adey et al., 2010) to simultaneously cleave regions of accessible chromatin and ligate high-throughput sequencing adaptors. ATAC-seq has rapidly become a widely used method for chromatin accessibility (Buenrostro et al., 2013, 2015a). Factors that have likely contributed to the popularity of ATAC-seq include a relatively short protocol that requires few cells (between 500 and 50,000 cells per assay, with single-cell assays available (see below)) and a robust signal over background that is comparable with DNase-seq data. Similar to DNase I, Tn5 transposase cleaves accessible regions of chromatin that are free of both nucleosomes and other DNA-bound proteins (Buenrostro et al., 2013). ATAC-seq has generated high-resolution chromatin accessibility profiles for a range of cells, which has permitted mechanistic insights into both normal and disease cell regulation (Kundaje et al., 2015; ENCODE Project Consortium et al., 2012). For instance, ATAC-seq profiling was used to identify a key transcriptional regulator that promotes metastasis of small-cell lung cancer through the alteration of chromatin accessibility at distal regulatory elements (Denny et al., 2016).

Initially, the Tn5 transposase was only available for purchase from a single commercial source, which increased costs associated with this assay (Adey et al., 2010). Recently, however, two versions of DNA encoding hyperactive Tn5 enzyme mutants have become available for in-house enzyme purification, which significantly decreases the cost for multisample assays (Hennig et al., 2018; Picelli et al., 2014). Besides costs associated with Tn5 enzyme, a major consideration in performing ATAC-seq was nonspecific detection of mitochondrial DNA, which could account for 50%–75% of all sequencing reads, thereby requiring significant and costly sequencing read depth to obtain accessible chromatin data (Buenrostro et al., 2013). To address this issue, methods to remove mitochondrial DNA were recently developed that use either CRISPR technology (Fast-ATAC) or a series of detergent washes (Omni-ATAC) (Corces et al., 2016, 2017; Montefiori et al., 2017). When compared with previous ATAC-seq protocols, the Omni-ATAC method had up to 13-fold fewer sequencing reads that mapped to mitochondrial DNA across multiple cell types (Corces et al., 2017).

Nucleosome Occupancy and Methylome Sequencing (NOMe-seq)

NOMe-seq assesses both nucleosome positioning and DNA methylation to generate a snapshot of chromatin architecture (Kelly et al., 2012; Lay et al., 2018; Rhie et al., 2018a). The assay was developed based on the work that showed a DNA methyltransferase enzyme could be utilized to determine nucleosome positioning in cells (Jessen et al., 2004, 2006; Kilgore et al., 2007; Pardo et al., 2011; Xu et al., 1998). NOMe-seq uses the GpC methyltransferase isolated from the *Chlorella* virus, M.CviPI (Xu et al., 1998). The presence of nucleosomes prevents GpC methylation. After M.CviPI treatment, DNA is treated with bisulfite to convert unmethylated cytosines to thymidine. High-throughput sequencing is then used to identify cytosines in the CpG dinucleotide context that indicate cellular DNA methylation and those in the GpC dinucleotide context, which indicate accessible chromatin (Kelly et al., 2012). NOMe-seq has been used to map DNA methylation and nucleosome positioning in a number of cell lines (Kelly et al., 2012; Rhie et al., 2018b; Taberlay et al., 2014; Statham et al., 2015; Lay et al., 2015). This technique is especially useful for separating inactive gene promoters that have densely packed nucleosomes without DNA methylation (poised or repressed promoters) and those with DNA methylation (silenced promoters). For example, NOMe-seq was used to simultaneously demonstrate that the loss of methylation at promoters with CpG islands led to the acquisition of positioned nucleosomes that promoted either a poised (positioned nucleosomes across the promoter) or an active (positioned nucleosomes that flanked the promoter) chromatin state (Lay et al., 2015).

NOMe-seq was recently adapted to include a formaldehyde cross-linking step. It was noted in earlier studies that dynamic nucleosome movement during isolation of cell nuclei could result in methylation of GpC sites that were otherwise inaccessible (Lay et al., 2018). This cross-linking step, however, makes the chromatin resistant to M.CviPI activity, such that titrating enzyme concentration becomes critical to avoid enzyme bias (Lay et al., 2018). Another important consideration for the NOMe-seq protocol is the size of DNA fragments selected for analysis. Selection of small sequencing library fragments will lead to enrichment of CpG islands but likely lose most information for other regulatory elements such as distal enhancers. If the fragments are large, coverage of CpG islands will be low and will impair the ability to identify nucleosome-depleted regions (Rhie et al., 2018a).

Single Cell Assays for Determining Chromatin Accessibility in Heterogenous Cell Populations

Recently, several single-cell assays have been developed to interrogate the role of chromatin accessibility in normal cell development or disease. Single-cell assays are particularly useful for samples with significant cellular heterogeneity such as developing embryos, immune cells, or blood tumors (Corces and Corces, 2016; Cusanovich et al., 2018; Satpathy et al., 2018). DNase-seq, NOMe-seq, and ATAC-seq have all been adapted for single-cell analyses (Buenrostro et al., 2015b; Cusanovich et al., 2015; Jin et al., 2015; Pott, 2017). In each case, the protocols require specialized cell processing to minimize perturbations to the epigenome, with individual cells being sorted either by flow cytometry or on microfluidic platforms. The yield of recovered accessible chromatin regions for single-cell ATAC-seq methods ranges from 500 to 70,000 tags per cell (Buenrostro et al., 2015b; Cusanovich et al., 2015; Schwartzman and Tanay, 2015). Caution must be taken in analyzing chromatin accessibility data from individual cells. Sufficient sequencing read depth of typically 500,000 to 1.5 million reads per cell must be achieved to accurately assess chromatin accessibility patterns. Without statistical pooling of individual cell data, it is challenging to distinguish true-positive from false-positive signals at individual genetic loci (Schwartzman and Tanay, 2015).

CHROMATIN ACCESSIBILITY AND DISEASE

Through normal development, cells maintain a strict epigenetic program to ensure correct cell lineage decisions. For this reason, these cells have limited epigenetic plasticity. There is growing evidence that dysregulation of epigenetic programs plays a central role in the transition of cells from normal to diseased state. Loss of epigenetic constraints alters cellular signaling and transcription pathways, potentially resulting in large-scale changes in chromatin organization, which can occur during development, somatic mutation, or as a result of external influences such as lifestyle or exposure to environmental toxicants (Flavahan et al., 2017; Lewis et al., 2017; Bowers and McCullough, 2017). Since chromatin accessibility offers an unbiased overview of changes to cell regulatory pathways, it is possible that chromatin signature may inform biomarkers of both extrinsic and intrinsic changes to normal cell function and predict disease susceptibility or resistance to treatment.

Understanding How Environmental Factors Influence Chromatin Accessibility

Established connections between chromatin accessibility patterns and susceptibility to environmental influences remain limited, compared with studies of normal human development and diseases, in particular cancer. Chromatin accessibility assays can be used to identify regional chromatin changes associated with sensitivity to exposure only if a baseline chromatin signature (accessibility prior to exposure) can be established (Lewis et al., 2017). Environmental influences during development and after birth have been shown to shape an individual's genome and epigenome. For instance, altered developmental outcomes persisted in the grandchildren of women who experienced severe maternal starvation during the Dutch famine of 1944–45, including persistent shifts in DNA methylation patterns

(Tobi et al., 2014, 2015). Another study of elderly twins over 10 years could attribute ~90% of the changes in CpG DNA methylation over the course of the study to environmental influences and only 10% to familial factors (Tan et al., 2016). Therefore, both epigenomic and genomic interindividual variabilities are significant factors to consider when identifying a baseline chromatin signature for subsequent comparison (Fig. 3). As a result, epigenome-wide association studies in humans are complicated by the need to follow a large populations of individuals over time to achieve sufficient statistical power to provide a meaningful link between a cellular chromatin signature and specific environmental exposures (Mill and Heijmans, 2013). Collection of specific tissue samples from healthy or exposed individuals is limited mainly to noninvasive methods such as blood collection, which significantly decreases the range of tissues available for analysis (Mill and Heijmans, 2013; Tobi et al., 2014). Sex differences must also be considered. For example, several studies have demonstrated sex-specific changes to DNase I accessibility in the rodent livers (Sugathan and Waxman, 2013; Zhang et al., 2012; Ling et al., 2010; Chia and Rotwein, 2010).

It was recently proposed that chromatin accessibility studies in combination with gene expression data would provide a cost-effective, powerful, unbiased overview of the complex genetic and epigenetic changes associated with interindividual variability in response to environmental toxicant exposures (Lewis et al., 2017). DNase-seq has been used to demonstrate that the genetic variants that influence chromatin accessibility are significant determinants of gene expression variation in humans (Degner et al., 2012). These findings are supported by studies in genetically divergent mice, which show that strain-specific variation in chromatin accessibility profiles is a consequence of interindividual differences in DNA sequence (Leung

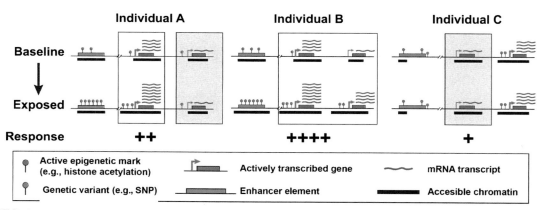

FIG. 3   Interindividual variability and toxicant-induced response. A combination of genetic and epigenetic analysis provides a snapshot of individual response to toxicant exposure and can possibly predict susceptibility to the negative effects of the toxicant. In this simple example, each individual has a different genetic and epigenetic profile that results in varying degrees of gene transcription. Following toxicant exposure, the ability to upregulate genes required to respond (e.g., upregulation of gene involved in efflux of toxicant from cells) is dependent on increasing both chromatin accessibility and gene expression. Individual B has the highest response due to the upregulation of gene expression and enhancer activation as indicated by increased chromatin accessibility, gene transcription, and activating histone posttranslational modifications (PTMs) at the enhancer. Individual C has the lowest response since enhancer activity is decreased by a genetic variation and a lack of activating histone PTMs, which blocks the upregulation of transcription. Individual A has a moderate response since transcription is increased at one of the response genes.

et al., 2014; Chappell et al., 2017; Israel et al., 2018) and that this variation significantly influences transcriptional changes in response to environmental stimuli. Animal models offer a controlled system that compares a baseline with postexposure chromatin signature. Unlike humans, genetic and epigenetic variation is limited in model organisms permitting robust detection of responses to environmental stimuli within a single strain or between strains. FAIRE-seq performed in the livers of genetically divergent mouse strains fed a high-fat diet indicated that up to 50% of chromatin accessibility changes between the strains could be attributed to local sequence variability (Leung et al., 2014). In contrast, ATAC-seq profiles of lung cells from genetically diverse mouse strains exposed to the inhaled genotoxic chemical, 1,3-butadiene, were correlated to severity of response to exposure (Israel et al., 2018). Subsequent work discovered that the mouse strain with the least DNA damage from 1,3-butadiene exposure also showed the greatest increase in chromatin accessibility around genes involved in detoxification pathways (Chappell et al., 2017). These studies demonstrate the feasibility of assaying chromatin accessibility to determine individual and population responses to toxicant exposures. Therefore, translation of these types of exposure studies to human populations may be possible by combining chromatin accessibility and gene expression data to identify the appropriate baselines prior to exposure. Also, the role of interindividual genetic and epigenetic variability in modulating exposure response must be assessed by using enough individuals to obtain the appropriate statistical power to achieve meaningful conclusions (Bowers and McCullough, 2017; McCullough et al., 2016; Lewis et al., 2017; Mill and Heijmans, 2013).

Chromatin Accessibility and Cancer

Changes in chromatin accessibility have been linked to a wide range of diseases, with cancer being one of the most widely studied. For this reason, cancer biology constitutes a prime example of how information obtained from chromatin accessibility studies can be utilized in both therapeutic discovery and diagnostic applications.

Aberrant changes in the epigenome can lead to alterations in gene expression, which result in regulatory disruption of key cancer-associated genes. Approximately 50% of human cancers have mutations in chromatin-associated proteins, with a growing number of these genetic changes being recognized as driver mutations across a diverse range of tumor types (Flavahan et al., 2017; Shen and Laird, 2013; Timp and Feinberg, 2013; Van Rechem and Whetstine, 2014; You and Jones, 2012). Changes in chromatin accessibility patterns in tumors have been linked to decreased genomic integrity and consequently a greater frequency of genetic mutation rates (Papamichos-Chronakis and Peterson, 2013). Some tumors, however, may result primarily from epigenetic alterations. Many pediatric tumors have a low mutation frequency. For instance, genome-wide analysis of pediatric ependymoma across a broad cohort of patients did not reveal any significantly or recurrently mutated genes (Mack et al., 2014). Instead, DNA hypermethylation combined with a loss of histone H3 lysine 27 trimethylation drives the biology of these tumors (Bayliss et al., 2016; Mack et al., 2014). Ewing's sarcoma, a pediatric soft tissue and bone sarcoma, is dependent on a chromosomal translocation that results in the expression of an oncogenic fusion protein. This oncoprotein drives the disease and is directly linked to regions of aberrant chromatin accessibility at simple GGAA repeat regions. Loss of the oncoprotein decreases chromatin accessibility at these GGAA repeats and halts proliferation of Ewing's sarcoma cells (Boulay et al., 2017; Patel et al., 2012; Gangwal et al., 2008). Thus, as the result of an interplay between genetic and epigenetic mechanisms or mainly epigenetic drivers, tumor cells acquire a unique chromatin signature that is different from both normal differentiated cells and their embryonic progenitors. These disease-specific chromatin signatures have both diagnostic and therapeutic potential.

Chromatin Signature as a Diagnostic and Therapeutic Target in Cancer

Unlike genetic mutations, changes in chromatin organization associated with disease are dynamic. As a result, therapies that target chromatin-based mechanisms are emerging with several receiving FDA approval or in clinical trials, including histone deacetylase, DNA methyltransferase, histone methyltransferase, histone demethylase, and BET domain inhibitors (Mair et al., 2014; Rodriguez-Paredes and Esteller, 2011; Shortt et al., 2017). Chromatin-associated proteins frequently occur as modular subunits of various protein complexes that have nonoverlapping activities within the cell (Shortt et al., 2017; Dawson, 2017). For this reason, hit compounds derived from typical in vitro screens based on a single domain of a chromatin-associated protein are more likely to have limited efficacy or significant off-target effects when tested in vivo (Cai et al., 2015; Flavahan et al., 2017; Rodriguez-Paredes and Esteller, 2011). In addition, since tumor development is associated with a broad range of genetic events, therapeutic target selection can be difficult (Garraway and Lander, 2013). For instance, while gain-of-function mutants can be compatible with direct target inhibition, transcription factors represent an important exception since they often lack enzymatic activity or binding pockets with targetable molecular features. Transcription factor activity, however, can be directly linked to tumor-specific chromatin signatures. As discussed earlier, the oncoprotein that drives Ewing's sarcoma is responsible for maintaining disease-specific accessible chromatin that is linked to cell proliferation (Boulay et al., 2017; Patel et al., 2012; Gangwal et al., 2008). FOXA1, a "pioneer" transcription factor, is required for chromatin accessibility to enable binding of estrogen receptor-α (ERα). Reducing FOXA1 expression halts the proliferation of tamoxifen-resistant ER-positive luminal breast cancer cells (Hurtado et al., 2011). Based on their protein structure, neither of these transcription factors are good candidates for direct targeting. By targeting chromatin accessibility instead of the transcription factor itself, the limitations associated with in vitro screening and single protein target identification can be overcome. For example, a high-throughput FAIRE (HT-FAIRE) screen has been developed that successfully identified small-molecule inhibitors of a disease-specific aberrant chromatin accessibility in Ewing's sarcoma (Pattenden et al., 2016).

Tumor-specific chromatin signatures have also been linked to metastatic and therapeutic resistance mechanisms. Metastatic small-cell lung cancer cells were discovered by ATAC-seq to have a unique chromatin signature, which revealed that the prometastatic pathway was controlled by a single transcription factor (Denny et al., 2016). Like metastasis, drug resistance is a multistep process involving wide-scale changes in chromatin accessibility (Knoechel et al., 2014; Sharma et al., 2010; Shaffer et al., 2017). These changes can result in cellular heterogeneity and the emergence of small populations of cells that persist following therapeutic treatment. These transitions can be accompanied by redistribution of chromatin accessibility (Liau et al., 2017; Zawistowski et al., 2017). Future application of chromatin accessibility assays to therapeutically resistant tumor cells could potentially unlock mechanisms associated with tumor heterogeneity and drug resistance and identify possible therapeutic targets. Thus, chromatin accessibility offers a general, unbiased approach to identify common output from many cellular pathways without an a priori understanding of mechanism. Since diseased cells exhibit a chromatin signature that differs from their more epigenetically constrained normal cell developmental lineage, chromatin accessibility offers incredible potential as both a biomarker and therapeutic target.

References

Adey, A., Morrison, H.G., Asan, X.X., Kitzman, J.O., Turner, E.H., Stackhouse, B., Mackenzie, A.P., Caruccio, N.C., Zhang, X., Shendure, J., 2010. Rapid, low-input, low-bias construction of shotgun fragment libraries by high-density in vitro transposition. Genome Biol. 11, R119.

Bayliss, J., Mukherjee, P., Lu, C., Jain, S.U., Chung, C., Martinez, D., Sabari, B., Margol, A.S., Panwalkar, P., Parolia, A., Pekmezci, M., Mceachin, R.C., Cieslik, M., Tamrazi, B., Garcia, B.A., La Rocca, G., Santi, M., Lewis, P.W., Hawkins, C., Melnick, A., David Allis, C., Thompson, C.B., Chinnaiyan, A.M., Judkins, A.R., Venneti, S., 2016. Lowered H3K27me3 and DNA hypomethylation define poorly prognostic pediatric posterior fossa ependymomas. Sci. Transl. Med. 8, 366ra161.

Boulay, G., Sandoval, G.J., Riggi, N., Iyer, S., Buisson, R., Naigles, B., Awad, M.E., Rengarajan, S., Volorio, A., McBride, M.J., Broye, L.C., Zou, L., Stamenkovic, I., Kadoch, C., Rivera, M.N., 2017. Cancer-specific retargeting of BAF complexes by a prion-like domain. Cell 171, 163–178.e19.

Bowers, E.C., McCullough, S.D., 2017. Linking the epigenome with exposure effects and susceptibility: the epigenetic seed and soil model. Toxicol. Sci. 155, 302–314.

Boyle, A.P., Song, L., Lee, B.K., London, D., Keefe, D., Birney, E., Iyer, V.R., Crawford, G.E., Furey, T.S., 2011. High-resolution genome-wide in vivo footprinting of diverse transcription factors in human cells. Genome Res. 21, 456–464.

Buenrostro, J.D., Giresi, P.G., Zaba, L.C., Chang, H.Y., Greenleaf, W.J., 2013. Transposition of native chromatin for fast and sensitive epigenomic profiling of open chromatin, DNA-binding proteins and nucleosome position. Nat. Methods 10, 1213–1218.

Buenrostro, J.D., Wu, B., Chang, H.Y., Greenleaf, W.J., 2015a. ATAC-seq: a method for assaying chromatin accessibility genome-wide. Curr. Protoc. Mol. Biol. 109 (21.29), 1–9.

Buenrostro, J.D., Wu, B., Litzenburger, U.M., Ruff, D., Gonzales, M.L., Snyder, M.P., Chang, H.Y., Greenleaf, W.J., 2015b. Single-cell chromatin accessibility reveals principles of regulatory variation. Nature 523, 486–490.

Byrum, S., Mackintosh, S.G., Edmondson, R.D., Cheung, W.L., Taverna, S.D., Tackett, A.J., 2011. Analysis of histone exchange during chromatin purification. J. Integr. OMICS 1, 61–65.

Cai, S.F., Chen, C.W., Armstrong, S.A., 2015. Drugging chromatin in cancer: recent advances and novel approaches. Mol. Cell 60, 561–570.

Chappell, G.A., Israel, J.W., Simon, J.M., Pott, S., Safi, A., Eklund, K., Sexton, K.G., Bodnar, W., Lieb, J.D., Crawford, G.E., Rusyn, I., Furey, T.S., 2017. Variation in DNA-damage responses to an inhalational carcinogen (1,3-butadiene) in relation to strain-specific differences in chromatin accessibility and gene transcription profiles in C57BL/6J and CAST/EiJ mice. Environ. Health Perspect. 125, 107006.

Chereji, R.V., Ocampo, J., Clark, D.J., 2017. MNase-sensitive complexes in yeast: nucleosomes and non-histone barriers. Mol. Cell 65, 565–577. e3.

Chia, D.J., Rotwein, P., 2010. Defining the epigenetic actions of growth hormone: acute chromatin changes accompany GH-activated gene transcription. Mol. Endocrinol. 24, 2038–2049.

Corces, M.R., Corces, V.G., 2016. The three-dimensional cancer genome. Curr. Opin. Genet. Dev. 36, 1–7.

Corces, M.R., Buenrostro, J.D., Wu, B., Greenside, P.G., Chan, S.M., Koenig, J.L., Snyder, M.P., Pritchard, J.K., Kundaje, A., Greenleaf, W.J., Majeti, R., Chang, H.Y., 2016. Lineage-specific and single-cell chromatin accessibility charts human hematopoiesis and leukemia evolution. Nat. Genet. 48, 1193–1203.

Corces, M.R., Trevino, A.E., Hamilton, E.G., Greenside, P.G., Sinnott-Armstrong, N.A., Vesuna, S., Satpathy, A.T., Rubin, A.J., Montine, K.S., Wu, B., Kathiria, A., Cho, S.W., Mumbach, M.R., Carter, A.C., Kasowski, M., Orloff, L.A., Risca, V.I., Kundaje, A., Khavari, P.A., Montine, T.J., Greenleaf, W.J., Chang, H.Y., 2017. An improved ATAC-seq protocol reduces background and enables interrogation of frozen tissues. Nat. Methods 14, 959–962.

Cui, K., Zhao, K., 2012. Genome-wide approaches to determining nucleosome occupancy in metazoans using MNase-Seq. Methods Mol. Biol. 833, 413–419.

Cusanovich, D.A., Daza, R., Adey, A., Pliner, H.A., Christiansen, L., Gunderson, K.L., Steemers, F.J., Trapnell, C., Shendure, J., 2015. Multiplex single-cell profiling of chromatin accessibility by combinatorial cellular indexing. Science 348, 910–914.

Cusanovich, D.A., Reddington, J.P., Garfield, D.A., Daza, R.M., Aghamirzaie, D., Marco-Ferreres, R., Pliner, H.A., Christiansen, L., Qiu, X., Steemers, F.J., Trapnell, C., Shendure, J., Furlong, E.E.M., 2018. The cis-regulatory dynamics of embryonic development at single-cell resolution. Nature 555, 538–542.

Davie, K., Jacobs, J., Atkins, M., Potier, D., Christiaens, V., Halder, G., Aerts, S., 2015. Discovery of transcription factors and regulatory regions driving in vivo tumor development by ATAC-seq and FAIRE-seq open chromatin profiling. PLoS Genet. 11, e1004994.

Dawson, M.A., 2017. The cancer epigenome: concepts, challenges, and therapeutic opportunities. Science 355, 1147–1152.

Degner, J.F., Pai, A.A., Pique-Regi, R., Veyrieras, J.B., Gaffney, D.J., Pickrell, J.K., De Leon, S., Michelini, K., Lewellen, N., Crawford, G.E., Stephens, M., Gilad, Y., Pritchard, J.K., 2012. DNase I sensitivity QTLs are a major determinant of human expression variation. Nature 482, 390–394.

Denny, S.K., Yang, D., Chuang, C.H., Brady, J.J., Lim, J.S., Gruner, B.M., Chiou, S.H., Schep, A.N., Baral, J., Hamard, C., Antoine, M., Wislez, M., Kong, C.S., Connolly, A.J., Park, K.S., Sage, J., Greenleaf, W.J., Winslow, M.M., 2016. Nfib promotes metastasis through a widespread increase in chromatin accessibility. Cell 166, 328–342.

Dingwall, C., Lomonossoff, G.P., Laskey, R.A., 1981. High sequence specificity of micrococcal nuclease. Nucleic Acids Res. 9, 2659–2673.

ENCODE Project Consortium, 2004. The ENCODE (ENCyclopedia of DNA elements) project. Science 306, 636–640.

ENCODE Project Consortium, Bernstein, B.E., Birney, E., Dunham, I., Green, E.D., Gunter, C., Snyder, M., 2012. An integrated encyclopedia of DNA elements in the human genome. Nature 489, 57–74.

Felsenfeld, G., 1978. Chromatin. Nature 271, 115–122.

Fittler, F., Zachau, H.G., 1979. Subunit structure of alpha-satellite DNA containing chromatin from African green monkey cells. Nucleic Acids Res. 7, 1–13.

Flavahan, W.A., Gaskell, E., Bernstein, B.E., 2017. Epigenetic plasticity and the hallmarks of cancer. Science 357, 256–266.

Gaffney, D.J., Mcvicker, G., Pai, A.A., Fondufe-Mittendorf, Y.N., Lewellen, N., Michelini, K., Widom, J., Gilad, Y., Pritchard, J.K., 2012. Controls of nucleosome positioning in the human genome. PLoS Genet. 8, e1003036.

Gangwal, K., Sankar, S., Hollenhorst, P.C., Kinsey, M., Haroldsen, S.C., Shah, A.A., Boucher, K.M., Watkins, W.S., Jorde, L.B., Graves, B.J., Lessnick, S.L., 2008. Microsatellites as EWS/FLI response elements in Ewing's sarcoma. Proc. Natl. Acad. Sci. USA 105, 10149–10154.

Garraway, L.A., Lander, E.S., 2013. Lessons from the cancer genome. Cell 153, 17–37.

Giresi, P.G., Lieb, J.D., 2009. Isolation of active regulatory elements from eukaryotic chromatin using FAIRE (formaldehyde assisted isolation of regulatory elements). Methods 48, 233–239.

Giresi, P.G., Kim, J., McDaniell, R.M., Iyer, V.R., Lieb, J.D., 2007. FAIRE (formaldehyde-assisted isolation of regulatory elements) isolates active regulatory elements from human chromatin. Genome Res. 17, 877–885.

Gomez, N.C., Hepperla, A.J., Dumitru, R., Simon, J.M., Fang, F., Davis, I.J., 2016. Widespread chromatin accessibility at repetitive elements links stem cells with human Cancer. Cell Rep. 17, 1607–1620.

Goryshin, I.Y., Miller, J.A., Kil, Y.V., Lanzov, V.A., Reznikoff, W.S., 1998. Tn5/IS50 target recognition. Proc. Natl. Acad. Sci. USA 95, 10716–10721.

Groudine, M., Weintraub, H., 1975. Rous sarcoma virus activates embryonic globin genes in chicken fibroblasts. Proc. Natl. Acad. Sci. USA 72, 4464–4468.

Groudine, M., Weintraub, H., 1980. Activation of cellular genes by avian RNA tumor viruses. Proc. Natl. Acad. Sci. USA 77, 5351–5354.

Groudine, M., Weintraub, H., 1982. Propagation of globin DNAase I-hypersensitive sites in absence of factors required for induction: a possible mechanism for determination. Cell 30, 131–139.

1. HISTONE MODIFICATIONS AND CHROMATIN STRUCTURE

He, H.H., Meyer, C.A., Hu, S.S., Chen, M.W., Zang, C., Liu, Y., Rao, P.K., Fei, T., Xu, H., Long, H., Liu, X.S., Brown, M., 2014. Refined DNase-seq protocol and data analysis reveals intrinsic bias in transcription factor footprint identification. Nat. Methods 11, 73–78.

Hennig, B.P., Velten, L., Racke, I., Tu, C.S., Thoms, M., Rybin, V., Besir, H., Remans, K., Steinmetz, L.M., 2018. Large-scale low-cost NGS library preparation using a robust Tn5 purification and tagmentation protocol. G3 (Bethesda) 8, 79–89.

Hesselberth, J.R., Chen, X., Zhang, Z., Sabo, P.J., Sandstrom, R., Reynolds, A.P., Thurman, R.E., Neph, S., Kuehn, M.S., Noble, W.S., Fields, S., Stamatoyannopoulos, J.A., 2009. Global mapping of protein-DNA interactions in vivo by digital genomic footprinting. Nat. Methods 6, 283–289.

Hogan, G.J., Lee, C.K., Lieb, J.D., 2006. Cell cycle-specified fluctuation of nucleosome occupancy at gene promoters. PLoS Genet. 2, e158.

Horz, W., Altenburger, W., 1981. Sequence specific cleavage of DNA by micrococcal nuclease. Nucleic Acids Res. 9, 2643–2658.

Hurtado, A., Holmes, K.A., Ross-Innes, C.S., Schmidt, D., Carroll, J.S., 2011. FOXA1 is a key determinant of estrogen receptor function and endocrine response. Nat. Genet. 43, 27–33.

Israel, J.W., Chappell, G.A., Simon, J.M., Pott, S., Safi, A., Lewis, L., Cotney, P., Boulos, H.S., Bodnar, W., Lieb, J.D., Crawford, G.E., Furey, T.S., Rusyn, I., 2018. Tissue- and strain-specific effects of a genotoxic carcinogen 1,3-butadiene on chromatin and transcription. Mamm. Genome 29, 153–167.

Iwafuchi-Doi, M., Donahue, G., Kakumanu, A., Watts, J.A., Mahony, S., Pugh, B.F., Lee, D., Kaestner, K.H., Zaret, K.S., 2016. The Pioneer transcription factor FoxA maintains an accessible nucleosome configuration at enhancers for tissue-specific gene activation. Mol. Cell 62, 79–91.

Jessen, W.J., Dhasarathy, A., Hoose, S.A., Carvin, C.D., Risinger, A.L., Kladde, M.P., 2004. Mapping chromatin structure in vivo using DnA methyltransferases. Methods 33, 68–80.

Jessen, W.J., Hoose, S.A., Kilgore, J.A., Kladde, M.P., 2006. Active PHO5 chromatin encompasses variable numbers of nucleosomes at individual promoters. Nat. Struct. Mol. Biol. 13, 256–263.

Jin, W., Tang, Q., Wan, M., Cui, K., Zhang, Y., Ren, G., Ni, B., Sklar, J., Przytycka, T.M., Childs, R., Levens, D., Zhao, K., 2015. Genome-wide detection of DNase I hypersensitive sites in single cells and FFPE tissue samples. Nature 528, 142–146.

John, S., Sabo, P.J., Canfield, T.K., Lee, K., Vong, S., Weaver, M., Wang, H., Vierstra, J., Reynolds, A.P., Thurman, R.E., Stamatoyannopoulos, J.A., 2013. Genome-scale mapping of DNase I hypersensitivity. Curr. Protoc. Mol. Biol. (Chapter 27, Unit 21.27).

Kaplan, N., Moore, I.K., Fondufe-Mittendorf, Y., Gossett, A.J., Tillo, D., Field, Y., Leproust, E.M., Hughes, T.R., Lieb, J.D., Widom, J., Segal, E., 2009. The DNA-encoded nucleosome organization of a eukaryotic genome. Nature 458, 362–366.

Keene, M.A., Elgin, S.C., 1981. Micrococcal nuclease as a probe of DNA sequence organization and chromatin structure. Cell 27, 57–64.

Kelly, T.K., Liu, Y., Lay, F.D., Liang, G., Berman, B.P., Jones, P.A., 2012. Genome-wide mapping of nucleosome positioning and DNA methylation within individual DNA molecules. Genome Res. 22, 2497–2506.

Kilgore, J.A., Hoose, S.A., Gustafson, T.L., Porter, W., Kladde, M.P., 2007. Single-molecule and population probing of chromatin structure using DNA methyltransferases. Methods 41, 320–332.

Knoechel, B., Roderick, J.E., Williamson, K.E., Zhu, J., Lohr, J.G., Cotton, M.J., Gillespie, S.M., Fernandez, D., Ku, M., Wang, H., Piccioni, F., Silver, S.J., Jain, M., Pearson, D., Kluk, M.J., Ott, C.J., Shultz, L.D., Brehm, M.A., Greiner, D.L., Gutierrez, A., Stegmaier, K., Kung, A.L., Root, D.E., Bradner, J.E., Aster, J.C., Kelliher, M.A., Bernstein, B.E., 2014. An epigenetic mechanism of resistance to targeted therapy in T cell acute lymphoblastic leukemia. Nat. Genet. 46, 364–370.

Kundaje, A., Meuleman, W., Ernst, J., Bilenky, M., Yen, A., Heravi-Moussavi, A., Kheradpour, P., Zhang, Z., Wang, J., Ziller, M.J., Amin, V., Whitaker, J.W., Schultz, M.D., Ward, L.D., Sarkar, A., Quon, G., Sandstrom, R.S., Eaton, M.L., Wu, Y.C., Pfenning, A.R., Wang, X., Claussnitzer, M., Liu, Y., Coarfa, C., Harris, R.A., Shoresh, N., Epstein, C.B., Gjoneska, E., Leung, D., Xie, W., Hawkins, R.D., Lister, R., Hong, C., Gascard, P., Mungall, A.J., Moore, R., Chuah, E., Tam, A., Canfield, T.K., Hansen, R.S., Kaul, R., Sabo, P.J., Bansal, M.S., Carles, A., Dixon, J.R., Farh, K.H., Feizi, S., Karlic, R., Kim, A.R., Kulkarni, A., Li, D., Lowdon, R., Elliott, G., Mercer, T.R., Neph, S.J., Onuchic, V., Polak, P., Rajagopal, N., Ray, P., Sallari, R.C., Siebenthall, K.T., Sinnott-Armstrong, N.A., Stevens, M., Thurman, R.E., Wu, J., Zhang, B., Zhou, X., Beaudet, A.E., Boyer, L.A., De Jager, P.L.,

Farnham, P.J., Fisher, S.J., Haussler, D., Jones, S.J., Li, W., Marra, M.A., McManus, M.T., Sunyaev, S., Thomson, J.A., Tlsty, T.D., Tsai, L.H., Wang, W., Waterland, R.A., Zhang, M.Q., Chadwick, L.H., Bernstein, B.E., Costello, J.F., Ecker, J.R., Hirst, M., Meissner, A., Milosavljevic, A., Ren, B., Stamatoyannopoulos, J.A., Wang, T., Kellis, M., 2015. Integrative analysis of 111 reference human epigenomes. Nature 518, 317–330.

Lay, F.D., Liu, Y., Kelly, T.K., Witt, H., Farnham, P.J., Jones, P.A., Berman, B.P., 2015. The role of DNA methylation in directing the functional organization of the cancer epigenome. Genome Res. 25, 467–477.

Lay, F.D., Kelly, T.K., Jones, P.A., 2018. Nucleosome occupancy and Methylome sequencing (NOME-seq). Methods Mol. Biol. 1708, 267–284.

Leung, A., Parks, B.W., Du, J., Trac, C., Setten, R., Chen, Y., Brown, K., Lusis, A.J., Natarajan, R., Schones, D.E., 2014. Open chromatin profiling in mice livers reveals unique chromatin variations induced by high fat diet. J. Biol. Chem. 289, 23557–23567.

Levy, A., Noll, M., 1981. Chromatin fine structure of active and repressed genes. Nature 289, 198–203.

Lewis, L., Crawford, G.E., Furey, T.S., Rusyn, I., 2017. Genetic and epigenetic determinants of inter-individual variability in responses to toxicants. Curr. Opin. Toxicol. 6, 50–59.

Liau, B.B., Sievers, C., Donohue, L.K., Gillespie, S.M., Flavahan, W.A., Miller, T.E., Venteicher, A.S., Hebert, C.H., Carey, C.D., Rodig, S.J., Shareef, S.J., Najm, F.J., Van Galen, P., Wakimoto, H., Cahill, D.P., Rich, J.N., Aster, J.C., Suva, M.L., Patel, A.P., Bernstein, B.E., 2017. Adaptive chromatin remodeling drives glioblastoma stem cell plasticity and drug tolerance. Cell Stem Cell 20, 233–246 e7.

Ling, G., Sugathan, A., Mazor, T., Fraenkel, E., Waxman, D.J., 2010. Unbiased, genome-wide in vivo mapping of transcriptional regulatory elements reveals sex differences in chromatin structure associated with sex-specific liver gene expression. Mol. Cell. Biol. 30, 5531–5544.

Mack, S.C., Witt, H., Piro, R.M., Gu, L., Zuyderduyn, S., Stutz, A.M., Wang, X., Gallo, M., Garzia, L., Zayne, K., Zhang, X., Ramaswamy, V., Jager, N., Jones, D.T., Sill, M., Pugh, T.J., Ryzhova, M., Wani, K.M., Shih, D.J., Head, R., Remke, M., Bailey, S.D., Zichner, T., Faria, C.C., Barszczyk, M., Stark, S., Seker-Cin, H., Hutter, S., Johann, P., Bender, S., Hovestadt, V., Tzaridis, T., Dubuc, A.M., Northcott, P.A., Peacock, J., Bertrand, K.C., Agnihotri, S., Cavalli, F.M., Clarke, I., Nethery-Brokx, K., Creasy, C.L., Verma, S.K., Koster, J., Wu, X., Yao, Y., Milde, T., Sin-Chan, P., Zuccaro, J., Lau, L., Pereira, S., Castelo-Branco, P., Hirst, M., Marra, M.A., Roberts, S.S., Fults, D., Massimi, L., Cho, Y.J., Van Meter, T., Grajkowska, W., Lach, B., Kulozik, A.E., Von Deimling, A., Witt, O., Scherer, S.W., Fan, X., Muraszko, K.M., Kool, M., Pomeroy, S.L., Gupta, N., Phillips, J., Huang, A., Tabori, U., Hawkins, C., Malkin, D., Kongkham, P.N., Weiss, W.A., Jabado, N., Rutka, J.T., Bouffet, E., Korbel, J.O., Lupien, M., Aldape, K.D., Bader, G.D., Eils, R., Lichter, P., Dirks, P.B., Pfister, S.M., Korshunov, A., Taylor, M.D., 2014. Epigenomic alterations define lethal CIMP-positive ependymomas of infancy. Nature 506, 445–450.

Mair, B., Kubicek, S., Nijman, S.M., 2014. Exploiting epigenetic vulnerabilities for cancer therapeutics. Trends Pharmacol. Sci. 35, 136–145.

McCullough, S.D., Bowers, E.C., On, D.M., Morgan, D.S., Dailey, L.A., Hines, R.N., Devlin, R.B., Diaz-Sanchez, D., 2016. Baseline chromatin modification levels may predict interindividual variability in ozone-induced gene expression. Toxicol. Sci. 150, 216–224.

McDaniell, R., Lee, B.K., Song, L., Liu, Z., Boyle, A.P., Erdos, M.R., Scott, L.J., Morken, M.A., Kucera, K.S., Battenhouse, A., Keefe, D., Collins, F.S., Willard, H.F., Lieb, J.D., Furey, T.S., Crawford, G.E., Iyer, V.R., Birney, E., 2010. Heritable individual-specific and allele-specific chromatin signatures in humans. Science 328, 235–239.

Mieczkowski, J., Cook, A., Bowman, S.K., Mueller, B., Alver, B.H., Kundu, S., Deaton, A.M., Urban, J.A., Larschan, E., Park, P.J., Kingston, R.E., Tolstorukov, M.Y., 2016. MNase titration reveals differences between nucleosome occupancy and chromatin accessibility. Nat. Commun.. 711485.

Mill, J., Heijmans, B.T., 2013. From promises to practical strategies in epigenetic epidemiology. Nat. Rev. Genet. 14, 585–594.

Montefiori, L., Hernandez, L., Zhang, Z., Gilad, Y., Ober, C., Crawford, G., Nobrega, M., Jo Sakabe, N., 2017. Reducing mitochondrial reads in Atac-seq using Crispr/Cas9. Sci. Rep. 7, 2451.

Mueller, B., Mieczkowski, J., Kundu, S., Wang, P., Sadreyev, R., Tolstorukov, M.Y., Kingston, R.E., 2017. Widespread changes in nucleosome accessibility without changes in nucleosome occupancy during a rapid transcriptional induction. Genes Dev. 31, 451–462.

Noll, M., 1974. Internal structure of the chromatin subunit. Nucleic Acids Res. 1, 1573–1578.

Obier, N., Bonifer, C., 2016. Chromatin programming by developmentally regulated transcription factors: Lessons from the study of haematopoietic stem cell specification and differentiation. FEBS Lett. 590, 4105–4115.

Omidbakhshfard, M.A., Winck, F.V., Arvidsson, S., Riano-Pachon, D.M., Mueller-Roeber, B., 2014. A step-by-step protocol for formaldehyde-assisted isolation of regulatory elements from Arabidopsis thaliana. J. Integr. Plant Biol. 56, 527–538.

Papamichos-Chronakis, M., Peterson, C.L., 2013. Chromatin and the genome integrity network. Nat. Rev. Genet. 14, 62–75.

Pardo, C.E., Carr, I.M., Hoffman, C.J., Darst, R.P., Markham, A.F., Bonthron, D.T., Kladde, M.P., 2011. MethylViewer: computational analysis and editing for bisulfite sequencing and methyltransferase accessibility protocol for individual templates (MAPit) projects. Nucleic Acids Res. 39, e5.

Patel, M., Simon, J.M., Iglesia, M.D., Wu, S.B., McFadden, A.W., Lieb, J.D., Davis, I.J., 2012. Tumor-specific retargeting of an oncogenic transcription factor chimera results in dysregulation of chromatin and transcription. Genome Res. 22, 259–270.

Pattenden, S.G., Simon, J.M., Wali, A., Jayakody, C.N., Troutman, J., McFadden, A.W., Wooten, J., Wood, C.C., Frye, S.V., Janzen, W.P., Davis, I.J., 2016. High-throughput small molecule screen identifies inhibitors of aberrant chromatin accessibility. Proc. Natl. Acad. Sci. USA 113, 3018–3023.

Picelli, S., Bjorklund, A.K., Reinius, B., Sagasser, S., Winberg, G., Sandberg, R., 2014. Tn5 transposase and tagmentation procedures for massively scaled sequencing projects. Genome Res. 24, 2033–2040.

Pott, S., 2017. Simultaneous measurement of chromatin accessibility, DNA methylation, and nucleosome phasing in single cells. Elife (June), 6.

Rhie, S.K., Schreiner, S., Farnham, P.J., 2018a. Defining regulatory elements in the human genome using nucleosome occupancy and Methylome sequencing (NOMe-Seq). Methods Mol. Biol. 1766, 209–229.

Rhie, S.K., Yao, L., Luo, Z., Witt, H., Schreiner, S., Guo, Y., Perez, A.A., Farnham, P.J., 2018b. ZFX acts as a transcriptional activator in multiple types of human tumors by binding downstream of transcription start sites at the majority of CpG island promoters. Genome Res. 28, 310–320.

Rivera, C.M., Ren, B., 2013. Mapping human epigenomes. Cell 155, 39–55.

Rizzo, J.M., Bard, J.E., Buck, M.J., 2012. Standardized collection of MNase-seq experiments enables unbiased dataset comparisons. BMC Mol. Biol. 13, 15.

Rodriguez-Paredes, M., Esteller, M., 2011. Cancer epigenetics reaches mainstream oncology. Nat. Med. 17, 330–339.

Satpathy, A.T., Saligrama, N., Buenrostro, J.D., Wei, Y., Wu, B., Rubin, A.J., Granja, J.M., Lareau, C.A., Li, R., Qi, Y., Parker, K.R., Mumbach, M.R., Serratelli, W.S., Gennert, D.G., Schep, A.N., Corces, M.R., Khodadoust, M.S., Kim, Y.H., Khavari, P.A., Greenleaf, W.J., Davis, M.M., Chang, H.Y., 2018. Transcript-indexed ATAC-Seq for precision immune profiling. Nat Med. 24 (5), 580–590.

Schwartzman, O., Tanay, A., 2015. Single-cell epigenomics: techniques and emerging applications. Nat. Rev. Genet. 16, 716–726.

Shaffer, S.M., Dunagin, M.C., Torborg, S.R., Torre, E.A., Emert, B., Krepler, C., Beqiri, M., Sproesser, K., Brafford, P.A., Xiao, M., Eggan, E., Anastopoulos, I.N., Vargas-Garcia, C.A., Singh, A., Nathanson, K.L., Herlyn, M., Raj, A., 2017. Rare cell variability and drug-induced reprogramming as a mode of cancer drug resistance. Nature 546, 431–435.

Sharma, S.V., Lee, D.Y., Li, B., Quinlan, M.P., Takahashi, F., Maheswaran, S., McDermott, U., Azizian, N., Zou, L., Fischbach, M.A., Wong, K.K., Brandstetter, K., Wittner, B., Ramaswamy, S., Classon, M., Settleman, J., 2010. A chromatin-mediated reversible drug-tolerant state in cancer cell subpopulations. Cell 141, 69–80.

Shen, H., Laird, P.W., 2013. Interplay between the cancer genome and epigenome. Cell 153, 38–55.

Shortt, J., Ott, C.J., Johnstone, R.W., Bradner, J.E., 2017. A chemical probe toolbox for dissecting the cancer epigenome. Nat. Rev. Cancer 17, 268.

Simon, J.M., Giresi, P.G., Davis, I.J., Lieb, J.D., 2013. A detailed protocol for formaldehyde-assisted isolation of regulatory elements (FAIRE). Curr. Protoc. Mol. Biol. (Chapter 21, Unit21 26).

Song, L., Crawford, G.E., 2010. DNase-Seq: a high-resolution technique for mapping active gene regulatory elements across the genome from mammalian cells. Cold Spring Harb. Protoc. pdb.prot5384.

Song, L., Zhang, Z., Grasfeder, L.L., Boyle, A.P., Giresi, P.G., Lee, B.K., Sheffield, N.C., Graf, S., Huss, M., Keefe, D., Liu, Z., London, D., McDaniell, R.M., Shibata, Y., Showers, K.A., Simon, J.M., Vales, T., Wang, T., Winter, D., Zhang, Z., Clarke, N.D., Birney, E., Iyer, V.R., Crawford, G.E., Lieb, J.D., Furey, T.S., 2011. Open chromatin defined by DNaseI and FAIRE identifies regulatory elements that shape cell-type identity. Genome Res. 21, 1757–1767.

1. HISTONE MODIFICATIONS AND CHROMATIN STRUCTURE

Southern, E.M., 1975. Detection of specific sequences among DNA fragments separated by gel electrophoresis. J. Mol. Biol. 98, 503–517.

Stalder, J., Groudine, M., Dodgson, J.B., Engel, J.D., Weintraub, H., 1980. Hb switching in chickens. Cell 19, 973–980.

Statham, A.L., Taberlay, P.C., Kelly, T.K., Jones, P.A., Clark, S.J., 2015. Genome-wide nucleosome occupancy and DNA methylation profiling of four human cell lines. Genom. Data 3, 94–96.

Stergachis, A.B., Neph, S., Reynolds, A., Humbert, R., Miller, B., Paige, S.L., Vernot, B., Cheng, J.B., Thurman, R.E., Sandstrom, R., Haugen, E., Heimfeld, S., Murry, C.E., Akey, J.M., Stamatoyannopoulos, J.A., 2013. Developmental fate and cellular maturity encoded in human regulatory DNA landscapes. Cell 154, 888–903.

Sugathan, A., Waxman, D.J., 2013. Genome-wide analysis of chromatin states reveals distinct mechanisms of sex-dependent gene regulation in male and female mouse liver. Mol. Cell. Biol. 33, 3594–3610.

Taberlay, P.C., Statham, A.L., Kelly, T.K., Clark, S.J., Jones, P.A., 2014. Reconfiguration of nucleosome-depleted regions at distal regulatory elements accompanies DNA methylation of enhancers and insulators in cancer. Genome Res. 24, 1421–1432.

Tan, Q., Heijmans, B.T., Hjelmborg, J.V., Soerensen, M., Christensen, K., Christiansen, L., 2016. Epigenetic drift in the aging genome: a ten-year follow-up in an elderly twin cohort. Int. J. Epidemiol. 45, 1146–1158.

Teytelman, L., Ozaydin, B., Zill, O., Lefrancois, P., Snyder, M., Rine, J., Eisen, M.B., 2009. Impact of chromatin structures on DNA processing for genomic analyses. PLoS One 4, e6700.

Thurman, R.E., Rynes, E., Humbert, R., Vierstra, J., Maurano, M.T., Haugen, E., Sheffield, N.C., Stergachis, A.B., Wang, H., Vernot, B., Garg, K., John, S., Sandstrom, R., Bates, D., Boatman, L., Canfield, T.K., Diegel, M., Dunn, D., Ebersol, A.K., Frum, T., Giste, E., Johnson, A.K., Johnson, E.M., Kutyavin, T., Lajoie, B., Lee, B.K., Lee, K., London, D., Lotakis, D., Neph, S., Neri, F., Nguyen, E.D., Qu, H., Reynolds, A.P., Roach, V., Safi, A., Sanchez, M.E., Sanyal, A., Shafer, A., Simon, J.M., Song, L., Vong, S., Weaver, M., Yan, Y., Zhang, Z., Zhang, Z., Lenhard, B., Tewari, M., Dorschner, M.O., Hansen, R.S., Navas, P.A., Stamatoyannopoulos, G., Iyer, V.R., Lieb, J.D., Sunyaev, S.R., Akey, J.M., Sabo, P.J., Kaul, R., Furey, T.S., Dekker, J., Crawford, G.E., Stamatoyannopoulos, J.A., 2012. The accessible chromatin landscape of the human genome. Nature 489, 75–82.

Timp, W., Feinberg, A.P., 2013. Cancer as a dysregulated epigenome allowing cellular growth advantage at the expense of the host. Nat. Rev. Cancer 13, 497–510.

Tobi, E.W., Goeman, J.J., Monajemi, R., Gu, H., Putter, H., Zhang, Y., Slieker, R.C., Stok, A.P., Thijssen, P.E., Muller, F., Van Zwet, E.W., Bock, C., Meissner, A., Lumey, L.H., Eline Slagboom, P., Heijmans, B.T., 2014. DNA methylation signatures link prenatal famine exposure to growth and metabolism. Nat. Commun. 5, 5592.

Tobi, E.W., Slieker, R.C., Stein, A.D., Suchiman, H.E., Slagboom, P.E., Van Zwet, E.W., Heijmans, B.T., Lumey, L.H., 2015. Early gestation as the critical time-window for changes in the prenatal environment to affect the adult human blood methylome. Int. J. Epidemiol. 44, 1211–1223.

Tsompana, M., Buck, M.J., 2014. Chromatin accessibility: A window into the genome. Epigenetics Chromatin 7, 33.

Van Rechem, C., Whetstine, J.R., 2014. Examining the impact of gene variants on histone lysine methylation. Biochim. Biophys. Acta 1839, 1463–1476.

Weinreich, M.D., Gasch, A., Reznikoff, W.S., 1994. Evidence that the cis preference of the Tn5 transposase is caused by nonproductive multimerization. Genes Dev. 8, 2363–2374.

Weintraub, H., Groudine, M., 1976. Chromosomal subunits in active genes have an altered conformation. Science 193, 848–856.

Weintraub, H., Larsen, A., Groudine, M., 1981. Alpha-globin-gene switching during the development of chicken embryos: expression and chromosome structure. Cell 24, 333–344.

Wu, C., Gilbert, W., 1981. Tissue-specific exposure of chromatin structure at the 5′ terminus of the rat preproinsulin II gene. Proc. Natl. Acad. Sci. USA 78, 1577–1580.

Wu, C., Bingham, P.M., Livak, K.J., Holmgren, R., Elgin, S.C., 1979a. The chromatin structure of specific genes: I. Evidence for higher order domains of defined DNA sequence. Cell 16, 797–806.

Wu, C., Wong, Y.C., Elgin, S.C., 1979b. The chromatin structure of specific genes: II. Disruption of chromatin structure during gene activity. Cell 16, 807–814.

Xu, M., Kladde, M.P., Van Etten, J.L., Simpson, R.T., 1998. Cloning, characterization and expression of the gene coding for a cytosine-5-DNA methyltransferase recognizing GpC. Nucleic Acids Res. 26, 3961–3966.

You, J.S., Jones, P.A., 2012. Cancer genetics and epigenetics: two sides of the same coin? Cancer Cell 22, 9–20.

Yu, V.W., Yusuf, R.Z., Oki, T., Wu, J., Saez, B., Wang, X., Cook, C., Baryawno, N., Ziller, M.J., Lee, E., Gu, H., Meissner, A., Lin, C.P., Kharchenko, P.V., Scadden, D.T., 2016. Epigenetic memory underlies cell-autonomous heterogeneous behavior of hematopoietic stem cells. Cell 167, 1310–1322. e17.

Zawistowski, J.S., Bevill, S.M., Goulet, D.R., Stuhlmiller, T.J., Beltran, A.S., Olivares-Quintero, J.F., Singh, D., Sciaky, N., Parker, J.S., Rashid, N.U., Chen, X., Duncan, J.S., Whittle, M.C., Angus, S.P., Velarde, S.H., Golitz, B.T., He, X., Santos, C., Darr, D.B., Gallagher, K., Graves, L.M., Perou, C.M., Carey, L.A., Earp, H.S., Johnson, G.L., 2017. Enhancer remodeling during adaptive bypass to MEK inhibition is attenuated by pharmacologic targeting of the P-TEFb complex. Cancer Discov. 7, 302–321.

Zhang, Z., Pugh, B.F., 2011. High-resolution genome-wide mapping of the primary structure of chromatin. Cell 144, 175–186.

Zhang, Y., Laz, E.V., Waxman, D.J., 2012. Dynamic, sex-differential STAT5 and BCL6 binding to sex-biased, growth hormone-regulated genes in adult mouse liver. Mol. Cell. Biol. 32, 880–896.

Zhou, M., Reznikoff, W.S., 1997. Tn5 transposase mutants that alter DNA binding specificity. J. Mol. Biol. 271, 362–373.

Zhou, M., Bhasin, A., Reznikoff, W.S., 1998. Molecular genetic analysis of transposase-end DNA sequence recognition: cooperativity of three adjacent base-pairs in specific interaction with a mutant Tn5 transposase. J. Mol. Biol. 276, 913–925.

Further Reading

Tucker, P.W., Hazen Jr., E.E., Cotton, F.A., 1978. Staphylococcal nuclease reviewed: a prototypic study in contemporary enzymology. I. Isolation; physical and enzymatic properties. Mol. Cell. Biochem. 22, 67–77.

Tucker, P.W., Hazen Jr., E.E., Cotton, F.A., 1979a. Staphylococcal nuclease reviewed: a prototypic study in contemporary enzymology. II. Solution studies of the nucleotide binding site and the effects of nucleotide binding. Mol. Cell. Biochem. 23, 3–16.

Tucker, P.W., Hazen Jr., E.E., Cotton, F.A., 1979b. Staphylococcal nuclease reviewed: a prototypic study in contemporary enzymology. III. Correlation of the three-dimensional structure with the mechanisms of enzymatic action. Mol. Cell. Biochem. 23, 67–86.

Tucker, P.W., Hazen Jr., E.E., Cotton, F.A., 1979c. Staphylococcal nuclease reviewed: a prototypic study in contemporary enzymology. IV. The nuclease as a model for protein folding. Mol. Cell. Biochem. 23, 131–141.

1-4

Implications for Chromatin Biology in Toxicology

Tiffany A. Katz, Cheryl L. Walker

Center for Precision Environmental Health, Department of Molecular and Cellular Biology, Baylor College of Medicine, Houston, TX, United States

O U T L I N E

CHANGES IN CHROMATIN MODIFICATION STATES ARE IMPORTANT IN AGING AND THE DEVELOPMENTAL ORIGINS OF HEALTH AND DISEASE

Chromatin modifications that comprise the epigenome include methylation, acetylation, phosphorylation, sumoylation, and ubiquitination and occur on both DNA and the associated histone proteins of chromatin. These modifications and the enzymes and mechanisms responsible for their creation and maintenance are highly conserved across species. As a result, a variety of model systems are amenable to epigenetic toxicology studies, including rodents, flies, and nematodes. We focus here on histone modifications, which are conserved across species and cell type (Woo and Li, 2012). Studies across a large number of different species have demonstrated that both histone modifications themselves and colocalization of specific histone marks are conserved across humans, mice, and pigs (Xiao et al., 2012). The degree of conservation is dependent on the genomic feature; for example, histone acetylations located at transcriptional start sites are highly conserved, as is the (activating) function of this mark (Roh et al., 2007). Additionally, the enzymes that "read, write, and erase" histone marks are also highly conserved across species (Table 1). This high degree of conservation allows for the use of model systems when studying the effects of toxicants on chromatin biology.

As we age, human tissues undergo several periods of essential epigenomic plasticity, with embryonic development, fetal development, puberty, pregnancy, and lactation being the most prominent (Kanherkar et al., 2014). A massive erasure of parental epigenetic marks occurs prior to implantation, after which unique epigenomic patterns are programmed into specific cell types, culminating in cellular differentiation and decreased epigenomic plasticity, and ultimately tissue specification and organ system development (Kanherkar et al., 2014; Ho et al., 2017; Zhu et al., 2013). Consequently, the environment has a large influence on how the epigenetic landscape is constructed during development. The developmental origins of health and disease (DOHaD) hypothesis was first described in a 1989 report showing that birth weight was correlated with several adulthood diseases including heart disease, diabetes, and hypertension (Barker et al., 1989a,b; Barker, 2007). These early observations were further supported by other studies in which maternal environmental exposures were correlated with adult disease in the offspring, including the studies of the Dutch and Chinese famines (Schulz, 2010; Wang et al., 2017a,b,c). These early DOHaD studies have given rise to an entire field of research devoted to investigating developmental exposures and risk for diseases later in life including cancer, metabolic disorders such as diabetes, and neurological disorders (Hilakivi-Clarke and de Assis, 2006; Godfrey et al., 2016; Mochizuki et al., 2017; Foulds et al., 2017).

Reprogramming of the epigenome is thought to be one of the major mechanisms by which early environmental exposures influence health and disease susceptibility across the life course. During development prior to cell-fate determination, the epigenome is erased and subsequently reapplied allowing cell-specific profiles to influence cell fate (Kanherkar et al., 2014). While cells are undergoing migration, proliferation, and ultimately differentiation, the epigenomic landscape is highly fluid and amenable to the environment. Importantly, exposure to xenoestrogens during development can be accompanied by the reprogramming of epigenetic histone methyl marks, which, while persistent into later life, may not manifest as aberrant gene expression until challenged by subsequent exposures to dietary factors, hormones, or chemicals. Thus, epigenetic reprogramming at genes in this setting may not appear to have immediate effects on gene expression, but can subsequently lead to an array of effects later in life including increased susceptibility to metabolic disease, altered neurological function, cancer, and fertility (Howdeshell et al., 1999; Schonfelder et al., 2002; Durando et al., 2007).

Early examples of this include reprogramming of lactoferrin gene expression and DNA methylation, which was induced in response to early-life DES treatment but was only

TABLE 1 Conservation of Proteins Involved in Histone Modification Activity

Complex		Function	Histone Mark	Human	Mouse	Rat	Fly	Zebrafish	Frog
Polycomb group proteins	PRC2	Methyltransferase	H3K27me3	EZH1/2	EZH2	EZH2	EZ	EZH2	EZH2
				Suz12	Suz12	Suz12	Su(z)12	Suz12b	Suz12
				EED	EE+D	EED	ESC	EED	EED
				RbBP4	RbBP4	RbBP4	PCL/Caf1	RbBP4	RbBP4
	PRC1	Ubiquitinase	K119u	CBX	CBX1	CBX1	HP1b	CBX1b	CBX1
				PHC1/2	PHC1/2	PHC1/2	PH	PHC2a/b	PHC1
				RING1	RING1	RING1	dRING		RING1
				BMI1	BMI1	BMI1	PSC	BMI1a	BMI1
COMPASS complex	MLL1/2	Methyltransferase	H3K4me3	MLL1/2	MLL	MLL	TRX	MLL	MLL
	MLL3/4	Methyltransferase	H3K4me1	MLL3/4	MLL3/4	MLL3/4	TRR/LPT	MLL3a/b	MLL3
	SET1a/b	Methyltransferase	H3K4me3	SET1a/b	SET1a	ASH2l	SET1B	ASH2l	ASH2
G9a	G9a	Methyltransferase	H3K9me1/2	G9a	G9a	G9a	G9a	G9a	G9a
Histone acetylation	Sirtuin1	Histone deacetylase	H4K16, H3K9, H1K26	SIRT1	SIRT1	SIRT1	SIRT1	SIRT1	SIRT1
	p300	Histone acetyltransferase	H3K27, H3K122	p300	p300	p300	NEJ	p300	p300
SETD2	SETD2	Methyltransferase	H3K36me3	SETD2	SETD2	SETD2	SETD2	SETD2	SETD2

Homologs were identified using the NCBI database Homologene.

observed in adult animals that had undergone puberty (Li et al., 1997). More recently, BPA-induced reprogramming of androgen-responsive genes in the prostate was shown to remain silent until exposure to testosterone and activation of androgen receptor (AR) signaling. In those studies, early-life BPA reprogramming increased trimethylation of lysine 4 on the tail of histone 3 (H3K4me3) at promoters of AR target genes and did not cause any aberrant gene expression but rather made these genes hyperresponsive to later-life testosterone exposure. The activation of the PI3K/Akt pathway preceded H3K4me3 gene occupancy and was responsible for the cleavage and activation of the mixed lineage leukemia (MLL) 1/2, a writer of the H3K4me3 mark. MLL activation was necessary for the trimethylation of H3K4, as H3K4me3 induction was blocked by inhibiting MLL itself or the cleavage and activation of MLL by taspase 1 (Wang et al., 2016). As a result, reprogrammed genes in BPA-exposed animals became hyperresponsive to androgen, exhibiting significantly more expression than in similarly challenged animals that had not been exposed to BPA during development (Wang et al., 2016).

Aging itself is also associated with changes in the epigenetic landscape. In *Caenorhabditis elegans,* the activity of several histone methyltransferases and demethylases has been shown to influence life span (Hamilton et al., 2005). Trimethylation of lysine 27 on the tail of histone 3 (H3K27me3) has been associated with effects on life span in both *C. elegans* and *Drosophila melanogaster* (Jin et al., 2011; Siebold et al., 2010). In *C. elegans,* the knockdown of the H3K27me3 demethylase UTX-1 increased life span, whereas the loss of the H3K27me3 methyltransferase polycomb repressive complex 2 (PRC2) increased the life span of *D. melanogaster* (Siebold et al., 2010). These marks have also been shown to play an important role in aging in mammals. In muscle tissue from old mice, H3K27me3 is increased compared with muscle tissue from young mice (Liu et al., 2013), and the loss of UTX-1 in male mice reduced life span (Welstead et al., 2012). In human adipose mesenchymal stem cells (MSCs), young cells lost H3K27me3 upon differentiation, whereas old MSCs retained the mark upon differentiation, which could be explained by their higher levels of EZH2, the enzymatic component of PRC2 (Noer et al., 2009).

The histone marks written by SET-domain-containing methyltransferases are also associated with aging. Knockdown of SETD1A, an H3K4 methyltransferase, increased life span in *C. elegans* (Greer et al., 2010), whereas the loss of KDM5B, an H3K4me3 demethylase, shortened the life span in *D. melanogaster* and *C. elegans* (Greer et al., 2010; Li et al., 2010). Conversely, the knockdown of the H3K4me3 demethylases KDM5B or KDM1A extended the life span in sterile *C. elegans* (Lee et al., 2003; Ni et al., 2012; McColl et al., 2008).

Additionally, the loss of trimethylated lysine 36 on histone 3 (H3K36me3) methyltransferase, SET2 (SETD2 homolog), shortened the life span in *C. elegans* (Pu et al., 2015) and was associated with reduced life span in yeast (Sen et al., 2015). Similarly, the loss of the H3K36me3 demethylase, JMJD2, extended life span in both *S. cerevisiae* and *C. elegans* (Ni et al., 2012; Sen et al., 2015).

Given that histone modifications and the proteins and enzymes that govern them play major roles in development and aging, studies evaluating the impact of chemical exposures on histone modifications and chromatin state must carefully consider exposure timing. The next section of the chapter presents studies investigating the effects of both direct and developmental exposure to toxicants and the mechanisms by which they alter the epigenome.

MECHANISMS BY WHICH TOXICANT EXPOSURES ALTER THE CHROMATIN LANDSCAPE

The epigenome as a target and determinant of response to toxicant exposures has emerged as an important theme in toxicological research. Toxicant-, cell-, tissue-, sex-, and species-specific effects on the epigenomic landscape have been observed for exposures that impact chromatin modifications, which depend on toxicant-specific mechanism(s) of action, and host determinants such as age, sex, and genetic background. In some instances, these differences are determined by the toxicant class, but even within a single class, individual toxicants may exhibit dramatically different effects on chromatin modifications. Here, we overview toxicant-epigenome alterations in which exposures have a direct effect on chromatin modifications, as distinct from toxicant exposures that may indirectly impact the epigenome, for example, via DNA repair inhibition, that could lead to underlying changes in the DNA and secondary reprogramming of the epigenome.

Repressive Histone Marks and Polycomb Group Proteins

The polycomb group proteins function as two major complexes, polycomb repressive complex 1 and 2 (PRC1 and PRC2). These complexes include epigenetic readers, writers, and erasers. PRC2 binds chromatin and, using the catalytic EZH2 histone methyltransferase "writer," adds a trimethyl group to lysine 27 on histone 3 (H3K27me3). This mark is then recognized (read) by the CBX subunit of the PRC1 complex to facilitate the recruitment of PRC1. Once recruited, the RING1/2 subunit of PRC1 monoubiquitinates K119 on histone 2A causing the pausing of RNA pol II and gene repression (Francis et al., 2004). Both PRC1 and PRC2 play key roles in development (Sparmann and van Lohuizen, 2006), and perturbation of these complexes can have detrimental effects on development and risk for diseases, including cancer (Sparmann and van Lohuizen, 2006). This section outlines mechanisms by which environmental toxicants can disrupt the activity of polycomb group proteins to reprogram the epigenome.

Xenoestrogens

Bisphenol A (BPA) is a broadly used plasticizer to which most humans are ubiquitously exposed; detectible levels of BPA are found in the urine of over 90% of Americans (Calafat et al., 2008; Ye et al., 2015; Vandenberg et al., 2007; Centers for Disease Control and Prevention, 2009). Routes of BPA exposure include contaminated food and drink, as BPA can migrate or be leached from plastic containers or the lining of cans into food or beverages (Vandenberg et al., 2007; Kang et al., 2006), and contact with BPA-containing materials such as cash register receipts and tubing in medical applications (Hehn, 2016; Babu et al., 2015; Biedermann et al., 2010; Hope et al., 2016; Kanno et al., 2007). It is now widely accepted that BPA and many of its analogs activate estrogen receptors (ERs), estrogen-related receptor-γ, and pregnane X receptors by direct binding, causing endocrine disruption (Rubin, 2011; Stossi et al., 2014; Szafran et al., 2017; Gore et al., 2015a,b). In animal studies, direct exposure to BPA affects fertility by altering ovarian function and estradiol metabolism (Lee et al., 2013; Kim et al., 2014a; Ziv-Gal et al., 2013; Peretz et al., 2011) and can alter development and promote cancer in estrogen-sensitive tissues such as mammary, vaginal, and uterine tissues (Jirtle and Skinner, 2007; Weidman et al., 2007; Gao et al., 2015; Seachrist et al., 2016; Soto et al., 2013; Wang et al., 2017d).

The impact of BPA on DNA methylation has been widely studied in animal models of BPA exposure and will be covered in subsequent chapters of this book. In contrast to DNA methylation, much less is known regarding the effects of BPA on histone modifications, despite the known cross talk between estrogen signaling and regulation of proteins involved in histone modification activity (Weinhouse et al., 2015; Ma et al., 2013; Kim et al., 2014b; van Esterik et al., 2015; Anderson et al., 2012, 2017; Bhan et al., 2014a; Bredfeldt et al., 2010; Greathouse et al., 2012). BPA reportedly induces the RNA and protein expression of the histone methyltransferase EZH2 (Bhan et al., 2014a), which, as mentioned above, "writes" the H3K27me3 mark, silencing gene transcription. The overexpression of EZH2 has been shown to influence carcinogenesis in a variety of organs (Greathouse et al., 2012; Croonquist and Van Ness, 2005; Kleer et al., 2003; Varambally et al., 2002; Bachmann et al., 2006; Collett et al., 2006; Cha et al., 2005). Direct exposures to BPA, estradiol, and diethylstilbestrol (DES) were all able to induce EZH2 expression both in MCF7 breast cancer cells and in adult rat mammary glands (Bhan et al., 2014a), and the EZH2 promoter contains three functional estrogen-response elements (EREs). Additionally, all three chemicals were able to induce luciferase activity and ER-α and ER-β enrichment in two out of the three EZH2 promoter EREs in an *in vitro* activity assay and a chromatin immunoprecipitation coupled with quantitative PCR (ChIP-qPCR) assay (Bhan et al., 2014a).

Xenoestrogens have been shown to increase EZH2 activity and H3K27me3 enrichment and repress gene expression through mechanisms that involve ER interacting with the COMPASS complex. Specifically, ER colocalizes with MLL, an epigenetic "writer" responsible for the H3K4me mark associated with both active promoters and enhancers (Sze and Shilatifard, 2016; Herz et al., 2013). This methyltransferase also acts as a coactivator with ERs (Bhan et al., 2014a), and both MLL2 and MLL3 colocalized to EREs within the EZH2 promoter when ERs are liganded and activated by BPA, estradiol, or DES. Subsequently, RNApolII, CBP, p300, and histone trimethylation and acetylation were enriched at EREs in the EZH2 promoter, which can be blocked by MLL2 siRNA in MCF7 cells (Bhan et al., 2014a). These data imply that ER and MLL are corecruited to ER target genes, including the EZH2 promoter, in response to BPA, estradiol, or DES, to transactivate gene and protein expression.

Additionally, we have identified rapid ER signaling to lead to phosphorylation and inactivation of EZH2 and subsequent reductions in H3K27me3 after developmental exposure to DES (Bredfeldt et al., 2010), which presumably contributes to reprogramming of the rodent uterine tissue by developmental DES exposure (Greathouse et al., 2008). We have also identified genistein, a soy phytoestrogen, to induce nongenomic ER signaling and lead to phosphorylation and repression of EZH2 (Greathouse et al., 2012). Developmental genistein exposure also leads estrogen target genes to become hyperresponsive to hormone stimulation and the promotion of uterine leiomyomas (Greathouse et al., 2012).

Arsenic

Arsenic, which is widespread in the environment, is classified as a group I carcinogen. In addition to cancer, arsenic is associated with many other adverse health outcomes including cardiovascular disease, lung disease, and neuronal development. Exposure typically occurs through contaminated drinking water, diet, or occupationally (Mantha et al., 2017; Minatel et al., 2018; Shakoor et al., 2017; Smeester and Fry, 2018). Arsenic is not directly mutagenic, but is rather thought to act through epigenetic mechanisms to induce carcinogenesis (Rossman, 2003). Several studies have identified changes in histone modifications in response to arsenic exposure in vitro and in rodents and human tissues (Howe and Gamble, 2016). In this chapter, we focus on studies that have elucidated mechanisms by which arsenic exposures induce changes in histone modifications.

In a mouse model of arsenic-induced carcinogenesis, epigenetic mechanisms were found to play a major role in malignant transformation (Kim et al., 2012). In BALB/c 3T3 cells exposed directly to arsenic or vehicle then transplanted into Swiss nude mice, only the arsenic-exposed cells formed tumors. The tumors exhibited reduced protein expression of the tumor suppressors p16INK4a and p19ARF. In these studies, a key component of the PRC1 and PRC2 complexes, BMI1, and the corresponding histone mark H3K27me2/3 were increased in cells exposed to arsenic. Further, knockdown of BMI1 or Suz12 (another member of the PRC complexes) suppressed the arsenic-induced cell transformation and the accumulation of the H3K27me2/3 mark. BMI1 and Suz12 knockdown also inhibited the arsenic-induced reduction in p16INK4a and p19ARF (Kim et al., 2012). This research strongly suggests that the PRC complexes play a major role in arsenic-induced carcinogenic cell transformation in the mouse model.

COMPASS Complex

The COMPASS complex is largely responsible for the establishment and maintenance of transcriptional memory. The enzymatic activity of the COMPASS complex is carried out by SET-domain-containing subunits that methylate lysine 4 on histone 3 (H3K4me) leading to gene activation (Herz et al., 2013). There are three major COMPASS complexes largely defined by the enzymatic subunit, MLL1/2, MLL3/4, or Set1a/b (Sze and Shilatifard, 2016). Canonically, MLL1/2 trimethylates H3K4 marking promoters and bivalent genes, MLL3/4 monomethylates H3K4 at enhancers, and Set1a/b di- and trimethylates H3K4 at a variety of genomic locations (Sze and Shilatifard, 2016). Exposure to xenoestrogens or arsenic, both developmentally and during adulthood, has been shown to perturb COMPASS complex activity.

Xenoestrogens

Developmental BPA exposure induced prostate dysplasia during adulthood and genome-wide reprogramming of the H3K4me3 mark (Wang et al., 2016; Wong et al., 2015). In adulthood, animals exposed to BPA neonatally exhibited increased H3K4me3 at the promoters of several genes on rat chromosome 1 in the kallikrein and prostastatin gene clusters and increased expression of those genes. Promotion to neoplasia following hormonal challenge (testosterone and estradiol) was also more severe in animals exposed to BPA neonatally (Wong et al., 2015), demonstrating a hypersensitivity to a hormonal challenge.

Interestingly, another study demonstrated that direct exposure to BPA induced the expression of HOTAIR in MCF7 cells and that the induction occurs through a mechanism similar to that of EZH2 described above. Increased occupancy of MLL1, MLL3, CPB, p300, and H3K4me3 occurred at the ERE in the HOTAIR promoter, accompanied by increased histone acetylation, following direct treatment with BPA, estradiol, or DES (Bhan et al., 2014b).

Notably, the induction of HOTAIR likely leads to additional epigenetic consequences, as HOTAIR is known to interact with PRC2 and LDS1/CoREST, responsible for removing H3K4me1 and H3K4me2. Both PRC2 and CoREST are responsible for gene silencing and may be disrupted by the overexpression of HOTAIR.

Arsenic

Developmental arsenic exposure reportedly promoted hepatocellular carcinoma in a mouse model of spontaneous hepatic tumors (C3H/HeN mice) (Nohara et al., 2012). Fifty-one percent of the mice exposed to 85 ppm arsenic from day E8 to E18 developed tumors compared with 41% of the controls. Tumors in animals exposed during development to arsenic were also larger (92 mm^2) compared with tumors in the control animals (76 mm^2). Interestingly, developmental arsenic exposure increased the proportion of tumors containing mutations in the proto-oncogene Ha-*ras* (Nohara et al., 2012). Two genes (*Slc25a30* and *Fabp4*) were identified to be late-onset alterations as they were present at 74 but not 49 weeks of age. Remarkably, the trimethylation of H3K4, an active histone mark written by the COMPASS complex, was increased in arsenic-exposed animals at *Slc25a30*, which was transcriptionally upregulated, whereas the demethylation of H3K9, a repressive histone mark, was increased at *fabp4*, which was transcriptionally downregulated (Nohara et al., 2012), implying that epigenetic regulation influences the expression of the late-onset genes. Additional studies are needed to identify the upstream pathways leading to epigenetic reprogramming in this model; nonetheless, arsenic has been detected in human fetal liver tissue, indicating that developmental arsenic exposure occurs in humans and can travel to the fetal liver (Drake et al., 2015).

The Histone Methyltransferase G9a

The histone methyltransferase G9a is a SET-domain-containing enzyme that mono- and dimethylates lysine 9 on histone 3, effectively silencing gene transcription in euchromatic regions (Casciello et al., 2015). G9a also contains ankyrin repeats that recognize the H3K9me1/2, and therefore, G9a has two distinct functions as a reader and a writer of the H3K9me1/2 mark (Mozzetta et al., 2015). G9a dimerizes with G9a-like protein (GLP) to carry out its function (Casciello et al., 2015). Exposure to both phthalates and arsenic can disrupt the function of G9a leading to epigenomic alterations and disease phenotypes.

Phthalates

Phthalates are plasticizers found in a wide variety of products including food containers, pharmaceuticals, packaging, children's toys, medical devices, and household items (Chou and Wright, 2006; Schettler, 2006). Di(2-ethylhexyl) phthalate (DEHP) and benzyl butyl phthalate (BBP) are two of the most commonly used phthalates. Phthalates have been shown to cause male and female reproductive toxicities after both in utero and postnatal exposures in rodent models and human studies (Singh and Li, 2012; Martino-Andrade and Chahoud, 2010; Swan et al., 2005; Shelby, 2006). Male infants in the United States display an inverse association between anogenital distance and maternal urinary phthalate levels during gestation (Swan et al., 2005). This finding has been confirmed in rodent models, which have consistently exhibited a higher sensitivity to phthalates during development, not only in males in the primary testis but also in females in ovarian tissue (Martino-Andrade and Chahoud, 2010).

Direct phthalate exposure has been associated with altered histone modifications through endocrine disruption. One study investigating DEHP exposure in pubertal male mice found that sperm count, testis weight, testicular damage, plasma testosterone, and steroidogenic acute regulatory protein (StAR, the rate-limiting enzyme in steroid biosynthesis) RNA expression decreased in response to DEHP in a dose-dependent manner (Liu et al., 2016). Interestingly, there was a dramatic reduction in the expression of G9a in the testes of animals exposed to DEHP directly and a decrease in H3K9me1/2. Previously, G9a was shown to be essential for germ cell development, as mice lacking G9a are sterile and harbor germ cells that undergo apoptosis instead of meiosis during sperm development (Tachibana et al., 2007). DEHP-induced changes in G9a expression and H3K9 methylation correlated with altered expression of known G9a targets, as well as genes known to be involved in the observed testicular phenotype (Liu et al., 2016). The authors speculated that DEHP may work through the Nr0b2 receptor, as Nr0b2 was increased in a dose-responsive fashion by direct exposure to DEHP. It has been reported that G9a expression can be controlled by Nr0b2 (Volle et al., 2009), and further supporting their hypothesis, Nr5a2 (which is known to be negatively regulated by Nr0b2) decreased following direct DEHP exposure (Liu et al., 2016).

Data in adults and children have suggested that phthalates act as obesogens, although the data are not completely clear (Goodman et al., 2014). Direct exposure to BBP increased adipogenesis in stromal mesenchymal stem cells (MSCs) by activating PPAR-γ (Sonkar et al., 2016; Pereira-Fernandes et al., 2013a,b). BBP also increased the expression of the adipogenic transcription factors C/EBP-α and aP2, while G9a and H3K9me2 were decreased following direct BBP treatment. The authors also reported decreased HDACs 3 and 10 and increased H3K9 acetylation in BBP-treated MSCs (Sonkar et al., 2016). These epigenetic effects were reversed by blocking PPAR-γ, indicating an integral role of the receptor in the process of altering the epigenome. After the inhibition of PPAR-γ by siRNA, the adipogenic phenotype was lost, and BBP no longer caused reduction in G9a or increased H3K9Ac.

Arsenic

In mouse Leydig cells, direct arsenic exposure diminished steroidogenesis and decreased testosterone and progesterone, which was accompanied by the decreased expression of the important steroidogenic enzymes StAR, P450scc, P450c17, and Hsd17b, except for Hsd3b, which was increased (Alamdar et al., 2017). In these cells, global H3K9me2/3 was lost in response to direct arsenic exposure, most likely due to the downregulation of G9a and Suv39h1/2, H3K9 methyltransferases, and the upregulation of the histone demethylase Jmjd2a (Alamdar et al., 2017). The inhibition of Jmjd2a by quercetin resulted in the recovery of the H3K9me2/3 marks. Similarly, the arsenic-induced reduction in H3K9me2/3 levels at the Hsd3b promoter was rescued by quercetin. In testicular cells, 3β-hydroxysteroid dehydrogenase (3β-HSD) catalyzes steroid synthesis, activating 3β-HSD by reducing H3K9me2/3 at the Hsd3b promoter, and is thought to be a compensatory mechanism to combat the reduction in many other steroidogenic enzymes (Alamdar et al., 2017). In contrast to these findings, a separate report in A549 lung carcinoma cells found that direct arsenite (+3 oxidized arsenic) increased H3K9me2/3 levels and its writer G9a, whereas JHDM2A was not changed (Zhou et al., 2008), indicating that modulation of G9a and/or jumonji demethylases may be cell-type- or cancer-specific.

Histone Acetylation

Transcriptional regulation of gene expression is tightly controlled by numerous mechanisms, one of the most well studied being histone acetylation, which was first discovered to effect transcription in the 1960s (Allfrey et al., 1964). Histone acetylation is generally associated with euchromatin and was later appreciated to be a dynamic histone modification after the discovery of the first histone acetyltransferase (HAT), p55 (Brownell and Allis, 1996; Brownell et al., 1996). It is now widely understood that histone acetylation is controlled by both HATs that "write" this mark and histone deacetylases (HDACs) "erasers" that remove it. There are two families of HDACs: the histone deacetylase family, which includes HDACs 1–11, and the Sir2 regulator family, which includes sirtuins 1–7 (Seto and Yoshida, 2014). The HAT family is divided into classes based on where they reside in the cell. Type A HATs are nuclear and are involved in regulating gene transcription, and type B HATs can be found in the cytoplasm and acetylate newly formed histones before they are packaged into nucleosomes (Lee and Workman, 2007). The following sections describe how toxicant exposures alter histone acetylation.

Organotins

Tributyltin (TBT) and triphenyltin (TPT) were used as antifouling agents for aquatic vessels in the 1970s but had severe effects on marine life including inducing imposex, which resulted in reproductive deficiencies (Shimasaki et al., 2003; McAllister and Kime, 2003; Horiguchi et al., 2006; Horiguchi, 2006). In addition to the reproductive effects on marine life, TBT and TPT induce increased adiposity by binding to the PPAR-γ and retinoid X receptors (RXR), leading to investigations of the compounds as potential obesogens in mammals (Kanayama et al., 2005; Grun et al., 2006; Grun and Blumberg, 2006; Janesick and Blumberg, 2012). The United States banned TBT and TPT in the 1980s, but both compounds have since been found in ocean sediment and seawater in ports and harbors around the world including Europe, Asia, and North America, as well as in fish prepared for human consumption (Ashraf et al., 2017; Jadhav et al., 2011; Filipkowska et al., 2016; Antizar-Ladislao, 2008; Airaksinen et al., 2010; Kucuksezgin et al., 2011; Rastkari et al., 2012). TBT and TPT have likewise been detected in house dust and flooring materials in Germany, the Netherlands, the United Kingdom, and the United States (Albany, NY) (Fromme et al., 2005; Kannan et al., 2010). Consequently, TBT is detectable in human serum. In a study conducted in central Michigan, TBT was detected in all 32 subjects tested, with an average concentration of 8 ng/mL and a peak level of 85 ng/mL (Kannan et al., 1999).

TBT and TPT have both been shown to increase the activity of HATs GCN5, PCAF, P300, and CBP in a standard in vitro HAT activity assay using core histones or nucleosomes as the substrate in a dose-dependent manner (Osada et al., 2005). TBT and TPT metabolites (DBT, MBT, DPT, and MPT) also enhanced HAT activity, but to a lesser degree compared with the parent compounds. Neither TBT nor TPT affected HDAC activity. In these assays, tin was essential in augmenting HAT activity, as replacing tin with carbon or silicon abrogated the increase in HAT activity seen with TBT and TPT (Osada et al., 2005). Two potential mechanisms were proposed for the effect of TBT and TPT on HAT activity: (1) TBT and TPT may bind HAT active sites to increase activity, or (2) TBT and TPT may interact with histones, increasing activity by unmasking HAT binding sites.

Phthalates

Developmental exposure to DEHP has been shown to induce metabolic syndrome in rodent models (Heindel et al., 2017). Exposed rats displayed glucose and insulin intolerance and reduced insulin receptor expression and downstream signaling activity in muscle tissue (Rajesh and Balasubramanian, 2014). These animals also exhibited a dose-dependent decline in GLUT4 expression, the insulin responsive GLUT protein. Chromatin immunoprecipitation revealed increased occupancy of HDAC 2 at the *Glut4* promoter, which likely contributed to decreased expression of this gene.

Dioxin

Dioxin is a persistent environmental pollutant and potent carcinogen, which is primarily found in food, particularly animal fats, and is linked to a number of diseases including type 2 diabetes, heart disease, and skin disorders (Travis and Hattemer-Frey, 1991; Marinkovic et al., 2010). Dioxin is known to bind the aryl hydrocarbon receptor (AhR) and to induce the CYP1 family of metabolizing genes. Since the 1990s, dioxin has been thought to interact with epigenetic machinery to elicit its effects on Cyp1a1 (Xu et al., 1997); however, recently, more detailed mechanisms have been elucidated.

In mouse hepatoma cells, increased acetylation at H3K9, H3K14, and H4 and phosphorylation of H3S10 by IKK-α, a serine/threonine protein kinase, are essential for dioxin-induced upregulation of Cyp1a1. Direct treatment with 2,3,7,8-tetrachlorodibenzo-*p*-dioxin (TCDD) leads to increased occupancy of AhR and the phosphorylation of H2S10, and several kinases, including IKK-α, ARNT, MSK1, and MSK2, are enriched at the *Cyp1a1* enhancer region in response to direct TCDD (Kurita et al., 2014). Direct exposure to dioxin also induced the acetylation of H3K9, H3K14, and H4 and the trimethylation of H3K4, at *Cyp1a1* and *Cyp1b1* promoter regions, and recruited HATs p300 and PCAF to the *Cyp1a1* and *Cyp1b1* enhancers (Beedanagari et al., 2010).

Lead

Lead is a ubiquitous metal in our environment, with human exposure occurring occupationally or through contaminated water, paint, and dust (Patrick, 2006). Lead causes severe neurological toxicity and hematologic, skeletal, and cardiovascular system abnormalities (Patrick, 2006). In Sprague–Dawley rats, direct lead exposure increased both systolic and diastolic blood pressure (Xu et al., 2015), and increased global acetylated H3 was observed in both vascular and cardiac tissues, which was evident 12 days post exposure and persisted until 40 days post exposure.

Chromium

Chromium is a known carcinogen, with the most potent species being Cr(VI), although the most prevalent chromium, Cr(III), is not considered carcinogenic. Administering dichromate directly to cells was shown to be reduced to Cr(VI) and subsequently Cr(III) within 2 h. This treatment blocked the release of HDAC 1 and prevented p300 from binding the AhR complex at the *Cyp1a1* promoter in hepatoma cells (Wei et al., 2004).

Arsenic

In primary human keratinocytes (NHEKs), direct arsenic exposure induced H4K16 acetylation (Herbert et al., 2014a), whereas arsenic reduced the acetylation of H4K16 in human embryonic kidney (HEK293t) and human cervical carcinoma (HeLa) cells (Liu et al., 2015). At low doses, arsenic induced the expression of SIRT1 and p53 proteins in NHEK cells. The same group previously determined that *miR-34a* was involved in the upregulation of SIRT1 (Herbert et al., 2014b), and therefore investigated if *miR-34a* was involved in arsenic-induced SIRT1 upregulation. Arsenic induced p53 occupancy at both the *SIRT1* and *miR-34a* promoters. H4K16 acetylation was mixed at the *SIRT1* promoter and enriched at the *miR-34a* promoter, possibly due to SIRT1 downregulation.

As previously noted, arsenic reduces global H4K16 acetylation in HEK293t and HeLa cell lines (Liu et al., 2015); the histone acetyltransferase hMOF plays a key role in arsenic-induced epigenetic alterations. In an in vitro activity assay, the catalytic activity of hMOF was blocked by arsenic, and arsenic was shown to bind hMOF directly, as revealed by matrix-assisted laser desorption/ionization time-of-flight mass spectrometry (MALDI-TOF-MS) (Liu et al., 2015). Additionally, overexpression of hMOF attenuated arsenic-induced cytotoxicity and global H4K16 deacetylation (Liu et al., 2015). Further contributing to the loss of histone acetylation, direct exposure to arsenic upregulated HDAC 4 was reversed by overexpressing hMOF. Occupancy of hMOF upstream of the HDAC 4 TSS was blocked by arsenic exposure, although HDAC 4 inhibition did not substantially affect global H4K16 acetylation (Liu et al., 2015). These findings represent a distinct mechanism for arsenic regulation of H4K16 acetylation.

APPLICATIONS OF CHROMATIN BIOLOGY TO TOXICOLOGICAL STUDIES

Chromatin modifications can be meiotically and/or mitotically heritable, persistent, or transient. This inherent plasticity makes the epigenome highly responsive to environmental cues, both physiological (such as puberty) and environmental (such as chemical exposures). As a result, persistent and heritable chromatin modifications can serve as sensitive indicators of past environmental exposures, with the potential to act as molecular "signatures" or "fingerprints" for toxicant exposures. DNA- and RNA-based biomarkers of exposure have proved to be robust and reproducible (Waring et al., 2004; Hamadeh et al., 2002; Tennant, 2002). The National Institute of Environmental Health Sciences (NIEHS) created the National Center for Toxicogenomics (NCT) in 2000 to coordinate efforts in developing a "worldwide reference system of genome-wide gene expression data and to develop a knowledge base of chemical effects in biological systems" (Tennant, 2002). Since those initial efforts led by the NCT to link gene expression signatures with toxicant exposures, gene expression signatures have been extensively used to develop toxicant exposure identifiers (Lettieri, 2006). However, the use of chromatin modifications in signatures for toxicant exposure, whether in response to an acute exposure or due to the heritable nature of epigenetic alterations, as a signature for exposures that occurred distantly in time, is just beginning.

Importantly, chromatin modifications can both direct and respond to changes in gene expression. Toxicant exposure can lead to chromatin entering a "poised" state and thus becoming more sensitive to transcriptional activation via transcription factors. Data are emerging from many studies that chemical (and other) exposures can induce changes in chromatin modifications, which persist across the life course but which may remain silent (i.e., cause no change in gene expression) until triggered by later-life exposures such as high-fat diet or hormones (endogenous such as puberty or pharmacology), resulting in increased expression of the reprogrammed genes (Li et al., 1997; Wang et al., 2016; Polanco et al., 2010; Hilakivi-Clarke et al., 1997, 2004; Walker and Ho, 2012). Therefore, while changes in gene expression can be reliably used to detect response to toxicant exposure, chromatin modifications may be an even more sensitive biomarker for such exposures, as they may precede both the development of an adverse phenotype and any observable changes in gene expression (Fig. 1).

Analysis of chromatin modifications and detection of toxicant-induced epigenetic alterations has become more sensitive and cost-effective over time. Chromatin-based analyses for epigenetic marks can be performed in virtually any tissue and cell type, often even at the single-cell level, which opens the possibility for their use in easily accessed tissues such as blood and skin, for biomarker studies in both animal models and humans.

The NIEHS is again leading the field by supporting an effort to determine how environmental exposures induce epigenomic alterations in surrogate tissues, such as blood, correlate with changes that occur in tissues that are the targets for adverse health outcomes associated with those exposures. This effort is supported by the Toxicant Exposure and Responses by Genomic and Epigenomic Regulators of Transcription (TaRGET) II consortium (Wang et al., 2018). TaRGET II is a multi-institutional collaboration working to identify epigenetic signatures across various organs and tissues produced in response to a wide variety of environmental exposures and developing robust methods for epigenomic analyses of surrogate tissues (blood or skin) that reflect the impact of exposures on specific target tissues (e.g., liver and brain). This important undertaking will enable the identification of chromatin-based biomarkers of exposure that can be investigated with a simple blood draw or skin biopsy and that may be able to inform us about exposure-related risk of disease prior to onset of any adverse effects or even toxicant-induced changes in gene expression.

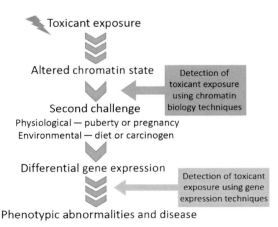

FIG. 1 The use of chromatin biology to assess exposure status enables earlier detection of exposure-related health outcomes compared with the use of gene expression techniques.

1. HISTONE MODIFICATIONS AND CHROMATIN STRUCTURE

TABLE 2 Summary of Epigenetic Consequences of Toxicant Exposures

	Xenoestrogens	Arsenic	Phthalates	Organotins	Dioxin	Other Metals
PRC2/1	In mammary tissue and breast cancer cell lines, xenoestrogens induce EZH2 expression (Stossi et al., 2014), while in uterine tissue, DES induces EZH2 phosphorylation and inactivation leading to reduced H3K27me3 (Jirtle and Skinner, 2007)	Arsenic increases BMI1 and H3K27me2/3 in BALB/c 3T3 cells that was associated with a reduction in tumor suppressors and tumor development in allograft assays (Anderson et al., 2012)	ND	ND	ND	ND
COMPASS Complex	Induce recruitment of MML2 to the EZH2 and MLL3 to the HOTAIR promoters in MCF7 cells (Stossi et al., 2014; Bhan et al., 2014a) In rat prostate, BPA induces genome-wide reprogramming of H2K4me3 and subsequent increased severity of neoplasia (Anderson et al., 2017)	Developmental exposure to arsenic induces H3K4me3 at the Slc25a30 gene in murine hepatocellular carcinoma (Bredfeldt et al., 2010)	ND	ND	ND	ND
G9a	ND	H3K9me2/3 was lost after As exposure, and G9a was also reduced in mouse Leydig cells (Howe and Gamble, 2016), while G9a and H3K9me2/3 were increased in mouse lung tissue (Kim et al., 2012)	DEHP and BBP lead to a reduction in G9a in rodent models and a decrease in H3K9me1/2 (Greathouse et al., 2008; Shakoor et al., 2017)	ND	ND	ND
Histone acetylation	ND	As has opposite effects on H4K16 acetylation in NHEK and HEK293t cells (Ashraf et al., 2017; Jadhav et al., 2011) and blocks the acetyltransferase activity of hMOF while upregulating HDAC 4 contributing to global H4K16 deacetylation (Jadhav et al., 2011)	DEHP increases the occupancy of HDAC 2 at the GLUT4 promoter in mouse muscle tissue (Lee and Workman, 2007)	TBT and TPT increase the activity of HATs (Brownell et al., 1996)	Dioxin induced the acetylation of H3K9, H3K14, and H4 and increasing the recruitment of HATs to the Cyp1a1 and Cyp1b1 enhancers in mouse hepatoma cells (Kanayama et al., 2005)	Lead increases global H3 acetylation in vascular and cardiac tissues (Grun and Blumberg, 2006), and chromium HDAC 1 releases and prevents p300 binding at the Cyp1a1 promoter in hematoma cells (Janesick and Blumberg, 2012)

ND, not discussed.

SUMMARY

The epigenetic landscape can be reconstructed in response to physiological processes, such as development and aging, and in response to environmental factors, such as toxicant exposures. Epigenetic reprogramming in response to toxicant exposures, both direct and developmental, can lead to immediate physiological outcomes or instead poise the system to respond to exposures later in life, such as hormonal cues or high-fat diet. This enables chromatin modifications to serve as biomarkers for toxicant exposure or disease risk. This chapter has presented alterations in the chromatin landscape induced by specific environmental toxicants (Table 2). Continued research is warranted to further dissect the mechanisms by which these alterations occur, which can aid in the eventual development of therapeutics for those found to be at increased disease risk due to epigenomic reprogramming.

Funding

National Institute of Environmental Health Sciences R01ES023206 and U01ES026719.

Disclosure Statement

The authors have nothing to disclose.

References

Airaksinen, R., et al., 2010. Organotin intake through fish consumption in Finland. Environ. Res. 110 (6), 544–547.

Alamdar, A., et al., 2017. Arsenic activates the expression of 3beta-HSD in mouse Leydig cells through repression of histone H3K9 methylation. Toxicol. Appl. Pharmacol. 326, 7–14.

Allfrey, V.G., Faulkner, R., Mirsky, A.E., 1964. Acetylation and methylation of histones and their possible role in the regulation of RNA synthesis. Proc. Natl. Acad. Sci. USA 51, 786–794.

Anderson, O.S., et al., 2012. Epigenetic responses following maternal dietary exposure to physiologically relevant levels of bisphenol a. Environ. Mol. Mutagen. 53 (5), 334–342.

Anderson, O.S., et al., 2017. Novel epigenetic biomarkers mediating bisphenol a exposure and metabolic phenotypes in female mice. Endocrinology 158 (1), 31–40.

Antizar-Ladislao, B., 2008. Environmental levels, toxicity and human exposure to tributyltin (TBT)-contaminated marine environment. A review. Environ. Int. 34 (2), 292–308.

Ashraf, M.W., Salam, A., Mian, A., 2017. Levels of organotin compounds in selected fish species from the Arabian gulf. Bull. Environ. Contam. Toxicol. 98 (6), 811–816.

Babu, S., et al., 2015. Unusually high levels of bisphenol A (BPA) in thermal paper cash register receipts (CRs): development and application of a robust LC-UV method to quantify BPA in CRs. Toxicol. Mech. Methods 25 (5), 410–416.

Bachmann, J., et al., 2006. Pancreatic resection for pancreatic cancer. HPB (Oxford) 8 (5), 346–351.

Barker, D.J., 2007. The origins of the developmental origins theory. J. Intern. Med. 261 (5), 412–417.

Barker, D.J., et al., 1989a. Growth in utero, blood pressure in childhood and adult life, and mortality from cardiovascular disease. BMJ 298 (6673), 564–567.

Barker, D.J., et al., 1989b. Weight in infancy and death from ischaemic heart disease. Lancet 2 (8663), 577–580.

Beedanagari, S.R., et al., 2010. Role of epigenetic mechanisms in differential regulation of the dioxin-inducible human CYP1A1 and CYP1B1 genes. Mol. Pharmacol. 78 (4), 608–616.

Bhan, A., et al., 2014a. Histone methyltransferase EZH2 is transcriptionally induced by estradiol as well as estrogenic endocrine disruptors bisphenol-A and diethylstilbestrol. J. Mol. Biol. 426 (20), 3426–3441.

Bhan, A., et al., 2014b. Bisphenol-A and diethylstilbestrol exposure induces the expression of breast cancer associated long noncoding RNA HOTAIR in vitro and in vivo. J. Steroid Biochem. Mol. Biol. 141, 160–170.

Biedermann, S., Tschudin, P., Grob, K., 2010. Transfer of bisphenol A from thermal printer paper to the skin. Anal. Bioanal. Chem. 398 (1), 571–576.

Bredfeldt, T.G., et al., 2010. Xenoestrogen-induced regulation of EZH2 and histone methylation via estrogen receptor signaling to PI3K/AKT. Mol. Endocrinol. 24 (5), 993–1006.

Brownell, J.E., Allis, C.D., 1996. Special HATs for special occasions: linking histone acetylation to chromatin assembly and gene activation. Curr. Opin. Genet. Dev. 6 (2), 176–184.

Brownell, J.E., et al., 1996. Tetrahymena histone acetyltransferase A: a homolog to yeast Gcn5p linking histone acetylation to gene activation. Cell 84 (6), 843–851.

Calafat, A.M., et al., 2008. Exposure of the U.S. population to bisphenol A and 4-tertiary-octylphenol: 2003-2004. Environ. Health Perspect. 116 (1), 39–44.

Casciello, F., et al., 2015. Functional role of G9a histone methyltransferase in cancer. Front. Immunol. 6, 487.

Centers for Disease Control and Prevention, 2009. Fourth Report on Human Exposure to Environmental Chemicals. U.S. Department of Health and Human Services, Centers for Disease Control and Prevention.

Cha, T.L., et al., 2005. Akt-mediated phosphorylation of EZH2 suppresses methylation of lysine 27 in histone H3. Science 310 (5746), 306–310.

Chou, K., Wright, R.O., 2006. Phthalates in food and medical devices. J. Med. Toxicol. 2 (3), 126–135.

Collett, K., et al., 2006. Expression of enhancer of zeste homologue 2 is significantly associated with increased tumor cell proliferation and is a marker of aggressive breast cancer. Clin. Cancer Res. 12 (4), 1168–1174.

Croonquist, P.A., Van Ness, B., 2005. The polycomb group protein enhancer of zeste homolog 2 (EZH 2) is an oncogene that influences myeloma cell growth and the mutant ras phenotype. Oncogene 24 (41), 6269–6280.

Drake, A.J., et al., 2015. In utero exposure to cigarette chemicals induces sex-specific disruption of one-carbon metabolism and DNA methylation in the human fetal liver. BMC Med. 13, 18.

Durando, M., et al., 2007. Prenatal bisphenol A exposure induces preneoplastic lesions in the mammary gland in Wistar rats. Environ. Health Perspect. 115 (1), 80–86.

Filipkowska, A., et al., 2016. Organotins in fish muscle and liver from the Polish coast of the Baltic Sea: is the total ban successful? Mar. Pollut. Bull. 111 (1–2), 493–499.

Foulds, C.E., et al., 2017. Endocrine-disrupting chemicals and fatty liver disease. Nat. Rev. Endocrinol. 13 (8), 445–457.

Francis, N.J., Kingston, R.E., Woodcock, C.L., 2004. Chromatin compaction by a polycomb group protein complex. Science 306 (5701), 1574–1577.

Fromme, H., et al., 2005. Occurrence of organotin compounds in house dust in Berlin (Germany). Chemosphere 58 (10), 1377–1383.

Gao, H., et al., 2015. Bisphenol A and hormone-associated cancers: current progress and perspectives. Medicine (Baltimore). 94(1):e211.

Godfrey, K.M., Costello, P.M., Lillycrop, K.A., 2016. Development, epigenetics and metabolic programming. Nestle Nutr. Inst. Workshop Ser. 85, 71–80.

Goodman, M., Lakind, J.S., Mattison, D.R., 2014. Do phthalates act as obesogens in humans? A systematic review of the epidemiological literature. Crit. Rev. Toxicol. 44 (2), 151–175.

Gore, A.C., et al., 2015a. EDC-2: the Endocrine Society's second scientific statement on endocrine-disrupting chemicals. Endocr. Rev. 36 (6), E1–E150.

Gore, A.C., et al., 2015b. Executive summary to EDC-2: The Endocrine Society's second scientific statement on endocrine-disrupting chemicals. Endocr. Rev. 36 (6), 593–602.

Greathouse, K.L., et al., 2008. Identification of uterine leiomyoma genes developmentally reprogrammed by neonatal exposure to diethylstilbestrol. Reprod. Sci. 15 (8), 765–778.

Greathouse, K.L., et al., 2012. Environmental estrogens differentially engage the histone methyltransferase EZH2 to increase risk of uterine tumorigenesis. Mol. Cancer Res. 10 (4), 546–557.

Greer, E.L., et al., 2010. Members of the H3K4 trimethylation complex regulate lifespan in a germline-dependent manner in C. elegans. Nature 466 (7304), 383–387.

Grun, F., Blumberg, B., 2006. Environmental obesogens: organotins and endocrine disruption via nuclear receptor signaling. Endocrinology 147 (6 Suppl), S50–S55.

Grun, F., et al., 2006. Endocrine-disrupting organotin compounds are potent inducers of adipogenesis in vertebrates. Mol. Endocrinol. 20 (9), 2141–2155.

1. HISTONE MODIFICATIONS AND CHROMATIN STRUCTURE

Hamadeh, H.K., et al., 2002. Methapyrilene toxicity: anchorage of pathologic observations to gene expression alterations. Toxicol. Pathol. 30 (4), 470–482.

Hamilton, B., et al., 2005. A systematic RNAi screen for longevity genes in C. elegans. Genes Dev. 19 (13), 1544–1555.

Hehn, R.S., 2016. NHANES data support link between handling of thermal paper receipts and increased urinary bisphenol A excretion. Environ. Sci. Technol. 50 (1), 397–404.

Heindel, J.J., et al., 2017. Metabolism disrupting chemicals and metabolic disorders. Reprod. Toxicol. 68, 3–33.

Herbert, K.J., et al., 2014a. Arsenic exposure disrupts epigenetic regulation of SIRT1 in human keratinocytes. Toxicol. Appl. Pharmacol. 281 (1), 136–145.

Herbert, K.J., Cook, A.L., Snow, E.T., 2014b. SIRT1 inhibition restores apoptotic sensitivity in p53-mutated human keratinocytes. Toxicol. Appl. Pharmacol. 277 (3), 288–297.

Herz, H.M., Garruss, A., Shilatifard, A., 2013. SET for life: biochemical activities and biological functions of SET domain-containing proteins. Trends Biochem. Sci. 38 (12), 621–639.

Hilakivi-Clarke, L., de Assis, S., 2006. Fetal origins of breast cancer. Trends Endocrinol. Metab. 17 (9), 340–348.

Hilakivi-Clarke, L., et al., 1997. A maternal diet high in n-6 polyunsaturated fats alters mammary gland development, puberty onset, and breast cancer risk among female rat offspring. Proc. Natl. Acad. Sci. USA 94 (17), 9372–9377.

Hilakivi-Clarke, L., et al., 2004. In utero alcohol exposure increases mammary tumorigenesis in rats. Br. J. Cancer 90 (11), 2225–2231.

Ho, S.M., et al., 2017. Environmental factors, epigenetics, and developmental origin of reproductive disorders. Reprod. Toxicol. 68, 85–104.

Hope, E., Reed, D.R., Moilanen, L.H., 2016. Potential confounders of bisphenol-A analysis in dental materials. Dent. Mater. 32 (8), 961–967.

Horiguchi, T., 2006. Masculinization of female gastropod mollusks induced by organotin compounds, focusing on mechanism of actions of tributyltin and triphenyltin for development of imposex. Environ. Sci. 13 (2), 77–87.

Horiguchi, T., et al., 2006. Impact of tributyltin and triphenyltin on ivory shell (Babylonia japonica) populations. Environ. Health Perspect. 114 (Suppl. 1), 13–19.

Howdeshell, K.L., et al., 1999. Exposure to bisphenol A advances puberty. Nature 401 (6755), 763–764.

Howe, C.G., Gamble, M.V., 2016. Influence of arsenic on global levels of histone posttranslational modifications: a review of the literature and challenges in the field. Curr. Environ. Health Rep. 3 (3), 225–237.

Jadhav, S., Bhosale, D., Bhosle, N., 2011. Baseline of organotin pollution in fishes, clams, shrimps, squids and crabs collected from the west coast of India. Mar. Pollut. Bull. 62 (10), 2213–2219.

Janesick, A., Blumberg, B., 2012. Obesogens, stem cells and the developmental programming of obesity. Int. J. Androl. 35 (3), 437–448.

Jin, C., et al., 2011. Histone demethylase UTX-1 regulates C. elegans life span by targeting the insulin/IGF-1 signaling pathway. Cell Metab. 14 (2), 161–172.

Jirtle, R.L., Skinner, M.K., 2007. Environmental epigenomics and disease susceptibility. Nat. Rev. Genet. 8 (4), 253–262.

Kanayama, T., et al., 2005. Organotin compounds promote adipocyte differentiation as agonists of the peroxisome proliferator-activated receptor gamma/retinoid X receptor pathway. Mol. Pharmacol. 67 (3), 766–774.

Kang, J.H., Kondo, F., Katayama, Y., 2006. Human exposure to bisphenol A. Toxicology 226 (2–3), 79–89.

Kanherkar, R.R., Bhatia-Dey, N., Csoka, A.B., 2014. Epigenetics across the human lifespan. Front. Cell Dev. Biol. 2, 49.

Kannan, K., et al., 1999. Butyltin compounds in river otters (Lutra canadensis) from the northwestern United States. Arch. Environ. Contam. Toxicol. 36 (4), 462–468.

Kannan, K., et al., 2010. Organotin compounds, including butyltins and octyltins, in house dust from Albany, New York, USA. Arch. Environ. Contam. Toxicol. 58 (4), 901–907.

Kanno, Y., et al., 2007. Effects of endocrine disrupting substance on estrogen receptor gene transcription in dialysis patients. Ther. Apher. Dial. 11 (4), 262–265.

Kim, H.G., et al., 2012. Polycomb (PcG) proteins, BMI1 and SUZ12, regulate arsenic-induced cell transformation. J. Biol. Chem. 287 (38), 31920–31928.

Kim, E.J., et al., 2014a. Association between urinary levels of bisphenol-A and estrogen metabolism in Korean adults. Sci. Total Environ. 470–471, 1401–1407.

Kim, J.H., et al., 2014b. Perinatal bisphenol A exposure promotes dose-dependent alterations of the mouse methylome. BMC Genomics 15, 30.

Kleer, C.G., et al., 2003. EZH2 is a marker of aggressive breast cancer and promotes neoplastic transformation of breast epithelial cells. Proc. Natl. Acad. Sci. USA 100 (20), 11606–11611.

Kucuksezgin, F., et al., 2011. Assessment of organotin (butyltin species) contamination in marine biota from the eastern Aegean Sea, Turkey. Mar. Pollut. Bull. 62 (9), 1984–1988.

Kurita, H., et al., 2014. The Ah receptor recruits IKKalpha to its target binding motifs to phosphorylate serine-10 in histone H3 required for transcriptional activation. Toxicol. Sci. 139 (1), 121–132.

Lee, K.K., Workman, J.L., 2007. Histone acetyltransferase complexes: one size doesn't fit all. Nat. Rev. Mol. Cell Biol. 8 (4), 284–295.

Lee, S.S., et al., 2003. DAF-16 target genes that control C. elegans life-span and metabolism. Science 300 (5619), 644–647.

Lee, S.G., et al., 2013. Bisphenol A exposure during adulthood causes augmentation of follicular atresia and luteal regression by decreasing 17beta-estradiol synthesis via downregulation of aromatase in rat ovary. Environ. Health Perspect. 121 (6), 663–669.

Lettieri, T., 2006. Recent applications of DNA microarray technology to toxicology and ecotoxicology. Environ. Health Perspect. 114 (1), 4–9.

Li, S., et al., 1997. Developmental exposure to diethylstilbestrol elicits demethylation of estrogen-responsive lactoferrin gene in mouse uterus. Cancer Res. 57 (19), 4356–4359.

Li, L., et al., 2010. Essential functions of the histone demethylase lid. PLoS Genet. 6 (11), e1001221.

Liu, L., et al., 2013. Chromatin modifications as determinants of muscle stem cell quiescence and chronological aging. Cell Rep. 4 (1), 189–204.

Liu, D., et al., 2015. Arsenic trioxide reduces global histone H4 acetylation at lysine 16 through direct binding to histone acetyltransferase hMOF in human cells. PLoS One 10 (10): e0141014.

Liu, C., et al., 2016. Pubertal exposure to di-(2-ethylhexyl)-phthalate inhibits G9a-mediated histone methylation during spermatogenesis in mice. Arch. Toxicol. 90 (4), 955–969.

Ma, Y., et al., 2013. Hepatic DNA methylation modifications in early development of rats resulting from perinatal BPA exposure contribute to insulin resistance in adulthood. Diabetologia 56 (9), 2059–2067.

Mantha, M., et al., 2017. Estimating inorganic arsenic exposure from U.S. rice and total water intakes. Environ. Health Perspect. 125 (5), 057005.

Marinkovic, N., et al., 2010. Dioxins and human toxicity. Arh. Hig. Rada. Toksikol. 61 (4), 445–453.

Martino-Andrade, A.J., Chahoud, I., 2010. Reproductive toxicity of phthalate esters. Mol. Nutr. Food Res. 54 (1), 148–157.

McAllister, B.G., Kime, D.E., 2003. Early life exposure to environmental levels of the aromatase inhibitor tributyltin causes masculinisation and irreversible sperm damage in zebrafish (Danio rerio). Aquat. Toxicol. 65 (3), 309–316.

McColl, G., et al., 2008. Pharmacogenetic analysis of lithium-induced delayed aging in Caenorhabditis elegans. J. Biol. Chem. 283 (1), 350–357.

Minatel, B.C., et al., 2018. Environmental arsenic exposure: from genetic susceptibility to pathogenesis. Environ. Int. 112, 183–197.

Mochizuki, K., et al., 2017. Relationship between epigenetic regulation, dietary habits, and the developmental origins of health and disease theory. Congenit. Anom. (Kyoto) 57 (6), 184–190.

Mozzetta, C., Pontis, J., Ait-Si-Ali, S., 2015. Functional crosstalk between lysine methyltransferases on histone substrates: the case of G9A/GLP and polycomb repressive complex 2. Antioxid. Redox Signal. 22 (16), 1365–1381.

Ni, Z., et al., 2012. Two SET domain containing genes link epigenetic changes and aging in Caenorhabditis elegans. Aging Cell 11 (2), 315–325.

Noer, A., Lindeman, L.C., Collas, P., 2009. Histone H3 modifications associated with differentiation and long-term culture of mesenchymal adipose stem cells. Stem Cells Dev. 18 (5), 725–736.

Nohara, K., et al., 2012. Late-onset increases in oxidative stress and other tumorigenic activities and tumors with a Ha-ras mutation in the liver of adult male C3H mice gestationally exposed to arsenic. Toxicol. Sci. 129 (2), 293–304.

Osada, S., et al., 2005. Some organotin compounds enhance histone acetyltransferase activity. Toxicol. Lett. 155 (2), 329–335.

Patrick, L., 2006. Lead toxicity, a review of the literature. Part 1: exposure, evaluation, and treatment. Altern. Med. Rev. 11 (1), 2–22.

Pereira-Fernandes, A., et al., 2013a. Evaluation of a screening system for obesogenic compounds: screening of endocrine disrupting compounds and evaluation of the PPAR dependency of the effect. PLoS One 8 (10): e77481.

Pereira-Fernandes, A., et al., 2013b. Unraveling the mode of action of an obesogen: mechanistic analysis of the model obesogen tributyltin in the 3T3-L1 cell line. Mol. Cell. Endocrinol. 370 (1–2), 52–64.

Peretz, J., et al., 2011. Bisphenol A impairs follicle growth, inhibits steroidogenesis, and downregulates rate-limiting enzymes in the estradiol biosynthesis pathway. Toxicol. Sci. 119 (1), 209–217.

Polanco, T.A., et al., 2010. Fetal alcohol exposure increases mammary tumor susceptibility and alters tumor phenotype in rats. Alcohol. Clin. Exp. Res. 34 (11), 1879–1887.

Pu, M., et al., 2015. Trimethylation of Lys36 on H3 restricts gene expression change during aging and impacts life span. Genes Dev. 29 (7), 718–731.

Rajesh, P., Balasubramanian, K., 2014. Phthalate exposure in utero causes epigenetic changes and impairs insulin signalling. J. Endocrinol. 223 (1), 47–66.

Rastkari, N., et al., 2012. Butyltin compounds in fish commonly sold in north of Iran. Bull. Environ. Contam. Toxicol. 88 (1), 74–77.

Roh, T.Y., et al., 2007. Genome-wide prediction of conserved and nonconserved enhancers by histone acetylation patterns. Genome Res. 17 (1), 74–81.

Rossman, T.G., 2003. Mechanism of arsenic carcinogenesis: an integrated approach. Mutat. Res. 533 (1–2), 37–65.

Rubin, B.S., 2011. Bisphenol A: an endocrine disruptor with widespread exposure and multiple effects. J. Steroid Biochem. Mol. Biol. 127 (1–2), 27–34.

Schettler, T., 2006. Human exposure to phthalates via consumer products. Int. J. Androl. 29 (1), 134–139 (discussion 181-5).

Schonfelder, G., et al., 2002. In utero exposure to low doses of bisphenol A lead to long-term deleterious effects in the vagina. Neoplasia 4 (2), 98–102.

Schulz, L.C., 2010. The Dutch hunger winter and the developmental origins of health and disease. Proc. Natl. Acad. Sci. USA 107 (39), 16757–16758.

Seachrist, D.D., et al., 2016. A review of the carcinogenic potential of bisphenol A. Reprod. Toxicol. 59, 167–182.

Sen, P., et al., 2015. H3K36 methylation promotes longevity by enhancing transcriptional fidelity. Genes Dev. 29 (13), 1362–1376.

Seto, E., Yoshida, M., 2014. Erasers of histone acetylation: the histone deacetylase enzymes. Cold Spring Harb. Perspect. Biol. 6 (4): a018713.

Shakoor, M.B., et al., 2017. Human health implications, risk assessment and remediation of as-contaminated water: a critical review. Sci. Total Environ. 601–602, 756–769.

Shelby, M.D., 2006. NTP-CERHR monograph on the potential human reproductive and developmental effects of di (2-ethylhexyl) phthalate (DEHP). NTP CERHR MON. (18). v, vii-7, II-iii-xiii passim.

Shimasaki, Y., et al., 2003. Tributyltin causes masculinization in fish. Environ. Toxicol. Chem. 22 (1), 141–144.

Siebold, A.P., et al., 2010. Polycomb repressive complex 2 and trithorax modulate Drosophila longevity and stress resistance. Proc. Natl. Acad. Sci. USA 107 (1), 169–174.

Singh, S., Li, S.S., 2012. Epigenetic effects of environmental chemicals bisphenol A and phthalates. Int. J. Mol. Sci. 13 (8), 10143–10153.

Smeester, L., Fry, R.C., 2018. Long-term health effects and underlying biological mechanisms of developmental exposure to arsenic. Curr. Environ. Health Rep. 5 (1), 134–144.

Sonkar, R., Powell, C.A., Choudhury, M., 2016. Benzyl butyl phthalate induces epigenetic stress to enhance adipogenesis in mesenchymal stem cells. Mol. Cell. Endocrinol. 431, 109–122.

Soto, A.M., et al., 2013. Does cancer start in the womb? Altered mammary gland development and predisposition to breast cancer due to in utero exposure to endocrine disruptors. J. Mammary Gland Biol. Neoplasia 18 (2), 199–208.

Sparmann, A., van Lohuizen, M., 2006. Polycomb silencers control cell fate, development and cancer. Nat. Rev. Cancer 6 (11), 846–856.

Stossi, F., et al., 2014. Defining estrogenic mechanisms of bisphenol A analogs through high throughput microscopy-based contextual assays. Chem. Biol. 21 (6), 743–753.

Swan, S.H., et al., 2005. Decrease in anogenital distance among male infants with prenatal phthalate exposure. Environ. Health Perspect. 113 (8), 1056–1061.

Szafran, A.T., et al., 2017. Characterizing properties of non-estrogenic substituted bisphenol analogs using high throughput microscopy and image analysis. PLoS One 12 (7), e0180141.

Sze, C.C., Shilatifard, A., 2016. MLL3/MLL4/COMPASS family on epigenetic regulation of enhancer function and cancer. Cold Spring Harb. Perspect. Med. 6 (11), 1–15

Tachibana, M., et al., 2007. Functional dynamics of H3K9 methylation during meiotic prophase progression. EMBO J. 26 (14), 3346–3359.

1. HISTONE MODIFICATIONS AND CHROMATIN STRUCTURE

Tennant, R.W., 2002. The National Center for Toxicogenomics: using new technologies to inform mechanistic toxicology. Environ. Health Perspect. 110 (1), A8–10.

Travis, C.C., Hattemer-Frey, H.A., 1991. Human exposure to dioxin. Sci. Total Environ. 104 (1–2), 97–127.

van Esterik, J.C., et al., 2015. Liver DNA methylation analysis in adult female C57BL/6JxFVB mice following perinatal exposure to bisphenol A. Toxicol. Lett. 232 (1), 293–300.

Vandenberg, L.N., et al., 2007. Human exposure to bisphenol A (BPA). Reprod. Toxicol. 24 (2), 139–177.

Varambally, S., et al., 2002. The polycomb group protein EZH2 is involved in progression of prostate cancer. Nature 419 (6907), 624–629.

Volle, D.H., et al., 2009. The orphan nuclear receptor small heterodimer partner mediates male infertility induced by diethylstilbestrol in mice. J. Clin. Invest. 119 (12), 3752–3764.

Walker, C.L., Ho, S.M., 2012. Developmental reprogramming of cancer susceptibility. Nat. Rev. Cancer 12 (7), 479–486.

Wang, Q., et al., 2016. Reprogramming of the epigenome by MLL1 links early-life environmental exposures to prostate cancer risk. Mol. Endocrinol. 30 (8), 856–871.

Wang, N., et al., 2017a. The famine exposure in early life and metabolic syndrome in adulthood. Clin. Nutr. 36 (1), 253–259.

Wang, Z., et al., 2017b. Fetal and infant exposure to severe Chinese famine increases the risk of adult dyslipidemia: Results from the China health and retirement longitudinal study. BMC Public Health 17 (1), 488.

Wang, Z., et al., 2017c. Association between exposure to the Chinese famine during infancy and the risk of self-reported chronic lung diseases in adulthood: a cross-sectional study. BMJ Open 7 (5), e015476.

Wang, Z., Liu, H., Liu, S., 2017d. Low-dose bisphenol a exposure: a seemingly instigating carcinogenic effect on breast cancer. Adv. Sci. (Weinh.) 4 (2), 1600248.

Wang, T., et al., 2018. The NIEHS TaRGET II consortium and environmental epigenomics. Nat. Biotechnol. 36 (3), 225–227.

Waring, J.F., et al., 2004. Interlaboratory evaluation of rat hepatic gene expression changes induced by methapyrilene. Environ. Health Perspect. 112 (4), 439–448.

Wei, Y.D., et al., 2004. Chromium inhibits transcription from polycyclic aromatic hydrocarbon-inducible promoters by blocking the release of histone deacetylase and preventing the binding of p300 to chromatin. J. Biol. Chem. 279 (6), 4110–4119.

Weidman, J.R., et al., 2007. Cancer susceptibility: epigenetic manifestation of environmental exposures. Cancer J. 13 (1), 9–16.

Weinhouse, C., et al., 2015. Stat3 is a candidate epigenetic biomarker of perinatal bisphenol A exposure associated with murine hepatic tumors with implications for human health. Epigenetics 10 (12), 1099–1110.

Welstead, G.G., et al., 2012. X-linked H3K27me3 demethylase Utx is required for embryonic development in a sex-specific manner. Proc. Natl. Acad. Sci. USA 109 (32), 13004–13009.

Wong, R.L., et al., 2015. Identification of secretaglobin Scgb2a1 as a target for developmental reprogramming by BPA in the rat prostate. Epigenetics 10 (2), 127–134.

Woo, Y.H., Li, W.H., 2012. Evolutionary conservation of histone modifications in mammals. Mol. Biol. Evol. 29 (7), 1757–1767.

Xiao, S., et al., 2012. Comparative epigenomic annotation of regulatory DNA. Cell 149 (6), 1381–1392.

Xu, L., Ruh, T.S., Ruh, M.F., 1997. Effect of the histone deacetylase inhibitor trichostatin A on the responsiveness of rat hepatocytes to dioxin. Biochem. Pharmacol. 53 (7), 951–957.

Xu, L.H., et al., 2015. Lead induces apoptosis and histone hyperacetylation in rat cardiovascular tissues. PLoS One 10 (6): e0129091.

Ye, X., et al., 2015. Urinary concentrations of bisphenol A and three other bisphenols in convenience samples of U.S. adults during 2000-2014. Environ. Sci. Technol. 49 (19), 11834–11839.

Zhou, X., et al., 2008. Arsenite alters global histone H3 methylation. Carcinogenesis 29 (9), 1831–1836.

Zhu, J., et al., 2013. Genome-wide chromatin state transitions associated with developmental and environmental cues. Cell 152 (3), 642–654.

Ziv-Gal, A., et al., 2013. Bisphenol A inhibits cultured mouse ovarian follicle growth partially via the aryl hydrocarbon receptor signaling pathway. Reprod. Toxicol. 42, 58–67.

1. HISTONE MODIFICATIONS AND CHROMATIN STRUCTURE

DNA METHYLATION

C H A P T E R

2-1

The Role of DNA Methylation in Gene Regulation

Paige A. Bommarito*, Rebecca C. Fry*,†

*Department of Environmental Sciences and Engineering, Gillings School of Global Public Health, University of North Carolina at Chapel Hill, Chapel Hill, NC, United States
†Curriculum in Toxicology, School of Medicine, University of North Carolina at Chapel Hill, Chapel Hill, NC, United States

O U T L I N E

CHAPTER OVERVIEW

DNA methylation represents the best characterized form of epigenetic modification. DNA methylation is written onto the genome by two methyltransferases, known as Dnmt3a and Dnmt3b. These enzymes catalyze the addition of a methyl group to the fifth carbon of the cytosine DNA base. Once DNA methylation is established, it must be maintained when the cell divides. The maintenance of DNA methylation is carried out by a separate enzyme, Dnmt1. Together, these enzymes ensure that DNA methylation patterns are established and carried on into subsequent cellular generations. In this way, DNA methylation is a mechanism of cellular memory and it carries important information about the programming of gene expression along with it.

Patterns of DNA methylation are established early during development, with two critical waves of methylation and demethylation occurring during embryogenesis. This process is tightly regulated, and inhibition of the machinery responsible for establishing these patterns results in abnormal development and, eventually, death of the organism. Important examples of the role of DNA methylation in gene expression during early development include X-chromosome inactivation, genomic imprinting, and the repression of transposable elements.

DNA methylation has the potential to alter gene expression through both direct and indirect mechanisms. When DNA methylation occurs within gene promoters, it can impede the transcriptional machinery from accessing DNA and initiating transcription. However, not all transcription factors are sensitive to DNA methylation occurring within their binding sites. Therefore, it cannot be the only mechanism by which DNA methylation regulates gene expression. Alternatively, methyl-CpG-binding proteins have been implicated in the DNA-methylation-mediated regulation of gene expression. These proteins bind to methylated DNA and mediate interactions between DNA methylation and histone modifications, producing a repressive chromatin structure. Thus, DNA methylation can also regulate gene expression by altering chromatin structure. This intimate link between DNA methylation and other epigenetic modifications ensures the stable maintenance of methylation-dependent changes in gene expression across cell replications.

The methylome has been characterized in a wide variety of human diseases, including cancers, metabolic disorders, and cardiovascular diseases. As the understanding of the writing and erasure of DNA methylation is improved, clinical applications of such technologies in disease identification and treatment are becoming a reality. DNA methylation has been used as a biomarker in clinical settings to identify imprinting diseases, such as Prader-Willi and Angelman syndromes. Moreover, drugs targeting DNA methylation have also been approved by the Food and Drug Administration (FDA) or are undergoing clinical trials for the treatment of cancers, where inhibition of methylation may lead to a reactivation of tumor-suppressing genes.

This chapter highlights the mechanisms of establishing and maintaining DNA methylation, with a discussion of critical developmental processes in mammals that require DNA methylation. It further focuses on the mechanisms by which DNA methylation can repress or activate gene expression, with a focus on how DNA methylation interacts with chromatin structure. Moreover, this chapter considers the mechanisms for DNA demethylation, which plays an important role in embryonic development. Lastly, with understanding of the mechanisms by which DNA is methylated and demethylated, future clinical applications of DNA methylation in understanding and treating disease are considered.

INTRODUCTION

DNA methylation refers to the process by which a methyl group is covalently added to the C5 position of cytosine, creating the modified base 5-methylcytosine (Fig. 1). Methylation occurs primarily at locations of the genome that have a cytosine proximal to a guanine base, otherwise known as CpG dinucleotides. 5-Methylcytosine was first discovered in 1925 after having been isolated from *Mycobacterium tuberculosis* and again, in 1948, when it was identified in DNA extracted from a calf thymus (Johnson and Coghill, 1925; Hotchkiss, 1948). At that time, the function of this methylated cytosine remained unknown. In 1975, two independent laboratories proposed that DNA methylation was an important mechanism in the regulation of gene expression and cellular memory (Holliday and Pugh, 1975; Riggs, 1975).

These researchers recognized that the CpG dinucleotide is a palindrome, with both strands of DNA reading the same sequence in the 5′ to 3′ direction. Therefore, they proposed that methylation of CpG dinucleotides could be maintained after cell division. More specifically, a methylated strand of DNA would become two hemimethylated strands following mitosis. Each daughter cell could then use the methylated template strand of DNA as a reference to copy methylation marks onto the newly synthesized strand (Holliday and Pugh, 1975; Riggs, 1975).

This theory required that cells have "methyltransferases" responsible for creating methylation marks (de novo methylation) and "methyltransferases" responsible for maintaining methylation (maintenance methylation). They concluded that a semiconservative method of replicating DNA methylation would also provide a mechanism for cells to carry and preserve information related to gene expression. Moreover, they proposed that DNA methylation in promoter regions of DNA could disrupt DNA-binding proteins involved in transcription and affect downstream gene expression (Holliday and Pugh, 1975; Riggs, 1975). In particular, they recognized the potential for DNA methylation to, in part, control the tight regulation of gene expression during embryogenesis (Holliday and Pugh, 1975; Riggs, 1975). Since this time, their hypotheses have been supported by a growing body of evidence that demonstrates the critical role of DNA methylation in embryonic development, the regulation of gene expression, and as a mechanism of cellular memory.

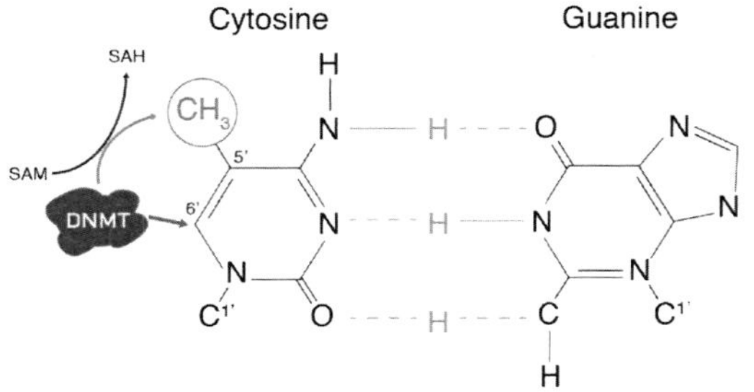

FIG. 1 DNA methyltransferases (DNMTs) are responsible for catalyzing the addition of a methyl group to the C5 position of cytosine. Specifically, DNMTs perform a nucleophilic attack on C6 and, using the cofactor S-adenosyl-L-methionine (SAM) as a methyl donor, covalently add a methyl group to the C5 position via a second nucleophilic attack. The methyl group does not impact hydrogen bonding of cytosine to guanine.

2. DNA METHYLATION

MECHANISMS OF DNA METHYLATION

As suggested by Holliday and Pugh, DNA methylation is carried out by a set of enzymes, known as methyltransferases, which control de novo and maintenance methylation (Holliday and Pugh, 1975). The enzymes that control these processes in mammals and their mechanisms are discussed in the following section.

De Novo DNA Methylation

De novo methylation refers to the methylation of previously unmethylated DNA (Fig. 2). Early evidence for de novo methylation comes from experiments where mouse embryos were infected with a virus, known as the Moloney murine leukemia virus (M-MuLV). After infection, this virus was integrated into the genome of the mouse embryo. Using methylation-sensitive enzymes, which are unable to cleave methylated DNA, it was shown that proviral DNA became methylated after incorporation into the mouse genome (Jahner et al., 1982).

In mammals, de novo methylation is carried out by two DNA methyltransferases (DNMT) of the DNMT3 family, known as DNMT3A and DNMT3B. The discovery of the *DNMT3A* and *DNMT3B* genes followed a targeted search for sequences in mammalian DNA with close similarity to known sequences of DNMTs in bacteria (Okano et al., 1998). Okano et al. (1998)

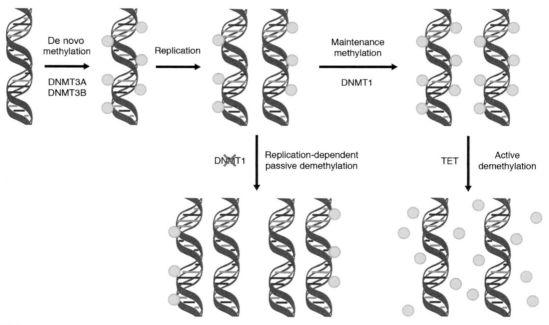

FIG. 2 DNA methylation is written onto the genome by DNMT3A and DNMT3B. However, after semiconservative replication, DNA methylation patterns must be maintained. This maintenance occurs primarily through the actions of DNMT1. Without DNMT1, the cell experiences replication-dependent, passive demethylation. DNA can also undergo active demethylation, catalyzed by ten-eleven translocation (TET) proteins.

found that these genes were highly expressed in embryonic stem (ES) cells, but lowly expressed in differentiated tissues. Given that methylation patterns are first established during embryogenesis, embryogenesis is a period of intense de novo methylation. This suggested that the proteins encoded by *DNMT3A* and *DNMT3B* were likely responsible for de novo methylation. Further evidence for the role of DNMT3A and DNMT3B in de novo methylation comes from studies using murine ES cells lacking either functional *Dnmt3a*, *Dnmt3b*, or both. Cells with either nonfunctional *Dnmt3a* or *Dnmt3b* retained the ability to methylate proviral DNA, while cells lacking both *Dnmt3a* and *Dnmt3b* did not (Okano et al., 1999). De novo methylation is also impaired in model organisms with *Dnmt3* knockouts, resulting in death of the organism (Okano et al., 1999). In humans, a disorder known as immunodeficiency, centromeric instability, and facial anomalies (ICF) syndrome results from the loss of function mutations in just one copy of *DNMT3B*(Table 1) (Ueda et al., 2006).

One additional member of the DNMT3 family has been characterized: DNMT 3-like protein (DNMT3L). On its own, DNMT3L does not have methyltransferase activity. While DNMT3L shares some conserved domains with DNMT3A and DNMT3B, it features a truncated catalytic domain and only weakly binds DNA (Hata et al., 2002; Kareta et al., 2006; Aapola et al., 2000). Instead, DNMT3L stimulates the activity of DNMT3A and DNMT3B, increasing their affinity for the cofactor and universal methyl donor, S-adenosyl-L-methionine (SAM) (Suetake et al., 2004).

Maintenance DNA Methylation

Early on, researchers realized that, if DNA methylation created stable changes in gene expression, a "maintenance methyltransferase" would be required in order to fully methylate hemimethylated DNA generated during mitosis (Holliday and Pugh, 1975; Riggs, 1975). Indeed, this "maintenance methyltransferase" is now known as DNMT1 (Fig. 2).

The scientific evidence for the activity of DNMT1 included studies of crude cellular extracts with DNMT (Sneider et al., 1975). These results were confirmed by others and eventually, using recombinant human protein purified from insect cells, researchers identified and confirmed DNMT1's preference for hemimethylated DNA (Bestor and Ingram, 1983; Pradhan et al., 1999; Gruenbaum et al., 1982; Kalousek and Morris, 1968). Interestingly, studies using purified recombinant DNMT1 demonstrated that the enzyme has a 15-fold preference for hemimethylated DNA compared with unmethylated DNA (Pradhan et al., 1999). Thus, DNMT1 is preferentially responsible for maintenance, rather than de novo, methylation. Moreover, *Dnmt1*$^{-/-}$ murine ES cells had significant decreases in total DNA methylation content, with an increase in the proportion of hemimethylated DNA (Liang et al., 2002). Taken together, this evidence suggests that DNMT1 is responsible for maintaining DNA methylation patterns. The role of this enzyme is essential for normal development, with *Dnmt1*$^{-/-}$ mice dying during embryogenesis (Table 1) (Rougier et al., 1998).

It should be noted that DNMT1 is not the only enzyme responsible for the maintenance of DNA methylation. In fact, DNMT3A and DNMT3B also contribute to maintenance methylation. Evidence for this comes from murine ES cells, where the inactivation of *Dnmt3a* and *Dnmt3b* resulted in the gradual loss of methylation that cannot be rescued by overexpression of *Dnmt1*. Instead, normal methylation levels could only be restored by

TABLE 1   Summary of Mammalian DNMTs and the Loss of Function Phenotypes

Gene	Function	Species	Loss of Function Phenotype	Nature of the Loss of Function
Dnmt1	Maintenance DNA methylation	Mouse	Embryonic lethality prior to embryonic day 10.5. Embryos have reduced mass and stunted development and lack a normal yolk sac and visible vasculature Embryos and ES cells have ~70% reduction in methylcytosine content and a significant increase in hemimethylated DNA strands	Homozygous deletions (Li et al., 1992)
Dnmt3a	De novo DNA methylation	Mouse	Mice appear normal at birth but have stunted postnatal growth and die at ~4 weeks of age	Homozygous deletions (Okano et al., 1999)
Dnmt3b	De novo DNA methylation	Mouse	Embryonic lethality between embryonic days 13.5 and 16.5. Recovered embryos featured a variety of developmental anomalies, including liver abnormalities, hemorrhaging in the head, failure of the ventricular septum to close, and necrosis	Homozygous deletions (Okano et al., 1999)
		Human	ICF syndrome—reduction of immunoglobulin production and T cells; demethylation of the satellites of chromosomes 1, 9, and 16; facial abnormalities; stunted growth; and mental disabilities	Heterozygous missense substitution, heterozygous insertion (Okano et al., 1999)
Dnmt3L	Stimulates de novo DNA methylation	Mouse	Embryos are viable but sterile. Males have hypogonadism, and females produce offspring that die by embryonic day 9.5 with defects of the both neural tube and extraembryonic tissues Oocytes of $Dnmt3L^{-/-}$ females feature demethylation at maternally imprinted sites Male germ cells have significant expression of transposable elements and meiotic failure	Homozygous deletions (Bourc'his et al., 2001; Hata et al., 2002; Arima et al., 2006)

introducing functional *Dnmt3a* or *Dnmt3b* back into the cells (Chen et al., 2003; Liang et al., 2002). Thus, there do not appear to be strict divisions of labor between DNMT1 and DNMT3 (Chédin, 2011).

DNA METHYLATION PATTERNS AND THE DEVELOPING ORGANISM

DNA methylation during embryogenesis is a critical developmental process, as evidenced by the embryonic lethality of DNMT1 and DNMT3 knockouts (Li et al., 1992; Okano et al., 1999). DNA methylation and demethylation occur in two distinct waves during development (Fig. 3). Prior to fertilization, the oocyte is relatively undermethylated compared to the

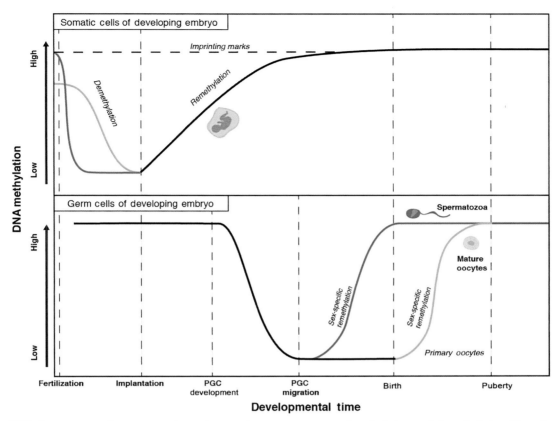

FIG. 3 During embryogenesis, the embryo undergoes two waves of methylation and demethylation. First, the paternal genome is rapidly demethylated following fertilization, while the maternal genome undergoes more gradual, passive demethylation when cell division starts. The DNA in the somatic cells of the embryo is remethylated during development in a tissue-specific manner. The second wave of demethylation occurs in the primordial germ cells (PGCs). Mature PGCs undergo demethylation as they migrate to the genital ridge in order to erase the imprinting marks from the parental genomes. DNA methylation is then reestablished to produce germ cells that contain the appropriate sex-specific methylation patterns for future generations.

spermatocyte (Monk et al., 1987). However, in the first genome-wide demethylation event occurring immediately after fertilization, the paternal genome is rapidly demethylated prior to the first cell division. On the other hand, the maternal genome undergoes more gradual, replication-dependent demethylation during the first few cell divisions (Rougier et al., 1998). Notably, imprinted loci within the somatic portions of the embryo escape this wave of demethylation. DNA methylation patterns are reestablished after implantation and remain stable after differentiation into somatic tissues (Monk et al., 1987). During this remethylation, DNA is not differentially methylated depending on the parent of origin. The second wave of demethylation and methylation occurs within primordial germ cells (PGCs). Cells fated to become PGCs initially contain the same epigenetic marks as other cells in the embryo. However, as these PGCs mature and migrate to the genital ridge, these epigenetic marks are erased

(Monk et al., 1987; Hajkova et al., 2002). The remainder of this section will discuss important patterns of DNA methylation that are established during and maintained after development. This includes X-chromosome inactivation, genetic imprinting, and methylation of transposable elements.

X-Chromosome Inactivation

In his initial hypothesis, Riggs recognized the potential role that DNA methylation could play in X-chromosome inactivation, a process in female mammals where one of the two X chromosomes in a nucleus is silenced during development (Riggs, 1975). X-chromosome inactivation, which is a mechanism of dosage compensation, prevents the difference in copy number of X-linked genes from leading to developmental differences between the sexes (Lyon, 1961, 1962). There are two types of X-chromosome inactivation: imprinted and random. After fertilization, the paternal X chromosome is inactivated in a process known as imprinted X-chromosome inactivation (Okamoto et al., 2004). The paternal X chromosome is later reactivated in the inner cell mass of the blastocyst and the embryonic tissues to undergo random X-chromosome inactivation (Okamoto et al., 2004). During random X-chromosome inactivation, a noncoding RNA contained within the X-inactivation center, Xist, is transcribed from the future inactive X chromosome. Xist progressively coats the inactive X chromosome, recruits repressive histone modifications and leads to changes in DNA methylation. At the end of this process, the inactive X chromosome, known as the Barr body, can be seen as a condensed body of DNA in the nucleus (Gendrel and Heard, 2014). It is worth noting that there are important species differences in X-chromosome inactivation, and while it is best understood in mice, the process differs from that of humans and other mammals (Migeon, 2016). For example, in humans, extraembryonic tissues do not appear to have imprinted X-chromosome inactivation (Okamoto et al., 2011). Instead, the human placenta appears mosaic and has random inactivation of the X chromosome (Moreira de Mello et al., 2010).

DNA methylation is necessary for X-chromosome inactivation. However, the understanding of the role of DNA methylation in X-chromosome inactivation is limited. In the inactive X chromosome, there is hypermethylation of CpG islands within promoter regions of the chromosome (Weber et al., 2005). This hypermethylation corresponds to the transcriptional repression of genes on the inactive X chromosome. Early research demonstrating that DNA methylation is necessary for X-chromosome inactivation used mouse-human hybrid somatic cell clones. These clones were created using mouse fibroblasts deficient in the X-linked gene, hypoxanthine-guanine phosphoribosyltransferase (*HPRT*), and human fibroblasts in order to generate *HPRT*-deficient cells with an inactive human X chromosome (Mohandas et al., 1981). These clones were then treated with an inhibitor of DNA methylation, 5-azacytidine. In cells treated with 5-azacytidine, the human-derived HPRT was detected 1000 times more frequently than in untreated cells, indicating a reactivation of the inactive human X chromosome (Mohandas et al., 1981). Thus, inhibiting methylation lead to a reactivation of the previously inactive X chromosome. Further evidence for the necessity of DNA methylation in X-chromosome inactivation comes from studies utilizing Dnmt1 null mice. In such mice, random X-chromosome inactivation begins normally. However, the inactive

X chromosome reactivates and gene expression from the chromosome resumes at a later stage of development (Sado et al., 2000). That X-chromosome inactivation begins normally may be attributed to abundant levels of maternal Dnmt1 protein in the oocyte, which becomes depleted as the cell divisions progress (Sado et al., 2000). While the precise role that DNA methylation plays in X-chromosome inactivation remains a mystery, it is clear that the process is required for stably maintained repression of genes on the inactive X chromosome.

Imprinted Genes and Development

In the developing female embryo, the X chromosome is inactivated at random. However, there are other loci within the genome where genes are expressed depending on the parent of origin. This is known as genomic imprinting. DNA methylation was first posited to play a role in genomic imprinting because it was clear that the underlying mechanism had several requirements that DNA methylation could satisfy. First, it needed to be linked to the pronucleus (Surani et al., 1984; McGrath and Solter, 1984a). Second, the imprint needed to be persistent and maintained through subsequent cell divisions (Surani et al., 1986). Third, the imprint had to affect gene expression (McGrath and Solter, 1984b). Lastly, the imprint needed to be able to change from one sex to the other in order to produce successive generations (Sapienza et al., 1987). Evidence supporting the role of DNA methylation in imprinting comes from experiments utilizing cells from *Dnmt1*-deficient mice, which have aberrant expression of imprinted genes, such as *H19* and insulin-like growth factor 2 (*Igf2*) (Li et al., 1993). The disrupted transcriptional control of these imprinted loci corresponds disruptions in patterns of DNA methylation at the promoter regions of these genes, providing some of the first evidence that DNA methylation is responsible for imprinting (Li et al., 1993).

Interestingly, DNMT3L plays a special role in establishing genomic imprinting. $Dnmt3L^{-/-}$ mice develop normally and survive into adulthood. However, female $Dnmt3L^{-/-}$ mice do not produce embryos with methylation at maternally imprinted sites and their offspring typically die during embryogenesis (Arima et al., 2006; Bourc'his et al., 2001). Similarly, *male $Dnmt3L^{-/-}$* mice are infertile, with incomplete spermatogenesis and hypomethylated spermatocyte DNA (Bourc'his and Bestor, 2004; Hata et al., 2002). Looking more specifically, at offspring with a maternal *Dnmt3L* deficiency, genes at imprinted loci are aberrantly expressed. Such embryos feature gross anatomical abnormalities, such as neural tube defects and abnormalities of the extraembryonic tissues (Bourc'his et al., 2001; Hata et al., 2002). In humans, the loss of imprinting and the subsequent overexpression of imprinted genes later in life have been linked to a variety of diseases, including fetal growth restriction; diabetes; and cancers of the breast, lung, and colon, among others (Robertson, 2005; Smith et al., 2006). Thus, the importance of establishing and maintaining appropriate imprinting via DNA methylation remains critical throughout the life span.

Transposable Elements

Transposable elements are a class of DNA sequences that can range anywhere from hundreds to thousands of nucleotides in length. These elements make up over half of the genome and are known as "jumping genes" because they have the ability to move around in the genome (Smit, 1996; Lander et al., 2001). Because of this unique capability, transposons can create insertional mutations and alter the size of the genome. Therefore, transposons contribute to genomic instability (Yoder et al., 1997). Insertional mutations caused by transposon expression have been noted in diseases such as hemophilia, cancers, Duchenne muscular dystrophy, and X-linked severe combined immunodeficiency syndrome (Chen et al., 2005).

Hypermethylation of transposons represses their transcription and prevents their movement around the genome (Hata and Sakaki, 1997; Kochanek et al., 1993). Transposable elements are highly methylated compared with the rest of the genome. In fact, the majority of methylated cytosines can be found within transposons (Kochanek et al., 1993). Loss of *Dnmt3L* results in the significant demethylation of the promoter regions of transposons, such as the long interspersed nuclear element 1 (LINE1) retrotransposon. Expression of LINE1 has also been detected in *Dnmt3L* null mice (Bourc'his and Bestor, 2004). Methylation of LINE1 is significantly reduced in multiple types of human cancers, suggesting that transposon methylation is related to cancer incidence (Barchitta et al., 2014). Thus, DNA methylation is required for the repression of transposable elements, both during development and into adulthood. Disruptions in this transcriptional silencing may also be related to disease.

DNA METHYLATION AS A REGULATOR OF GENE EXPRESSION

The original hypotheses put forth by Riggs, Holliday, and Pugh suggested that DNA methylation could be responsible for the tight regulation of gene expression during development. Based on experimental evidence that DNA methylation affects the expression of (1) imprinted genes, (2) genes on the inactive X chromosome, and (3) transposable elements, DNA methylation does, indeed, play an important role in regulating gene expression. DNA methylation does so by both directly and indirectly affecting the accessibility of DNA to the gene expression machinery. Importantly, while DNA methylation is typically thought of as a repressive epigenetic mark, there is an abundance of evidence that suggests DNA methylation, particularly within the gene body, can serve as an activating mechanism (Suzuki and Bird, 2008).

DNA Methylation in Gene Promoters Affects Transcription Factor Binding and Vice Versa

DNA is methylated primarily at CpG dinucleotides, which are underrepresented in the genome (Antequera and Bird, 1993). However, there are GC-enriched regions, known as CpG islands, which are scattered throughout the genome. In humans, approximately 70% of gene promoters have CpG islands (Saxonov et al., 2006). One theory of how DNA methylation regulates gene expression is that CpG methylation blocks transcription factors from binding to gene promoters (Fig. 4A). Therefore, DNA methylation prevents the initiation of gene expression. This theory provides an explanation for the inverse relationship between gene expression and DNA methylation. In vitro studies have demonstrated that even methylation at a single CpG dinucleotide within a transcription factor binding site can inhibit transcription factor binding (Kovesdi et al., 1987; Watt and Molloy, 1988). To demonstrate this, Watt and Molloy (1988) synthesized oligonucleotides of the adenovirus major late promoter

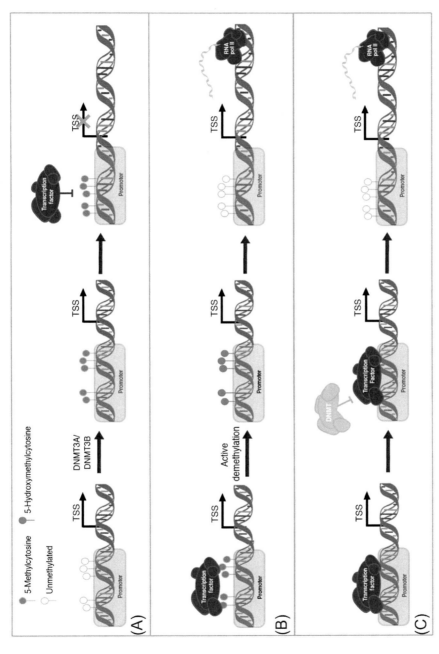

FIG. 4 DNA methylation is hypothesized, in part, to repress gene expression by blocking transcription factors from accessing DNA promoter regions (Part A). However, the actions of transcription factors are also responsible for shaping DNA methylation patterns. This is possible through active (Part B) and static mechanisms (Part C). *TSS*, transcription start site; *RNA pol II*, RNA polymerase II.

2. DNA METHYLATION

(AdMLP), a promoter bound by upstream transcription factor 1 (USF1). Some of these synthesized oligonucleotides featured a single methylated CpG dinucleotide, while others were free of methylation. Using an in vitro assay to separate unbound DNA from the DNA-USF1 complex, they found that CpG methylation within the transcription factor binding site resulted in a 15-fold reduction in transcription factor binding of AdMLP (Watt and Molloy, 1988). Moreover, methylation-induced reductions in transcription factor binding led to a reduction in gene expression in vivo (Kovesdi et al., 1987). Taken together, DNA methylation regulates gene expression, at least in part, by disrupting the access of the DNA to transcription factors. It is important to note, however, that this cannot be the only method for DNA methylation to control gene expression because some transcription factors bind DNA regardless of its methylation status. Additionally, CpG methylation within the promoter, but outside of a transcription factor binding site, can also affect gene expression (Tate and Bird, 1993). Thus, interfering with transcription factor binding is one of several ways that DNA methylation can regulate gene expression.

While it is recognized that DNA methylation patterns impact transcription factor binding, there is also evidence for the reciprocal relationship. That is, transcription factor binding may also influence DNA methylation patterns. Transcription factor binding is necessary to create areas of low methylation in the genome in neuronal stem cells (Stadler et al., 2011). At bound loci, transcription factors can block DNMTs from accessing the DNA (Fig. 4C). However, it has also been shown that transcription factor binding is linked to active demethylation of CpG islands within regulatory regions of the DNA (Fig. 4B) (Feldmann et al., 2013). More evidence for the role of transcription factor binding as a determinant of DNA methylation patterns comes from studies of environmental exposures on the human methylome. Using aggregated data from human populations with environmental exposure to heavy metals, it has been noted that differentially methylated genes within these populations are enriched for similar transcription factor binding sites. This suggests that contaminant exposures may induce activities of particular response-related transcription factors. This binding activity may result in remodeling of the human methylome at a number of sites throughout the human genome (Martin and Fry, 2016). In essence, DNA methylation patterns influence gene expression, through blocking transcription factor binding, and transcription factor binding influences DNA methylation patterns.

DNA Methylation Recruits Methyl-CpG Binding Proteins and Remodels Chromatin

A second mechanism by which DNA methylation negatively regulates gene expression is through the recruitment of methyl-CpG-binding proteins. The understanding of this comes, in part, from experiments in cellular extracts demonstrating that methylation-dependent inhibition of a specific gene's expression could be overcome by adding more of that methylated DNA sequence or by adding an alternative, "competitor" methylated construct (Boyes and Bird, 1991). These results suggested that a protein may mediate the methylation-dependent repression of gene expression. The first of these, a protein complex known as methyl-CpG-binding protein 1 (MeCP1), was observed to mediate this methylation-dependent repression by directing histone deacetylation to areas of DNA methylation (Boyes and Bird, 1991). A second methyl-CpG-binding protein, MeCP2, was later identified (Fig. 5A). MeCP2

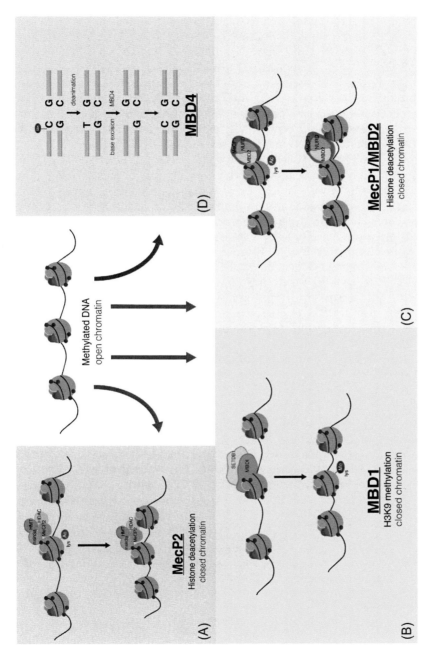

FIG. 5 Methyl-CpG-binding proteins are responsible for mediating a link between DNA methylation and chromatin structure, as demonstrated by the "core" MBD proteins. MeCP2 (Part A) and MeCP1 (Part C) are responsible for histone deacetylation of methylated sites, while MBD1 complexes with proteins methylate H3K9 (Part B). Altogether, these proteins ensure stable transcriptional repression by creating a closed chromatin state. Interestingly, MBD4 plays a role in DNA repair of mutations occurring at methylated CpG sites (Part D).

preferentially binds methylated CpG dinucleotides and interacts with and recruits a corepressor complex containing mSin3A and histone deacetylase (Nan et al., 1997, 1998). Thus, MeCP2 creates colocalization of CpG methylation and deacetylated histones, which have known repressive functions in gene expression. Following the discovery and characterization of MeCP2, four additional proteins were discovered that contained the same methyl-CpG-binding domain (MBD). These proteins—MBD1, MBD2, MBD3, and MBD4—have been further characterized for their role in methylation-dependent repression of gene transcription (Hendrich and Bird, 1998; Cross et al., 1997). Importantly, these proteins play a critical role in mediating interactions between DNA methylation, histone modifications, and chromatin structure that affect gene expression. While other methyl-CpG-binding proteins have since been discovered, these "core" MBDs remain the best characterized (Du et al., 2015).

Methyl-CpG-binding protein 1 (MBD1) creates an important link between DNA methylation and histone methylation. Specifically, MBD1 forms a complex with the SET domain bifurcated 1 (SETDB1), a methyl-CpG-binding protein that methylates histones. Together, the MBD1-SETDB1 complex directs histone methylation to areas of methylated DNA. Specifically, SETDB1 methylates H3K9, producing a repressive histone mark characteristic of heterochromatin (Fig. 5B) (Schultz et al., 2002). Additionally, during DNA replication, the MBD1-SETDB1 complex interacts with the chromatin assembly factor-1 (CAF-1). This ensures that newly assembled, methylated DNA also has H3K9 methylation. HeLa cells without functional MBD1 or SETDB1 lack stable transcriptional repression at methylated promoters and, instead, have histone markers of transcriptional activation at methylated loci (Sarraf and Stancheva, 2004). Therefore, MBD1 is responsible for the methylation-dependent maintenance of heterochromatin across cell divisions and contributes to methylation-dependent transcriptional repression.

MBD2 is the DNA-binding protein found within the MeCP1 protein complex and mediates the link between DNA methylation and histone acetylation (Ng et al., 1999). MBD2 binds to DNA and recruits the nucleosome remodeling and histone deacetylase (NuRD) complex to DNA, which together form MeCP1. Interestingly, MBD3 can be interchanged for MBD2 in complex with NuRD (Le Guezennec et al., 2006). While MBD2-NuRD and MBD3-NuRD have some overlapping functions, these complexes appear to play separate roles within the cell. First, MBD2 preferentially binds methylated DNA, directing NuRD to regions of DNA methylation (Wade et al., 1999). MBD2-NuRD is localized to densely methylated regions of the genome, particularly promoter regions (Chatagnon et al., 2011; Gunther et al., 2013). NuRD then deacetylates the histones within these highly methylated regions and produces more compact chromatin, leading to transcriptional repression (Fig. 5C). While MBD3 was likely produced as a duplication of MBD2, it does not bind methylated DNA because it contains mutations within its MBD (Saito and Ishikawa, 2002). Instead, several studies have linked MBD3 to unmethylated or hydroxymethylated CpGs, although it is still not clear what the distribution of MBD3-NuRD is throughout the genome (Gunther et al., 2013). Although its functions are not well characterized, MBD3, not MBD2, has been shown to be embryonic lethal in mice (Hendrich et al., 2001).

In contrast, MBD4 appears to be more involved in DNA repair than in epigenetic gene regulation. While MBD4 contains the methyl-CpG-binding domain, it also bears sequence homology with known bacterial DNA repair enzymes (Hendrich and Bird, 1998). Thus, instead of serving as a link between DNA methylation and other epigenetic modifications,

MBD4 is a part of the base excision repair pathway, repairing mismatches that result from the deamination of 5-methylcytosine to thymine, which occur relatively frequently throughout the genome (Fig. 5D) (Duncan and Miller, 1980). In $Mbd4^{-/-}$ mice, there is an increased rate of CT mutation at CpG sites (Millar et al., 2002). Importantly, mutations of methylated CpG dinucleotides are known to give rise to human disease (Jones et al., 1992). Since the discovery of the MBD family, several other methyl-CpG-binding proteins have been discovered. Moreover, our understanding of the roles that these methyl-CpG-binding proteins play in DNA methylation, chromatin structure, and a variety of other cellular processes is expanding. Nevertheless, the important takeaway is that these proteins serve as a critical link between DNA methylation, histone modifications, and the overall chromatin structure (Du et al., 2015).

DNA Methylation in the Gene Body Impacts Transcription

Typically, DNA methylation is believed to be a marker of transcriptional repression. However, DNA methylation within the gene body appears to serve a different function than DNA methylation within the promoter region. While there is some evidence that intragenic DNA methylation is related to transcriptional repression (Lorincz et al., 2004), the bulk of evidence suggests that it is associated with gene activation.

For instance, consider X-chromosome inactivation. After X-chromosome inactivation is complete, the active and inactive X chromosomes have disparate amounts of DNA methylation. While the inactive X chromosome relies on DNA methylation to maintain its transcriptional repression, it actually has lower 5-methylcytosine content than the active X chromosome. This is because, while the inactive X chromosome is hypermethylated within promoters, the active X chromosome is densely methylated within its gene bodies (Hellman and Chess, 2007). Moreover, gene body methylation on the active X chromosome corresponds to an increase in gene expression. This observation is not unique to X-chromosome inactivation. In fact, a positive relationship between gene body methylation and gene expression has been noted elsewhere in the genome, in a range of cell and tissue types, and in both plants and animals (Jjingo et al., 2012; Rauch et al., 2009; Aran et al., 2011; Smith et al., 2006; Baubec et al., 2015). Thus, there is a role for DNA methylation in active gene transcription.

A few theories have been proposed to explain how intragenic methylation promotes gene expression. One suggests that intragenic methylation represses transcription of the antisense sequence, which may interfere with the expression of the sense gene (Tran et al., 2005; Suzuki and Bird, 2008). An alternative theory proposes that intragenic methylation increases transcriptional efficiency through the repression of cryptic promoters, which are normally silenced promoters found within genes (Fig. 6A) (Shenker and Flanagan, 2012). Looking specifically at cryptic promoters found within the SH3 and multiple ankyrin repeat domains 3 (SHANK3) locus, intragenic CpG methylation did not impair the expression of SHANK3 in human cortical samples. Instead, it reduced the expression of transcripts produced by the induction of gene transcription from cryptic promoters found within the gene (Fig. 6B) (Maunakea et al., 2010). Others have also demonstrated that $Dnmt3b^{-/-}$ murine ES cells have a significant amount of RNA produced from transcription start sites found within gene bodies at the locations of cryptic promoters (Suzuki and Bird, 2008). These studies suggest that intragenic DNA methylation activates gene expression by repressing transcription initiated

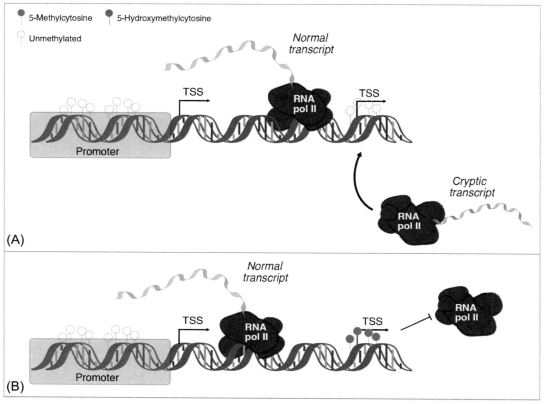

FIG. 6 Intragenic DNA methylation is associated with highly expressed genes. When unmethylated, RNA polymerase II may begin transcription at cryptic promoters. This activity may reduce the transcriptional efficiency of the gene product (Part A). Therefore, intragenic methylation at cryptic promoters may improve transcriptional efficiency (Part B). *TSS*, transcription start site; *RNA pol II*, RNA polymerase II.

at cryptic promoters. While the precise role that DNA methylation plays in active gene expression is still being uncovered, it is clear that it does not merely serve as a marker distinguishing active and inactive genes.

DNA DEMETHYLATION

During embryogenesis, cells in the developing organism undergo several waves of DNA methylation and demethylation. The patterns of demethylation exhibit the characteristics of both active and passive DNA demethylation. The first genome-wide demethylation event, occurring after fertilization, demonstrates the difference between active and passive DNA methylation (Fig. 3). In the maternal genome of the zygote, 5-methylcytosine content gradually declines in a replication-dependent manner (Mayer et al., 2000). This is known as passive demethylation. In contrast, the paternal genome undergoes rapid demethylation prior to DNA replication and cell division. This demethylation cannot be explained by passive demethylation and, instead, requires an active process (Mayer et al., 2000).

TET-Mediated Oxidation and Demethylation

Initially, the search for an enzyme responsible for active demethylation is centered on identifying a 5-methylcytosine demethylase (Ooi and Bestor, 2008). No such enzyme was discovered. Instead, researchers discovered the modified base 5-hydroxymethylcytosine in the Purkinje neurons of mice (Kriaucionis and Heintz, 2009). The relatively high frequency of 5-hydroxymethylcytosine in the genome suggested that it played a key role in gene expression in the brain. Based on this observation, researchers suggested that 5-hydroxymethylcytosine could play a role in demethylation and was either (1) an intermediate base produced via oxidative demethylation or (2) the end product of demethylation (Kriaucionis and Heintz, 2009).

With this new hypothesis centered on the oxidation of 5-methylcytosine, researchers refined their search for an enzyme that was instead capable of modifying methylated bases. One such enzyme was already known to oxidize the methyl group found on thymine in trypanosomes. Trypanosomes contain an additional modified base that is produced by enzymes that catalyze the hydroxylation and glycosylation of the methyl group on thymine. Using this information to search for genes with sequence homology, the ten-eleven translocation 1 (TET1) protein, which has been shown to oxidize 5-methylcytosine to 5-hydroxymethylcytosine in vitro, was discovered (Tahiliani et al., 2009). Subsequently, all three proteins in the TET family—TET1, TET2, and TET3—were shown to oxidize 5-methylcytosine to 5-hydroxymethylcytosine (Ito et al., 2010). Additionally, the TET enzymes can further oxidize 5-hydroxymethylcytosine to 5-formylcytosine and 5-carboxylcytosine (He et al., 2011). These intermediate bases are all part of the active demethylation pathway back to an unmodified cytosine (Fig. 7).

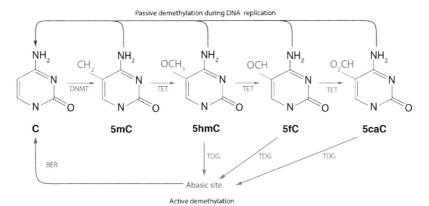

FIG. 7   DNA is actively demethylated via TET-mediated oxidation. 5-Methylcytosine is oxidized to 5-hydroxymethylcytosine, followed by 5-formylcytosine and 5-carboxylcytosine. Following oxidation, TDG excises the base, creating an abasic site that is repaired to an unmodified cytosine via base excision repair.

DNMT1 is sensitive to the oxidation of 5-methylcytosine. TET-mediated oxidation of 5-methylcytosine to 5-hydroxymethylcytosine reduces the activity of DNMT1 by over 90% and prevents the maintenance of DNA methylation (Valinluck and Sowers, 2007). Without the maintenance provided by DNMT1, TET-mediated oxidation may result in passive demethylation following DNA replication (Fig. 7). Additionally, the oxidation of 5-methylcytosine alters the activity of methyl-CpG-binding proteins, such as MeCP2 (Valinluck et al., 2004). Therefore, oxidation of 5-methylcytosine also disrupts methylation-dependent changes in chromatin structure, possibly reversing repressive heterochromatin markers.

In addition to passive demethylation following the TET-mediated oxidation, the enzyme thymine DNA glycosylase (TDG) can also excise these oxidized bases. TDG is a member of the larger uracil DNA glycosylase (UDG) family of DNA repair enzymes. Interestingly, TDG is the only member of the UDG family that has been shown to be embryonic lethal (Cortazar et al., 2011). This observation, in part, suggested that TDG is involved in DNA demethylation, a process known to be necessary for embryonic development. TDG has been shown to cleave 5-formylcytosine and 5-carboxylcytosine in vitro (He et al., 2011; Maiti and Drohat, 2011). Cleavage by TDG creates an abasic site, which is repaired with a cytosine via the base excision repair pathway. Thus, active demethylation proceeds via TET-mediated oxidation followed by TDG cleavage and base excision repair (Fig. 7) (He et al., 2011).

Notably, recent research suggests that, instead of being a transient intermediate, 5-hydroxymethylcytosine may be a stable epigenetic mark (Hahn et al., 2014; Bachman et al., 2014). 5-Hydroxymethylcytosine has been shown to be enriched in several tissues, including the brain, where it accumulates over development (Globisch et al., 2010; Hahn et al., 2014). In these tissues, 5-hydroxymethylcytosine appears stable and may play a role in regulating active gene expression and proliferation (Bachman et al., 2014). However, in other developmental contexts, 5-hydroxymethylcytosine still appears transient (Inoue and Zhang, 2011; Iqbal et al., 2011). Thus, it may serve different roles depending on the tissue and timing of development.

APPLICATIONS OF DNA METHYLATION IN UNDERSTANDING AND TREATING DISEASE

Despite DNA methylation being the most heavily studied epigenetic modification, there are still gaps in understanding precisely how DNA methylation interacts with other epigenetic modifications and affects gene expression. Nevertheless, research on DNA methylation has demonstrated that the epigenetic mark plays an important role in processes underlying disease development and may offer an explanation for how the environment can make lasting impacts on an organism. Moreover, while DNA methylation can be stable and carried on into subsequent cellular generations, it is also reversible. The prospect of manipulating DNA methylation patterns has ignited interest in potential clinical applications of such therapies in disease treatments.

Environmental Links to DNA Methylation

DNA methylation is a mechanism of cellular memory. That memory includes both past and present environmental exposures. DNA methylation allows the body to modify gene expression in response to environmental exposures. Environmental exposures may also manipulate the methylome in ways that can contribute to the development of disease. One famous example of the relationship between DNA methylation and environmental exposures originates from studies of the Dutch Hunger Winter, a severe famine in the Netherlands during the winter of 1944–45 (Roseboom et al., 2006; Schulz, 2010). Adults who were conceived during the Dutch Hunger Winter were exposed to malnutrition during the prenatal period. This prenatal malnutrition was associated with differential methylation of the imprinted IGF2 locus in adults, demonstrating that environmental exposures can lead to long-lasting epigenomic changes (Heijmans et al., 2008). Aside from malnutrition and other dietary exposures, a wide range of environmental factors have been linked to changes in the methylome, such as emotional distress, pesticides, and heavy metals (Provençal and Binder, 2015; Collotta et al., 2013; Arita and Costa, 2009). However, it is often unclear how such environmental exposures alter the epigenome and whether observed changes in DNA methylation lead to functional changes in gene expression.

Within environmental epigenetics, an important consideration is the possibility for transgenerational inheritance of altered methylation patterns. While environmental exposures can lead to changes in the exposed organism, it can also impact that organism's offspring through direct effects of exposure on the gametes or embryos. In addition, subsequent generations that did not directly experience the exposure may also be affected. Following exposure to vinclozolin in gestating female mice (F0 generation), changes in DNA methylation were observed in the great-grandchildren (F3 offspring) of the exposed mice. These changes in DNA methylation were also related to changes in gene expression (Skinner et al., 2013). It is also possible that such epigenetic differences may be related to transgenerational effects that vinclozolin has on the behavior and reproductive function noted in the offspring of exposed mice (Anway et al., 2005, 2006; Crews et al., 2007). Research into these effects may prove critical in understanding persistent health disparities in populations with historical exposures to stress and other environmental disturbances.

DNA Methylation as a Biomarker for Identifying and Understanding Disease

Epigenetic patterns, including DNA methylation, are critical for the understanding of both normal and disease states, particularly for complex diseases that cannot be explained by genetic factors alone. Methylation biomarkers have been examined extensively in human disease, and the increasing availability of technologies capable of profiling the epigenome has encouraged the search for differentially methylated loci that are predictive of human disease (Rakyan et al., 2011). One class of diseases for which methylation biomarkers have proved useful is genomic imprinting disorders, such as Prader-Willi syndrome or Angelman syndrome, which result in the loss of the expression of imprinted genes (Ramsden et al., 2010). For instance, Prader-Willi syndrome can arise from the silencing of the paternal alleles of a cluster of imprinted genes on the proximal long arm of human chromosome 15. As a disease that results from the disruption of genomic imprinting, it can be reliably diagnosed by methylation analysis of the imprinted gene *SNRPN*, which is found within the affected region of chromosome 15 (Ramsden et al., 2010). These biomarkers may allow for diseases related to methylation to be identified more reliably and faster than they otherwise could be.

The increasing popularity of epigenome-wide association studies has also resulted in the increased study of epigenetic modifications in relation to disease, particularly those that may include early-life reprogramming of gene expression, such as childhood asthma, depression, and metabolic disorders (Ji et al., 2016; Jakovcevski and Akbarian, 2012; Gluckman et al., 2009). Detailing the patterns of DNA methylation that are associated with disease states is critical for the identification and understanding of the processes that underlie disease. Such advances may be seen in cancer research, where an extensive comparison between normal and tumor tissues has revealed typical DNA methylation signatures underlying the disease. Generally speaking, it has been observed that cancerous tissues tend to have genome-wide demethylation and tumor-suppressor-specific hypermethylation (Jones et al., 2016; Das and Singal, 2004). Genome-wide demethylation, which has often been measured by assaying DNA methylation levels within transposable elements, is indicative of genomic instability and the expression of oncogenes. Hypermethylation of tumor suppressor genes, on the other hand, suggests that the ability to regulate cell survival and proliferation is compromised (Jones et al., 2016; Das and Singal, 2004). This research has enhanced the understanding of cancer progression, the search for prognostic and diagnostic biomarkers, and has informed the development of drugs that may counteract the epigenetic changes occurring in cancer.

DNA Methylation in Disease Therapy

Perhaps, one of the most exciting aspects of DNA methylation is its reversibility, which creates the possibility of modifying it in a clinical setting. Notably, some methylation-inhibiting drugs are already in use for the treatment of cancers, such as acute myelogenous leukemia, with the understanding that inhibiting methylation could lead to the reexpression of tumor-suppressing genes and activation of the patient's own immune response (Jones et al., 2016). This includes 5-azacytidine, which serves as a cytidine analog and is incorporated into DNA. Unlike cytidine, 5-azacytidine forms an irreversible bond with DNMT1 and causes the degradation of the enzyme and an overall inhibition of DNA methylation (Heerboth et al., 2014). Anticancer drugs targeting other components of the epigenome, including histone deacetylation and histone methylation, have also been approved by the FDA, and many others are going through clinical trials (Jones et al., 2016).

References

Aapola, U., Shibuya, K., Scott, H.S., Ollila, J., Vihinen, M., Heino, M., Shintani, A., Kawasaki, K., Minoshima, S., Krohn, K., Antonarakis, S.E., Shimizu, N., Kudoh, J., Peterson, P., 2000. Isolation and initial characterization of a novel zinc finger gene, DNMT3L, on 21q22.3, related to the Cytosine-5-methyltransferase 3 gene family. Genomics 65, 293–298.

Antequera, F., Bird, A., 1993. Number of CpG islands and genes in human and mouse. Proc. Natl. Acad. Sci. 90, 11995–11999.

Anway, M.D., Cupp, A.S., Uzumcu, M., Skinner, M.K., 2005. Epigenetic transgenerational actions of endocrine disruptors and male fertility. Science 308, 1466–1469.

Anway, M.D., Memon, M.A., Uzumcu, M., Skinner, M.K., 2006. Transgenerational effect of the endocrine disruptor vinclozolin on male spermatogenesis. J. Androl. 27, 868–879.

Aran, D., Toperoff, G., Rosenberg, M., Hellman, A., 2011. Replication timing-related and gene body-specific methylation of active human genes. Hum. Mol. Genet. 20, 670–680.

Arima, T., Hata, K., Tanaka, S., Kusumi, M., Li, E., Kato, K., Shiota, K., Sasaki, H., Wake, N., 2006. Loss of the maternal imprint in Dnmt3Lmat$^{-/-}$ mice leads to a differentiation defect in the extraembryonic tissue. Dev. Biol. 297, 361–373.

Arita, A., Costa, M., 2009. Epigenetics in metal carcinogenesis: nickel, arsenic, chromium and cadmium. Metallomics 1, 222–228.

Bachman, M., Uribe-Lewis, S., Yang, X., Williams, M., Murrell, A., Balasubramanian, S., 2014. 5-Hydroxymethylcytosine is a predominantly stable DNA modification. Nat. Chem. 6, 1049–1055.

Barchitta, M., Quattrocchi, A., Maugeri, A., Vinciguerra, M., Agodi, A., 2014. LINE-1 Hypomethylation in blood and tissue samples as an epigenetic marker for Cancer risk: a systematic review and meta-analysis. PLoS One. 9: e109478.

Baubec, T., Colombo, D.F., Wirbelauer, C., Schmidt, J., Burger, L., Krebs, A.R., Akalin, A., Schubeler, D., 2015. Genomic profiling of DNA methyltransferases reveals a role for DNMT3B in genic methylation. Nature 520, 243–247.

Bestor, T.H., Ingram, V.M., 1983. Two DNA methyltransferases from murine erythroleukemia cells: purification, sequence specificity, and mode of interaction with DNA. Proc. Natl. Acad. Sci. U. S. A. 80, 5559–5563.

Bourc'his, D., Bestor, T.H., 2004. Meiotic catastrophe and retrotransposon reactivation in male germ cells lacking Dnmt3L. Nature 431, 96–99.

Bourc'his, D., Xu, G.-L., Lin, C.-S., Bollman, B., Bestor, T.H., 2001. Dnmt3L and the establishment of maternal genomic imprints. Science 294, 2536–2539.

Boyes, J., Bird, A., 1991. DNA methylation inhibits transcription indirectly via a methyl-CpG binding protein. Cell 64, 1123–1134.

Chatagnon, A., Perriaud, L., Nazaret, N., Croze, S., Benhattar, J., Lachuer, J., Dante, R., 2011. Preferential binding of the methyl-CpG binding domain protein 2 at methylated transcriptional start site regions. Epigenetics 6, 1295–1307.

Chédin, F., 2011. The DNMT3 family of mammalian de novo DNA methyltransferases. Prog. Mol. Biol. Transl. Sci. 101, 255–285.

Chen, J.M., Stenson, P.D., Cooper, D.N., Ferec, C., 2005. A systematic analysis of LINE-1endonuclease-dependent retrotranspositional events causing human genetic disease. Hum. Genet. 117, 411–427.

Chen, T., Ueda, Y., Dodge, J.E., Wang, Z., Li, E., 2003. Establishment and maintenance of genomic methylation patterns in mouse embryonic stem cells by Dnmt3a and Dnmt3b. Mol. Cell. Biol. 23, 5594–5605.

Collotta, M., Bertazzi, P.A., Bollati, V., 2013. Epigenetics and pesticides. Toxicology 307, 35–41.

Cortazar, D., Kunz, C., Selfridge, J., Lettieri, T., Saito, Y., Macdougall, E., Wirz, A., Schuermann, D., Jacobs, A.L., Siegrist, F., Steinacher, R., Jiricny, J., Bird, A., Schar, P., 2011. Embryonic lethal phenotype reveals a function of TDG in maintaining epigenetic stability. Nature 470, 419–423.

Crews, D., Gore, A.C., Hsu, T.S., Dangleben, N.L., Spinetta, M., Schallert, T., Anway, M.D., Skinner, M.K., 2007. Transgenerational epigenetic imprints on mate preference. Proc. Natl. Acad. Sci. U. S. A. 104, 5942–5946.

Cross, S.H., Meehan, R.R., Nan, X., Bird, A., 1997. A component of the transcriptional repressor MeCP1 shares a motif with DNA methyltransferase and HRX proteins. Nat. Genet. 16, 256–259.

Das, P.M., Singal, R., 2004. DNA methylation and Cancer. J. Clin. Oncol. 22, 4632–4642.

Du, Q., Luu, P.-L., Stirzaker, C., Clark, S.J., 2015. Methyl-CpG-binding domain proteins: readers of the epigenome. Epigenomics 7, 1051–1073.

Duncan, B.K., Miller, J.H., 1980. Mutagenic deamination of cytosine residues in DNA. Nature 287, 560–561.

Feldmann, A., Ivanek, R., Murr, R., Gaidatzis, D., Burger, L., Schübeler, D., 2013. Transcription factor occupancy can mediate active turnover of DNA methylation at regulatory regions. PLoS Genet. 9e1003994.

Gendrel, A.-V., Heard, E., 2014. Noncoding RNAs and epigenetic mechanisms during X-chromosome inactivation. Annu. Rev. Cell Dev. Biol. 30, 561–580.

Globisch, D., Munzel, M., Muller, M., Michalakis, S., Wagner, M., Koch, S., Bruckl, T., Biel, M., Carell, T., 2010. Tissue distribution of 5-hydroxymethylcytosine and search for active demethylation intermediates. PLoS One 5:e15367.

Gluckman, P.D., Hanson, M.A., Buklijas, T., Low, F.M., Beedle, A.S., 2009. Epigenetic mechanisms that underpin metabolic and cardiovascular diseases. Nat. Rev. Endocrinol. 5, 401–408.

Gruenbaum, Y., Cedar, H., Razin, A., 1982. Substrate and sequence specificity of a eukaryotic DNA methylase. Nature 295, 620–622.

Gunther, K., Rust, M., Leers, J., Boettger, T., Scharfe, M., Jarek, M., Bartkuhn, M., Renkawitz, R., 2013. Differential roles for MBD2 and MBD3 at methylated CpG islands, active promoters and binding to exon sequences. Nucleic Acids Res. 41, 3010–3021.

Hahn, M.A., Szabo, P.E., Pfeifer, G.P., 2014. 5-Hydroxymethylcytosine: a stable or transient DNA modification? Genomics 104, 314–323.

Hajkova, P., Erhardt, S., Lane, N., Haaf, T., El-Maarri, O., Reik, W., Walter, J., Surani, M.A., 2002. Epigenetic reprogramming in mouse primordial germ cells. Mech. Dev. 117, 15–23.

Hata, K., Sakaki, Y., 1997. Identification of critical CpG sites for repression of L1 transcription by DNA methylation. Gene 189 (2), 227–234. https://dx.doi.org/10.1016/S0378-1119(96)00856-6.

Hata, K., Okano, M., Lei, H., Li, E., 2002. Dnmt3L cooperates with the Dnmt3 family of de novo DNA methyltransferases to establish maternal imprints in mice. Development 129, 1983–1993.

He, Y.-F., Li, B.-Z., Li, Z., Liu, P., Wang, Y., Tang, Q., Ding, J., Jia, Y., Chen, Z., Li, L., Sun, Y., Li, X., Dai, Q., Song, C.-X., Zhang, K., He, C., Xu, G.-L., 2011. Tet-mediated formation of 5-Carboxylcytosine and its excision by TDG in mammalian DNA. Science (New York, N.Y.) 333, 1303–1307.

Heerboth, S., Lapinska, K., Snyder, N., Leary, M., Rollinson, S., Sarkar, S., 2014. Use of epigenetic drugs in disease: an overview. Genet. Epigenet. 6, 9–19.

Heijmans, B.T., Tobi, E.W., Stein, A.D., Putter, H., Blauw, G.J., Susser, E.S., Slagboom, P.E., Lumey, L.H., 2008. Persistent epigenetic differences associated with prenatal exposure to famine in humans. Proc. Natl. Acad. Sci. 105, 17046–17049.

Hellman, A., Chess, A., 2007. Gene body-specific methylation on the active X chromosome. Science 315, 1141–1143.

Hendrich, B., Bird, A., 1998. Identification and characterization of a family of mammalian methyl-CpG binding proteins. Mol. Cell. Biol. 18, 6538–6547.

Hendrich, B., Guy, J., Ramsahoye, B., Wilson, V.A., Bird, A., 2001. Closely related proteins MBD2 and MBD3 play distinctive but interacting roles in mouse development. Genes Dev. 15, 710–723.

Holliday, R., Pugh, J.E., 1975. DNA modification mechanisms and gene activity during development. Cold Spring Harbor Monogr. Ser. 32, 639–645.

Hotchkiss, R.D., 1948. The quantitative separation of purines, pyrimidines, and nucleosides by paper chromatography. J. Biol. Chem. 175, 315–332.

Inoue, A., Zhang, Y., 2011. Replication-dependent loss of 5-hydroxymethylcytosine in mouse preimplantation embryos. Science 334, 194.

Iqbal, K., Jin, S.G., Pfeifer, G.P., Szabo, P.E., 2011. Reprogramming of the paternal genome upon fertilization involves genome-wide oxidation of 5-methylcytosine. Proc. Natl. Acad. Sci. U. S. A. 108, 3642–3647.

Ito, S., D'alessio, A.C., Taranova, O.V., Hong, K., Sowers, L.C., Zhang, Y., 2010. Role of Tet proteins in 5mC to 5hmC conversion, ES-cellself-renewal and inner cell mass specification. Nature 466, 1129–1133.

Jahner, D., Stuhlmann, H., Stewart, C.L., Harbers, K., Lohler, J., Simon, I., Jaenisch, R., 1982. De novo methylation and expression of retroviral genomes during mouse embryogenesis. Nature 298, 623–628.

Jakovcevski, M., Akbarian, S., 2012. Epigenetic mechanisms in neurological disease. Nat. Med. 18, 1194–1204.

Ji, H., Biagini Myers, J.M., Brandt, E.B., Brokamp, C., Ryan, P.H., Khurana Hershey, G.K., 2016. Air pollution, epigenetics, and asthma. Allergy Asthma Clin. Immunol. 12, 51.

Jjingo, D., Conley, A.B., Yi, S.V., Lunyak, V.V., Jordan, I.K., 2012. On the presence and role of human gene-body DNA methylation. Oncotarget 3, 462–474.

Johnson, T.B., Coghill, R.D., 1925. Researches on pyrimidines. C111. The discovery of 5-methyl-cytosine in tuberculinic acid, the nucleic acid of the tubercle bacillus1. J. Am. Chem. Soc. 47, 2838–2844.

Jones, P.A., Issa, J.-P.J., Baylin, S., 2016. Targeting the cancer epigenome for therapy. Nat. Rev. Genet. 17, 630–641.

Jones, P.A., Rideout 3rd, W.M., Shen, J.C., Spruck, C.H., Tsai, Y.C., 1992. Methylation, mutation and cancer. BioEssays 14, 33–36.

Kalousek, F., Morris, N.R., 1968. Deoxyribonucleic acid methylase activity in rat spleen. J. Biol. Chem. 243, 2440–2442.

Kareta, M.S., Botello, Z.M., Ennis, J.J., Chou, C., Chédin, F., 2006. Reconstitution and mechanism of the stimulation of de novo methylation by human DNMT3L. J. Biol. Chem. 281, 25893–25902.

Kochanek, S., Renz, D., Doerfler, W., 1993. DNA methylation in the Alu sequences of diploid and haploid primary human cells. EMBO J. 12, 1141–1151.

Kovesdi, I., Reichel, R., Nevins, J.R., 1987. Role of an adenovirus E2 promoter binding factor in E1A-mediated coordinate gene control. Proc. Natl. Acad. Sci. U. S. A. 84, 2180–2184.

Kriaucionis, S., Heintz, N., 2009. The nuclear DNA base 5-hydroxymethylcytosine is present in Purkinje neurons and the brain. Science 324, 929–930.

Lander, E.S., Linton, L.M., Birren, B., Nusbaum, C., Zody, M.C., Baldwin, J., Devon, K., Dewar, K., Doyle, M., Fitzhugh, W., Funke, R., Gage, D., Harris, K., Heaford, A., Howland, J., Kann, L., Lehoczky, J., Levine, R., Mcewan, P., Mckernan, K., Meldrim, J., Mesirov, J.P., Miranda, C., Morris, W., Naylor, J., Raymond, C., Rosetti, M., Santos, R., Sheridan, A., Sougnez, C., Stange-Thomann, Y., Stojanovic, N., Subramanian, A., Wyman, D., Rogers, J., Sulston, J., Ainscough, R., Beck, S., Bentley, D., Burton, J., Clee, C., Carter, N., Coulson, A., Deadman, R., Deloukas, P., Dunham, A., Dunham, I., Durbin, R., French, L., Grafham, D., Gregory, S., Hubbard, T., Humphray, S., Hunt, A., Jones, M., Lloyd, C., Mcmurray, A., Matthews, L., Mercer, S., Milne, S., Mullikin, J.C., Mungall, A., Plumb, R., Ross, M., Shownkeen, R., Sims, S., Waterston, R.H., Wilson, R.K., Hillier, L.W., Mcpherson, J.D., Marra, M.A., Mardis, E.R., Fulton, L.A., Chinwalla, A.T., Pepin, K.H., Gish, W.R., Chissoe, S.L., Wendl, M.C., Delehaunty, K.D., Miner, T.L., Delehaunty, A., Kramer, J.B., Cook, L.L., Fulton, R.S., Johnson, D.L., Minx, P.J., Clifton, S.W., Hawkins, T., Branscomb, E., Predki, P., Richardson, P., Wenning, S., Slezak, T., Doggett, N., Cheng, J.F., Olsen, A., Lucas, S., Elkin, C., Uberbacher, E., Frazier, M., et al., 2001. Initial sequencing and analysis of the human genome. Nature 409, 860–921.

Le Guezennec, X., Vermeulen, M., Brinkman, A.B., Hoeijmakers, W.A., Cohen, A., lasonder, E., Stunnenberg, H.G., 2006. MBD2/NuRD and MBD3/NuRD, two distinct complexes with different biochemical and functional properties. Mol. Cell. Biol. 26, 843–851.

Li, E., Beard, C., Jaenisch, R., 1993. Role for DNA methylation in genomic imprinting. Nature 366, 362–365.

Li, E., Bestor, T.H., Jaenisch, R., 1992. Targeted mutation of the DNA methyltransferase gene results in embryonic lethality. Cell 69, 915–926.

Liang, G., Chan, M.F., Tomigahara, Y., Tsai, Y.C., Gonzales, F.A., Li, E., Laird, P.W., Jones, P.A., 2002. Cooperativity between DNA methyltransferases in the maintenance methylation of repetitive elements. Mol. Cell. Biol. 22, 480–491.

Lorincz, M.C., Dickerson, D.R., Schmitt, M., Groudine, M., 2004. Intragenic DNA methylation alters chromatin structure and elongation efficiency in mammalian cells. Nat. Struct. Mol. Biol. 11, 1068–1075.

Lyon, M., 1961. Gene action in the X-chromosome of the mouse (*Mus musculus* L.). Nature 190, 372–373.

Lyon, M., 1962. Sex chromatin and gene action in the mammalian X-chromosome. Am. J. Hum. Genet. 14, 135–148.

Maiti, A., Drohat, A.C., 2011. Thymine DNA glycosylase can rapidly excise 5-formylcytosine and 5-carboxylcytosine: potential implications for active demethylation of CpG sites. J. Biol. Chem. 286, 35334–35338.

Martin, E.M., Fry, R.C., 2016. A cross-study analysis of prenatal exposures to environmental contaminants and the epigenome: support for stress-responsive transcription factor occupancy as a mediator of gene-specific CpG methylation patterning. Environ. Epigenet. 2.

Maunakea, A.K., Nagarajan, R.P., Bilenky, M., Ballinger, T.J., D'souza, C., Fouse, S.D., Johnson, B.E., Hong, C., Nielsen, C., Zhao, Y., Turecki, G., Delaney, A., Varhol, R., Thiessen, N., Shchors, K., Heine, V.M., Rowitch, D.H., Xing, X., Fiore, C., Schillebeeckx, M., Jones, S.J., Haussler, D., Marra, M.A., Hirst, M., Wang, T., Costello, J.F., 2010. Conserved role of intragenic DNA methylation in regulating alternative promoters. Nature 466, 253–257.

Mayer, W., Niveleau, A., Walter, J., Fundele, R., Haaf, T., 2000. Demethylation of the zygotic paternal genome. Nature 403, 501–502.

McGrath, J., Solter, D., 1984a. Completion of mouse embryogenesis requires both the maternal and paternal genomes. Cell 37, 179–183.

McGrath, J., Solter, D., 1984b. Maternal Thp lethality in the mouse is a nuclear, not cytoplasmic, defect. Nature 308, 550–551.

Migeon, B.R., 2016. An overview of X inactivation based on species differences. Semin. Cell Dev. Biol. 56, 111–116.

Millar, C.B., Guy, J., Sansom, O.J., Selfridge, J., Macdougall, E., Hendrich, B., Keightley, P.D., Bishop, S.M., Clarke, A.R., Bird, A., 2002. Enhanced CpG mutability and tumorigenesis in MBD4-deficient mice. Science 297, 403–405.

Mohandas, T., Sparkes, R.S., shapiro, L.J., 1981. Reactivation of an inactive human X chromosome: evidence for X inactivation by DNA methylation. Science 211, 393–396.

Monk, M., Boubelik, M., Lehnert, S., 1987. Temporal and regional changes in DNA methylation in the embryonic, extraembryonic and germ cell lineages during mouse embryo development. Development 99, 371–382.

Moreira de Mello, J.C., De Araujo, E.S., Stabellini, R., Fraga, A.M., De Souza, J.E., Sumita, D.R., Camargo, A.A., Pereira, L.V., 2010. Random X inactivation and extensive mosaicism in human placenta revealed by analysis of allele-specific gene expression along the X chromosome. PLoS One 5:e10947.

Nan, X., Campoy, F.J., Bird, A., 1997. MeCP2 is a transcriptional repressor with abundant binding sites in genomic chromatin. Cell 88, 471–481.

Nan, X., Ng, H.H., Johnson, C.A., Laherty, C.D., Turner, B.M., Eisenman, R.N., Bird, A., 1998. Transcriptional repression by the methyl-CpG-binding protein MeCP2 involves a histone deacetylase complex. Nature 393, 386–389.

Ng, H.H., Zhang, Y., Hendrich, B., Johnson, C.A., Turner, B.M., Erdjument-Bromage, H., Tempst, P., Reinberg, D., Bird, A., 1999. MBD2 is a transcriptional repressor belonging to the MeCP1 histone deacetylase complex. Nat. Genet. 23, 58–61.

Okamoto, I., Otte, A.P., Allis, C.D., Reinberg, D., Heard, E., 2004. Epigenetic dynamics of imprinted X inactivation during early mouse development. Science 303, 644–649.

Okamoto, I., Patrat, C., Thépot, D., Peynot, N., Fauque, P., Daniel, N., Diabangouaya, P., Wolf, J.-P., Renard, J.-P., Duranthon, V., Heard, E., 2011. Eutherian mammals use diverse strategies to initiate X-chromosome inactivation during development. Nature 472, 370.

Okano, M., Bell, D.W., Haber, D.A., Li, E., 1999. DNA methyltransferases Dnmt3a and Dnmt3b are essential for De novo methylation and mammalian development. Cell 99, 247–257.

Okano, M., Xie, S., Li, E., 1998. Cloning and characterization of a family of novel mammalian DNA (cytosine-5) methyltransferases. Nat. Genet. 19, 219–220.

Ooi, S.K.T., Bestor, T.H., 2008. The colorful history of active DNA demethylation. Cell 133, 1145–1148.

Pradhan, S., Bacolla, A., Wells, R.D., Roberts, R.J., 1999. Recombinant human DNA (cytosine-5) methyltransferase. I. Expression, purification, and comparison of de novo and maintenance methylation. J. Biol. Chem. 274, 33002–33010.

Provençal, N., Binder, E.B., 2015. The effects of early life stress on the epigenome: from the womb to adulthood and even before. Exp. Neurol. 268, 10–20.

Rakyan, V.K., Down, T.A., Balding, D.J., Beck, S., 2011. Epigenome-wide association studies for common human diseases. Nat. Rev. Genet. 12, 529–541.

Ramsden, S.C., Clayton-Smith, J., Birch, R., Buiting, K., 2010. Practice guidelines for the molecular analysis of Prader-Willi and Angelman syndromes. BMC Med. Genet. 11, 70.

Rauch, T.A., Wu, X., Zhong, X., Riggs, A.D., Pfeifer, G.P., 2009. A human B cell methylome at 100-base pair resolution. Proc. Natl. Acad. Sci. U. S. A. 106, 671–678.

Riggs, A.D., 1975. X inactivation, differentiation, and DNA methylation. Cytogenet. Genome Res. 14, 9–25.

Robertson, K.D., 2005. DNA methylation and human disease. Nat. Rev. Genet. 6, 597–610.

Roseboom, T., De Rooij, S., Painter, R., 2006. The Dutch famine and its long-term consequences for adult health. Early Hum. Dev. 82, 485–491.

Rougier, N., Bourc'his, D., Gomes, D.M., Niveleau, A., Plachot, M., Pàldi, A., Viegas-Péquignot, E., 1998. Chromosome methylation patterns during mammalian preimplantation development. Genes Dev. 12, 2108–2113.

Sado, T., Fenner, M.H., Tan, S.-S., Tam, P., Shioda, T., Li, E., 2000. X inactivation in the mouse embryo deficient for Dnmt1: distinct effect of hypomethylation on imprinted and random X inactivation. Dev. Biol. 225, 294–303.

Saito, M., Ishikawa, F., 2002. The mCpG-binding domain of human MBD3 does not bind to mCpG but interacts with NuRD/Mi2 components HDAC1 and MTA2. J. Biol. Chem. 277, 35434–35439.

Sapienza, C., Peterson, A.C., Rossant, J., Balling, R., 1987. Degree of methylation of transgenes is dependent on gamete of origin. Nature 328, 251–254.

Sarraf, S.A., Stancheva, I., 2004. Methyl-CpG binding protein MBD1 couples histone H3 methylation at lysine 9 by SETDB1 to DNA replication and chromatin assembly. Mol. Cell 15, 595–605.

Saxonov, S., Berg, P., Brutlag, D.L., 2006. A genome-wide analysis of CpG dinucleotides in the human genome distinguishes two distinct classes of promoters. Proc. Natl. Acad. Sci. U. S. A. 103, 1412–1417.

Schultz, D.C., Ayyanathan, K., Negorev, D., Maul, G.G., Rauscher 3rd, F.J., 2002. SETDB1: a novel KAP-1-associated histone H3, lysine 9-specific methyltransferase that contributes to HP1-mediated silencing of euchromatic genes by KRAB zinc-finger proteins. Genes Dev. 16, 919–932.

Schulz, L.C., 2010. The Dutch hunger winter and the developmental origins of health and disease. Proc. Natl. Acad. Sci. U. S. A. 107, 16757–16758.

2. DNA METHYLATION

Shenker, N., Flanagan, J.M., 2012. Intragenic DNA methylation: implications of this epigenetic mechanism for cancer research. Br. J. Cancer 106, 248–253.

Skinner, M.K., Haque, C.G.-B.M., Nilsson, E., Bhandari, R., Mccarrey, J.R., 2013. Environmentally induced transgenerational epigenetic reprogramming of primordial germ cells and the subsequent germ line. PLoS One 8:e66318.

Smit, A.F., 1996. The origin of interspersed repeats in the human genome. Curr. Opin. Genet. Dev. 6, 743–748.

Smith, F.M., Garfield, A.S., Ward, A., 2006. Regulation of growth and metabolism by imprinted genes. Cytogenet. Genome Res. 113, 279–291.

Sneider, T., Teague, W., Rogachevsky, L., 1975. S-adenosylmethionine: DNA-cytosine5-methyltransferase from a Novikoff rat hepatoma cell line. Nucleic Acids Res. 2, 1685–1700.

Stadler, M.B., Murr, R., Burger, L., Ivanek, R., Lienert, F., Scholer, A., Van Nimwegen, E., Wirbelauer, C., Oakeley, E.J., Gaidatzis, D., Tiwari, V.K., Schubeler, D., 2011. DNA-binding factors shape the mouse methylome at distal regulatory regions. Nature 480, 490–495.

Suetake, I., Shinozaki, F., Miyagawa, J., Takeshima, H., Tajima, S., 2004. DNMT3L stimulates the DNA methylation activity of Dnmt3a and Dnmt3b through a direct interaction. J. Biol. Chem. 279, 27816–27823.

Surani, M.A., Barton, S.C., Norris, M.L., 1984. Development of reconstituted mouse eggs suggests imprinting of the genome during gametogenesis. Nature 308, 548–550.

Surani, M.A., Barton, S.C., Norris, M.L., 1986. Nuclear transplantation in the mouse: heritable differences between parental genomes after activation of the embryonic genome. Cell 45, 127–136.

Suzuki, M.M., Bird, A., 2008. DNA methylation landscapes: provocative insights from epigenomics. Nat. Rev. Genet. 9, 465–476.

Tahiliani, M., Koh, K.P., Shen, Y., Pastor, W.A., Bandukwala, H., Brudno, Y., Agarwal, S., Iyer, L.M., Liu, D.R., Aravind, L., Rao, A., 2009. Conversion of 5-methylcytosine to 5-hydroxymethylcytosine in mammalian DNA by MLL partner TET1. Science 324, 930–935.

Tate, P.H., Bird, A.P., 1993. Effects of DNA methylation on DNA-binding proteins and gene expression. Curr. Opin. Genet. Dev. 3, 226–231.

Tran, R.K., Henikoff, J.G., Zilberman, D., Ditt, R.F., Jacobsen, S.E., Henikoff, S., 2005. DNA methylation profiling identifies CG methylation clusters in *Arabidopsis* genes. Curr. Biol. 15, 154–159.

Ueda, Y., Okano, M., Williams, C., Chen, T., Georgopoulos, K., Li, E., 2006. Roles for Dnmt3b in mammalian development: a mouse model for the ICF syndrome. Development 133, 1183–1192.

Valinluck, V., Sowers, L.C., 2007. Endogenous cytosine damage products alter the site selectivity of human DNA maintenance methyltransferase DNMT1. Cancer Res. 67, 946–950.

Valinluck, V., Tsai, H.-H., Rogstad, D.K., Burdzy, A., Bird, A., Sowers, L.C., 2004. Oxidative damage to methyl-CpG sequences inhibits the binding of the methyl-CpG binding domain (MBD) of methyl-CpG binding protein 2 (MeCP2). Nucleic Acids Res. 32, 4100–4108.

Wade, P.A., Gegonne, A., Jones, P.L., Ballestar, E., Aubry, F., Wolffe, A.P., 1999. Mi-2 complex couples DNA methylation to chromatin remodelling and histone deacetylation. Nat. Genet. 23, 62–66.

Watt, F., Molloy, P.L., 1988. Cytosine methylation prevents binding to DNA of a HeLa cell transcription factor required for optimal expression of the adenovirus major late promoter. Genes Dev. 2, 1136–1143.

Weber, M., Davies, J.J., Wittig, D., Oakeley, E.J., Haase, M., Lam, W.L., Schubeler, D., 2005. Chromosome-wide and promoter-specific analyses identify sites of differential DNA methylation in normal and transformed human cells. Nat. Genet. 37, 853–862.

Yoder, J.A., Walsh, C.P., Bestor, T.H., 1997. Cytosine methylation and the ecology of intragenomic parasites. Trends Genet. 13, 335–340.

Implications of DNA Methylation in Toxicology

Christopher Faulk

Department of Animal Science, College of Food, Agricultural, and Natural Resource Sciences, University of Minnesota, Saint Paul, MN, United States

OUTLINE

Toxicoepigenetics
https://doi.org/10.1016/B978-0-12-812433-8.00006-X

Abbreviations

5hmC	5-hydroxymethylcytosine
5mC	5-methylcytosine
BPA	bisphenol A
CpG site	cytosine-phosphate-guanine
DOHaD	developmental origins of health and disease
ELISA	enzyme-linked immunosorbent assay
HF	high fat
LC–MS	liquid chromatography-mass spectrometry
LF	low fat
LUMA	luminometric pyrosequencing assay
MS-PCR	methylation-specific PCR
SAH	S-adenosylhomocysteine
SAM	S-adenosylmethionine
SMRT seq	single-molecule real-time sequencing

INTRODUCTION

Measurement and functional analysis of DNA methylation in toxicology studies is still an emerging field with multiple technologies maturing to probe the epigenome. In this chapter, emphasis will be on the modulation of DNA methylation by environmental exposures and modifiable risk factors. It is important to note that risk factors are not exclusive to traditional toxic agents, but inclusive of any environmental factor that can induce delayed manifestation of an altered phenotype without sequence mutations. Professor Moshe Szyf eloquently stated, "DNA function and health could be stably altered by exposure to environmental agents without changing the sequence, just by changing the state of DNA methylation. Our current screening tests do not detect agents that have long-term impact on the phenotype without altering the genotype. The realization that long-range damage could be caused without changing the DNA sequence has important implications on the way we assess the safety of chemicals, drugs, and food and broadens the scope of definition of toxic agents" (Szyf, 2011). In this vein, studies are presented with a range of "agents" that affect health via alteration of methylation. Emphasis will be on studies in which developmental exposure leads to persistent changes in DNA methylation through adulthood, since the perinatal window is the developmental stage most susceptible to environmental exposure (Faulk and Dolinoy, 2011). The surveyed examples provided are not meant to be exhaustive, but representative of exposures that modify methylation.

DETECTION OF DNA METHYLATION

Bisulfite Methods of DNA Methylation Detection

5-Methylcytosine (5mC) is the most abundant modified nucleotide in eukaryotes, accounting for up to 5% of cytosines in vertebrates. Detecting cytosine DNA methylation remained a challenge until the development of the bisulfite conversion technique in 1970 by Hayatsu

(2008) and Hayatsu et al. (1970). This reaction selectively deaminates cytosines to uracil, read by DNA polymerases as thymine, allowing for the differentiation between methylated and unmethylated cytosines when sequenced. The first use of sodium bisulfite to sequence 5-methylcytosine within DNA was performed by Marianne Frommer et al. in 1992 (Frommer et al., 1992). Numerous refinements and techniques have been developed to make use of this method of DNA methylation identification. Broadly speaking, these techniques fall into enzymatic, pyrosequencing, or short-read sequencing methods and can give global, genome-wide, or site-specific values of methylation (Fig. 1). A recent review by Kurdyukov and Bullock provides a guide to choosing a method for any type of DNA methylation analysis (Kurdyukov and Bullock, 2016).

Direct Methods of Locus-Specific or Global DNA Methylation Detection

Few methods of querying DNA methylation natively, that is, without bisulfite conversion, have historically been available, and all with caveats (Fig. 1). The difference between the terms "global" and "genome wide" is that the former means a method of detecting CpG methylation in aggregate of all or nearly all of the CpG sites, while "genome wide" means the ability to know the DNA methylation status of all or most CpG sites on an individual basis. A global method uses methylation-sensitive enzymatic recognition and cutting of the DNA, commonly referred to as the luminometric methylation assay (LUMA). A similar method using methylation-specific isoschizomers combined with PCR gives site-specific methylation values (MS-PCR). Both LUMA and MS-PCR have low resolution and low throughput, which limits their application to using samples with large amounts of high-quality DNA. Second, liquid chromatography-mass spectrometry (LC-MS) analysis can only give a global percentage of 5-methylcytosine within a sample and is relatively laborious. Third, ELISA-based assays also give a global measure, and are inexpensive, but give only a rough estimate of 5-methylcytosine. Nevertheless, ELISA assays can be used with low-quality DNA and done without specialized equipment.

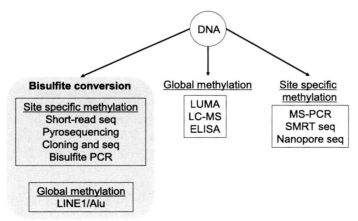

FIG. 1 Methods of DNA methylation detection. Bisulfite-converted DNA can be queried to determine the methylation of specific CpG sites or via pyrosequencing of the transposons, LINE or Alu. Native DNA can be queried for global levels of DNA methylation via liquid chromatography-mass spectrometry (LC-MS), enzyme-linked immunosorbent assay (ELISA), or luminometric pyrosequencing assay (LUMA). Site-specific levels of CpG methylation at one gene can be determined with native DNA via methylation-specific PCR (MS-PCR) or genome- wide with single-molecule real-time sequencing (SMRT seq) or nanopore sequencing.

Direct Methods of Genome-Wide DNA Methylation Detection

Recent developments in DNA sequencing are giving rise to methods of direct detection of 5-methylcytosine on native DNA strands with genome-wide scale and single-base resolution. In single-molecule real-time (SMRT) sequencing developed by Pacific Biosciences, a bound DNA polymerase incorporates fluorescently labeled nucleotides to a complementary strand, with pulse characteristics used to infer base composition. This technique can detect 5-methylcytosine as well as 5-hydroxymethylcytosine (5hmC) and N6-methyladenosine (Flusberg et al., 2010). Similarly, nanopore sequencing developed by Oxford Nanopore Technologies has demonstrated the ability to directly detect 5-methylcytosine and 6-methyladenine (Simpson et al., 2017). While both of these methods are still experimental and under development, their potential is to provide high-throughput long-read sequence reads with direct assessment of these epigenetic DNA modifications.

Principles of DNA Methylation Modification

DNA methylation in eukaryotes is mediated by the one-carbon metabolism pathway. This complex and ancient pathway has dietary choline as the primary input in humans, accounting for 60% of our methyl group substrates (Niculescu and Zeisel, 2002). Other methyl donors are methionine (20% of daily supply); methylfolate (10%–20% of our daily supply); or the choline metabolite, betaine (trimethylglycine). The metabolism of these donors interacts at the point of homocysteine-to-methionine conversion. Methionine is the primary substrate to S-adenosylmethionine, the active methylating agent used to couple the methyl group to the cytosine of a CpG site, which is the most common form of DNA methylation in vertebrates. Importantly, these methyl donors are fungible sources of methyl groups and can be interconverted when substrates are restricted due to either diet or mutation in critical enzymes. For example, rodents fed diets deficient in several methyl donors had hypomethylated DNA (Dizik et al., 1991). While choline deficiency is unlikely to be prevalent, folate deficiency is relatively common in between 22% and 52% of Indians (Kapil and Bhadoria, 2014). Folate deficiency alone or with concurrent B12 deficiency can induce hypomethylation in both rodents and humans (Brunaud et al., 2003; Agodi et al., 2015). Indirect modification of DNA methylation machinery by definition does not affect the one-carbon metabolism pathway directly, but is the result of influence on enzymes and pathways that feed into the pathway.

Methylation occurs through the activity of a family of enzymes called DNA methyltransferases (DNMTs). In this family, there are three functional members, DNMT1, the most abundant enzyme that primarily maintains existing DNA methylation patterns, and DNMT3a and DNMT3b that de novo methylate regions and act on hemimethylated DNA. Active demethylation occurs through the activity of the ten-eleven translocation enzymes (TETs) that oxidize 5-methylcytosine to 5-hydroxymethylcytosine (Fig. 2). TET can further oxidize this base to 5-formylcytosine and then to 5-carboxycytosine. The latter two versions of cytosine can be removed by thymine DNA glycosylase (TDG), leaving an abasic position. Restoration of an unmodified cytosine is completed by the base excision repair pathway, completing the cycle (Kohli and Zhang, 2013). Inhibition of either the methylation or demethylation mechanisms will alter methylation patterns. For example, disruption of the DNMT enzymes will prevent establishment or cell-division maintenance of DNA

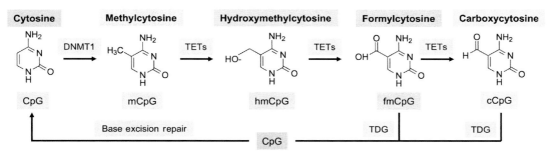

FIG. 2 Pathway of CpG site demethylation. DNA methyltransferase (DNMT1) adds a methyl group to the 5-carbon. Demethylation occurs via oxidation of the methyl group through ten-eleven translocation enzymes (TETs) and continues through hydroxymethylation to formylcytosine and reduction to carboxycytosine. From this and the prior substrate, DNA glycosylase (TDG) can remove the base leaving an abasic site targeted by the base excision repair (BER) pathway to replace the formylcytosine or carboxycytosine with an unmodified cytosine.

methylation, leading to a lower level of methylation. Likewise, dysregulation of the TET enzymes will interrupt the oxidation pathway leading to demethylation and resulting in the retention of DNA methylation.

Conservation of DNA Methylation Dynamics

DNA methylation appears in all domains of life as a functional modification of DNA for various reasons. It is believed that methylation in bacteria first evolved as a method of self-defense used by restriction enzymes to distinguish native DNA from nonmethylated phage DNA. Plants use methylation for gene suppression like other eukaryotes, and it is even speculated that methylation plays a role in plant memory (Latzel et al., 2016). Even archaea species contain functional DNA methyltransferases and methylated nucleotides, both cytosine and adenine (Blow et al., 2016). Here, we focus primarily on animals where methylation is limited to the context of palindromic CpG dinucleotides. Unlike the underlying sequence, methylation is modified throughout the life of an individual. That is to say, individuals are born and die with the same genome in all their cells, yet the methylome changes as cells replicate over time. Mammals have an innovative method of resetting DNA methylation not generally present in insects or basal vertebrates such as the frog, *Xenopus* spp. (Veenstra and Wolffe, 2001). In mammals, the male pronucleus undergoes rapid global demethylation, and methylation is reestablished as the embryo develops (Mayer et al., 2000; Reik et al., 2001). This effect is seen in pigs, mice, and humans and is the basis for the window of susceptibility in methylation dysregulation during early embryogenesis. Importantly, this effect is not conserved in magnitude or timing and is reduced in species such as rabbit and sheep or cloned embryos (Beaujean et al., 2004; Dean et al., 2001). Likewise, not all regions of the genome are fully reset, with some regions escaping full demethylation, particularly transposons and imprinted genes (Li et al., 2004). A second wave of methylation resetting occurs in the primordial germ cells of mammals (Yamazaki et al., 2003; Faulk and Dolinoy, 2011). It is during reestablishment of methylation that exposures are most impactful. The environment at this state, particularly chemical and nutritional exposures, can most strongly shape the epigenome.

DIET-INDUCED CHANGES TO DNA METHYLATION

It is sometimes overlooked that the largest exogenous chemical exposure comes daily throughout life of kilograms of uncharacterized material in the form of our diet. The role of nutrition in toxicology has received attention for decades (Parke and Ioannides, 1981). Coupled with our changing diet, toxicological studies must chase direct effects and dietary background effects. While dietary deficiencies in methyl donors are known to be tumorigenic alone, deficient diets exacerbate the tumorigenic activities of several classes of compounds such as barbiturates (phenobarbital), pesticides (DDT), and other carcinogens (2-acetylaminofluorene) (Poirier, 1994). Conversely, methyl supplementation can ameliorate these effects and the hypomethylating effect of bisphenol A (BPA) (Dolinoy et al., 2007). Long-term effects on epigenetic drift act in concert with different diet backgrounds and BPA exposure to effect the phenotype (Kochmanski et al., 2016). In fact, the term "environmental deflection" has been coined to describe the toxicological and dietary synergistic activity on the epigenome and physiology (Kochmanski et al., 2017). So, it is important to remember the influence of nutrition on toxicology and epigenetics.

Evolutionary Conservation in Diet and DNA Methylation

The most dramatic example of a diet-induced change in the epigenome occurs in insects, where honey bee larvae fed royal jelly, containing the *royalactin* protein, develop as queens while bees who do not consume the royal jelly have no chance of being queen. The underlying mechanism to this unique and complex social hierarchy appears to be DNA methylation (Kucharski et al., 2008). Queens average a 4-year life span, while workers only live a few months, and genes such as *dynactin p62* differ by at least 10% DNA methylation between the castes. Indeed, siRNA-induced silencing of *Dnmt3* causes a majority of larvae to develop morphologically as queens (Shi et al., 2011). Most fascinatingly, even the evolutionarily distant fruit fly, *Drosophila melanogaster*, exhibits a "queen" phenotype when fed *royalactin*, despite not possessing a functional DNA methylation system and last sharing a common ancestor over 325 million years ago (Kamakura, 2012). On the mammalian side, two populations of baboons (*Papio cynocephalus)* one feeding in the wild and one feeding on human scraps were found to have 0.20% of CpG sites tested as differentially methylated (Lea et al., 2016). Intriguingly, the males, who disperse between groups, were found to maintain the methylation pattern of their childhood diet throughout adulthood, even when switching food sources as adults. Further, the differentially methylated loci were frequently in promoters and enhancers of metabolism-related genes. These examples speak to a broadly conserved mechanism of developmental plasticity built upon variation in the methylome (Champagne, 2013). Indeed, differentially methylated CpG sites, without any differences in amino acid sequences, appear to underlie species-specific differences in morphology, neurology, and development between humans and great apes (Hernando-Herraez et al., 2015).

Background Diet Considerations in Rodents

The background or "base" diet is an underappreciated issue in rodent studies of toxicological exposures. Ten years ago, Warden and Fisler stated, "many papers using animal models draw conclusions about dietary effects from comparison of natural ingredient chow with defined diets, despite marked differences in micro- and macronutrient content," and this problem continues today (Warden and Fisler, 2008). Even when using defined diets, the choice of diet can be crucial. For example, the most widely used rodent defined diet, AIN-93, was preceded 17 years prior by the AIN-76 defined diet (Reeves et al., 1993). The updated AIN-93 diet increases B-12 from 10.1 to 25 µg/kg and choline bitartrate from 0.20% to 0.25%, both components of the DNA methylation pathway. Still, today, studies continue to be published using AIN-76. Unfortunately, there remains a paucity of data regarding epigenetic differences resulting from differences between these two diets.

Caloric Restriction and DNA Methylation

Nutrition in nature is never guaranteed or consistent. Deprivation during early development disproportionately affects animals in the most vulnerable stage, prior to and during organogenesis, laying the groundwork for diseases later in life. The transmission of nutritional status during development to the adult organism is, by definition, epigenetic (Laubach et al., 2017). Deprivation can take the form of total caloric restriction, macronutrient restriction (e.g., the lack of protein but sufficient calories overall), or micronutrient deficiencies. The most famous study of caloric deficits in humans has been with the Dutch hunger winter fetal cohort that had reduced DNA methylation at the insulin-like growth factor 2 (*IGF2*) locus in offspring that persisted for 60 years (Heijmans et al., 2008). As adults, this cohort suffered from increased coronary heart disease, lipids, and obesity (Painter et al., 2005). Seasonally induced nutritional challenge in rural Gambians has similarly been shown to be associated with methylation shifts in several genes (Waterland et al., 2010). In rodents, the evidence for caloric restriction inducing adult changes in methylation is robust. Caloric restriction in male mice reduces global sperm methylation and increases adiposity and dyslipidemia in offspring of paternally restricted mice (McPherson et al., 2016). In mice, in utero undernourishment leads to locus-specific hypomethylation and enrichment of retained nucleosomes in the sperm of adult males (Radford et al., 2014). These results are especially interesting given that nearly all histones are removed from the genome during sperm maturation and replaced with small protamine proteins. Any information transmitted by histone modifications would normally be lost during the histone-protamine transition, so preferential retention of histones is an interesting vector for potential information transfer from one generation to the next.

Macronutrient Restriction

Macronutrient restriction is independent of caloric restriction and, during susceptible developmental windows, can also induce changes in DNA methylation in the adult animal. Protein restriction in mouse dams yields changes in DNA methylation in their adult offspring in at least one gene promoter, phosphoenolpyruvate carboxykinase, used in gluconeogenesis and found to be maintained in two further generations (Hoile et al., 2011). Offspring of protein-restricted rats have increased expression of glucocorticoid receptor and peroxisomal proliferator-activated receptor, inversely associated with DNA methylation level at these genes (Lillycrop et al., 2005a, b). Both direct and indirect pathways of altered methylation can intersect, for example, with macronutrient protein restriction and methyl-donor supplementation. Coadministration with dietary folate rescues methylation and gene expression effects at these loci in rats (Lillycrop et al., 2005a, b). This combination of specific micronutrient depletion in pregnancy with dietary challenge for the adult animal was also studied by Langie et al. who found that low versus normal folate in dams resulted in altered DNA methylation in 200–300 promoters (Langie et al., 2013). When the offspring were challenged with low- versus high-fat diets, the diet effect was synergistic with the maternal low folate effect, yielding the largest expression change for most of the differentially expressed genes. Rats that were protein-restricted in utero also secreted more leptin and gained body fat faster when challenged with a high-fat diet (Giudicelli et al., 2013). Simultaneous oversupply of a methyl-donor diet, however, can rescue the leptin and weight gain response by increasing the methylation of the leptin gene promoter. As with many perinatal exposure studies, this one found sex-specific effects. In swine, a similar effect is seen, with protein restriction during pregnancy inducing global changes in methylation that also coincide with excess protein (Altmann et al., 2012). Protein restriction in pregnant mice, with the use of isocaloric diets, results in robust, sustainably changed methylation profiles of differentially methylated regions in imprinted genes of offspring and has become an established model (Ivanova et al., 2012). While imprinted genes are not directly involved in protein metabolism, many are growth regulators.

Macronutrient Shifts

Macronutrient ratio changes have been extensively studied in light of the obesity epidemic, and a body of work indicates that isocaloric shifts (e.g., change from low fat to high fat without a significant change in the number of calories consumed) or changes in carbohydrate ratios can alter DNA methylation. It is important to distinguish the effects of isocaloric changes in composition from macronutrient restriction as discussed previously. Here, we are concerned with diets in which no component is truly restricted, only changed in relative dietary contribution. Also, this section focuses on developmental exposures and transgenerational maintenance of DNA methylation.

In a careful study, Cannon et al. performed a parent-offspring dietary switching experiment wherein dams were fed either high-fat (HF) or low-fat (LF) diets and half of each resulting litter were either kept on the same diet or switched. All pups fed low-fat diets were normal weight and indistinguishable (Cannon et al., 2014). However, while the HF–HF pups became obese, the LF-HF pups became even more relatively obese. These findings indicate developmental programming resulted in a metabolic mismatch of the offspring between their in utero diet cues and their postweaning diet exposure. While they found many differences in gene expression, they did not find any differential DNA methylation. Interestingly, in a reanalysis of their data, Stubbs et al. did indeed find numerous differentially methylated sites; moreover, these sites correlated with increased biological age according to the epigenetic clock (Stubbs et al., 2017).

Studies of maternal exposure to high-fat diets repeatedly find a developmental effect of maternal programming, most likely mediated by epigenetic marks (Samuelsson et al., 2008; Ashino et al., 2012; Masuyama and Hiramatsu, 2012). The most convincing evidence that epigenetic marks are the mechanism by which parental exposures influence offspring phenotype comes from an elegant study by Huypens et al. who used in vitro fertilization to remove the confounding effect of direct in utero exposure (Huypens et al., 2016). In their experiment, high-fat- and low-fat-fed dams and sires had gametes collected for use in in vitro fertilization. The resulting embryos were implanted in surrogates who ate only chow diet through pregnancy and weaning. Despite a controlled chow diet in utero, the offspring from either high-fat-fed parental donor gametes reacted to a high-fat diet challenge by gaining weight faster than low-fat-fed parental donors. The effect was synergistic, with contributions from both parents. So, even prior to conception, the environment can alter the epigenome, particularly the methylome, resulting in phenotypic changes in offspring.

The above sections taken together give clear implications for toxicological studies in that macronutrient availability and composition must be determined in conjunction with any exposures to accurately assess relative risks of toxicants. While many studies focus on deficiencies, even "normal" diet backgrounds should be carefully assessed when used as base diets for toxic exposure studies. If both diet deficiencies and toxicants are present, the epigenetic consequences could be amplified leading to confounding results.

Micronutrient or Trace Mineral Deficiencies

Micronutrient restriction or deficiency can alter DNA methylation independently of caloric sufficiency. Similar to mice (McPherson et al., 2016), sheep restricted in the micronutrients B12, folate, and methionine have normal birthweights; however, as the offspring age, the restricted offspring are heavier, fatter, insulin-resistant and have altered immune function (Sinclair et al., 2007). These changes were concordant with shifts in 4% of CpG islands examined. Conversely, mouse maternal diet supplementation to create a methyl-donor-rich diet increases methylation and suppresses transcription of the *Runx3* gene in offspring lung tissue, a transcription factor that has suppressor and activating activity and tumor suppressor activity (Hollingsworth et al., 2008). This methyl-supplemented diet also alters the coat color pattern of the agouti viable yellow strain mouse (Waterland and Jirtle, 2003), a common biosensor model (Dolinoy, 2008).

Even micronutrient deficiency not directly implicated in the one-carbon metabolism pathway can influence DNA methylation. Neonatal iron deficiency in pigs results in altered hippocampal DNA methylation in >800 CpG sites within CpG islands and ~100 outside of CpG islands (Schachtschneider et al., 2016). Twelve of these sites were associated with differentially expressed genes. Fetal iron deficiency in rats induces epigenetic changes in both DNA methylation and histones (Tran et al., 2015). Calcium deficiency in perinatal rats induces hypomethylation in a gene-specific manner in pups (Takaya et al., 2013). Even chromium deficiency in dams induces hypomethylation in offspring mitogen-activated protein kinase (*Mapk*) signaling pathway genes (Zhang et al., 2016). There are no reports of cobalt deficiency directly impacting DNA methylation; however, in cases of extreme restriction, it can induce B12 deficiency owing to B12's incorporation of a cobalt atom in its structure;

however, in cattle, it appeared to have little practical effect on tissue metabolism (Kennedy et al., 1995). Copper is an especially crucial micronutrient involved in numerous catalytic activities. While copper deficiency is so rare that moderate restriction alone has not been studied for epigenetic effects, there is a body of evidence linking copper accumulation with Wilson's disease altering the SAM/SAH ratio, methionine metabolism, and chelation to prevent liver damage in mouse (Medici et al., 2014). Global methylation of the placenta is reduced, coincident with fetal growth restriction in some strains of rat dams fed a copper-deficient diet (Ergaz et al., 2014). Magnesium deficiency appears to also generate consistent gene-specific changes in methylation, though its epigenetic effect is only reported in a single study so far. In this case, deficiency in pregnant rats led to offspring with altered glucocorticoid metabolism and large methylation changes in at least one promoter, 11β-hydroxysteroid dehydrogenase-2 (*Hsd11b2*) as measured by pyrosequencing (Takaya et al., 2011). Selenium's effects on the methylome are under investigation with reports indicating that selenium deficiency, as seen in Keshan disease, is associated with reduced methylation in several genes, *Gadd45α*, *Tlr2*, and *Icam1* (Yang et al., 2014). Both selenium deficiency and supplementation appear to induce a similar myocardial dysfunction, but only the supplemented group exhibited global hypomethylation (Metes-Kosik et al., 2012). In rats, selenium deficiency was seen to be tissue-specific (Uthus et al., 2006). Importantly, there is an inverse interaction of selenium and arsenic on global methylation at physiological levels of exposure seen in an Indian cohort (Richard Pilsner et al., 2011). Zinc deficiency in utero induces hypomethylation at the metallothionein-2 promoter in mice, which persists to adulthood (Kurita et al., 2013). In contrast to nearly all the other effects of deficiencies or toxicant exposures, zinc deficiency increases methylation at one locus, the brain-derived neurotrophic factor (*Bdnf*) exon IX, concurrently with decreased expression. This change in the hippocampus of zinc-deficient rats 0–2 months old is likely the mechanism behind their impaired learning and memory (Hu et al., 2016). There are no reports of boron, manganese, or molybdenum deficiency affecting methylation.

Micronutrients and trace metals are highlighted here for similar reasons as the importance of dietary background. Many studies of altered macronutrient composition (e.g., the "Western diet") retain the same mineral or nutrient mix that is found in a standard rodent chow. It is important to note that in human diets, many trace nutrients are of low concentration or borderline deficient, so rodent diets must not be limited to altered macronutrient without the concomitant effect of micronutrient supply. The same concern is noted with toxicant exposure. The comorbidity of nutrient restriction and toxicant exposure is common and probably synergistic in their effects on the epigenome.

PHARMACEUTICALS

Iatrogenic effects of pharmaceutical treatments on the methylome are increasingly being studied. Some drugs are specifically designed to act upon the methylome, most commonly, azacytidine; deoxyazacytidine; zebularine; or the next-generation drug, guadecitabine (Sato et al., 2017). Originally, azacytidine was developed as a nucleoside analogue whose cytotoxic activity was thought to underlie its anticancer activity by killing faster dividing cancer cells. Subsequently, it was determined that its main effect derived from its effect on the methylome. Deoxyazacytidine is the deoxynucleoside analogue that specifically acts on DNA by inhibiting DNMT enzymes. Both drugs are now used in a low but repeated dose schedule to induce epigenetic effects without direct cytotoxicity. The newer forms nucleoside analogues, zebularine and guadecitabine, have lower toxicity profiles while still acting on the epigenome. Repurposing of drugs for use or recognition of their methylation altering ability is under active examination, as with the anti-inflammatory drug olsalazine (Méndez-Lucio et al., 2014). Olsalazine has similar DNMT inhibition activity as deoxyazacytidine but is better tolerated and is already approved for human use. Unfortunately, no studies have yet determined in vivo demethylation activity. Since this chapter is primarily focused on incidental environmental exposures, the epigenetic activity of pharmaceuticals is out of scope. See Heerboth et al. for an in-depth review of the methylation activities of pharmaceuticals (Heerboth et al., 2014).

METALS

Metals as Toxicants

What distinguishes metallic micronutrients from toxicants? If a substance is necessary but absent, we can be deficient in it and so consider it a nutrient, while if it is present but damaging, it is a toxicant. Some compounds can be both, as Paracelsus, the father of toxicology said, "the dose makes the poison." In the case of some metals, it can be the oxidation state that makes the poison. For instance, chromium(III) was discussed as a necessary micronutrient whose deficiency causes dysregulation of methylation. At the same time, chromium(VI) is a well-known genotoxic carcinogen that forms DNA adducts and is generally considered to be toxic. Chromium(VI), along with nickel, and arsenic, all cause changes in DNA methylation that may be causal to their carcinogenicity (Salnikow and Zhitkovich, 2008).

High Density Metals

"Heavy metal" is an ambiguous term (Duffus, 2002), so this section focuses on examples of metallic elements that influence DNA methylation, especially via developmental exposure. Some metals are essential micronutrients (e.g., iron, cobalt, copper, and zinc). Others seem to be harmless, for example, gold, silver, platinum, gallium, and tungsten. Historically, some high-density metals were considered toxic, via either acute exposure (arsenic) or chronic exposure (lead). Only recently are the epigenetic effects of metal exposure becoming elucidated. Arsenic is a common exposure in contaminated water around the world that can induce transient and permanent changes to DNA methylation (Eckstein et al., 2017). It influences DNA methylation globally and gene specifically in both high and low doses (Intarasunanont et al., 2012; Reichard and Puga, 2010). Even lead, whose toxic properties have been well known since Roman times, is only now being probed for epigenetic effects (Faulk et al., 2013, 2014). Since lead does not directly act on the one-carbon metabolism pathway, the mechanism by which it alters the epigenome remains unclear, though some work in perinatally exposed rhesus macaques suggests that permanent alteration of DNMT expression level may explain these effects (Wu et al., 2008). Copper falls into both the "micronutrient" and "toxic" categories, depending on dose. An endogenous form of copper toxicity manifests in Wilson's disease and interacts with the one-carbon metabolism pathway via inhibition of S-adenosylhomocysteine hydrolase causing a buildup of that substrate. A mouse model of Wilson's disease also exhibits altered methionine metabolism, leading to global hypomethylation (Medici et al., 2013). When treated with either betaine or copper chelator, *DNMT3b* is upregulated, homocysteine levels are normalized, and global methylation is restored.

Light Metals

Lithium has a long history in neuropsychological management of bipolar disorder; however, the mechanism of its effects is still unclear. Recent work in human neuroblastoma cell culture with lithium treatment finds surprisingly gene-specific effects on DNA methylation (Asai et al., 2013). Specifically, lithium induced changes at the *BDNF* promoter IV locus, which has been repeatedly shown to be a target of dysregulated methylation after acute developmental psychological stress (Moser et al., 2015). Indeed, lithium appears to directly upregulate *BDNF* via hypomethylation of the promoter locus in a defined manner (Dwivedi and Zhang, 2015). The most surprising aspect of this activity is that it can occur in mature adults, in nondividing neural tissue, and in a locus-specific way, despite being merely a monoatomic metal ion. For these reasons, lithium and *BDNF* are likely to become the best studied example of a specific epigenetic modifying agent that affects behavior through reproducible methylation at a gene locus. This author notes that natural levels of lithium vary throughout watersheds offering the opportunity for a natural epidemiological experiment. Regarding magnesium, there are no reports of hypermagnesemia associated with methylation changes, and given strong mechanisms of homeostasis, epigenetic changes are of low concern in cases of magnesium overload (Ayuk and Gittoes, 2014). Aluminum in blood serum accumulates primarily though airborne occupational exposure and is inversely correlated with global DNA methylation (Yang et al., 2015). Beryllium sensitization is still under investigation, but at least one study found no association with methylation at multiple promoters with beryllium exposure (Tooker et al., 2016). The final light metal under examination is titanium. Due to its inert nature, titanium-induced changes to the methylome are restricted to its use in nanoparticle testing and exposures, not necessarily via the chemical interaction of the metal with any biological processes. Nevertheless, caution in future space travelers may be warranted as radioactive Ti(48), found primarily in space, was found to induce global hypermethylation in exposed mice (Jangiam et al., 2015).

ORGANIC TOXICANTS

Perhaps the best studied organic toxicant thought to affect the epigenome is bisphenol A (BPA), a high-volume chemical that can act as an endocrine disrupter (De Coster and Van Larebeke, 2012). Studies of perinatal BPA exposure in both mouse and human have found altered global, gene-specific, and imprinted gene altered methylation (Nahar et al., 2015;

Faulk et al., 2016). Perinatal BPA has been used to probe epigenetic drift, the concept that natural changes in the epigenome over the life course can be altered by toxicant exposure (Kochmanski et al., 2016). Agricultural use chemicals such as herbicides, pesticides, and fungicides are also increasingly under scrutiny for epigenetic effects. Water fleas (*Daphnia magna*) show decreased DNA methylation when reared in the presence of vinclozolin, a fungicide and endocrine disruptor (Vandegehuchte et al., 2010). Vinclozolin also alters methylation in outbred mice (Guerrero-Bosagna et al., 2012). Notably, this effect is transgenerational, with the altered state persisting until at least the third generation beyond the exposed dam (Skinner and Guerrero-Bosagna, 2014). Industrial solvents like trichloroethylene at levels below the limit for drinking water increases methylation variability and epigenetic drift in mice (Gilbert et al., 2016). The airborne organic carcinogen formaldehyde causes global hypomethylation (Liu et al., 2011). Several studies have linked perinatal phthalate plasticizers to altered human methylation at either specific genes or genome wide (Solomon et al., 2017; Zhao et al., 2016).

All of the above chemicals are ubiquitous and present in human populations to varying degrees. The toxicological impact of their alteration of DNA methylation is under intense scrutiny. However, even in studies where these organics are not the primary focus, care must be taken to account for their confounding effects.

NONCHEMICAL STRESS

Early-life stress is a common and potent programmer of methylation state, often correlated with phenotype or behavior as animals move to adulthood. A vivid example is the pea aphid (*Acyrthosiphon pisum*), which has a functional DNA methylation system and produces more winged offspring and DNMT expression when under predator stress or crowded conditions, a clear adaptive advantage (Weisser et al., 1999; Walsh et al., 2010). Even the invasion of a new species into the territory of another species can drive changes to DNA methylation and ultimately phenotype. For instance, red imported fire ants (*Solenopsis invicta*) were recently introduced to the territory of the eastern fence lizard (*Sceloporus undulatus*). Subsequently, methylation variation increased, and epiallele frequencies have changed within the lizard genomes, correlated with behavioral, morphological, and physiological changes (Schrey et al., 2016). In mammals, *Bdnf* gene promoter methylation is especially susceptible to stress, as seen in the light metal section. Rats can experience a phenomenon similar to post-traumatic stress disorder after being in proximity to cats, concomitant with increases in hippocampal *Bdnf* methylation (Roth et al., 2011). The same effect on *BDNF* is seen in human suicidal ideation and completion (Kang et al., 2013; Keller et al., 2010) and victims of abuse (Thaler et al., 2014). Other genes are affected similarly, with early-life stress in mice increasing the methylation of the arginine vasopressin (*Avp*) gene in neurons (Murgatroyd et al., 2009). Maternal separation in mice also causes increased methylation at *Mecp2* and *Cb1*, in conjunction with decreases in other genes, and depressive behavior (Franklin et al., 2010). Intriguingly, these methylation changes are also transmitted transgenerationally. Notably, in these cases, stress causes increased methylation, a rarely observed effect from a detrimental exposure as is seen in most metal or nutrient overloads or deficiencies.

Nonchemical stress can be a confounder when studying chemical exposures. For example, movement of mice between rooms is sufficient to cause changes in food intake, which can subsequently affect the programming of offspring of pregnant mice, independent of any chemical exposure. Therefore, it is crucial to design controls that match the nonchemical stress as closely as possible.

Modulation of these stress-induced methylation changes is still an area in need of further study.

FUTURE DIRECTIONS

The importance of DNA methylation in toxicology is being increasingly recognized. We now have extensive evidence from nutritional studies, deficiencies, and exposures of metals and organic compounds that DNA methylation can be altered in consistent ways that lead to detrimental physical effects. In the beginning of this chapter, diet background was emphasized. Rodents, in particular, are given highly nutritious diets that do not reflect human diets that are borderline deficient in many nutrients. Our diets leave little room for buffering capacity, especially when we are challenged by toxicants. Since in utero development window is the most crucial time point for toxicant exposure to affect the epigenome, dietary deficiencies that a mother could recover from may be most impactful to the fetus. These impacts on the fetal epigenome may not be immediately evident but can increase the odds of late-life disease. Thus, the most crucial time for epigenome dysregulation overlaps the most sensitive time for chemical exposure, early life. Most of our experimental evidence of changes to the epigenome comes from perinatal exposures, with developmental plasticity of the methylome corresponding to phenotypic outcomes in the adult. This evidence fits with the developmental origins of health and disease paradigm. In the case of organic toxicants, effects are occasionally found to be transgenerational, arguing for investigators of other exposures to extend their observations beyond one generation. Overall, toxicological studies that modulate the epigenome both increase our knowledge of risk factors altering the epigenome and open new avenues for potential treatments.

References

Agodi, A., et al., 2015. Low fruit consumption and folate deficiency are associated with LINE-1 hypomethylation in women of a cancer-free population. Genes Nutr. 10(5).

Altmann, S., et al., 2012. Maternal dietary protein restriction and excess affects offspring gene expression and methylation of non-SMC subunits of condensin I in liver and skeletal muscle. Epigenetics 7 (3), 239–252.

Asai, T., et al., 2013. Effect of mood stabilizers on DNA methylation in human neuroblastoma cells. Int. J. Neuropsychopharmacol. 16 (10), 2285–2294. Available at: http://ijnp.oxfordjournals.org/content/16/10/2285.abstract.

Ashino, N.G., et al., 2012. Maternal high-fat feeding through pregnancy and lactation predisposes mouse offspring to molecular insulin resistance and fatty liver. J. Nutr. Biochem. 23 (4), 341–348.

Ayuk, J., Gittoes, N.J.L., 2014. Contemporary view of the clinical relevance of magnesium homeostasis. Ann. Clin. Biochem. 51 (Pt 2), 179–188. Available at: http://www.ncbi.nlm.nih.gov/pubmed/24402002.

Beaujean, N., et al., 2004. Non-conservation of mammalian preimplantation methylation dynamics. , 14(7), pp.R266–R267. Available at. Curr. Biol. 14 (7), R266–R267. Available at: http://www.ncbi.nlm.nih.gov/pubmed/15062117. [(Accessed 4 August 2017)].

Blow, M.J., et al., 2016. The epigenomic landscape of prokaryotes. PLoS Genet. 12(2).

Brunaud, L., et al., 2003. Effects of vitamin B12 and folate deficiencies on DNA methylation and carcinogenesis in rat liver. Clin. Chem. Lab. Med. 41 (8), 1012–1019. Available at: http://www.degruyter.com.proxy.wexler.hunter.cuny.edu/view/j/cclm.2003.41.issue-8/cclm.2003.155/cclm.2003.155.xml.

Cannon, M.V., et al., 2014. Maternal nutrition induces pervasive gene expression changes but no detectable DNA methylation differences in the liver of adult offspring. PLoS One 9(3).

Champagne, F.A., 2013. Epigenetics and developmental plasticity across species. Dev. Psychobiol. 55 (1), 33–41.

De Coster, S., Van Larebeke, N., 2012. Endocrine-disrupting chemicals: associated disorders and mechanisms of action. J. Environ. Publ. Health 2012.

Dean, W., et al., 2001. Conservation of methylation reprogramming in mammalian development: aberrant reprogramming in cloned embryos. Proc. Natl. Acad. Sci. U. S. A. 98, 13734–13738.

Dizik, M., Christman, J.K., Wainfan, E., 1991. Alterations in expression and methylation of specific gene in livers of rats fed a cancer promoting methyl-deficient diet. Carcinogenesis 12 (7), 1307–1312.

Dolinoy, D.C., 2008. The agouti mouse model: an epigenetic biosensor for nutritional and environmental alterations on the fetal epigenome. In: Nutrition Reviews.

Dolinoy, D.C., Huang, D., Jirtle, R.L., 2007. Maternal nutrient supplementation counteracts bisphenol A-induced DNA hypomethylation in early development. Proc. Natl. Acad. Sci. U. S. A. 104 (32), 13056–13061. Available at: http://www.pubmedcentral.nih.gov/articlerender.fcgi?artid=1941790&tool=pmcentrez&rendertype=abstract.

Duffus, J.H., 2002. Heavy metals—a meaningless term? Pure Appl. Chem. 74 (5), 793–807.

Dwivedi, T., Zhang, H., 2015. Lithium-induced neuroprotection is associated with epigenetic modification of specific BDNF gene promoter and altered expression of apoptotic-regulatory proteins. Front. Neurosci. 9 (January).

Eckstein, M., Rea, M., Fondufe-Mittendorf, Y.N., 2017. Transient and permanent changes in DNA methylation patterns in inorganic arsenic-mediated epithelial-to-mesenchymal transition. Toxicol. Appl. Pharmacol. 331, 6–17.

Ergaz, Z., et al., 2014. Placental oxidative stress and decreased global DNA methylation are corrected by copper in the Cohen diabetic rat. Toxicol. Appl. Pharmacol. 276 (3), 220–230.

Faulk, C., et al., 2016. Detection of differential DNA methylation in repetitive DNA of mice and humans perinatally exposed to bisphenol A. Epigenetics.

Faulk, C., et al., 2013. Early-life lead exposure results in dose- and sex-specific effects on weight and epigenetic gene regulation in weanling mice. Epigenomics 5 (5), 487–500. Available at: http://www.pubmedcentral.nih.gov/articlerender.fcgi?artid=3873735&tool=pmcentrez&rendertype=abstract%5Cnhttp://www.ncbi.nlm.nih.gov/pubmed/24059796.

Faulk, C., et al., 2014. Longitudinal epigenetic drift in mice perinatally exposed to lead. Epigenomics 9 (7), 934–941. Available at: http://www.ncbi.nlm.nih.gov/pubmed/24786859.

Faulk, C., Dolinoy, D.C., 2011. Timing is everything: the when and how of environmentally induced changes in the epigenome of animals. Epigenetics 6 (7), 791–797.

Flusberg, B.A., et al., 2010. Direct detection of DNA methylation during single-molecule, real-time sequencing. Nat. Methods 7 (6), 461–465. Available at: https://doi.org/10.1038/nmeth.1459.

Franklin, T.B., et al., 2010. Epigenetic transmission of the impact of early stress across generations. Biol. Psychiatr. 68 (5), 408–415.

Frommer, M., et al., 1992. A genomic sequencing protocol that yields a positive display of 5-methylcytosine residues in individual DNA strands. Proc. Natl. Acad. Sci. 89 (5), 1827–1831. Available at: http://www.pnas.org/cgi/doi/10.1073/pnas.89.5.1827.

Gilbert, K.M., et al., 2016. Chronic exposure to trichloroethylene increases DNA methylation of the Ifng promoter in CD4+ T cells. Toxicol. Lett. 260, 1–7.

Giudicelli, F., et al., 2013. Excess of methyl donor in the perinatal period reduces postnatal leptin secretion in rat and interacts with the effect of protein content in diet. PLoS One. 8(7).

Guerrero-Bosagna, C., et al., 2012. Epigenetic transgenerational inheritance of vinclozolin induced mouse adult onset disease and associated sperm epigenome biomarkers. Reprod. Toxicol. 34 (4), 694–707.

Hayatsu, H., 2008. Discovery of bisulfite-mediated cytosine conversion to uracil, the key reaction for DNA methylation analysis–a personal account. Proc. Jpn. Acad. Ser. B Phys. Biol. Sci. 84 (8), 321–330. Available at: http://www.ncbi.nlm.nih.gov/pubmed/18941305%5Cnhttp://www.pubmedcentral.nih.gov/articlerender.fcgi?artid=PMC3722019.

Hayatsu, H., et al., 1970. Reaction of sodium bisulfite with uracil, cytosine, and their derivatives. Biochemistry 9 (14), 2858–2865.

Heerboth, S., et al., 2014. Use of epigenetic drugs in disease: an overview. Genet. Epigenet. 1 (6), 9–19.

2. DNA METHYLATION

Heijmans, B.T., et al., 2008. Persistent epigenetic differences associated with prenatal exposure to famine in humans. Proc. Natl. Acad. Sci. U. S. A. 105 (44), 17046–17049. Available at: http://www.pubmedcentral.nih.gov/articlerender.fcgi?artid=2579375&tool=pmcentrez&rendertype=abstract.

Hernando-Herraez, I., et al., 2015. DNA methylation: insights into human evolution. PLoS Genet. 11(12).

Hoile, S.P., et al., 2011. Dietary protein restriction during F0 pregnancy in rats induces transgenerational changes in the hepatic transcriptome in female offspring. PLoS One 6(7).

Hollingsworth, J.W., et al., 2008. In utero supplementation with methyl donors enhances allergic airway disease in mice. J. Clin. Investig. 118 (10), 3462–3469.

Hu, Y.-D., et al., 2016. The cognitive impairment induced by zinc deficiency in rats aged 0–2 months related to BDNF DNA methylation changes in the hippocampus. Nutr. Neurosci. 1–7. Available at: https://www.scopus.com/inward/record.uri?eid=2-s2.0-84978474547&partnerID=40&md5=57e4919cd53291d3e38bb87512084df4.

Huypens, P. et al., 2016. Epigenetic germline inheritance of diet-induced obesity and insulin resistance. Nat. Genet., 48(5), pp. 497–499. https://doi.org/10.1038/ng.3527 (Accessed 15 March 2016).

Intarasunanont, P., et al., 2012. Effects of arsenic exposure on DNA methylation in cord blood samples from newborn babies and in a human lymphoblast cell line. Environ. Health: A Global Access Sci. Source. 11(1).

Ivanova, E., et al., 2012. DNA methylation at differentially methylated regions of imprinted genes is resistant to developmental programming by maternal nutrition. Epigenetics 7 (10), 1200–1210.

Jangiam, W., Tungjai, M., Rithidech, K.N., 2015. Induction of chronic oxidative stress, chronic inflammation and aberrant patterns of DNA methylation in the liver of titanium-exposed CBA/CaJ mice. Int. J. Radiat. Biol. 91 (5), 389–398. Available at: http://informahealthcare.com/doi/abs/10.3109/09553002.2015.1001882.

Kamakura, M., 2012. Royalactin induces queen differentiation in honeybees. Seikagaku 84 (12), 994–1003.

Kang, H.J., et al., 2013. BDNF promoter methylation and suicidal behavior in depressive patients. J. Affect. Disord. 151 (2), 679–685.

Kapil, U., Bhadoria, A., 2014. Prevalence of folate, ferritin and cobalamin deficiencies amongst adolescent in India. J. Family Med. Prim. Care 3 (3), 247–249. Available at: http://www.jfmpc.com/text.asp?2014/3/3/247/141619.

Keller, S., et al., 2010. Increased BDNF promoter methylation in the Wernicke area of suicide subjects. Arch. Gen. Psychiatr. 67 (3), 258–267. Available at: http://www.ncbi.nlm.nih.gov/pubmed/20194826.

Kennedy, D.G., et al., 1995. Cobalt-vitamin B12 deficiency and the activity of methylmalonyl CoA mutase and methionine synthase in cattle. Int. J. Vitam. Nutr. Res. 65 (4), 241–247.

Kochmanski, J., et al., 2017. Environmental deflection: the impact of toxicant exposures on the aging epigenome. Toxicol. Sci. 156 (2), 325–335.

Kochmanski, J., et al., 2016. Longitudinal effects of developmental bisphenol A and variable diet exposures on epigenetic drift in mice. Reprod. Toxicol.

Kohli, R.M., Zhang, Y., 2013. TET enzymes, TDG and the dynamics of DNA demethylation. Nature 502 (7472), 472–479. Available at: http://www.ncbi.nlm.nih.gov/pubmed/24153300.

Kucharski, R., et al., 2008. Nutritional control of reproductive status in honeybees via DNA methylation. Science 319 (5871), 1827–1830. Available at: http://www.sciencemag.org/cgi/doi/10.1126/science.1153069.

Kurdyukov, S., Bullock, M., 2016. DNA methylation analysis: choosing the right method. Biology 5 (1), 3. Available at: http://www.mdpi.com/2079-7737/5/1/3.

Kurita, H., et al., 2013. Prenatal zinc deficiency-dependent epigenetic alterations of mouse metallothionein-2 gene. J. Nutr. Biochem. 24 (1), 256–266.

Langie, S.A.S., et al., 2013. Maternal folate depletion and high-fat feeding from weaning affects DNA methylation and DNA repair in brain of adult offspring. FASEB J. 27 (8), 3323–3334.

Latzel, V., Rendina González, A.P., Rosenthal, J., 2016. Epigenetic memory as a basis for intelligent behavior in clonal plants. Front. Plant Sci. 7 (August), 1–7.

Laubach, Z., et al., 2017. Nutrition, DNA methylation, and developmental origins of cardiometabolic disease: a signal systems approach. In: Preedy, , Patel, (Eds.), Handbook of Nutrition, Diet, and Epigenetics. Springer International Publishing AG, Gewerbestrasse 11, Switzerland.

Lea, A.J., et al., 2016. Resource base influences genome-wide DNA methylation levels in wild baboons (Papio cynocephalus). Mol. Ecol. 25 (8), 1681–1696.

Li, J.Y., et al., 2004. Timing of establishment of paternal methylation imprints in the mouse. Genomics 84 (6), 952–960.

Lillycrop, K.A., et al., 2005a. Dietary protein restriction of pregnant rats induces and folic acid supplementation prevents epigenetic modification of hepatic gene expression in the offspring. J. Nutr. 135 (6), 1382–1386. Available at: http://www.ncbi.nlm.nih.gov/entrez/query.fcgi?cmd=Retrieve&db=PubMed&dopt=Citation&list_uids=15930441.

Lillycrop, K.A., et al., 2005b. Nutrient-gene interactions dietary protein restriction of pregnant rats induces and folic acid supplementation prevents epigenetic modification of hepatic gene expression in the offspring 1. J. Nutr. 135 (December 2004), 1382–1386.

Liu, Q., et al., 2011. Effects of long-term low-dose formaldehyde exposure on global genomic hypomethylation in 16HBE cells. Toxicol. Lett. 205 (3), 235–240.

Masuyama, H., Hiramatsu, Y., 2012. Effects of a high-fat diet exposure in utero on the metabolic syndrome-like phenomenon in mouse offspring through epigenetic changes in adipocytokine gene expression. Endocrinology 153 (6), 2823–2830.

Mayer, W., et al., 2000. Demethylation of the zygotic paternal genome. Nature 403 (6769), 501–502. Available at: http://www.ncbi.nlm.nih.gov/pubmed/10676950.

McPherson, N.O., et al., 2016. Paternal under-nutrition programs metabolic syndrome in offspring which can be reversed by antioxidant/vitamin food fortification in fathers. Sci. Rep. 6 (1), 27010. Available at: http://www.nature.com/articles/srep27010.

Medici, V., et al., 2014. Maternal choline modifies fetal liver copper, gene expression, DNA methylation, and neonatal growth in the tx-j mouse model of Wilson disease. Epigenetics 9 (2), 286–296.

Medici, V., et al., 2013. Wilson's disease: changes in methionine metabolism and inflammation affect global DNA methylation in early liver disease. Hepatology 57 (2), 555–565.

Méndez-Lucio, O., et al., 2014. Toward drug repurposing in epigenetics: olsalazine as a hypomethylating compound active in a cellular context. ChemMedChem 9 (3), 560–565.

Metes-Kosik, N., et al., 2012. Both selenium deficiency and modest selenium supplementation lead to myocardial fibrosis in mice via effects on redox-methylation balance. Mol. Nutr. Food Res. 56 (12), 1812–1824.

Moser, D.A., et al., 2015. BDNF methylation and maternal brain activity in a violence-related sample. PLoS One 10(12).

Murgatroyd, C., et al., 2009. Dynamic DNA methylation programs persistent adverse effects of early-life stress. Nat. Neurosci. 12 (12), 1559–1566. Available at: http://www.ncbi.nlm.nih.gov/pubmed/19898468%5Cnhttp://www.nature.com/neuro/journal/v12/n12/pdf/nn.2436.pdf.

Nahar, M.S., et al., 2015. In utero bisphenol A concentration, metabolism, and global DNA methylation across matched placenta, kidney, and liver in the human fetus. Chemosphere 124, 54–60. Available at: http://www.ncbi.nlm.nih.gov/pubmed/25434263.

Niculescu, M.D., Zeisel, S.H., 2002. Diet, methyl donors and DNA methylation: interactions between dietary folate, methionine and choline. J. Nutr. 132 (8 Suppl), 2333S–2335S. Available at: http://www.ncbi.nlm.nih.gov/pubmed/12163687.

Painter, R.C., Roseboom, T.J., Bleker, O.P., 2005. Prenatal exposure to the Dutch famine and disease in later life: an overview. Reprod. Toxicol. 20 (3), 345–352.

Parke, D.V., Ioannides, C., 1981. The role of nutrition in toxicology. Annu. Rev. Nutr. 1 (3), 207–234.

Poirier, L.A., 1994. Methyl group deficiency in hepatocarcinogenesis. Drug Metab. Rev. 26 (1–2), 185–199.

Radford, E.J., et al., 2014. In utero undernourishment perturbs the adult sperm methylome and intergenerational metabolism. Science 345 (6198), 1255903. Available at: http://www.sciencemag.org/cgi/doi/10.1126/science.1255903%5Cnhttp://www.sciencemag.org/content/345/6198/1255903.abstract.

Reeves, P.G., Nielsen, F.H., Fahey, G.C., 1993. AIN-93 purified diets for laboratory rodents: final report of the American Institute of Nutrition ad hoc Writing Committee on the reformulation of the AIN-76A rodent diet. J. Nutr. 123 (11), 1939–1951.

Reichard, J.F., Puga, A., 2010. Effects of arsenic exposure on DNA methylation and epigenetic gene regulation. Epigenomics 2 (1), 87–104.

Reik, W., Dean, W., Walter, J., 2001. Epigenetic reprogramming in mammalian development. Science (New York, NY) 293 (5532), 1089–1093. Available at: http://www.ncbi.nlm.nih.gov/pubmed/11498579.

Richard Pilsner, J., et al., 2011. Associations of plasma selenium with arsenic and genomic methylation of leukocyte DNA in Bangladesh. Environ. Health Perspect. 119 (1), 113–118.

Roth, T.L., et al., 2011. Epigenetic modification of hippocampal Bdnf DNA in adult rats in an animal model of post-traumatic stress disorder. J. Psychiatr. Res. 45 (7), 919–926.

Salnikow, K., Zhitkovich, A., 2008. Genetic and epigenetic mechanisms in metal carcinogenesis and cocarcinogenesis: nickel, arsenic, and chromium. Chem. Res. Toxicol. 21 (1), 28–44.

Samuelsson, A.M., et al., 2008. Diet-induced obesity in female mice leads to offspring hyperphagia, adiposity, hypertension, and insulin resistance: a novel murine model of developmental programming. Hypertension 51 (2), 383–392.

2. DNA METHYLATION

Sato, T., Issa, J.-P.J., Kropf, P., 2017. DNA hypomethylating drugs in cancer therapy. Cold Spring Harbor Perspect. Med. 7 (5), a026948. Available at: http://www.ncbi.nlm.nih.gov/pubmed/28159832. [(Accessed 11 August 2017)].

Schachtschneider, K.M., et al., 2016. Impact of neonatal iron deficiency on hippocampal DNA methylation and gene transcription in a porcine biomedical model of cognitive development. BMC Genomics 17 (856), 1–14. Available at: http://bmcgenomics.biomedcentral.com/articles/10.1186/s12864-016-3216-y.

Schrey, A.W., et al., 2016. Epigenetic response to environmental change: DNA methylation varies with invasion status. Environ. Epigenet. 2 (2), dvw008. Available at: https://academic.oup.com/eep/article-lookup/doi/10.1093/eep/dvw008.

Shi, Y.Y., et al., 2011. Diet and cell size both affect queen-worker differentiation through DNA methylation in honey bees (apis mellifera, apidae). PLoS One 6(4).

Simpson, J.T., et al., 2017. Detecting DNA cytosine methylation using nanopore sequencing. Nat. Methods 14 (4), 407–410. Available at: http://www.nature.com/doifinder/10.1038/nmeth.4184.

Sinclair, K.D., et al., 2007. DNA methylation, insulin resistance, and blood pressure in offspring determined by maternal periconceptional B vitamin and methionine status. Proc. Natl. Acad. Sci. U. S. A. 104 (49), 19351–19356. Available at: http://www.scopus.com/inward/record.url?eid=2-s2.0-37649011327&partnerID=tZOtx3y1.

Skinner, M., Guerrero-Bosagna, C., 2014. Role of CpG deserts in the epigenetic transgenerational inheritance of differential DNA methylation regions. BMC Genomics 15 (1), 692. Available at: http://www.biomedcentral.com/1471-2164/15/692.

Solomon, O., et al., 2017. Prenatal phthalate exposure and altered patterns of DNA methylation in cord blood. Environ. Mol. Mutagen.

Stubbs, T.M., et al., 2017. Multi-tissue DNA methylation age predictor in mouse. Genome Biol. 18 (1), 68. Available at: http://genomebiology.biomedcentral.com/articles/10.1186/s13059-017-1203-5.

Szyf, M., 2011. The implications of DNA methylation for toxicology: toward toxicomethylomics, the toxicology of DNA methylation. Toxicol. Sci. 120 (2), 235–255.

Takaya, J., et al., 2013. A calcium-deficient diet in pregnant, nursing rats induces hypomethylation of specific cytosines in the 11β-hydroxysteroid dehydrogenase-1 promoter in pup liver. Nutr. Res. 33 (11), 961–970.

Takaya, J., et al., 2011. Magnesium deficiency in pregnant rats alters methylation of specific cytosines in the hepatic hydroxysteroid dehydrogenase-2 promoter of the offspring. Epigenetics 6 (5), 573–578.

Thaler, L., et al., 2014. Methylation of BDNF in women with bulimic eating syndromes: associations with childhood abuse and borderline personality disorder. Prog. Neuro-Psychopharmacol. Biol. Psychiatry 54, 43–49.

Tooker, B.C., Ozawa, K., Newman, L.S., 2016. {CpG} promoter methylation status is not a prognostic indicator of gene expression in beryllium challenge. J. Immunotoxicol. 13 (3), 417–427. Available at: http://www.embase.com/search/results?subaction=viewrecord&from=export&id=L607288622.

Tran, P.V., et al., 2015. Fetal iron deficiency induces chromatin remodeling at the Bdnf locus in adult rat hippocampus. Am. J. Physiol. Regul. Integr. Comparat. Physiol. 308 (4), R276–R282. Available at: http://www.ncbi.nlm.nih.gov/pubmed/25519736%5Cnhttp://www.pubmedcentral.nih.gov/articlerender.fcgi?artid=PMC4329464.

Uthus, E.O., Ross, S.A., Davis, C.D., 2006. Differential effects of dietary selenium (Se) and folate on methyl metabolism in liver and colon of rats. Biol. Trace Elem. Res. 109 (3), 201–214. Available at: http://www.ncbi.nlm.nih.gov/entrez/query.fcgi?cmd=Retrieve&db=PubMed&dopt=Citation&list_uids=16632891.

Vandegehuchte, M.B., et al., 2010. Direct and transgenerational impact on *Daphnia magna* of chemicals with a known effect on DNA methylation. Comparat. Biochem. Physiol. C Toxicol. Pharmacol. 151 (3), 278–285.

Veenstra, G.J.C., Wolffe, A.P., 2001. Constitutive genomic methylation during embryonic development of Xenopus. Biochim. Biophys. Acta Gene Struct. Expr. 1521 (1–3), 39–44.

Walsh, T.K., et al., 2010. A functional DNA methylation system in the pea aphid, *Acyrthosiphon pisum*. Insect Mol. Biol. 19 (Suppl. 2), 215–228.

Warden, C.H., Fisler, J.S., 2008. Comparisons of diets used in animal models of high-fat feeding. *Cell metabolism*, 7(4), p.277. Available at. Cell Metab. 7 (4), 277. Available at: http://www.ncbi.nlm.nih.gov/pubmed/18396128. [(Accessed 4 August 2017)].

Waterland, R.A., et al., 2010. Season of conception in rural Gambia affects DNA methylation at putative human metastable epialleles. PLoS Genet. 6 (12), 1–10.

Waterland, R.A., Jirtle, R.L., 2003. Transposable elements: targets for early nutritional effects on epigenetic gene regulation. Mol. Cell. Biol. 23 (15), 5293–5300. Available at: http://www.pubmedcentral.nih.gov/articlerender.fcgi?artid=165709&tool=pmcentrez&rendertype=abstract.

Weisser, W.W., Braendle, C., Minoretti, N., 1999. Predator-induced morphological shift in the pea aphid. Proc. R. Soc. B: Biol. Sci. 266 (1424), 1175.

Wu, J., et al., 2008. Alzheimer's disease (AD)-like pathology in aged monkeys after infantile exposure to environmental metal lead (Pb): evidence for a developmental origin and environmental link for AD. J. Neurosci. 28 (1), 3–9. Available at: http://www.ncbi.nlm.nih.gov/pubmed/18171917%5Cnhttp://www.pubmedcentral.nih.gov/articlerender.fcgi?artid=PMC2486412.

Yamazaki, Y., et al., 2003. Reprogramming of primordial germ cells begins before migration into the genital ridge, making these cells inadequate donors for reproductive cloning. Proc. Natl. Acad. Sci. 100 (21), 12207–12212. Available at: http://www.pnas.org/content/100/21/12207%5Cnhttp://www.ncbi.nlm.nih.gov/pubmed/14506296%5Cnhttp://www.pnas.org/content/100/21/12207.full.pdf%5Cnhttp://www.pnas.org/content/100/21/12207.long.

Yang, G., et al., 2014. TLR2-ICAM1-Gadd45α axis mediates the epigenetic effect of selenium on DNA methylation and gene expression in Keshan disease. Biol. Trace Elem. Res. 159 (1–3), 69–80.

Yang, X., et al., 2015. The relationship between cognitive impairment and global DNA methylation decrease among aluminum Potroom workers. J. Occup. Environ. Med. 57 (7), 713–717. Available at: http://content.wkhealth.com/linkback/openurl?sid=WKPTLP:landingpage&an=00043764-201507000-00002.

Zhang, Q., et al., 2016. Effects of maternal chromium restriction on the long-term programming in MAPK signaling pathway of lipid metabolism in mice. Nutrients. 8(8).

Zhao, Y., et al., 2016. Third trimester phthalate exposure is associated with DNA methylation of growth-related genes in human placenta. Sci. Rep. 6 (August), 33449. Available at: http://www.nature.com/articles/srep33449.

2-3

DNA Methylation as a Biomarker in Environmental Epidemiology

Maya A. Deyssenroth, Robert O. Wright

Department of Environmental Medicine and Public Health, Icahn School of Medicine at Mount Sinai, New York, NY, United States

O U T L I N E

METHYLOMICS AS A POTENTIAL BIOMARKER FOR BOTH DISEASE AND TOXIC EXPOSURES

Biomarkers can be cellular, biochemical, or molecular alterations in human tissues that are directly or indirectly in the causal pathway of disease, or they may be indicators of external environmental factors that entered the body—such as chemicals. Biomarkers are often used to reflect the dose of a chemical, its biological effect, or both. Epigenomic biomarkers hold considerable promise as indicators of past toxic exposures or predictors of disease risk. While much work has focused on disease biomarkers, there is also a growing need for exposure-specific biomarkers, as in the absence of costly and time-consuming prospective data collection, past exposures to environmental agents are difficult to reconstruct and quantify. Past exposures to environmental factors such as air particles and metals (arsenic, cadmium, mercury, and lead) have been associated with multiple diseases. Past exposures to these agents and others that cause disease may leave epigenomic signatures on the human methylome. If so, identifying these signatures could allow researchers to reconstruct an individual's past environment. Since recent observations in experimental models and humans have shown alterations in both global and gene promoter DNA methylation from toxicants (Belinsky et al., 2002; Hou et al., 1999; Chanda et al., 2005; Chen et al., 2001, 2004; Sciandrello et al., 2004; Takiguchi et al., 2003; Sutherland and Costa, 2003; Zhao et al., 1997) exposures to these agents may leave unique signatures on the DNA methylome that can estimate past internal dose, particularly in the context of chronic dosing patterns.

Given above, DNA methylomic changes could in theory serve as several different types of biomarkers. They might be indicators of the biological pathway between exposure and disease; they might be a biomarker of early, preclinical disease states or a biomarker of environmental exposure. When researchers use methylomic data as a biomarker, they should consider that any of these three relationships might be the underlying root cause of an association between methylomic patterns and health. In general, the use of DNA methylomics as an indicator of the underlying biological mechanism (which may be the most common motivation) that leads to disease states is fraught with potential for misinterpretation. This is because any study, whether experimental or observational and subcellular- or population-based, cannot measure all known biological processes simultaneously and their changes over time. DNA methylation changes following an environmental exposure might be due to a direct effect on methylation, but as all biological processes have networks of interrelationships, alternative explanations will exist as well. The DNA methylation change might reflect a downstream reaction to the true underlying mechanism or an attempt by the cell/body to reset homeostasis. Feedback loops may exist in which the methylation change is a response to counter the effect of the environmental exposure. Similarly, a measured change in methylation might just represent shifts in cell populations in the target tissue. The source of the DNA and whether it is truly part of the underlying pathology should be considered. Using white blood cell (WBC) DNA as a proxy for brain or other target tissues must be justifiable, and results interpreted carefully as changes in WBC DNA methylation may be unrelated to changes in the target tissue/cell. Interpreting the biological meaning of DNA methylation changes should be done in the context of other measures, such as gene expression, histone modifications, miRNA expression, and the tissue and cells collected, and not in isolation. For these reasons, using DNA methylation patterns as a *biomarker of exposure or the presence of preclinical disease* is perhaps simpler than trying to interpret its biological meaning with respect to underlying mechanisms of action of environmental exposure or disease pathology. When using methylomic patterns as preclinical disease indicators or as biomarkers of exposure to a specific agent, the reasons why the methylomic signature arose are less important. The primary issue is whether the pattern is consistent and reproducible.

Methylomics has properties that make it exceptionally well suited to biomarker development compared with other "omic" platforms. Genomic measures of DNA *sequence* are not altered by the environment and cannot provide information on the effects of past exposure. Conversely, protein and metabolite expression are extremely dynamic, with profound changes that may occur with time of day or recent meals, limiting their utility as biomarkers as the signal to noise ratio can be very low. In contrast, DNA methylation is altered by the environment, but unlike proteins and metabolites, methylation changes can persist over time, even in the absence of the conditions that established them (Anway et al., 2005; Richards, 2006; Dolinoy et al., 2006). Furthermore, DNA methylation patterns are passed on during mitosis. Thus, DNA methylomics may be ideal for determining the presence of early disease states, as both the etiologic process for that disease and the effect of disease itself on methylomic patterns can reinforce methylomic patterns toward stability. Perhaps less appreciated but equally important, DNA methylomics in WBCs may represent a signature of past environmental exposures. Evidence for the effects of toxic exposures on the DNA methylome and the molecular mechanisms determining them (oxidative stress generation, attrition of the methyl-donor pool, and altered DNA methyltransferases activity) have been previously reviewed (Baccarelli and Bollati, 2009).

Longitudinal designs combined with methylomic approaches to biomarker development hold particular promise for exposure biomarker development. The development of biomarkers for exposure and preclinical disease has been slowed by scientific, technological, and design issues that have limited our ability to identify good biomarkers, including (1) limitations in the ability to scan relevant candidate biomarkers at high throughput and in large numbers of samples across relevant omic fields (e.g., metabolomics, proteomics, and histone epigenomics); (2) the dynamic nature of expression for many candidate biomarkers (metabolites, proteins, and mRNA levels can change dramatically on the scale of hours to minutes); (3) the static nature of DNA sequence that is not impacted by environment; and (4) the use of case-control designs in which the temporal nature of disease/exposure relationships cannot be discerned. Ideally, changes in a biomarker should precede the disease and follow the environmental exposure. By utilizing methylomic technology together with a longitudinal study design in a population with high annual disease incidence, many of these barriers to biomarker development can be overcome. If there are signature changes in the methylome secondary to disease states or to past environmental exposure, we can exploit the modifiable yet stable nature of the DNA methylome to develop relevant *exposure* and *disease* specific biomarkers.

METHYLOMICS OF WBCS IN RELATION TO DISEASE CAUSATION AND TOXIC EXPOSURES

A critical issue in any study of epigenetics is whether the DNA being assayed is a target tissue for the disease. Even identifying cell types within the target tissue is likely an important issue. Many, if not most, biomarkers of disease and toxic exposures are blood-based. WBCs can be involved in disease etiology in many cases, and the interpretation of these biomarkers can be a mix of both mechanistic factors and early disease states. In very early preclinical states, results may be driven more by mechanistic considerations, but over time, as disease changes metabolism, preclinical disease properties may be the primary driver of findings. For example, with respect to cardiovascular disease, WBCs participate in the generation of atherosclerotic plaques and in the production of inflammation, oxidative, and proteolytic damage to endothelial cells and hypercoagulability through tissue factor release (Madjid et al., 2004). Inflammatory changes from disease are reflected in blood biomarkers, as noted by the common use of serum c-reactive protein (CRP) and vascular cell adhesion molecule-1 (VCAM-1) as risk biomarkers.

Of course, as previously noted, WBC methylomics may also reflect relevant environmental exposures that predispose to cardiovascular disease. Because blood is a conduit between the external environment and human tissues, environmental chemicals *must expose WBC DNA* prior to exposing the target tissue DNA. Therefore, long-term exposure to multiple factors in the environment that predispose to disease may alter DNA methylation marks within the methylome of WBCs. Changes in the WBC methylome may last long enough to serve as a signature of past environmental exposures (Flanagan et al., 2015). Indeed, the relative stability of the methylome is part of the basis of the theory of environmental programming. Environmental stressors could therefore leave signature changes in the WBC methylome allowing us to reconstruct past environmental exposures. However, because methylomic patterns are cell-type-specific, changes in DNA methylomics might reflect changes in cell populations in a tissue rather than direct changes in methylomic patterns. This is important when using methylation as a biomarker of effect but is arguably less critical when using methylomics as a biomarker of exposure. In the latter case, one might argue that regardless of whether the methylomic change is a direct effect or due to changes in cell populations is not relevant when using it to estimate exposure. The only critical issue is the reproducibility of the methylomic pattern with respect to exposure.

METHODOLOGICAL CONSIDERATIONS IN METHYLOME-WIDE ASSOCIATION STUDIES

Discovery of methylation biomarkers of exposure and/or disease requires reliable methods for methylation quantitation. 5-methylcytosine (5mC) in the context of CG dinucleotides (CpG) is the most frequently interrogated form of DNA methylation, although, methylation at non-CpG contexts is also known to occur (Patil et al., 2014). There are ∼28 million CpG sites across the genome (Stirzaker et al., 2014), which are fewer than the expected frequency of these dinucleotides based on the GC content on the genome. This depletion has been attributed to the elevated mutability of 5mC to transition to thymine (Bird, 1980; Sved and Bird, 1990). While deamination of an unmethylated cytosine to uracil is recognized and repaired, deamination of methylated cytosine to thymine persists and is replicated, resulting in a gradual depletion of 5mC in the germline over time. The distribution of CpG sites across the genome is also not random, with a general genome-wide CpG depletion interrupted by an enrichment of CpG sites near gene promoter regions in "islands" (Bird, 1986). Part of this distribution pattern is likely attributable to the bimodal methylation pattern observed across the genome, with hypomethylation localized to CpG-dense islands near gene promoters and elevated levels of 5mC (hypermethylation) in the CpG-sparse "open sea" (Saxonov et al., 2006). Widespread aberrations in this bimodal pattern, including gene-specific hypermethylation and genome-wide hypomethylation, are known hallmarks of cancer (Lister et al., 2009), and the extent to which these and other such aberrations are triggered by environmental agents is a question of great importance in toxicoepigenetics.

Technological advances have made high-throughput, genome-wide assessments of methylation at single-base resolution possible. Methodologies to conduct methylome-wide association studies typically employ bisulfite conversion of genomic DNA prior to quantitation to enable the distinction of the methylation state at CpG sites. During this deamination process, unmethylated cytosines are converted to uracil and ultimately thymine through subsequent amplification cycles, while methylated cytosines are protected from deamination and remain intact. The calculated cytosine-thymine ratio detected at a given CpG site is used to infer the level of methylation (Frommer et al., 1992). While existing platforms, including probe-based arrays and next-generation sequencing, have been adapted to detect the methylation of bisulfite-converted DNA, additional challenges are introduced by bisulfite conversion, including information loss due to DNA damage, reduction in genomic complexity (nucleotide pool reduced from four to three bases), and doubling of the genome size (the loss of symmetry across strands).

Platforms That Assess Single-Base Resolution Methylation

Array-Based: MethylationEPIC BeadChip (Infinium) Microarray

The Infinium HumanMethylation 450 K BeadChip microarray has been the platform of choice for large-scale epidemiological studies, with the recently updated 850 K MethylationEPIC BeadChip array soon to be in wide-scale application. The original platform included >99% RefSeq annotated genes; was enriched for CpG Islands (96% of UCSC database) and informed by known differentially methylated regions (DMRs); and, therefore, was biased toward promoter regions. The recently updated 850 K array covers 90% of the original platform and has additional coverage of sites at more distal regulatory regions.

The Infinium assay includes individual beads containing 50 bp probes designed to hybridize bisulfite-converted DNA, base-pairing with the target CpG of interest at the 3' end of the probe. The subsequent methylation-state-specific single-base extension of the probe with labeled ddNTPs enables "genotyping" the 3' CpG site to deduce the methylation state (Bibikova et al., 2011; Sandoval et al., 2011). Two types of probe designs are employed by Infinium arrays. Infinium I probes utilize two beads for each CpG locus, one bead containing probes specific to the methylated state of the targeted site and the other bead containing probes specific to the unmethylated state of the targeted site. The bisulfite-converted DNA fragment hybridizes to either the unmethylated or methylated bead type, depending on the methylation state of the CpG site, with subsequent single-base extension of labeled ddNTPs in bead-type-specific manner. Infinium II assays utilize one bead per CpG locus, targeting both the methylated and unmethylated state of the CpG site of interest. Here, the cytosine of the CpG site is the single-base extension locus. The incorporation of a labeled G occurs opposite a methylated C, whereas the incorporation of a labeled A occurs opposite an unmethylated T (Bibikova et al., 2011). Type II probes occupy half the space of type I probes; however, type I probes are better designed to target CpG-dense regions (Pidsley et al., 2016).

Next-Generation Sequencing

Whole-genome bisulfite sequencing (WGBS) is the gold standard in single-base methylation quantitation, in terms of accuracy, coverage, and resolution (Suzuki and Bird, 2008; Bibikova and Fan, 2010; Laird, 2010). The in-depth characterization afforded by this methodology has been instrumental in delineating the methylation profile of the human genome. While it provides the most comprehensive survey of genome-wide methylation, covering ~20 million sites, it is currently not cost-effective for population-wide studies. WGBS is also inefficient for population-wide studies given that ~70% of the sequencing reads contain uninformative CpG sites with invariant methylation levels (Stirzaker et al., 2014; Lister et al., 2009; Ziller et al., 2013).

In an effort to lower the sequencing burden, reduced representation bisulfite sequencing (RRBS) was introduced to allow a more focused analysis of randomly selected genomic subsets. Here, genomic DNA is digested at CCGG sites with methylation insensitive restriction enzymes to enrich for fragments containing CpG sites. These fragments undergo size selection (~200bp), reducing to a subset of the genome (~2%) that is enriched for CpG sites, CpG islands, gene promoters, and gene bodies (Chatterjee et al., 2017). The resulting fragment pool is adenylated and ligated with bisulfite-resistant adapters, treated with sodium bisulfite, PCR-amplified, cloned, and sequenced (Meissner et al., 2005). Approximately, 20% of CpG sites (10 × of 450K array) are covered. Unlike probe-based methods, coverage for each site across samples is not guaranteed. By design, RRBS is biased toward CpG-rich regions, while CpG poor intergenic regions are underrepresented.

Capture-based targeted bisulfite sequencing can be performed following bisulfite conversion of capture-selected native DNA or through capture selection of bisulfite-converted DNA. Prebisulfite capture systems require a higher amount of input DNA (3 µg) due to the bottleneck incurred from the reduced complexity output library than postbisulfite capture systems (1 µg). However, careful design of capture probes is required to enable the detection of all potential alleles generated as a result of conversion and intermediate methylation states.

The SureSelect Methyl-Seq (Agilent Technologies) platform is a prebisulfite capture system. This platform captures regions enriched for CpG islands, shores, promoters, hypomethylated regions, and differentially methylated regions. It includes the same set of genes covered by the Infinium array, with coverage of 3.7 million CpG sites. Genomic DNA is fragmented, ligated to adapters, and hybridized by target-specific probes. The bead-captured library is bisulfite-converted, PCR-amplified, and sequenced using the Illumina platform (Lee et al., 2011).

The SeqCap Epi Enrichment (NimbleGen) platform is a postbisulfite capture system. Similar to the SureSelect platform, SeqCap targets the same set of genes as the Infinium array and surveys 5.5 million CpG sites (Allum et al., 2015; Li et al., 2015; Ziller et al., 2016). Unlike the single-stranded design of the SureSelect platform, both strands are targeted in the SeqCap system. Genomic DNA is fragmented to ~180bp, and adapters are ligated. Bisulfite conversion of the adapter-ligated library is followed by PCR amplification. Hybridization of probes is followed by bead capture and PCR amplification to enrich for targeted regions that are subsequently sequenced using the Illumina platform.

Preprocessing and Normalization

MethylationEPIC BeadChip (Infinium) Microarray

The Infinium BeadChip arrays have emerged as the most popular tools to evaluate methylome-wide changes in population-wide studies due to the reliability of the measurements, ease of use, and cost efficiency (Pidsley et al., 2016). To derive meaningful signals from the generated data, however, the unique technical aspects of the platform warrant consideration. Bioinformatics pipelines to address these issues have been developed and are now commonly applied in the analysis workflow.

In addition to removal of background fluorescent signals (Triche Jr. et al., 2013), a typical Infinium array preprocessing pipeline includes the removal of cross-reactive probes, probes overlaying polymorphic sites, and probe-type bias correction as described below.

A proportion of probes included on the array are known to align to multiple regions and repetitive sequences in the genome. These cross-reactive probes can lead to unreliable measurements of methylation at a given CpG site (Chen et al., 2013). Approximately 5% of the probes included on the 850K array have been observed to be cross-reactive (Pidsley et al., 2016). Current recommendations include the removal of these ambiguous probes from further analysis.

A proportion of the probes are also known to contain sites overlaying single-nucleotide polymorphisms (SNPs) (Naeem et al., 2014). Confounding can be introduced if the presence of these polymorphisms is not addressed, particularly in comparisons across individuals stemming from populations with varying allele frequencies. Current recommendations include cross-referencing the dataset with SNP databases of similar genetic structure as the study population to identify SNP sites residing within or near probes. Approximately 13% of probes included on the 850K array overlap SNPs (Pidsley et al., 2016). These probes should be considered for removal from further analysis.

The differing chemistries of the two probes used in the array result in a smaller distribution of methylation values among type II probes compared with type I probes. This difference in the dynamic range between the probe types lends itself to a potential bias in the determination of differentially methylated sites toward an enrichment of type I probes, particularly in the determination of DMRs that can include both types of probes. In addition to the reduced dynamic range, type II probes also demonstrate greater technical variability and lower reproducibility than type I probes (Dedeurwaerder et al., 2011). The Infinium II probe design predominates on the 850K platform, accounting for ~84% of the probes covered on the array (Pidsley et al., 2016; Moran et al., 2016). Bioinformatics-based approaches that apply peak-based correction to rescale type II probes on the basis of type I probes, including beta mixture quantile dilation (BMIQ), are now part of the preprocessing pipeline (Dedeurwaerder et al., 2011; Morris and Beck, 2015; Teschendorff et al., 2013).

Finally, removal of low-detection probes with detection P-values $>.05$ in more than a certain threshold of samples (e.g., 1%) and samples with certain threshold of probes failing detection (e.g., 10%) is typically included as additional processing steps (Wu and Kuan, 1708).

Bisulfite Sequencing

Similar to array-based approaches for methylation quantitation, bisulfite-sequencing-based methods require a unique set of preprocessing steps to facilitate the discovery of robust findings. The preprocessing workflow for bisulfite sequencing data is far less established than the Infinium array workflow, but general guidelines are provided below.

FastQ files are assessed for overall read level quality, using, for example, the fastQC pipeline (http://www.bioinformatics.babraham.ac.uk/projects/fastqc/). Important quality control assessments include determining base call quality and adaptor sequence contamination.

Mapping bisulfite-converted reads to a reference genome can be performed using methylation-aware alignment tools, where both cytosines and thymines are considered as potential matches to genomic cytosines. However, this approach is biased toward mapping the more informative, methylated allele over the unmethylated allele. In silico conversions prior to alignment using existing aligners such as Bowtie2 (Langmead and Salzberg, 2012), BLAT (Kent, 2002), SOAP (Li et al., 2008), and BWA (Li and Durbin, 2010) can be employed to minimize this bias (Tran et al., 2014). For example, prior to alignment, tools such as Bismark (Krueger and Andrews, 2011) and BS-Seeker2 (Guo et al., 2013) convert all residual cytosines in the sequencing reads and all cytosines in the reference genome to thymine, whereas BSMAP (Xi and Li, 2009) masks all thymines in the reads as potential matches to cytosines in the genome.

Additional postalignment processing steps may be warranted. Sites with low read depth (e.g., $< 10\times$) and reads with unusually high coverage (e.g., >95th percentile of read coverage distribution), which may indicate amplification bias, can be considered for removal. The presence of SNPs in the reads should be evaluated by referencing against available SNP databases, and sites overlapping polymorphisms should be considered for removal (Krueger et al., 2012; Baubec and Akalin, 2016). Bisulfite conversion rates should be evaluated, for example, by calculating the C:T ratio at non-CpG sites for each sample or referencing against unmethylated spike-in controls. Samples demonstrating low bisulfite conversion rates and samples with low call rates (e.g., $<75\%$) should be considered for removal (Wreczycka et al., 2017).

Addressing Confounding

Alterations in the epigenetic landscape is a hallmark of cell lineage commitment during development, determining the transcriptional profile specific to each cell type. Addressing differences in methylation due to cell-type composition is, therefore, of particular concern in environmental epidemiology studies, where methylation is typically assessed in biospecimens of mixed cell type. Indeed, whole blood, which is among the most commonly assessed tissue types in environmental epidemiology studies assessing methylomic biomarkers due to its ease in attainment, is particularly enriched with subclasses of immune cells. Observed differences in methylation may, therefore, reflect changes in cell-type populations due to a reactive inflammatory response to a pathological condition rather than a direct marker of a causal pathway. In addition to elevating false positives, not accounting for cell-type composition in analyses may also obscure true signals. This is especially the case in studies assessing exposure-related CpG sites since expected effect size estimates due to exposure are likely much smaller than cell-type-related differences in methylation. For the Infinium array, both reference-based and reference-free methods are available to address cell-type composition in methylome-wide studies. Reference-based methods use available datasets on a priori defined cell types known to be present in the analyzed tissue. CpG sites that maximally discriminate across these defined cell types are identified, and these sites are leveraged to estimate the cell-type composition in each sample. The estimated proportions can then be incorporated as covariates in models assessing differences in methylation (Teschendorff and Zheng, 2017). To date, such reference-based methods are only available in a select set of tissues, including whole blood (Jaffe and Irizarry, 2014; Houseman et al., 2012). An additional limitation of reference-based methods is that it requires a priori knowledge and reference databases for all relevant cell types of a tissue of interest. Reference free methods have been developed to address these issues. For example, given the assumption that cell-type composition accounts for the majority of the observed variability in methylation, several methods remove effects due to cell-type composition by identifying and removing the top components underlying the variability of the data (Houseman et al., 2014; Rahmani et al., 2016; Zou et al., 2014). Alternatively, surrogate variable analyses, as described below, have also been proposed (Leek and Storey, 2007; Teschendorff et al., 2011). To date, standardized methods for cell-type adjustments for bisulfite sequencing data are not yet established.

In addition to variation due to true biological signals, observed variability in methylation values likely also captures nonbiological factors, including technical artifacts. For example, global shifts in measurements may be introduced across batches of nucleic acid extraction, bisulfite conversion, plating, chip, and reagent lots. Subsequent analyses are likely confounded if these technical variables are additionally correlated with biological variables of interests. ComBat is a popular tool that returns batch-corrected methylation values based on user-provided data on methylation values, the outcome variable of interest, known batch variables, such as processing date and plate batch, and all other known covariates (Johnson et al., 2007). In practice, however, delineating all possible confounders in a study is not feasible. To account for unmodeled confounding, surrogate variable analysis (SVA) generates a residual matrix from data where the signal due to the primary variable has been removed. This residual matrix, representing any remaining systematic heterogeneity in the data, is decomposed to principal components reflecting surrogate variables of unmodeled confounding that can be included in models as covariates (Leek and Storey, 2007).

Inclusion of Technical Replicates

The inclusion of technical replicates facilitates further assessments of the quality of the generated data. Additional processing steps based on the evaluation of technical replicates can include the identification and removal of noninformative CpG sites with little interindividual variability across the assessed subjects to reduce the multiple testing burden (Meng et al., 2010). Modeling the intraclass correlation coefficient (ICC), which compares biological variability (between replicates) with total variability (sum of within and between replicates), facilitates the determination of CpG sites that are invariant (low interindividual variability) and sites with unstable (high intraindividual variability) methylation estimates (Bose et al., 2014; Chen et al., 2016). Studies have shown that the majority of genome-wide significant sites are among high ICC sites (Bose et al., 2014), highlighting the gain in statistical power that can be gained by restricting datasets to CpG sites demonstrating high ICC values.

2. DNA METHYLATION

Calculation for Differentially Methylated Sites

Methylation values derived from the microarray data are expressed in one of two forms. The beta value is the ratio of the methylated probe intensity and the overall intensity (methylated + unmethylated probe). This statistic ranges from 0 to 1 and follows a beta distribution. The M-value is the log2 ratio of the methylated and unmethylated probe intensities, with a data range consistent with the normal distribution (Du et al., 2010). Whether the beta or M-value is more appropriate should be evaluated based on the statistical framework selected to determine differentially methylated sites. For example, log-ratio-based M-values are a better fit when computing empirical Bayes-derived moderated t/F statistics as implemented in limma (Smyth, 2005; Robinson et al., 2014). R packages are available that wrap the detection of site-specific methylation differences into 450K preprocessing pipeline include RnBeads, which utilizes limma as the default method to conduct differential methylation analysis (Assenov et al., 2014).

Sequencing-derived methylation values are reported as counts of the methylated and unmethylated alleles. While a methylation ratio equivalent to the array-based beta value can be calculated, direct modeling of the count data is generally preferred. As sites with deep sequencing coverage have more reliable methylation value estimates than sites with sparse coverage, count-based models that account for the uncertainty associated with varying coverage, such as the beta-binomial model, are commonly applied in the analysis of sequencing data across replicate samples (Dolzhenko and Smith, 2014). R packages for the detection of differentially methylated sites in sequencing data based on the beta-binomial model [BiSeq (Hebestreit et al., 2013), methylSig (Park et al., 2014), and DSS (Feng et al., 2014)] and linear [limma (Smyth, 2005) and BSmooth (Hansen et al., 2012)] and logistic regression [methylkit (Akalin et al., 2012)] are available.

Calculation for Differentially Methylated Regions

Spatial correlations have been observed across neighboring CpG sites, likely reflecting a coordinated effort among multiple sites to exert a regulatory effect on gene activity. Site-specific detection of differential methylation may miss this contextual information provided by neighboring sites. The observed spatial correlation also has implications for the statistical power to detect significant sites, as the penalization due to multiple testing under an assumption of independence is likely too stringent. This is particularly pertinent in environmental epidemiology studies, where exposure-related effect sizes may be small at a given individual site, but persistent across a region. Incorporating spatial information into differential methylation analysis may, therefore, enhance statistical power to detect meaningful signals.

Bump hunting, which enables peak detection while accounting for potential unmeasured confounding using SVA, was among the earliest widely adopted data-driven methods to detect differentially methylated regions. Here, site-specific beta estimates are smoothed, and the area summary statistics of the smoothed regions are evaluated against a permutation-derived null distribution (Jaffe et al., 2012). Several additional methods are now available. For example, comb-p combines site-specific differential methylation analysis P-values of spatially correlated sites to generate aggregated and corrected P-values assigned to regions

(Pedersen et al., 2012). In DMRcate, the limma-based moderated *t*-statistics of individual sites are squared, and a gaussian kernel is applied to smooth the squared statistics within a given window (Peters et al., 2015). R packages are available that wrap detection of DMRs into 450 K preprocessing pipeline include minfi (Aryee et al., 2014), which utilizes bump hunting as its DMR detection method.

In addition to DMRcate, several additional R-based methods are available to detect differentially methylated regions in capture-based targeted sequencing and RRBS data, including methylKit (Akalin et al., 2012) and BiSeq (Hebestreit et al., 2013). In methylKit, the user specifies the tiling window parameters (e.g., 500 bp window and tiling at 100 bp steps), the methylated and unmethylated counts are then aggregated within the specified window, and the logistic regression framework is applied to identify differentially methylated sites. In BiSeq, a user-defined window (e.g., minimum 10 CpG sites, no >50 bp between neighboring sites) is first specified. CpG sites within the defined window are smoothed, and beta regression is applied on the smoothed methylation values at each CpG site. A hierarchical FDR correction is applied to detect clusters with a minimum of one differentially methylated site.

METHYLOMICS AS A POTENTIAL BIOMARKER FOR BOTH DISEASE AND TOXIC EXPOSURES

The development of cost-effective platforms to survey the methylome and the establishment of analytic pipelines to identify robust methylation signatures from the generated data has led to a steady growth of environmental epidemiology studies incorporating DNA methylation assessments. Below, we highlight several studies that showcase the utility of surveying the methylome for biomarkers of exposure and disease.

Example: Smoking and *AHRR* Methylation

Methylation patterns in retrotransposons are not the only reported relationships with the environment. In fact, gene-specific methylation patterns are likely more useful as presumably they may be more specific markers of environmental exposure. Recent studies demonstrate that smoke exposure in adults (Monick et al., 2012) and among fetuses (Joubert et al., 2012) is associated with altered leukocyte DNA methylation of the aryl-hydrocarbon receptor repressor (*AHRR*), located on chromosome 5 (Joubert et al., 2012; Philibert et al., 2013). The aryl-hydrocarbon receptor (AHR) is a cytoplasmic receptor, similar to steroid receptor, and mediates the toxicity of xenobiotics (Stockinger et al., 2014). The AHRR proteins inhibit AHR function by disrupting the AHR-ligand interaction. Methylation of the *AHRR* gene is inversely associated with other measures of smoking exposure such as serum cotinine and smoking questionnaire data. A recent metaanalysis conducted across 13 birth cohorts included in the Pregnancy and Childhood Epigenetics (PACE) Consortium also revealed an *AHRR* CpG site as the top most differentially methylated by maternal smoke status, highlighting the robustness of this exposure signal and the value of such collaborative efforts to replicate findings (Joubert et al., 2016).

Example: Air Pollution and DNA Methylation

Inhaled particulate pollutants have been shown to produce systemic changes in gene expression, which can be detected in peripheral blood of exposed individuals and might contribute to or at least reflect the toxic properties of particulate matter (PM) (Wang et al., 2005). Observations from in vitro and animal models have shown that air particles or air particle components such as toxic metals can induce changes in DNA methylation. Air particles are known to increase the production of reactive oxygen species, perhaps in a catalytic fashion via redox cycling (Donaldson et al., 2001). Oxidative DNA damage can interfere with the ability of methyltransferases to interact with DNA (Valinluck et al., 2004) and also alters the expression of genes belonging to DNA methylation machinery (Fratelli et al., 2005). Toxic metals have been demonstrated to colocalize in the DNA-binding domain of DNA methyltransferase and thus modify their activity (Takiguchi et al., 2003). Depletion of the cellular methyl pool has also been indicated as a contributor to the global hypomethylation observed in cells treated with toxic metals (Wright and Baccarelli, 2007). These observations suggest that PM exposure can modify the methylome and serve as signatures of PM exposures.

Environmental measures of air pollution and toxic metals are not easily quantified, and direct measures typically reflect only recent exposures. New omic biomarkers that can quantify the role of *long-term exposure* to air pollutants and metals would be a major scientific advancement. Such biomarkers would best be developed in a prospective cohort study and could then be replicated in a case-control design. A major challenge of epigenetics, as opposed to traditional genetics, is that precisely because epigenetic marks change, case-control studies, which offer more potential to analyze relevant target/diseased tissue samples, cannot provide evidence of the temporality of the association. Epigenetic differences could be a consequence of the disease, rather than on the causal pathway to the disease or a prediagnostic marker. In order to understand the role of DNA methylation as a biomarker of disease, one needs to understand the temporal relationship between the DNA methylation change and disease onset/severity. Prospective cohort studies allow the examination of whether disease preceded epigenetic changes or vice versa. If there are signature changes in the epigenome secondary to exposure, researchers could reconstruct past exposures based on these signatures, a process that would strengthen the design of future case-control studies.

Example: Biological Aging as Measured by the Epigenetic Clock

Several studies have demonstrated epigenetic drift, with a general trend toward global hypomethylation and region-specific hypermethylation, across the life span. The extent of this drift is shaped by life experiences, as showcased by studies that report similar methylation patterns among young monozygotic twins and more divergent methylation patterns among older monozygotic twins (Fraga et al., 2005). With the availability of base-resolution platforms, subsequent studies have resolved these overall trends to specific CpG sites that constitute an "epigenetic clock" (Horvath, 2013; Hannum et al., 2013). Acceleration of the epigenetic clock has been noted for several health outcomes, including obesity (Horvath et al., 2014), HIV infection (Horvath and Levine, 2015), cancer (Levine et al., 2015a), Alzheimer's disease (Levine et al., 2015b), and frailty (Breitling et al., 2016). Environmental exposures may trigger this process, as suggested by recent studies that report associations between DNA methylation age acceleration and traumatic stress (Wolf et al., 2018), as well as smoking (Gao et al., 2016).

Example: VTRNA2-1 as a Metastable Epiallele

While much of the genome exhibits little interindividual variation in tissue-specific methylation levels, specific regions, termed metastable epialleles, vary even across genetically identical individuals (Dolinoy et al., 2007). This was first exemplified through manipulations of the murine model agouti gene. These labile marks have since been shown to be responsive to gestational exposures with concomitant phenotypic alterations. Examples of such environmental biosensors are beginning to emerge in human studies as well. Most notably, differential methylation of the imprinted VTRNA2-1 locus has been identified across several epigenome-wide surveys, in association with environmental exposures including folic acid supplementation during pregnancy (Richmond et al., 2018), occupational pesticide exposure (van der Plaat et al., 2018), and maternal food access (Finer et al., 2016; Silver et al., 2015).

Example: Epigenetic Pathway Linking Prenatal Maternal Stress and Wheeze in Children

In addition to biomarkers of exposure, recent reports have also shown epigenetic marks that link environmental exposures to health outcomes as potential biomarkers of biological effect. A WGBS analysis revealed over 2000 prenatal stress-related DMRs, enriched for biological processes including Wnt signaling and calcium homeostasis, contrasting low versus high prenatal stress mother-child pairs. Expression changes of several DMR-associated genes, including *PPP3R1* and *NFATC3*, were additionally linked to the development of wheeze later in life (Trump et al., 2016).

CONCLUSION

Technological and analytic advances have enabled the application of base-resolution methylome-wide surveys in environmental epidemiology studies. Early evidence suggests that these methods can be leveraged to discover methylation signatures indicative of exposure and disease. Replication of existing findings is warranted to assess the robustness of these markers, and consortium-wide efforts using metaanalytic approaches will likely play an important role in this validation process. Additional characterizations of methylation markers are also warranted, including establishing the temporality of methylation signatures in relation to the exposure/outcome of interest. Finally, to fully realize the programmable aspect of these marks, such findings can inform intervention strategies that modify aberrant methylation marks in a targeted fashion to counteract exposure-induced deviations in health trajectories.

References

Akalin, A., Kormaksson, M., Li, S., et al., 2012. methylKit: a comprehensive R package for the analysis of genome-wide DNA methylation profiles. Genome Biol. 13, R87.

Allum, F., Shao, X., Guenard, F., et al., 2015. Characterization of functional methylomes by next-generation capture sequencing identifies novel disease-associated variants. Nat. Commun. 6, 7211.

Anway, M.D., Cupp, A.S., Uzumcu, M., Skinner, M.K., 2005. Epigenetic transgenerational actions of endocrine disruptors and male fertility. Science (New York, NY) 308, 1466–1469.

Aryee, M.J., Jaffe, A.E., Corrada-Bravo, H., et al., 2014. Minfi: a flexible and comprehensive bioconductor package for the analysis of Infinium DNA methylation microarrays. Bioinformatics (Oxford, England) 30, 1363–1369.

Assenov, Y., Muller, F., Lutsik, P., Walter, J., Lengauer, T., Bock, C., 2014. Comprehensive analysis of DNA methylation data with RnBeads. Nat. Methods 11, 1138–1140.

Baccarelli, A., Bollati, V., 2009. Epigenetics and environmental chemicals. Curr. Opin. Pediatr. 21, 243–251.

Baubec, T., Akalin, A., 2016. Genome-wide analysis of DNA methylation patterns by high-throughput sequencing. In: Aransay, A.M., Lavín Trueba, J.L. (Eds.), Field Guidelines for Genetic Experimental Designs in High-Throughput Sequencing. Springer International Publishing, Cham, pp. 197–221.

Belinsky, S.A., Snow, S.S., Nikula, K.J., Finch, G.L., Tellez, C.S., Palmisano, W.A., 2002. Aberrant CpG island methylation of the p16(INK4a) and estrogen receptor genes in rat lung tumors induced by particulate carcinogens. Carcinogenesis 23, 335–339.

Bibikova, M., Fan, J.B., 2010. Genome-wide DNA methylation profiling. Wiley Interdiscip. Rev. Syst. Biol. Med. 2, 210–223.

Bibikova, M., Barnes, B., Tsan, C., et al., 2011. High density DNA methylation array with single CpG site resolution. Genomics 98, 288–295.

Bird, A.P., 1980. DNA methylation and the frequency of CpG in animal DNA. Nucleic Acids Res. 8, 1499–1504.

Bird, A.P., 1986. CpG-rich islands and the function of DNA methylation. Nature 321, 209–213.

Bose, M., Wu, C., Pankow, J.S., et al., 2014. Evaluation of microarray-based DNA methylation measurement using technical replicates: the atherosclerosis risk in communities (ARIC) Study. BMC Bioinf. 15, 312.

Breitling, L.P., Saum, K.U., Perna, L., Schottker, B., Holleczek, B., Brenner, H., 2016. Frailty is associated with the epigenetic clock but not with telomere length in a German cohort. Clin. Epigenetics 8, 21.

Chanda, S., Dasgupta, U.B., Guhamazumder, D., et al., 2005. DNA hypermethylation of promoter of gene p53 and p16 in arsenic exposed people with and without malignancy. Toxicol. Sci.

Chatterjee, A., Rodger, E.J., Morison, I.M., Eccles, M.R., Stockwell, P.A., 2017. Tools and strategies for analysis of genome-wide and gene-specific DNA methylation patterns. Methods Mol. Biol. 1537, 249–277.

Chen, H., Liu, J., Zhao, C.Q., Diwan, B.A., Merrick, B.A., Waalkes, M.P., 2001. Association of c-myc overexpression and hyperproliferation with arsenite-induced malignant transformation. Toxicol. Appl. Pharmacol. 175, 260–268.

Chen, H., Li, S., Liu, J., Diwan, B.A., Barrett, J.C., Waalkes, M.P., 2004. Chronic inorganic arsenic exposure induces hepatic global and individual gene hypomethylation: implications for arsenic hepatocarcinogenesis. Carcinogenesis 25, 1779–1786.

Chen, Y.A., Lemire, M., Choufani, S., et al., 2013. Discovery of cross-reactive probes and polymorphic CpGs in the Illumina Infinium HumanMethylation450 microarray. Epigenetics 8, 203–209.

Chen, J., Just, A.C., Schwartz, J., et al., 2016. CpGFilter: model-based CpG probe filtering with replicates for epigenome-wide association studies. Bioinformatics 32, 469–471.

Dedeurwaerder, S., Defrance, M., Calonne, E., Denis, H., Sotiriou, C., Fuks, F., 2011. Evaluation of the Infinium Methylation 450K technology. Epigenomics 3, 771–784.

Dolinoy, D.C., Weidman, J.R., Jirtle, R.L., 2006. Epigenetic gene regulation: linking early developmental environment to adult disease. Reprod. Toxicol.

Dolinoy, D.C., Das, R., Weidman, J.R., Jirtle, R.L., 2007. Metastable epialleles, imprinting, and the fetal origins of adult diseases. Pediatr. Res. 61, 30r–37r.

Dolzhenko, E., Smith, A.D., 2014. Using beta-binomial regression for high-precision differential methylation analysis in multifactor whole-genome bisulfite sequencing experiments. BMC Bioinf. 15, 215.

Donaldson, K., Stone, V., Seaton, A., MacNee, W., 2001. Ambient particle inhalation and the cardiovascular system: potential mechanisms. Environ. Health Perspect. 109 (Suppl. 4), 523–527.

Du, P., Zhang, X., Huang, C.C., et al., 2010. Comparison of beta-value and M-value methods for quantifying methylation levels by microarray analysis. BMC Bioinf. 11, 587.

Feng, H., Conneely, K.N., Wu, H., 2014. A Bayesian hierarchical model to detect differentially methylated loci from single nucleotide resolution sequencing data. Nucleic Acids Res. 42, e69.

Finer, S., Iqbal, M.S., Lowe, R., et al., 2016. Is famine exposure during developmental life in rural Bangladesh associated with a metabolic and epigenetic signature in young adulthood? A historical cohort study. BMJ Open 6, e011768.

Flanagan, J.M., Brook, M.N., Orr, N., et al., 2015. Temporal stability and determinants of white blood cell DNA methylation in the breakthrough generations study. Cancer Epidemiol. Biomark. Prev. 24, 221–229.

Fraga, M.F., Ballestar, E., Paz, M.F., et al., 2005. Epigenetic differences arise during the lifetime of monozygotic twins. Proc. Natl. Acad. Sci. USA 102, 10604–10609.

Fratelli, M., Goodwin, L.O., Orom, U.A., et al., 2005. Gene expression profiling reveals a signaling role of glutathione in redox regulation. Proc. Natl. Acad. Sci. USA 102, 13998–14003.

Frommer, M., McDonald, L.E., Millar, D.S., et al., 1992. A genomic sequencing protocol that yields a positive display of 5-methylcytosine residues in individual DNA strands. Proc. Natl. Acad. Sci. USA 89, 1827–1831.

Gao, X., Zhang, Y., Breitling, L.P., Brenner, H., 2016. Relationship of tobacco smoking and smoking-related DNA methylation with epigenetic age acceleration. Oncotarget 7, 46878–46889.

Guo, W., Fiziev, P., Yan, W., et al., 2013. BS-Seeker2: a versatile aligning pipeline for bisulfite sequencing data. BMC Genomics 14, 774.

Hannum, G., Guinney, J., Zhao, L., et al., 2013. Genome-wide methylation profiles reveal quantitative views of human aging rates. Mol. Cell 49, 359–367.

Hansen, K.D., Langmead, B., Irizarry, R.A., 2012. BSmooth: from whole genome bisulfite sequencing reads to differentially methylated regions. Genome Biol. 13, R83.

Hebestreit, K., Dugas, M., Klein, H.U., 2013. Detection of significantly differentially methylated regions in targeted bisulfite sequencing data. Bioinformatics 29, 1647–1653.

Horvath, S., 2013. DNA methylation age of human tissues and cell types. Genome Biol. 14, R115.

Horvath, S., Levine, A.J., 2015. HIV-1 infection accelerates age according to the epigenetic clock. J. Infect. Dis. 212, 1563–1573.

Horvath, S., Erhart, W., Brosch, M., et al., 2014. Obesity accelerates epigenetic aging of human liver. Proc. Natl. Acad. Sci. USA 111, 15538–15543.

Hou, M., Morishita, Y., Iljima, T., et al., 1999. DNA methylation and expression of p16(INK4A) gene in pulmonary adenocarcinoma and anthracosis in background lung. Int. J. Cancer 84, 609–613.

Houseman, E.A., Accomando, W.P., Koestler, D.C., et al., 2012. DNA methylation arrays as surrogate measures of cell mixture distribution. BMC Bioinf. 13, 86.

Houseman, E.A., Molitor, J., Marsit, C.J., 2014. Reference-free cell mixture adjustments in analysis of DNA methylation data. Bioinformatics 30, 1431–1439.

Jaffe, A.E., Irizarry, R.A., 2014. Accounting for cellular heterogeneity is critical in epigenome-wide association studies. Genome Biol. 15, R31.

Jaffe, A.E., Murakami, P., Lee, H., et al., 2012. Bump hunting to identify differentially methylated regions in epigenetic epidemiology studies. Int. J. Epidemiol. 41, 200–209.

Johnson, W.E., Li, C., Rabinovic, A., 2007. Adjusting batch effects in microarray expression data using empirical Bayes methods. Biostatistics 8, 118–127.

Joubert, B.R., Haberg, S.E., Nilsen, R.M., et al., 2012. 450K epigenome-wide scan identifies differential DNA methylation in newborns related to maternal smoking during pregnancy. Environ. Health Perspect. 120, 1425–1431.

Joubert, B.R., Felix, J.F., Yousefi, P., et al., 2016. DNA methylation in newborns and maternal smoking in pregnancy: genome-wide consortium meta-analysis. Am. J. Hum. Genet. 98, 680–696.

Kent, W.J., 2002. BLAT–the BLAST-like alignment tool. Genome Res. 12, 656–664.

Krueger, F., Andrews, S.R., 2011. Bismark: a flexible aligner and methylation caller for Bisulfite-Seq applications. Bioinformatics 27, 1571–1572.

Krueger, F., Kreck, B., Franke, A., Andrews, S.R., 2012. DNA methylome analysis using short bisulfite sequencing data. Nat. Methods 9, 145–151.

Laird, P.W., 2010. Principles and challenges of genomewide DNA methylation analysis. Nat. Rev. Genet. 11, 191–203.

Langmead, B., Salzberg, S.L., 2012. Fast gapped-read alignment with Bowtie 2. Nat. Methods 9, 357–359.

Lee, E.J., Pei, L., Srivastava, G., et al., 2011. Targeted bisulfite sequencing by solution hybrid selection and massively parallel sequencing. Nucleic Acids Res. 39, e127.

2. DNA METHYLATION

Leek, J.T., Storey, J.D., 2007. Capturing heterogeneity in gene expression studies by surrogate variable analysis. PLoS Genet. 3, 1724–1735.

Levine, M.E., Hosgood, H.D., Chen, B., Absher, D., Assimes, T., Horvath, S., 2015a. DNA methylation age of blood predicts future onset of lung cancer in the women's health initiative. Aging 7, 690–700.

Levine, M.E., Lu, A.T., Bennett, D.A., Horvath, S., 2015b. Epigenetic age of the pre-frontal cortex is associated with neuritic plaques, amyloid load, and Alzheimer's disease related cognitive functioning. Aging 7, 1198–1211.

Li, H., Durbin, R., 2010. Fast and accurate long-read alignment with Burrows-Wheeler transform. Bioinformatics 26, 589–595.

Li, R., Li, Y., Kristiansen, K., Wang, J., 2008. SOAP: short oligonucleotide alignment program. Bioinformatics 24, 713–714.

Li, Q., Suzuki, M., Wendt, J., et al., 2015. Post-conversion targeted capture of modified cytosines in mammalian and plant genomes. Nucleic Acids Res. 43, e81.

Lister, R., Pelizzola, M., Dowen, R.H., et al., 2009. Human DNA methylomes at base resolution show widespread epigenomic differences. Nature 462, 315–322.

Madjid, M., Awan, I., Willerson, J.T., Casscells, S.W., 2004. Leukocyte count and coronary heart disease: implications for risk assessment. J. Am. Coll. Cardiol. 44, 1945–1956.

Meissner, A., Gnirke, A., Bell, G.W., Ramsahoye, B., Lander, E.S., Jaenisch, R., 2005. Reduced representation bisulfite sequencing for comparative high-resolution DNA methylation analysis. Nucleic Acids Res. 33, 5868–5877.

Meng, H., Joyce, A.R., Adkins, D.E., et al., 2010. A statistical method for excluding non-variable CpG sites in high-throughput DNA methylation profiling. BMC Bioinf. 11, 227.

Monick, M.M., Beach, S.R., Plume, J., et al., 2012. Coordinated changes in AHRR methylation in lymphoblasts and pulmonary macrophages from smokers. Am. J. Med. Genet. B Neuropsychiatr. Genet. 159B, 141–151.

Moran, S., Arribas, C., Esteller, M., 2016. Validation of a DNA methylation microarray for 850,000 CpG sites of the human genome enriched in enhancer sequences. Epigenomics 8, 389–399.

Morris, T.J., Beck, S., 2015. Analysis pipelines and packages for Infinium HumanMethylation450 BeadChip (450k) data. Methods (San Diego, CA) 72, 3–8.

Naeem, H., Wong, N.C., Chatterton, Z., et al., 2014. Reducing the risk of false discovery enabling identification of biologically significant genome-wide methylation status using the HumanMethylation450 array. BMC Genomics 15, 51.

Park, Y., Figueroa, M.E., Rozek, L.S., Sartor, M.A., 2014. MethylSig: a whole genome DNA methylation analysis pipeline. Bioinformatics 30, 2414–2422.

Patil, V., Ward, R.L., Hesson, L.B., 2014. The evidence for functional non-CpG methylation in mammalian cells. Epigenetics 9, 823–828.

Pedersen, B.S., Schwartz, D.A., Yang, I.V., Kechris, K.J., 2012. Comb-p: software for combining, analyzing, grouping and correcting spatially correlated P-values. Bioinformatics (Oxford, England) 28, 2986–2988.

Peters, T.J., Buckley, M.J., Statham, A.L., et al., 2015. De novo identification of differentially methylated regions in the human genome. Epigenetics Chromatin 8, 6.

Philibert, R.A., Beach, S.R., Lei, M.K., Brody, G.H., 2013. Changes in DNA methylation at the aryl hydrocarbon receptor repressor may be a new biomarker for smoking. Clin. Epigenetics 5, 19.

Pidsley, R., Zotenko, E., Peters, T.J., et al., 2016. Critical evaluation of the Illumina MethylationEPIC BeadChip microarray for whole-genome DNA methylation profiling. Genome Biol. 17, 208.

Rahmani, E., Zaitlen, N., Baran, Y., et al., 2016. Sparse PCA corrects for cell type heterogeneity in epigenome-wide association studies. Nat. Methods 13, 443–445.

Richards, E., 2006. Inherited epigenetic variation—revisiting soft inheritance. Nat. Rev. Genet. 7, 395–401.

Richmond, R.C., Sharp, G.C., Herbert, G., et al., 2018. The long-term impact of folic acid in pregnancy on offspring DNA methylation: follow-up of the Aberdeen Folic Acid Supplementation Trial (AFAST). Int. J. Epidemiol.

Robinson, M.D., Kahraman, A., Law, C.W., et al., 2014. Statistical methods for detecting differentially methylated loci and regions. Front. Genet. 5, 324.

Sandoval, J., Heyn, H., Moran, S., et al., 2011. Validation of a DNA methylation microarray for 450,000 CpG sites in the human genome. Epigenetics 6, 692–702.

Saxonov, S., Berg, P., Brutlag, D.L., 2006. A genome-wide analysis of CpG dinucleotides in the human genome distinguishes two distinct classes of promoters. Proc. Natl. Acad. Sci. USA 103, 1412–1417.

Sciandrello, G., Caradonna, F., Mauro, M., Barbata, G., 2004. Arsenic-induced DNA hypomethylation affects chromosomal instability in mammalian cells. Carcinogenesis 25, 413–417.

2. DNA METHYLATION

Silver, M.J., Kessler, N.J., Hennig, B.J., et al., 2015. Independent genomewide screens identify the tumor suppressor VTRNA2-1 as a human epiallele responsive to periconceptional environment. Genome Biol. 16, 118.

Smyth, G.K., 2005. limma: linear models for microarray data. In: Gentleman, R., Carey, V.J., Huber, W., Irizarry, R.A., Dudoit, S. (Eds.), Bioinformatics and Computational Biology Solutions Using R and Bioconductor. Springer New York, New York, NY, pp. 397–420.

Stirzaker, C., Taberlay, P.C., Statham, A.L., Clark, S.J., 2014. Mining cancer methylomes: prospects and challenges. Trends Genet. 30, 75–84.

Stockinger, B., Di Meglio, P., Gialitakis, M., Duarte, J.H., 2014. The aryl hydrocarbon receptor: multitasking in the immune system. Annu. Rev. Immunol. 32, 403–432.

Sutherland, J.E., Costa, M., 2003. Epigenetics and the environment. Ann. NY Acad. Sci. 983, 151–160.

Suzuki, M.M., Bird, A., 2008. DNA methylation landscapes: provocative insights from epigenomics. Nat. Rev. Genet. 9, 465–476.

Sved, J., Bird, A., 1990. The expected equilibrium of the CpG dinucleotide in vertebrate genomes under a mutation model. Proc. Natl. Acad. Sci. USA 87, 4692–4696.

Takiguchi, M., Achanzar, W.E., Qu, W., Li, G., Waalkes, M.P., 2003. Effects of cadmium on DNA-(cytosine-5) methyltransferase activity and DNA methylation status during cadmium-induced cellular transformation. Exp. Cell Res. 286, 355–365.

Teschendorff, A.E., Zheng, S.C., 2017. Cell-type deconvolution in epigenome-wide association studies: a review and recommendations. Epigenomics 9, 757–768.

Teschendorff, A.E., Zhuang, J., Widschwendter, M., 2011. Independent surrogate variable analysis to deconvolve confounding factors in large-scale microarray profiling studies. Bioinformatics 27, 1496–1505.

Teschendorff, A.E., Marabita, F., Lechner, M., et al., 2013. A beta-mixture quantile normalization method for correcting probe design bias in Illumina Infinium 450 k DNA methylation data. Bioinformatics 29, 189–196.

Tran, H., Porter, J., Sun, M.A., Xie, H., Zhang, L., 2014. Objective and comprehensive evaluation of bisulfite short read mapping tools. Adv. Bioinforma. 2014, 472045.

Triche Jr., T.J., Weisenberger, D.J., Van Den Berg, D., Laird, P.W., Siegmund, K.D., 2013. Low-level processing of Illumina Infinium DNA methylation BeadArrays. Nucleic Acids Res. 41, e90.

Trump, S., Bieg, M., Gu, Z., et al., 2016. Prenatal maternal stress and wheeze in children: novel insights into epigenetic regulation. Sci. Rep. 6, 28616.

Valinluck, V., Tsai, H.H., Rogstad, D.K., Burdzy, A., Bird, A., Sowers, L.C., 2004. Oxidative damage to methyl-CpG sequences inhibits the binding of the methyl-CpG binding domain (MBD) of methyl-CpG binding protein 2 (MeCP2). Nucleic Acids Res. 32, 4100–4108.

van der Plaat, D.A., de Jong, K., de Vries, M., et al., 2018. Occupational exposure to pesticides is associated with differential DNA methylation. Occup. Environ. Med.

Wang, Z., Neuburg, D., Li, C., et al., 2005. Global gene expression profiling in whole-blood samples from individuals exposed to metal fumes. Environ. Health Perspect. 113, 233–241.

Wolf, E.J., Maniates, H., Nugent, N., et al., 2018. Traumatic stress and accelerated DNA methylation age: a meta-analysis. Psychoneuroendocrinology 92, 123–134.

Wreczycka, K., Gosdschan, A., Yusuf, D., Gruning, B., Assenov, Y., Akalin, A., 2017. Strategies for analyzing bisulfite sequencing data. J. Biotechnol. 261, 105–115.

Wright, R.O., Baccarelli, A., 2007. Metals and neurotoxicology. J. Nutr. 137, 2809–2813.

Wu, M.C., Kuan, P.F., 1708. A guide to Illumina BeadChip data analysis. Methods Mol. Biol. (Clifton, NJ) 2018, 303–330.

Xi, Y., Li, W., 2009. BSMAP: whole genome bisulfite sequence MAPping program. BMC Bioinf. 10, 232.

Zhao, C.Q., Young, M.R., Diwan, B.A., Coogan, T.P., Waalkes, M.P., 1997. Association of arsenic-induced malignant transformation with DNA hypomethylation and aberrant gene expression. Proc. Natl. Acad. Sci. USA 94, 10907–10912.

Ziller, M.J., Gu, H., Muller, F., et al., 2013. Charting a dynamic DNA methylation landscape of the human genome. Nature 500, 477–481.

Ziller, M.J., Stamenova, E.K., Gu, H., Gnirke, A., Meissner, A., 2016. Targeted bisulfite sequencing of the dynamic DNA methylome. Epigenetics Chromatin 9, 55.

Zou, J., Lippert, C., Heckerman, D., Aryee, M., Listgarten, J., 2014. Epigenome-wide association studies without the need for cell-type composition. Nat. Methods 11, 309–311.

2. DNA METHYLATION

DNA Hydroxymethylation: Implications for Toxicology and Epigenetic Epidemiology

Jairus Pulczinski*, Bonnie H.Y. Yeung*, Qian Wu*,†, Robert Y.S. Cheng‡, Wan-yee Tang*

*Department of Environmental Health and Engineering, Johns Hopkins Bloomberg School of Public Health, Baltimore, MD, United States †Department of Hygienic Analysis and Detection and Ministry of Education Key Lab for Modern Toxicology, School of Public Health, Nanjing Medical University, Nanjing, China ‡Cancer and Inflammation Program, Center for Cancer Research, National Cancer Institute, Frederick, MD, United States

INTRODUCTION

Epigenetic mechanisms are critical for responding to environmental stressors, from famine to air pollution, as they allow fine tuning of gene expression patterns without changing the DNA sequence (Feil and Fraga, 2012). Therefore, epigenetic mechanisms can allow interpretation of environmental changes and interactions among genome, tissue development, and organ physiology. These mechanisms are implicated in phenotypic plasticity or the ability of the cells to change their behavior in response to internal and external environmental cues that can disrupt cellular homeostasis and give rise to the aberrant mechanisms implicated in disease development. Different lifestyle choices or environmental stressors render individuals more or less susceptible to chronic disease states through epigenetic reprogramming that can trigger either somatic or germ line changes in the epigenome during development and possibly be transmitted through multiple generations (Martos et al., 2015).

There are three major categories of epigenetic regulation: histone modification, noncoding RNA, and DNA methylation. One of the major types of DNA methylation in mammals is 5-methylcytosine (5-mC). DNA methylation, which is the addition of a methyl group to the 5-position of cytosine to produce 5-mC (Razin and Riggs, 1980), the "fifth base of DNA," is one of the most commonly studied epigenetic process in both experimental and epidemiological research (Bakulski and Fallin, 2014; Szyf, 2007) (Fig. 1). Further oxidation of cytosine (5-mC) to 5-hydroxymethylcytosine (5-hmC) results in the activation of gene transcription and provides new insight into how DNA demethylation occurs. The discovery of the ten-eleven translocation (TET) methylcytosine dioxygenase family raised substantial interest in 5-hmC-mediated DNA demethylation pathways and biological role of 5-hmC in development (Kriaucionis and Heintz, 2009; Tahiliani et al., 2009; Wu and Zhang, 2017). TET enzymes are responsible for catalyzing the oxidation of 5-mC to 5-hmC and the iterative oxidation of 5-hmC to 5-formylcytosine (5-fC) and 5-carboxycytosine (5-caC) DNA derivatives (He et al., 2011). 5-hmC is the most abundant catalytic product of TET-mediated 5-mC oxidation and has been proposed as the stable intermediate in DNA demethylation pathways (Bachman et al., 2014; Brazauskas and Kriaucionis, 2014; Hahn et al., 2014). Aberrant activities of TET proteins and their 5-hmC products are also implicated in different disease states, and these may have potential uses as epigenetic signatures for disease pathogenesis and exposure assessment (Yang et al., 2012; Shang et al., 2013; Cheng et al., 2014; Strand et al., 2015; Zhang et al., 2015; Saldanha et al., 2017; Tian et al., 2018). We therefore urge further investigation of the link between environmental cues and TET-mediated DNA hydroxymethylation (Dao et al., 2014).

CHEMISTRY OF DNA HYDROXYMETHYLATION

The Sixth-Base of Cytosine

The existence of noncanonical DNA bases has been known for decades, with methylated cytosine identified in calf thymus DNA in 1948 (Hotchkiss, 1948). As early as 1975, DNA methylation was postulated to play a role in gene regulation, serving as an epigenetic mechanism that could modify gene expression (Holliday and Pugh, 1975; Riggs, 1975). As the field

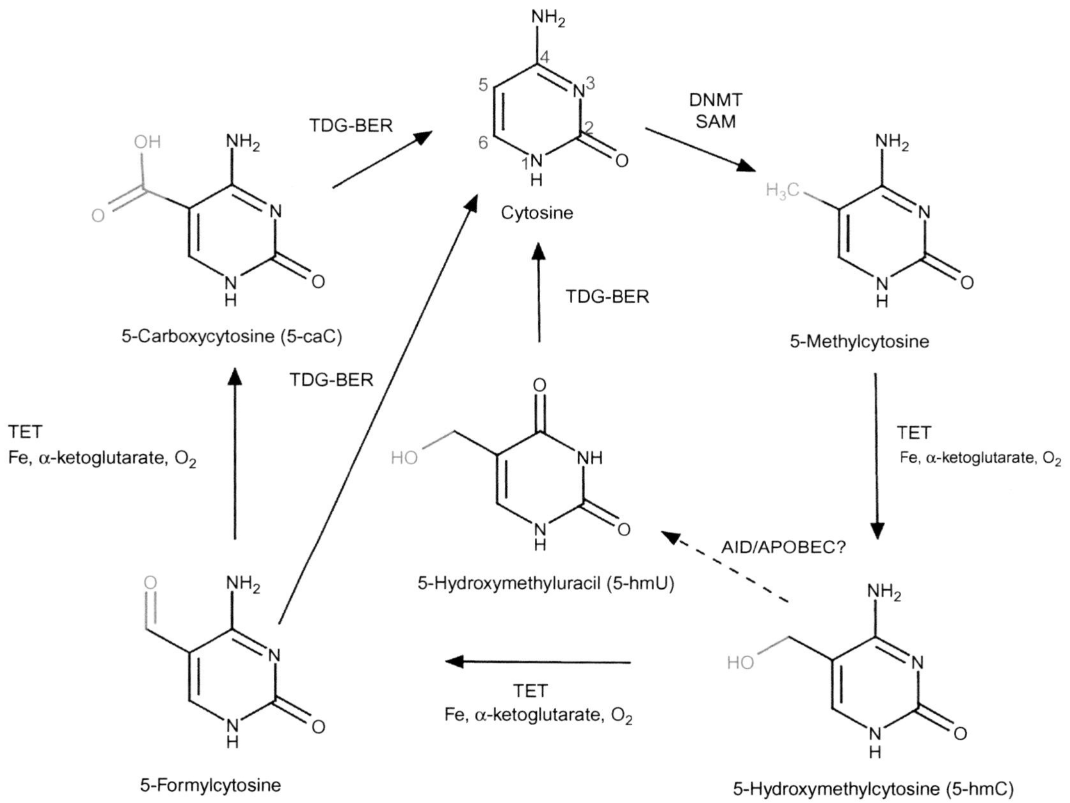

FIG. 1 Cytosine and its derivatives. Derivatives of cytosine are based on a pyrimidine ring, a six-membered aromatic heterocycle with nitrogen substitutions at the 1- and 3-positions. Additionally, cytosine has an ammonium side group on the 4-position and a keto side group at the 2-position. Methylation occurs at the 5-position to give 5-methylcytosine (5-mC). With hydroxyl addition occurring on the methyl side chain, it is known as 5-hydroxymethylcytosine (5-hmC). Additional modifications include the formation of a formyl group on the methyl side chain (known as 5-formylcytosine (5-fC)) or a carboxyl group at the 5-position (known as 5-carboxycytosine (5-caC)). The addition of the methyl group on cytosine is catalyzed by DNA methyltransferases (DNMT) that use S-adenosylmethionine (SAM) as a methyl donor. The ten-eleven translocation (TET) methylcytosine dioxygenases catalyze the hydroxylation of 5-mC, using an iron cofactor and α-ketoglutarate. It has been speculated that 5-hmC is either spontaneously or enzymatically converted to cytosine through DNA repair pathways. Activation-induced cytidine deaminases (AID/APOBEC) mediate deamination of 5-hmC to 5-hydroxymethlyuracil (5-hmU), generating an abasic site, which is recognized and subsequently removed by thymine DNA glycosylase (TDG) and base excision repair (BER) machinery, to restore an unmodified cytosine. TET can catalyze additional oxidative steps, leading sequentially to 5-fC and 5-caC. The 5-fC and 5-caC can then be excised by TDG, and a cytosine can be replaced at the resulting abasic site via BER. Software Marvin 17.24.0, 2017 (ChemAxon), was used for drawing, displaying, and characterizing chemical structures, substructures, and reactions.

has progressed, 5-mC has been shown to influence gene expression and regulation in a heritable manner and has since been considered the "fifth base" of DNA. Currently, DNA methylation is one of the best studied epigenetic mechanisms and has been implicated in transposon regulation, X inactivation, and gene imprinting, as well as in development and

2. DNA METHYLATION

disease (Jones, 2012). However, DNA methylation is not a static marker, as methylation levels vary considerably throughout early development (Guibert and Weber, 2013). Additionally, DNA methylation in the promoter regions of genes is generally a repressive marker; therefore, a mechanism must exist to regulate DNA methylation to remove it when needed.

One likely mechanism of demethylation involves 5-hmC, which was first identified in the DNA of T-even phages in 1953, and later in protozoans (Wyatt and Cohen, 1953). However, little thought was given to its possible existence in the genome of metazoans, although Penn et al. claimed to have identified 5-hmC in the brains of rats, mice, and frogs (Penn et al., 1972). Penn's findings were not replicated in two subsequent research studies, so 5-hmC was not believed to be a component of the DNA of higher organisms for many years (Gommers-Ampt and Borst, 1995; Kothari and Shankar, 1976). It was not until 2009, when 5-hmC was identified in human Purkinje cells and mouse embryonic stem cells, that the presence of 5-hmC in mammals was confirmed (Kriaucionis and Heintz, 2009; Tahiliani et al., 2009). Thereafter, the TET enzymes were identified as the dioxygenases that catalyze the hydroxyl addition to 5-mC (Tahiliani et al., 2009) via a reaction that uses an iron cofactor and α-ketoglutarate (α-KG) to produce 5-hmC. Furthermore, TET can catalyze additional oxidative steps, leading to sequentially 5-fC and 5-caC. 5-hmC, 5-fC, or 5-caC could act as an intermediate in both passive and active DNA demethylation pathways involving DNA repair enzymes like activation-induced cytidine deaminase (AID) and thymine DNA glycosylase (TDG) (He et al., 2011; Inoue and Zhang, 2011) (Fig. 1).

5-hmC is generally found on CpG dinucleotides (CpG) sites, as these are the sites where 5-mC is commonly present. The CpG sites are dispersed across the genome and occur at a much lower frequency than expected based on probability, implying that many of these sites have been lost via deamination of cytosine to uracil and replaced with thymine by DNA repair mechanisms or direct deamination of 5-mC to thymine, leading to a loss of cytosines in the genome (Walsh and Xu, 2006). Regions of the genome exist where these CpG sites occur at a high frequency, called CpG island (CGI). CGI is a region of at least 200 base pairs, of which the GC content is greater than 50% and the CpG ratio is greater than 60% (Gardiner-Garden and Frommer, 1987). CGIs occur mainly in the promoter regions, where ~70% of the CGIs are unmethylated except for those on the inactive X chromosome and some associated with imprinted genes (Deaton and Bird, 2011; Saxonov et al., 2006). Methylation of a CGI in a gene promoter generally represses the gene, likely due to impaired transcription factor binding at the promoter and via the recruitment of DNA methylation binding proteins that facilitate repression (Bogdanović and Veenstra, 2009). The CpG sites in the intragenic regions of genes are generally methylated, which is a permissive modification as it prevents initiation at alternative promoter regions (Neri et al., 2017). Unlike 5-mC, it has been suggested that the existence of 5-hmC associates with an increase in TET-mediated active DNA demethylation and gene transcription (reviewed by Wu and Zhang, 2017). Genomic distribution of 5-hmC is suggested to be cell-type specific. In embryonic stem cells (ESCs), 5-hmC is enriched at promoters and distal regulatory elements with low-to-intermediate CpG density (Xu et al., 2011; Wu et al., 2011). Strikingly, 5-hmC is enriched at bivalent promoters of genes that are repressed in ESCs but activated on cell differentiation (Wu et al., 2011; Pastor et al., 2011). 5-hmC also displays an increase in gene bodies, which may associate with the coupling between 5-mC oxidation and transcription elongation (Xu et al., 2011; Wu et al., 2011). In adult neurons, 5-hmC is depleted from transcription start sites, which is different from that in ESCs.

However, 5-hmC is highly enriched in gene bodies and enhancer regions, which is similar to that in ESCs (Wen et al., 2014). It is believed that more studies on genomic distribution of 5-hmC in different cell types or tissues will be completed as next-generation sequencing can provide whole-genome analysis of 5-hmC at base resolution. Several TET proteins are known that vary in catalytic activity and timing of expression (Wu and Zhang, 2017). For example, TET3 is highly expressed in the developing embryo, where it is believed to play a role in catalyzing genome-wide demethylation. TET1 and TET2 are expressed in early development and throughout adulthood and are believed to play a role in active demethylation, including primordial germ cell demethylation, and possibly in DNA repair (Pastor et al., 2013; An et al., 2015; Kafer et al., 2016). The three recognized TET family proteins, TET1, TET2, and TET3, have the catalytic domain, which coordinates the iron and α-KG with 5-mC. The long forms of TET1 and TET3 are characterized by the presence of a conserved cysteine-rich region (CXXC binding domain), which binds to unmethylated CGIs to stabilize the enzymes. TET2 lacks a CXXC site, which may influence its binding affinity to modified bases in the regulation of TET2 protein function (Ko et al., 2013). Taken together, the formation of this "sixth base of DNA (5-hmC)" represents a pathway that leads to passive or active demethylation, either by inhibition of the replicative DNA methyltransferase from identifying methyl markers on newly synthesized, hemimethylated DNA or by active, sequential oxidation until base excision that results in unmodified cytosine. However, the formation of 5-hmC appears to be more than simply a transient step toward demethylation, and in many cases, it may be a stable marker that itself mediates gene expression (Lin et al., 2017; Wu and Zhang, 2017).

Distribution of 5-hmC in the Genome

Unlike 5-mC, which represents roughly 4% of all cytosines in the genome, independent of somatic tissue type, the distribution of 5-hmC varies greatly by tissue type (Nestor et al., 2012). 5-hmC is most abundant in tissue of the central nervous system (CNS), where roughly 0.6% 5-hmC is detected (Globisch et al., 2010). In the CNS, the hypothalamus has the greatest abundance of 5-hmC and the spinal cord the least. In other tissues, like kidney, nasal epithelium, bladder, heart, muscle, and lung, roughly 0.15% 5-hmC is detected. The lowest levels (less than 0.1%) of 5-hmC are found in the testes, thyroid, liver, spleen, and pituitary (Globisch et al., 2010). Although 5-hmC can be involved in a pathway that leads to demethylation, emerging research suggests that 5-hmC may also be a stable epigenetic mark with transcriptional ramifications. For example, Bachman et al. (2014) radioactively labeled 5-hmC in vivo and later measured it via tandem liquid chromatography-mass spectrometry (LC/MS). They found that 5-hmC levels were unchanged during the cell cycle, and that 5-hmC addition occurs hours after DNA replication, unlike DNMT1-mediated DNA methylation, which is rapid. Due to this lag in 5-hmC addition, highly replicative cells are likely to lose 5-hmC over time (Bachman et al., 2014). A recent analysis of 5-hmC in mice identified a strong preference of 5-hmC for the transcriptional start site (TSS) in active genes in brain and liver tissue and a high association with gene expression. Strikingly, no association was noted between 5-hmC and the expression of housekeeping genes, leading the authors to speculate that 5-hmC may play a role in cell-specific gene expression (Lin et al., 2017). Furthermore, multiple studies reported enriched 5-hmC in enhancers and chromatin accessible regions and 5-hmC enrichment at lineage-specific transcription factor binding sites, providing credence to the belief that 5-hmC plays a crucial role in cell fate decisions leading to tissue-specific transcription differences (Bogdanović et al., 2016; Kim et al., 2018; Li et al., 2018; Lio et al., 2016).

Changes in 5-hmC Level During Development

5-hmC plays a pivotal and dynamic role in embryogenesis and development. Fig. 2 summarizes the dynamic changes in 5-hmC level during epigenetic reprogramming in embryo development and mammalian germ cells, as well as during puberty and aging. The germ cell from parental genome in mammals is condensed and highly methylated (80% of the CGI in sperm), and upon fertilization, it undergoes rapid demethylation. This demethylation

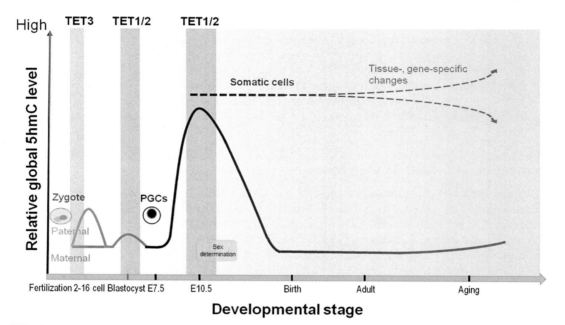

FIG. 2 Schematic diagram showing the dynamics of global 5-hmC levels in mammalian development (mouse model). After fertilization, the male pronucleus of the zygote undergoes TET3-mediated demethylation, which results in increased 5-hmC levels (Wossidlo et al., 2011). Imprinted genes or some repetitive sequences retain their methylation patterns during this epigenetic reprogramming. When the blastocyst implants into the uterus, the inner cell mass undergoes *de novo* DNA methylation. TET1 and TET2 proteins are highly expressed and are proposed to modify the methylation patterns (Pastor et al., 2013). The second TET1/TET2-dependent demethylation occurs in primordial germ cells (PGCs), leading to the induction of 5-hmC (López et al., 2017; Tan and Shi, 2012; Wu and Zhang, 2014). Male germ cells are gradually remethylated at E15, whereas the remethylation occurs after birth in females. DNA demethylation, which is associated with changes in global 5-hmC level, is involved in oocyte and sperm aging (Jenkins et al., 2013; Qian et al., 2015). In somatic cells, the 5-hmC marks are stabilized during development (López et al., 2017). However, tissue-, gene-, and time-specific changes in 5-hmC level may occur in response to environmental stimuli and aging. E, embryonic day; TET, ten-eleven translocation; PGCs, primordial germ cells; 5-hmC, 5-hydroxymethylcytosine.

coincides with a period of high TET3 expression and substantial formation 5-hmC, as well as 5-fC and 5-caC (Guibert and Weber, 2013; Wu and Zhang, 2017). Silencing of TET3 at this stage results in the retention of a considerable portion of the embryonic DNA methylation signature and aberrations in the resulting developmental pattern (Gu et al., 2011), indicating that paternal demethylation is an active event mediated by TET3 (Wossidlo et al., 2011). The maternal genome also undergoes demethylation, but this event was largely thought to be passive and due to DNA replication. Imprinted genes or some repetitive sequences retain their methylation patterns during this epigenetic reprogramming. After global demethylation, implantation follows, which coincides with genome-wide *de novo* DNA methylation. The inner cell mass that gives rise to all the somatic tissues becomes hypermethylated. The TET1 and TET2 proteins are highly expressed during this time and are proposed to modify the methylation patterns (Pastor et al., 2013). The second TET1/TET2-dependent demethylation occurs in primordial germ cells (PGCs), leading to the induction of 5-hmC (Hackett et al., 2013; Vincent et al., 2013; Tan and Shi, 2012; Wu and Zhang, 2014; López et al., 2017). Male germ cells are gradually remethylated at E15, whereas the remethylation occurs after birth in females. DNA demethylation, which is associated with changes in global 5-hmC level, is involved in oocyte and sperm aging (Jenkins et al., 2013; Qian et al., 2015). In somatic cells, the 5-hmC marks are stabilized during development (López et al., 2017). Nevertheless, tissue-, gene-, and time-specific changes in 5-hmC levels may occur in response to environmental stresses.

MEASUREMENT OF 5-hmC

The widely used techniques for the quantitation of 5-mC content, such as bisulfite sequencing and methylation-sensitive restriction enzyme digestion, cannot distinguish 5-hmC from 5-mC (Huang et al., 2010; Jin et al., 2010). Since the discovery of 5-hmC, numerous techniques have been developed to increase the accuracy, specificity, and sensitivity of 5-hmC detection. These techniques are mainly focused on the following approaches: (1) liquid chromatography and mass spectrometry (LC-MS/MS), (2) antibody affinity-based detection, (3) enzymatic and chemical modifications, and (4) single-molecule detection.

Global 5-hmC Measurement

LC-MS/MS with multiple reaction monitoring (MRM) provides a sensitive and accurate measurement of 5-hmC and 5-mC in the genome. The major drawbacks are the expense of the equipment and the requirement for considerable operator expertise and the need for a large amount of input DNA (Kriaucionis and Heintz, 2009; Le et al., 2011; Tang et al., 2013; Zhang et al., 2012). Antibody affinity-based techniques, such as dot blots (Koh et al., 2011) and enzyme-linked immunosorbent assays (ELISA) (Li and Liu, 2011), are commonly used, and they are fast and simple and require less input DNA when compared with LC-MS/MS, but these advantages come at the expense of sensitivity and accuracy. Enzymatic modification of 5-hmC allows labeling of the modified cytosines for quantitation of 5-hmC. For example, β-glucosyltransferase catalyzes the glucosylation of 5-hmC, yielding β-glucosyl-

5-hmC (5-ghmC) (Baskin et al., 2007; Song et al., 2011a). The 5-ghmC can be then labeled with radioisotopes or fluorescent dyes for detection (Chen et al., 2017; Gilat et al., 2017; Szwagierczak et al., 2010). Alternatively, glucosylation of 5-hmC can be conducted with a glucose moiety modified to contain an azide, resulting in azide-5ghmC. The azide-5ghmC can then initiate a click reaction with an alkyne-fluorescent tag to aid in detection and improve accuracy and sensitivity (Michaeli et al., 2013; Shahal et al., 2014).

Locus-Specific 5-hmC Measurement

DNA Immunoprecipitation (DIP)

Immune-based enrichment of DNA samples is a cost-effective method that captures DNA fragments containing the epitope for antibodies specific for 5-hmC, which can then be used for downstream analysis such as real-time PCR (qPCR) and high-throughput sequencing. For example, hydroxymethylated DNA immunoprecipitation (hMeDIP) uses an antibody specific for 5-hmC to enrich 5-hmC on sonicated genomic DNA (Ficz et al., 2011; Williams et al., 2011; Wu et al., 2011; Xu et al., 2011; Pastor et al., 2011; Huang et al., 2012). Unlike hMeDIP, 5-methylsulfonate (CMS)-DIP requires sodium bisulfite conversion of 5-hmC and then targets the resulting 5-hmC adduct via CMS antibody (Hayatsu and Shiragami, 1979; Pastor et al., 2011; Huang et al., 2012;). In a similar manner, JBP1-DIP utilizes an antibody specific to 5-ghmC to pull down modified bases after the T4 β-glucosyltransferase-mediated conversion of input DNA (Robertson et al., 2011, 2012). Each DIP method has its specific pros and cons; for instance, binding of 5-hmC-specific antibody is biased toward CpG-dense regions and simple repeats (Ficz et al., 2011; Nestor et al., 2012; Song et al., 2012; Stroud et al., 2011; Tan et al., 2013). CMS-DIP shows a PCR bias after bisulfite treatment, but less of a sequence bias, and has a lower background relative to hMeDIP (Huang et al., 2012; Pastor et al., 2011; Song et al., 2012). The major drawback of JBP1-DIP is that its data are incompatible with other techniques, although the labeling of 5-ghmC is highly efficient (Song et al., 2012; Thomson et al., 2013). The use of biotin/streptavidin is an alternative for enriching azide-5ghmC after a click reaction (Baskin et al., 2007; Song et al., 2011a). The GLIB method (glucosylation, periodate oxidation, and biotinylation) is a related technique that uses T4 β-glucosyltransferase to add glucose to 5-hmC. Sodium periodate is then added, which oxidized the glucose moiety forming aldehydes. Finally, an aldehyde-reactive probe is added, which adheres two biotin molecules to each 5-hmC-glucose molecule (Huang et al., 2012; Pastor et al., 2011).

Enzymatic and Chemical Modifications

In a similar manner to antibody detection, certain techniques use enzymes or chemicals to protect the 5-hmC on gene-specific loci. One such modification favors 5-ghmC, which is protected against cleavage by methylation-insensitive restriction enzymes (Borgaro and Zhu, 2013). This unique feature has led to the development of several techniques, such as combined glycosylation restriction analysis (CGRA) (Song et al., 2011b), reduced representation 5-hydroxymethylcytosine profiling (RRHP) (Petterson et al., 2014), AbaSI coupled with sequencing (Aba-seq) (Sun et al., 2013), and Pvu-Seal-seq (Sun et al., 2015). However, each restriction enzyme has limitations regarding the cleavage sequence, so utilization of multiple restriction enzymes is recommended to increase the genomic coverage (Davis and Vaisvila, 2011; Song et al., 2011b; Sun et al., 2016).

Selective chemical labeling with exonuclease digestion (SCL-exo) involves the azide glucosylation of 5-hmC, followed by treatment with biotin and streptavidin, in order to immobilize the DNA, and finally treatment with an exonuclease that is unable to cut at the modified base. The resulting strands can be sequenced and overlaid to identify the location of 5-hmC in the genome. In a similar manner, boronic acid can be used to inhibit the amplification of azide-glucosylated 5-hmC-containing DNA. Comparison with DNA not treated with boronic acid allows assessment of the density of 5-hmC in a gene- or fragment-specific manner (Zhao et al., 2014). TET-assisted bisulfite sequencing (TAB-seq) (Yu et al., 2012a, 2012b) and oxidative bisulfite sequencing (oxBS-seq) (Booth et al., 2013) are two techniques that can provide a single-base resolution with absolute abundance. TAB-seq involves glucosylation of 5-hmC, which protects the modified bases from oxidation during subsequent exposure to excess TET1. As 5-mC is not glucosylated, it is oxidized to 5-fC, which is converted to uracil upon subsequent bisulfite treatment. Thus, any cytosine identified via sequencing is 5-hmC protected in the first step by glucosylation. The main concerns about TAB-seq include the efficiency of TET enzyme-mediated conversion of 5-mC to 5-fC, protection efficiency of 5-hmC from glucosylation, abundance of 5-hmC at the modification site, and sequencing depths (Yu et al., 2012a,b). OxBS-seq selectively oxidizes 5-hmC to 5-fC with potassium perruthenate ($KRuO4$), followed by bisulfite conversion and sequencing. In parallel, samples without pretreatment of $KRuO4$ are subject to traditional bisulfite sequencing (BS-seq). The 5-hmC signal is calculated by the subtraction of 5-mC signals in ox-BS from those of BS-seq (Booth et al., 2013). OxBS-seq has similar limitations of that of TAB-seq, including the efficiency of $KRuO4$ on the conversion of 5-hmC to 5-fC and the sequencing depth.

Single Molecule Detection

Several emerging technologies utilize nanopore-based and single-molecule real-time (SMRT) sequencing to detect 5-hmC levels. The synthetic nanopore, which contains a small pore, often 1 nm in diameter, is inserted in an electrically resistant membrane that is then placed in between a salt solution in a well (Dekker, 2007). An electric current is applied to the membrane that causes the ions in the solution to pass through the pore while also driving DNA into the pore (Branton et al., 2008; Manrao et al., 2011). The electric current passing through the membrane is measured, and the unique perturbations caused by DNA driven through the pore allow for the identification of DNA bases. Modifications of this method reduce the speed at which DNA is drawn into the pore allowing single nucleotide detection of 5mC and 5hmC (Laszlo et al., 2013). SMRT sequencing is another single-molecule detection technique but one that relies on mimicking DNA replication to quantify DNA bases. SMRT sequencing relies on a technology called zero-model waveguide, where a plate is filled with tiny wells at which DNA polymerase paired with single-stranded DNA is immobilized (Ardui et al., 2018). An excitation beam is aimed into each well, and the wells are then filled with free phospholinked nucleotides bound to fluorescent tags (Kircher and Kelso, 2010). As the DNA polymerase incorporates the bases into a complementary strand, their fluorophores are excited by the excitation beam and give of specific colored light that can be detected

(Bayley, 2006). Cytosine modifications can be identified via azide glucosylation of 5-hmC and complexation with HS-N3-5 to create a large cytosine complex. This complex creates considerable lag time during DNA synthesis, which can then be interpreted as the presence of 5-hmC (Song et al., 2011a,b,c). Another SMRT-seq technique uses the T4 phage β-glucosyltransferase to add UDP-glucose to 5-hmC, followed by T6 phage β-glucosyl α-glucosyl transferase addition of a second UDP-glucose to the initial glucose moiety. This deglycosylation greatly enhances the kinetic signature of 5-hmC, improving detection (Chavez et al., 2014). Despite the ability of nanopore and SMRT techniques to process long reads rapidly, their major downside is a high error rate compared with the template strand (Matsuda et al., 2015; Ross et al., 2013).

EXPOSURES TO ENVIRONMENTAL TOXICANTS/STRESSORS AND CHANGES IN 5-hmC

5-hmC, much like its relative 5-mC, has been shown to be crucial in early development and disease pathogenesis (Tan and Shi, 2012). In addition, in certain contexts, 5-hmC appears to be a stable mark in the genome, rather than simply a transient marker of demethylation (Hahn et al., 2014). As our understanding of the multifaceted role of 5-hmC in the genome expands, so has the interest in the occurrence of dysregulation of 5-hmC due to exposures to environmental cues and disease states.

Heavy Metals

Most notable among the heavy metals are arsenic, lead, mercury, chromium, and cadmium, which are common environmental contaminants that have significant human toxicity and are linked to a range of negative outcomes, including cancer (Tchounwou et al., 2012). Current research points to dysregulation of DNA hydroxymethylation patterns or alteration of TET activity due to heavy metal exposure. In HEK293T cells, trivalent arsenic targets the zinc-finger domains of TET proteins, specifically the cysteine-rich region. This binding alters the catalytic efficiency of the TET proteins and reduces their oxygenase activity, resulting in decreased 5-hmC levels (Liu et al., 2015). Similarly, mouse embryonic stem cells (ESCs) treated with arsenic, cadmium, chromium, and antimony all displayed reduced 5-hmC content and reduced levels of the 5-hmC oxidative derivatives, 5-fC, and 5-caC (Xiong et al., 2017), pointing to potential stress responses due to heavy metal exposure. In human embryonic stem cells (hESCs), lead exposure altered the 5-hmC level at differentially hydroxymethylated regions (DHMRs). Exposure to lead decreased the 5-hmC content at the DHMRs but increased hydroxymethylation in certain regions outside of these DHMRs (Sen et al., 2015). An in vivo study demonstrated a significant alteration in 5-hmC content in an organ-specific manner in male rats exposed to sodium arsenite in drinking water (0.5, 2, or 10 part per million). The association between arsenic exposure and organ-specific distribution of 5-hmC was only found in the heart, spleen, and lung. The heart and spleen experienced an increase in 5-hmC content for all doses, while lung tissue showed a large increase in 5-hmC level only at the highest dose. By contrast, a loss of 5-hmC occurred in the pancreas of rats exposed to low-to-moderate level of arsenite. These findings suggest that organ-specific retention time of arsenite may modify the toxicity and epigenetic effect (i.e., increase in 5-hmC) of arsenite because both heart and spleen tissues are capable of accumulating high levels of arsenite (Zhang et al., 2014).

The effects of heavy metal exposure may also be modified by gender and time, according to an analysis of hydroxymethylation patterns in the circulating leukocytes of human populations. A cohort study in Bangladesh (Niedzwiecki et al., 2015) revealed a positive association between arsenic exposure and increased global 5-hmC and 5-mC contents in men. However, among women, arsenic exposure was negatively associated with a global 5-hmC level. In another human cohort, differential 5-hmC patterns at the DHMRs were identified in umbilical cord blood collected from children exposed to lead during gestation (Sen et al., 2015). Some changes at the DHMRs were gender-dependent, while some are not. The gender-independent ones could serve as biomarkers of early-life lead exposure. The effect of prenatal mercury exposure on global hydroxymethylation has been assessed in children in a large cohort study (Cardenas et al., 2017). An increase in maternal mercury exposure was associated with decreased global 5-hmC content in cord blood. This effect was maintained in early childhood (2.9–4.9 years), but it was not observed in the middle ages of childhood (6.7–10.5 years).

Environmental Estrogens

Environmental estrogens represent a broad class of compounds that have estrogen-like effects in vivo. These compounds range from phytoestrogens like genistein, commonly found in soy products, to plasticizers like phthalates, and epoxy resins precursors like bisphenol A (BPA). Given that these compounds mimic endogenous hormones, the bulk of the research into 5-hmC dysregulation has looked at fertility or the gonad-specific effects of exposure. Rats exposed to BPA or di(2-ethylhexyl)phthalate (DEHP) *in utero* showed a decrease in global 5-hmC in the testes (Abdel-Maksoud et al., 2015). Additionally, when BPA was administered to adult *Gobiocypris rarus*, a species of minnow, TET expression and global 5-hmC levels were reduced in the testes (Yuan et al., 2017). Mice exposed to diethylstilbestrol (DES) *in utero* had reduced TET1 expression that was associated with a reduction in global hydroxymethylation in adults but not at postnatal day 5 (Jefferson et al., 2013). In humans, BPA exposure reduces sperm quality in exposed men and alters the 5-hmC patterns in sperm. Assessment of men who were occupationally exposed to BPA for DHMRs via hMeDIP-seq (Zheng et al., 2017) revealed a vastly different 5-hmC profile for exposed men, who had 9610 DHMRs compared with controls, covering over 10% of all genes expressed in sperm and over 6% of the sperm genome. Of these DHMRs, over 90% was hyperhydroxymethylated. Many of the DHMRs were located in introns; however, multiple promoters in maternally imprinted regions showed increased 5-hmC levels. The association between phthalate exposure and 5-hmC content was assessed in a cohort of men undergoing reproductive counseling for low fertility (Pan et al., 2016). Total phthalate burdens were associated with the increased urinary 5-hmC levels and low sperm counts. However, the relationship between urinary 5-hmC and genomic 5-hmC has not been established yet.

Air Pollutants

Exposure to air pollution has been linked to respiratory and cardiovascular disease, including asthma and lung cancer (Kampa and Castanas, 2008). The effect of air pollutants, or its individual constituents, on DNA methylation has also been investigated for over a decade. However, few studies have examined 5-hmC changes in the context of air pollutant exposure. One notable study examined the relationship between exposure to particulate matter 2.5 microns in diameter and under (PM2.5) and particulate matter 10 microns in diameter and under (PM10) and global 5-hmC levels in office workers and truck drivers residing in Beijing (Sanchez-Guerra et al., 2015). PM10 exposure was associated with an increased global 5-hmC level in circulating white blood cells in both groups, with no association identified between PM2.5 exposures and 5-hmC contents. Another group investigated the role of short term PM2.5 and PM10 exposure on DNA methylation and hydroxymethylation in buccal cells as measured by UPLC-MS/MS. Despite considerable individual variability, exposure to PM2.5 and to PM10 was associated with a decrease in both global DNA methylation and DNA hydroxymethylation levels (De Nys et al., 2017).

A separate study investigated the relationship between traffic-related air pollution (TRAP) exposure and TET1 promoter methylation in the nasal epithelia from a cohort of asthmatic children and their nonasthmatic siblings, followed by further validation in a second pediatric cohort (Somineni et al., 2016). Children with asthma displayed reduced TET1 promoter methylation and increased global 5-hmC in saliva when compared with their nonasthmatic siblings, with 5-hmC levels correlated in the nasal epithelium, saliva, and peripheral blood mononuclear cells (PBMC). In addition, TET1 promoter methylation was associated with TRAP exposure only in the nonasthmatic children. Surprisingly, in in vitro model of TRAP exposure, in which human bronchial epithelial cells were exposed to diesel particulate matter, a common component of TRAP, the exposed cells displayed increased TET1 promoter methylation, reduced TET1 mRNA, and reduced global 5-hmC levels. The authors concluded that the role of TET1 in response to TRAP exposure may be time sensitive and thus vary based on the length of exposure.

Other than air pollution, there are studies that demonstrated the link between 5-hmC changes and exposures to cigarette smoke products and house allergens. Exposing thymocytes to nicotine induced TET2 expression and resulted in Fas promoter demethylation and upregulation of Fas apoptotic pathways (Liu et al., 2017). In mice, exposure to house dust mite (HDM), a ubiquitous indoor allergen, induced airway hyperresponsiveness and resulted in increased global DNA methylation and hydroxymethylation when compared with controls (Shang et al., 2013). In a similar model, mice that were exposed to HDM chronically developed airway inflammation and airway remodeling. These mice displayed reduced DNA methylation and increased DNA hydroxymethylation in the lung (Cheng et al., 2014).

Pesticides

Pesticide is often applied to insecticides, fungicides, herbicides, and other biocides but is most often used to describe insecticides. Many insecticides target the insect nervous system, via acetylcholine esterase inhibition; however, members of this broad class of compounds are not limited to targeting neurotransmitter pathways (Coman et al., 2013). In rat cortical neurons, treatment with rotenone, a natural insecticide and inhibitor of the mitochondrial complex I, increased 5-mC and 5-hmC levels. Treatment with lithium reversed the rotenone-induced changes in 5-hmC level (Scola et al., 2014). A separate rat model of Gulf War illness, in which rats were exposed to pyridostigmine bromide, permethrin, and N,N-diethyl-meta-toluamide (DEET), as well as restraint to induce stress (Pierce et al., 2016), revealed increased 5-mC content in the hippocampus and increased 5-hmC content in the cerebellum but decreased 5-hmC content in the cortex. However, the role of permethrin and DEET on the changes in 5-hmC is not fully understood.

Ionizing Radiation

Ionizing radiation, such as ultraviolet and high-frequency rays like x-rays, causes damage to living cells and is capable of inducing changes in the genome. Immortalized human keratinocyte (HaCaT) cells exposed to UVB radiation for 24 h showed a dose-dependent increase in global 5-hmC content and increased RNA and protein levels of TET1, TET2, and TET3 (Wang et al., 2017). Although the current research is limited, these data point toward the possibility of utilizing 5-hmC as a biosensor for skin UVB radiation exposure. Human fetal fibroblasts exposed to x-ray irradiation (2 and 4 Gy) showed no significant changes in global 5-hmC level (Wang et al., 2017), indicating that the effect of ionizing radiation on 5-hmC changes or TET protein expression may vary considerably among the types of ionizing radiation.

Lifestyle

The effects of lifestyle on disease development are well documented, as stress, high-fat diet, and sedentary lifestyle have all been linked to a myriad of diseases including obesity, insulin resistance, cardiovascular disease, and cancer (Hanson and Gluckman, 2015; Nielsen et al., 2016; Nilsson and Ling, 2017). Hyperhydroxymethylation of the glucocorticoid receptor gene Nr3c1 in the hippocampus was identified in acutely stressed mice using TAB-seq (Li et al., 2015). Further studies revealed that adult mice with acute stress and early-life stress had associated DHMRs in the hippocampus and hypothalamus, respectively (Li et al., 2016; Papale et al., 2017). The rat model of Gulf War illness showed persistent decreases in global 5-hmC levels in brain tissues (cortex and cerebellum) (Pierce et al., 2016). Conversely, a depression model of rats revealed that dietary butyrate, a four-carbon fatty acid that inhibits histone deacetylase activity, increased the level of 5-hmC, as measured by hMeDIP-qPCR, in promoter of the brain-derived neurotrophic factor gene in the brain, which may promote neurogenesis (Wei et al., 2015). The association between stress and DHMRs is also gender-specific. Stressed male C57BL/6J mice displayed significant hippocampal 5-hmC perturbations, including hyperhydroxymethylation of DHMRs in genes involved in structural development of the hippocampus and hypohydroxymethylation of DHMRs in genes involved in cell adhesion and communication (Li et al., 2016). Female mice exposed to acute stress had over 360 sex-specific DHMRs in the hippocampus (Papale et al., 2016). Female mice exposed to early-life stress also showed altered hydroxymethylation of DHMRs in genes in the hypothalamus involved in neurodevelopment and differentiation when compared with nonstressed controls (Papale et al., 2017). Taken together, these findings indicate that stress, in early life, and acute stress, during adulthood, could induce gender-specific 5-hmC alterations in brain tissue.

Dietary supplements and especially epigenetically active diets like those containing high levels of butyrate, vitamin C, choline, and folate are capable of altering the epigenome, including DNA hydroxymethylome (Lewis and Tollefsbol, 2017). Folate, betaine, choline, and methionine are dietary methyl donors and serve as the main sources of methyl groups in one-carbon metabolism. They donate methyl moieties during the synthesis of S-adenosylmethionine (SAM), which is used by DNMTs to methylate DNA. In vitro exposure to vitamin C (ascorbic acid) increased the global 5-hmC level in embryonic stem cells, likely through increased TET activity. Additionally, mice deficient in vitamin C synthesis showed increased global DNA hydroxymethylation in liver, lung, and cerebrum tissues following vitamin C supplementation (Yin et al., 2013). In humans, dietary intake of folate during pregnancy in women and dietary intake of betaine in men showed a positive correlation with the blood 5-hmC levels (Pauwels et al., 2016, 2017). Intake of methionine, choline, and betaine during pregnancy also resulted in a unique 5-hmC profile (Pauwels et al., 2016). Higher intake before and during early pregnancy resulted in lower global 5-hmC contents in the first and second trimesters, whereas higher intake in the second or third trimester resulted in higher global 5-hmC levels in the third trimester and at delivery.

The 5-hmC patterns can be modified by dietary intake of compounds other than epigenetically active substances. Rats given increased dietary salt (4% NaCl in AIN-76A diet for 7 days) showed modified 5-hmC patterns, as measured by oxidative reduced representation bisulfite sequencing, in the renal outer medulla (Liu et al., 2014). Newborn piglets exposed to a low-protein diet during gestation showed gender-specific 5-hmC patterns in the control region of mitochondrial DNA, as measured by hMeDIP-qPCR using a glucocorticoid receptor antibody. Specifically, males exhibited a higher degree of glucocorticoid receptor binding to the mitochondrial promoter and decreased 5-mC and 5-hmC levels, whereas females exhibited an opposite response (Jia et al., 2013). In rodent studies, mice on a calorie-restricted diet exhibited reduced 5-hmC levels in Purkinje cells of the cerebellum and in the cornu ammonis region CA3 of the hippocampus (Chouliaras et al., 2012; Lardenoije et al., 2015). Conversely, the offspring of rats fed with a high-fat diet during gestation showed a decrease in 5-hmC level in promoter of the hypothalamic anorexigenic neuropeptides, and the change was associated with an increased offspring body weight (Marco et al., 2016). In humans, normal weight and obese women showed a positive correlation of the global 5-hmC changes in blood to the body mass index, waist circumference, and total cholesterol and triglyceride levels. However, 5-hmC level in white blood cells was unaffected by an energy restricted diet in obese patients (Nicoletti et al., 2016).

Alcohol consumption primarily affects brain and liver tissue; however, when administered during gestation, alcohol can also function as a teratogen. Fetal alcohol exposure (E7–E16) in mice resulted in altered 5-hmC patterns in the embryonic frontal cortex (Öztürk et al., 2017; Tammen et al., 2014b) and hippocampus (Chen et al., 2013). Specifically, 5-hmC level was decreased in the ventricular zone, subventricular zone, and subplate region of the E17 frontal neocortex, while 5-hmC level was increased in the cortical plate (Öztürk et al., 2017; Tammen et al., 2014b). Mice exposed to alcohol during gestation showed decreased levels of 5-hmC in the hippocampus and an associated delay in hippocampal maturation (Chen et al., 2013).

In adult mice, alcohol reduced hepatic global 5-hmC levels in young mice (4 months) but not in old mice (18 months) (Tammen et al., 2014b). A similar finding was reported in adult rats chronically administered with alcohol, as these rats showed a reduced global 5-hmC content in the liver (Tammen et al., 2016). Several of the DHMRs identified via hMeDIP-array in young mice and old mice exposed to alcohol (Tammen et al., 2017) were highly involved in liver lipid metabolism and cell death pathways. Interestingly, aging was associated with a decrease in 5-hmC at the DHMRs, but alcohol reduced this effect. Overall, these results indicate that alcohol consumption *in utero* and in adulthood can affect 5-hmC levels, particularly in genes associated with neuronal development and metabolism.

Methamphetamine and cocaine are stimulant drugs of abuse with high addiction potential. Adult male C57BL/6J mice treated with cocaine for 7 days showed significantly decreased levels of TET1 expression and TET1 protein in the nucleus accumbens, one of the brain reward centers (Feng et al., 2015). The global 5-hmC level in the nucleus accumbens did not differ between controls and mice treated with cocaine, but oxBS-seq analysis revealed over 11,500 DHMRs in the exposed mice. Moreover, many of these DHMRs maintained their 5-hmC status for at least 1 month after the exposure (Feng et al., 2015). Similarly, rats in drug self-administration chambers addicted to methamphetamine showed DHMRs in genes encoding voltage- and calcium-gated potassium channels in the nucleus accumbens (Cadet et al., 2017). Rats administered methamphetamine daily showed reduced 5-hmC levels in promoter regions of the neurotransmitter receptors GluA1 and GluA2 in the striatum (Jayanthi et al., 2014). Together, these studies suggest that cocaine and methamphetamine addiction alters gene-specific hydroxymethylation, resulting in functional differences that can persist after drug cessation.

Aging

Aging is a complex biological process that can be affected by a host of factors, including lifestyle, stress, diet, and exposure to environmental stimuli. In mice, discrepancies are evident in the reported global 5-hmC expression among different brain regions, which may reflect the use of different detection methods, mouse strains, age, and other factors. Regardless, old mice predominantly show an increase in global 5-hmC levels in the brain (the hippocampus, frontal cortex, and cerebellum) (Chen et al., 2012; Chouliaras et al., 2012; Hadad et al., 2016; Kraus et al., 2016; Lardenoije et al., 2015; Münzel et al., 2010; Szulwach et al., 2011). In the liver, global 5-hmC levels are increased in old mice, and surprisingly, the expression of TET2 and TET3 is decreased when compared with young mice (Tammen et al., 2014a). A comparison of 5-hmC in the livers of young mice to old mice via hMeDIP-array identified 331 DHMRs with decreased 5-hmC at these regions with increased age. The DHMRs are related to the genes of hepatic system development, function, and lipid metabolism (Tammen et al., 2017). In germ cells, global 5-hmC was positively associated with aging in both mouse oocytes and sperms (Jenkins et al., 2013; Qian et al., 2015). In humans, age-dependent changes in 5-hmC level were identified in the brains of elderly individuals; specifically, global 5-hmC levels were increased in the frontal cortex and white matter, but not in the cerebellum (Kraus et al., 2015); additionally, elderly women showed decreases in global 5-hmC levels in the blood (Buscarlet et al., 2016). The global 5-hmC level and expression of TET1 and TET3 mRNA, but not TET2, were negatively associated with age in human peripheral blood T cells from 53 healthy volunteers, including both men and women (20–83 years of age, median 48.3) (Truong et al., 2015). These studies indicate that 5-hmC patterns are dynamic during aging and in certain contexts may be a marker of aging in specific tissues.

DISCUSSION AND FUTURE DIRECTIONS

In this review, we provide evidence to demonstrate a role for DNA hydroxymethylation in the response of mammals to environmental toxicants and its relationship to disease development. Similar to the changes in 5-mC, the changes in 5-hmC are critical for early development and appear to undergo modification throughout the life span. Although 5-hmC markers are indistinguishable from 5-mC by bisulfite modification, the use of enzymes, antibodies, or LC/MS allows the measurement of 5-hmC levels in the genome. In addition, platforms using high-throughput profiling, like 5-hMeDIP-array/seq, help to identify gene-specific 5-hmC changes. From cell lines to in vivo rodent models, we have seen the direct link between exposure to environmental stressors and changes in 5-hmC and the relationship of these changes to disease phenotypes. Independent human cohort studies have also revealed that 5-hmC changes are associated with environmental exposures to toxicants and with lifestyle factors. All in all, the evidence suggests that the 5-hmC profile could serve as a marker for toxicant exposures and/or as biomarkers for the diagnosis and prognosis of various diseases. Investigation of how DNA hydroxymethylation regulates gene transcription and phenotypes could help to improve our understanding of the etiology of disease and eventually aid in the design of new therapeutics or the recommendation of life choices to the public.

We have seen challenges in assessing 5-hmC profiles in toxicology and epigenetic epidemiology studies. For example, we have seen that 5-hmC changes can be exposure-, gender-, tissue-, and time-specific. Some discrepancies and limitations also complicate the assessment of 5-hmC changes using the currently available techniques, and these should be taken into consideration in future measurements and the analysis of 5-hmC profiles. First of all, new technologies are required to improve the sensitivity and specificity of the 5-hmC markers. Positive controls and proper standards are suggested for inclusion in any 5-hmC analysis. In order to demonstrate the gene-environment interaction with the epigenetics, big "omic" data from (hydroxy)methylome, transcriptome, proteome, and metabolome analysis are commonly applied. An integrated approach is important for data collection and data analysis. The TaRGET II consortium (National Institute of Environmental Health Sciences (NIEHS), 2016) was established in 2015 and aims to provide a guideline to identify and analyze the tissue-, exposure-, and time-specific epigenetic marks. In addition, comparing epigenetic marks in target tissues to those in surrogate tissues could improve the use of 5-hmC markers in epidemiology studies. Studies have demonstrated the potential use of 5-hmC signatures in circulating cell-free DNA as diagnostic biomarkers for human diseases such as cancer (Li et al., 2017). Furthermore, correlation analysis can be conducted for the 5-hmC changes with other 5-cytosine modifications. Identification of 5-fC and 5-caC markers could also be useful for monitoring stable or intermediate epigenetic changes in response to the environmental stressors (Valentini et al., 2016). When analyzing epidemiology data, proper adjustment of the exposures and other confounding factors is required (Bakulski and Fallin, 2014). Data validation from other independent cohorts is also highly recommended. GWAS has been well established to investigate the interaction between environment and genetic variations. Given the growing evidence of the methylome data, an interaction between genetic and epigenetic variations in disease susceptibility should be expected in the future (Ladd-Acosta and Fallin, 2016). Given the reversible nature of epigenetic changes, studies on removing the 5-hmC markers have been an increasing concern. The role of oxidative stress in regulating 5-hmC is also speculated (Delatte et al., 2015; Chia et al., 2011), so the use of antioxidants may be a new direction for modulating aberrant 5-hmC changes. Taken together, the current results indicate that we need more data and a systemic approach for analyzing the findings from epidemiology and toxicology studies.

Acknowledgments

National Institute of Environmental Health Sciences Grants ES024784 and ES028351 (WYT) and National Natural Science Foundation of China (81728018-WYT). The funders had no role in the decision to publish or the preparation of the manuscript. The authors declared that they have no potential competing financial interests.

References

Abdel-Maksoud, F.M., Leasor, K.R., Butzen, K., Braden, T.D., Akingbemi, B.T., 2015. Prenatal exposures of male rats to the environmental chemicals bisphenol A and di(2-ethylhexyl) phthalate impact the sexual differentiation process. Endocrinology 156, 4672–4683.

An, J., González-Avalos, E., Chawla, A., Jeong, M., López-Moyado, I.F., Li, W., Goodell, M.A., Chavez, L., Ko, M., Rao, A., 2015. Acute loss of TET function results in aggressive myeloid cancer in mice. Nat. Commun. 6,10071.

Ardui, S., Ameur, A., Vermeesch, J.R., Hestand, M.S., 2018. Single molecule real-time (SMRT) sequencing comes of age: applications and utilities for medical diagnostics. Nucleic Acids Res. 46, 2159–2168.

Bachman, M., Uribe-Lewis, S., Yang, X., Williams, M., Murrell, A., Balasubramanian, S., 2014. 5-Hydroxymethylcytosine is a predominantly stable DNA modification. Nat. Chem. 6, 1049–1055.

Bakulski, K.M., Fallin, M.D., 2014. Epigenetic epidemiology: promises for public health research. Environ. Mol. Mutagen. 55, 171–183.

Baskin, J.M., Prescher, J.A., Laughlin, S.T., Agard, N.J., Chang, P.V., Miller, I.A., Lo, A., Codelli, J.A., Bertozzi, C.R., 2007. Copper-free click chemistry for dynamic in vivo imaging. Proc. Natl. Acad. Sci. 104, 16793–16797.

Bayley, H., 2006. Sequencing single molecules of DNA. Curr. Opin. Chem. Biol. 10, 628–637.

Bogdanović, O., Veenstra, G.J.C., 2009. DNA methylation and methyl-CpG binding proteins: developmental requirements and function. Chromosoma 118, 549–565.

Bogdanović, O., Smits, A.H., de la Calle Mustienes, E., Tena, J.J., Ford, E., Williams, R., Senanayake, U., Schultz, M.D., Hontelez, S., van Kruijsbergen, I., Rayon, T., Gnerlich, F., Carell, T., Veenstra, G.J.C., Manzanares, M., Sauka-Spengler, T., Ecker, J.R., Vermeulen, M., Gómez-Skarmeta, J.L., Lister, R., 2016. Active DNA demethylation at enhancers during the vertebrate phylotypic period. Nat. Genet. 48, 417–426.

Booth, M.J., Ost, T.W.B., Beraldi, D., Bell, N.M., Branco, M.R., Reik, W., Balasubramanian, S., 2013. Oxidative bisulfite sequencing of 5-methylcytosine and 5-hydroxymethylcytosine. Nat. Protoc. 8, 1841–1851.

Borgaro, J.G., Zhu, Z., 2013. Characterization of the 5-hydroxymethylcytosine-specific DNA restriction endonucleases. Nucleic Acids Res. 41, 4198–4206.

Branton, D., Deamer, D.W., Marziali, A., Bayley, H., Benner, S.A., Butler, T., Di Ventra, M., Garaj, S., Hibbs, A., Huang, X., Jovanovich, S.B., Krstic, P.S., Lindsay, S., Ling, X.S., Mastrangelo, C.H., Meller, A., Oliver, J.S., Pershin, Y.V., Ramsey, J.M., Riehn, R., Soni, G.V., Tabard-Cossa, V., Wanunu, M., Wiggin, M., Schloss, J.A., 2008. The potential and challenges of nanopore sequencing. Nat. Biotechnol. 26, 1146–1153.

Brazauskas, P., Kriaucionis, S., 2014. Another stable base in DNA. Nat. Chem. 6, 1031–1033.

Buscarlet, M., Tessier, A., Provost, S., Mollica, L., Busque, L., 2016. Human blood cell levels of 5-hydroxymethylcytosine (5hmC) decline with age, partly related to acquired mutations in TET2. Exp. Hematol. 44, 1072–1084.

Cadet, J.L., Brannock, C., Krasnova, I.N., Jayanthi, S., Ladenheim, B., McCoy, M.T., Walther, D., Godino, A., Pirooznia, M., Lee, R.S., 2017. Genome-wide DNA hydroxymethylation identifies potassium channels in the nucleus accumbens as discriminators of methamphetamine addiction and abstinence. Mol. Psychiatry 22, 1196–1204.

Cardenas, A., Rifas-Shiman, S.L., Godderis, L., Duca, R.C., Navas-Acien, A., Litonjua, A.A., Demeo, D.L., Brennan, K.J., Amarasiriwardena, C.J., Hivert, M.F., Gillman, M.W., Oken, E., Baccarelli, A.A., 2017. Prenatal exposure to mercury: associations with global DNA methylation and hydroxymethylation in cord blood and in childhood. Environ. Health Perspect. 125,87022.

Chavez, L., Huang, Y., Luong, K., Agarwal, S., Iyer, L.M., Pastor, W.A., Hench, V.K., Frazier-Bowers, S.A., Korol, E., Liu, S., Tahiliani, M., Wang, Y., Clark, T.A., Korlach, J., Pukkila, P.J., Aravind, L., Rao, A., 2014. Simultaneous sequencing of oxidized methylcytosines produced by TET/JBP dioxygenases in Coprinopsis cinerea. Proc. Natl. Acad. Sci. 111, E5149–E5158.

Chen, H., Dzitoyeva, S., Manev, H., 2012. Effect of aging on 5-hydroxymethylcytosine in the mouse hippocampus. Restor. Neurol. Neurosci. 30, 237–245.

Chen, Y., Ozturk, N.C., Zhou, F.C., 2013. DNA methylation program in developing hippocampus and its alteration by alcohol. PLoS One 8.e60503.

Chen, H.y., Wei, J.R., Pan, J.X., Zhang, W., Dang, F.Q., Zhang, Z.Q., Zhang, J., 2017. Spectroscopic quantification of 5-hydroxymethylcytosine in genomic DNA using boric acid-functionalized nano-microsphere fluorescent probes. Biosens. Bioelectron. 91, 328–333.

Cheng, R.Y., Shang, Y., Limjunyawong, N., Dao, T., Das, S., Rabold, R., Sham, J.S., Mitzner, W., Tang, W.Y., 2014. Alterations of the lung methylome in allergic airway hyper-responsiveness. Environ. Mol. Mutagen. 55, 244–255.

Chia, N., Wang, L., Lu, X., Senut, M.C., Brenner, C., Ruden, D.M., 2011. Hypothesis: environmental regulation of 5-hydroxymethylcytosine by oxidative stress. Epigenetics 6, 853–856.

Chouliaras, L., van den Hove, D.L.A., Kenis, G., Keitel, S., Hof, P.R., van Os, J., Steinbusch, H.W.M., Schmitz, C., Rutten, B.P.F., 2012. Age-related increase in levels of 5-hydroxymethylcytosine in mouse hippocampus is prevented by caloric restriction. Curr. Alzheimer Res. 9, 536–544.

Coman, G., Farcas, A., Matei, A.V., Florian, C., 2013. Pesticides mechanisms of action in living organisms. In: NATO Science for Peace and Security Series C: Environmental Security. Springer, Dordrecht, pp. 173–184.

Dao, T., Cheng, R.Y.S., Revelo, M.P., Mitzner, W., Tang, W., 2014. Hydroxymethylation as a novel environmental biosensor. Curr. Environ. Health Rep. 37, 62–70.

Davis, T., Vaisvila, R., 2011. High sensitivity 5-hydroxymethylcytosine detection in Balb/C brain tissue. J. Vis. Exp., 48, 4–6.

De Nys, S., Duca, R.C., Nawrot, T., Hoet, P., Van Meerbeek, B., Van Landuyt, K.L., Godderis, L., 2017. Temporal variability of global DNA methylation and hydroxymethylation in buccal cells of healthy adults: association with air pollution. Environ. Int. 111, 301–308.

Deaton, A.M., Bird, A., 2011. CpG islands and the regulation of transcription. Genes Dev. 25, 1010–1022.

Dekker, C., 2007. Solid-state nanopores. Nat. Nanotechnol. 2, 209–215.

Delatte, B., Jeschke, J., Defrance, M., Bachman, M., Creppe, C., Calonne, E., Bizet, M., Deplus, R., Marroquí, L., Libin, M., Ravichandran, M., Mascart, F., Eizirik, D.L., Murrell, A., Jurkowski, T.P., Fuks, F., 2015. Genome-wide hydroxymethylcytosine pattern changes in response to oxidative stress. Sci. Rep. 5.12714.

Feil, R., Fraga, M.F., 2012. Epigenetics and the environment: emerging patterns and implications. Nat. Rev. Genet. 13, 97–109.

Feng, J., Shao, N., Szulwach, K.E., Vialou, V., Huynh, J., Zhong, C., Le, T., Ferguson, D., Cahill, M.E., Li, Y., Koo, J.W., Ribeiro, E., Labonte, B., Laitman, B.M., Estey, D., Stockman, V., Kennedy, P., Couroussé, T., Mensah, I., Turecki, G., Faull, K.F., Ming, G.L., Song, H., Fan, G., Casaccia, P., Shen, L., Jin, P., Nestler, E.J., 2015. Role of Tet1 and 5-hydroxymethylcytosine in cocaine action. Nat. Neurosci. 18, 536–544.

Ficz, G., Branco, M.R., Seisenberger, S., Santos, F., Krueger, F., Hore, T.A., Marques, C.J., Andrews, S., Reik, W., 2011. Dynamic regulation of 5-hydroxymethylcytosine in mouse ES cells and during differentiation. Nature 473, 398–404.

Gardiner-Garden, M., Frommer, M., 1987. CpG islands in vertebrate genomes. J. Mol. Biol. 196, 261–282.

Gilat, N., Tabachnik, T., Shwartz, A., Shahal, T., Torchinsky, D., Michaeli, Y., Nifker, G., Zirkin, S., Ebenstein, Y., 2017. Single-molecule quantification of 5-hydroxymethylcytosine for diagnosis of blood and colon cancers. Clin. Epigenetics 9, 70.

2. DNA METHYLATION

Globisch, D., Münzel, M., Müller, M., Michalakis, S., Wagner, M., Koch, S., Brückl, T., Biel, M., Carell, T., 2010. Tissue distribution of 5-hydroxymethylcytosine and search for active demethylation intermediates. PLoS One 5. e15367.

Gommers-Ampt, J.H., Borst, P., 1995. Hypermodified bases in DNA. FASEB J. 9, 1034–1042.

Gu, T.P., Guo, F., Yang, H., Wu, H.P., Xu, G.F., Liu, W., Xie, Z.G., Shi, L., He, X., Jin, S.G., Iqbal, K., Shi, Y.G., Deng, Z., Szabó, P.E., Pfeifer, G.P., Li, J., Xu, G.L., 2011. The role of Tet3 DNA dioxygenase in epigenetic reprogramming by oocytes. Nature 477, 606–612.

Guibert, S., Weber, M., 2013. Functions of DNA methylation and hydroxymethylation in mammalian development. Curr. Top. Dev. Biol. 104, 47–83.

Hackett, J.A., Sengupta, R., Zylicz, J.J., Murakami, K., Lee, C., Down, T.A., Surani, M.A., 2013. Germline DNA demethylation dynamics and imprint erasure through 5-hydroxymethylcytosine. Science 339, 448–452.

Hadad, N., Masser, D.R., Logan, S., Wronowski, B., Mangold, C.A., Clark, N., Otalora, L., Unnikrishnan, A., Ford, M.M., Giles, C.B., Wren, J.D., Richardson, A., Sonntag, W.E., Stanford, D.R., Freeman, W., 2016. Absence of genomic hypomethylation or regulation of cytosine-modifying enzymes with aging in male and female mice. Epigenetics Chromatin 9, 30.

Hahn, M.A., Szab, P.E., Pfeifer, G.P., 2014. 5-Hydroxymethylcytosine: a stable or transient DNA modification? Genomics 104, 314–323.

Hanson, M.A., Gluckman, P.D., 2015. Developmental origins of health and disease—global public health implications. Best Pract. Res. Clin. Obstet. Gynaecol. 29, 24–31.

Hayatsu, H., Shiragami, M., 1979. Reaction of bisulfite with the 5-hydroxymethyl group in pyrimidines and in phage DNAs. Biochemistry 18, 632–637.

He, Y.-F., Li, B.-Z., Li, Z., Liu, P., Wang, Y., Tang, Q., Ding, J., Jia, Y., Chen, Z., Li, L., Sun, Y., Li, X., Dai, Q., Song, C.-X., Zhang, K., He, C., Xu, G.-L., 2011. Tet-mediated formation of 5-carboxylcytosine and its excision by TDG in mammalian DNA. Science 333, 1303–1307.

Holliday, R., Pugh, J.E., 1975. DNA modification mechanisms and gene activity during development. Science 187 (80), 226–232.

Hotchkiss, R.D., 1948. The quantitative separation of purines, pyrimidines, and nucleosides by paper chromatography. J. Biol. Chem. 175, 315–332.

Huang, Y., Pastor, W.A., Shen, Y., Tahiliani, M., Liu, D.R., Rao, A., 2010. The behaviour of 5-hydroxymethylcytosine in bisulfite sequencing. PLoS One 5.e8888.

Huang, Y., Pastor, W.A., Zepeda-Martínez, J.A., Rao, A., 2012. The anti-CMS technique for genome-wide mapping of 5-hydroxymethylcytosine. Nat. Protoc. 7, 1897–1908.

Inoue, A., Zhang, Y., 2011. Replication-dependent loss of 5-hydroxymethylcytosine in mouse preimplantation embryos. Science 334, 194.

Jayanthi, S., McCoy, M.T., Chen, B., Britt, J.P., Kourrich, S., Yau, H.J., Ladenheim, B., Krasnova, I.N., Bonci, A., Cadet, J.L., 2014. Methamphetamine downregulates striatal glutamate receptors via diverse epigenetic mechanisms. Biol. Psychiatry 76, 47–56.

Jefferson, W.N., Chevalier, D.M., Phelps, J.Y., Cantor, A.M., Padilla-Banks, E., Newbold, R.R., Archer, T.K., Kinyamu, H.K., Williams, C.J., 2013. Persistently altered epigenetic marks in the mouse uterus after neonatal estrogen exposure. Mol. Endocrinol. 27, 1666–1677.

Jenkins, T.G., Aston, K.I., Cairns, B.R., Carrell, D.T., 2013. Paternal aging and associated intraindividual alterations of global sperm 5-methylcytosine and 5-hydroxymethylcytosine levels. Fertil. Steril. 100, 945–951.

Jia, Y., Li, R., Cong, R., Yang, X., Sun, Q., Parvizi, N., Zhao, R., 2013. Maternal low-protein diet affects epigenetic regulation of hepatic mitochondrial DNA transcription in a sex-specific manner in newborn piglets associated with GR binding to its promoter. PLoS One 8.e63855.

Jin, S.-G., Kadam, S., Pfeifer, G.P., 2010. Examination of the specificity of DNA methylation profiling techniques towards 5-methylcytosine and 5-hydroxymethylcytosine. Nucleic Acids Res. 38, e125.

Jones, P.A., 2012. Functions of DNA methylation: islands, start sites, gene bodies and beyond. Nat. Rev. Genet. 13, 484–492.

Kafer, G.R., Li, X., Horii, T., Suetake, I., Tajima, S., Hatada, I., Carlton, P.M., 2016. 5-Hydroxymethylcytosine marks sites of DNA damage and promotes genome stability. Cell Rep. 14, 1283–1292.

Kampa, M., Castanas, E., 2008. Human health effects of air pollution. Environ. Pollut. 151, 362–367.

2. DNA METHYLATION

Kim, H.S., Tan, Y., Ma, W., Merkurjev, D., Destici, E., Ma, Q., Suter, T., Ohgi, K., Friedman, M., Skowronska-Krawczyk, D., Rosenfeld, M.G., 2018. Pluripotency factors functionally premark cell-type-restricted enhancers in ES cells. Nature 556, 510–514.

Kircher, M., Kelso, J., 2010. High-throughput DNA sequencing—concepts and limitations. BioEssays 32, 524–536.

Ko, M., An, J., Bandukwala, H.S., Chavez, L., Äijö, T., Pastor, W.A., Segal, M.F., Li, H., Koh, K.P., Lähdesmäki, H., Hogan, P.G., Aravind, L., Rao, A., 2013. Modulation of TET2 expression and 5-methylcytosine oxidation by the CXXC domain protein IDAX. Nature 497, 122–126.

Koh, K.P., Yabuuchi, A., Rao, S., Huang, Y., Cunniff, K., Nardone, J., Laiho, A., Tahiliani, M., Sommer, C.A., Mostoslavsky, G., Lahesmaa, R., Orkin, S.H., Rodig, S.J., Daley, G.Q., Rao, A., 2011. Tet1 and Tet2 regulate 5-hydroxymethylcytosine production and cell lineage specification in mouse embryonic stem cells. Cell Stem Cell 8, 200–213.

Kothari, R.M., Shankar, V., 1976. 5-Methylcytosine content in the vertebrate deoxyribonucleic acids: species specificity. J. Mol. Evol. 7, 325–329.

Kraus, T.F.J., Guibourt, V., Kretzschmar, H.A., 2015. 5-Hydroxymethylcytosine, the "Sixth Base," during brain development and ageing. J. Neural Transm. 122, 1035–1043.

Kraus, T.F.J., Kilinc, S., Steinmaurer, M., Stieglitz, M., Guibourt, V., Kretzschmar, H.A., 2016. Profiling of methylation and demethylation pathways during brain development and ageing. J. Neural Transm. 123, 189–203.

Kriaucionis, S., Heintz, N., 2009. The nuclear DNA base 5-hydroxymethylcytosine is present in Purkinje neurons and the brain. Science 324, 929–930.

Ladd-Acosta, C., Fallin, M.D., 2016. The role of epigenetics in genetic and environmental epidemiology. Epigenomics 8, 271–283.

Lardenoije, R., van den Hove, D.L.A., Vaessen, T.S.J., Iatrou, A., Meuwissen, K.P.V., van Hagen, B.T.J., Kenis, G., Steinbusch, H.W.M., Schmitz, C., Rutten, B.P.F., 2015. Epigenetic modifications in mouse cerebellar Purkinje cells: effects of aging, caloric restriction, and overexpression of superoxide dismutase 1 on 5-methylcytosine and 5-hydroxymethylcytosine. Neurobiol. Aging 36, 3079–3089.

Laszlo, A.H., Derrington, I.M., Brinkerhoff, H., Langford, K.W., Nova, I.C., Samson, J.M., Bartlett, J.J., Pavlenok, M., Gundlach, J.H., 2013. Detection and mapping of 5-methylcytosine and 5-hydroxymethylcytosine with nanopore MspA. Proc. Natl. Acad. Sci. 110, 18904–18909.

Le, T., Kim, K.-P., Fan, G., Faull, K.F., 2011. A sensitive mass spectrometry method for simultaneous quantification of DNA methylation and hydroxymethylation levels in biological samples. Anal. Biochem. 412, 203–209.

Lewis, K.A., Tollefsbol, T.O., 2017. The influence of an epigenetics diet on the cancer epigenome. Epigenomics 9, 1153–1155.

Li, W., Liu, M., 2011. Distribution of 5-hydroxymethylcytosine in different human tissues. J. Nucleic Acids 2011, 1–5.

Li, S., Papale, L.A., Kintner, D.B., Sabat, G., Barrett-Wilt, G.A., Cengiz, P., Alisch, R.S., 2015. Hippocampal increase of 5-hmC in the glucocorticoid receptor gene following acute stress. Behav. Brain Res. 286, 236–240.

Li, S., Papale, L.A., Zhang, Q., Madrid, A., Chen, L., Chopra, P., Keleş, S., Jin, P., Alisch, R.S., 2016. Genome-wide alterations in hippocampal 5-hydroxymethylcytosine links plasticity genes to acute stress. Neurobiol. Dis. 86, 99–108.

Li, W., Zhang, X., Lu, X., You, L., Song, Y., Luo, Z., Zhang, J., Nie, J., Zheng, W., Xu, D., Wang, Y., Dong, Y., Yu, S., Hong, J., Shi, J., Hao, H., Luo, F., Hua, L., Wang, P., Qian, X., Yuan, F., Wei, L., Cui, M., Zhang, T., Liao, Q., Dai, M., Liu, Z., Chen, G., Meckel, K., Adhikari, S., Jia, G., Bissonnette, M.B., Zhang, X., Zhao, Y., Zhang, W., He, C., Liu, J., 2017. 5-Hydroxymethylcytosine signatures in circulating cell-free DNA as diagnostic biomarkers for human cancers. Cell Res. 27, 1243–1257.

Li, J., Wu, X., Zhou, Y., Lee, M., Guo, L., Han, W., Mo, W., Cao, W., Sun, D., Xie, R., Huang, Y., 2018. Decoding the dynamic DNA methylation and hydroxymethylation landscapes in endodermal lineage intermediates during pancreatic differentiation of hESC. Nucleic Acids Res. 46, 2883–2900.

Lin, I.H., Chen, Y.F., Hsu, M.T., 2017. Correlated 5-hydroxymethylcytosine (5hmC) and gene expression profiles underpin gene and organ-specific epigenetic regulation in adult mouse brain and liver. PLoS One 12, 1–25.

Lio, C.-W., Zhang, J., González-Avalos, E., Hogan, P.G., Chang, X., Rao, A., 2016. Tet2 and Tet3 cooperate with B-lineage transcription factors to regulate DNA modification and chromatin accessibility. eLife 5, e18290.

Liu, Y., Liu, P., Yang, C., Cowley, A.W., Liang, M., 2014. Base-resolution maps of 5-methylcytosine and 5-hydroxymethylcytosine in dahl S rats: effect of salt and genomic sequence. Hypertension 63, 827–838.

Liu, S., Jiang, J., Li, L., Amato, N.J., Wang, Z., Wang, Y., 2015. Arsenite targets the zinc finger domains of tet proteins and inhibits tet-mediated oxidation of 5-methylcytosine. Environ. Sci. Technol. 49, 11923–11931.

Liu, H.X., Liu, S., Qu, W., Yan, H.Y., Wen, X., Chen, T., Hou, L.F., Ping, J., 2017. α7 nAChR mediated Fas demethylation contributes to prenatal nicotine exposure-induced programmed thymocyte apoptosis in mice. Oncotarget 8, 93741–93756.

López, V., Fernández, A.F., Fraga, M.F., 2017. The role of 5-hydroxymethylcytosine in development, aging and age-related diseases. Ageing Res. Rev. 37, 28–38.

Manrao, E.A., Derrington, I.M., Pavlenok, M., Niederweis, M., Gundlach, J.H., 2011. Nucleotide discrimination with DNA immobilized in the MSPA nanopore. PLoS One 6.e25723.

Marco, A., Kisliouk, T., Tabachnik, T., Weller, A., Meiri, N., 2016. DNA CpG methylation (5-methylcytosine) and its derivative (5-hydroxymethylcytosine) alter histone posttranslational modifications at the Pomc promoter, affecting the impact of perinatal diet on leanness and obesity of the offspring. Diabetes 65, 2258–2267.

Martos, S.N., Tang, W.Y., Wang, Z., 2015. Elusive inheritance: transgenerational effects and epigenetic inheritance in human environmental disease. Prog. Biophys. Mol. Biol. 118, 44–54.

Matsuda, T., Matsuda, S., Yamada, M., 2015. Mutation assay using single-molecule real-time (SMRTTM) sequencing technology. Genes Environ. 37, 15.

Michaeli, Y., Shahal, T., Torchinsky, D., Grunwald, A., Hoch, R., Ebenstein, Y., 2013. Optical detection of epigenetic marks: sensitive quantification and direct imaging of individual hydroxymethylcytosine bases. Chem. Commun. 49, 8599.

Münzel, M., Globisch, D., Brückl, T., Wagner, M., Welzmiller, V., Michalakis, S., Müller, M., Biel, M., Carell, T., 2010. Quantification of the sixth DNA base hydroxymethylcytosine in the brain. Angew. Chem. Int. Ed. 49, 5375–5377.

National Institute of Environmental Health Sciences (NIEHS), 2016. Environmental Effects on the Epigenome are Focus of TaRGET II. [WWW Document]. Environ. Factor. URL, https://factor.niehs.nih.gov/2016/6/science-highlights/target/index.htm. [(Accessed 8 February 2018)].

Neri, F., Rapelli, S., Krepelova, A., Incarnato, D., Parlato, C., Basile, G., Maldotti, M., Anselmi, F., Oliviero, S., 2017. Intragenic DNA methylation prevents spurious transcription initiation. Nature 543, 72–77.

Nestor, C.E., Ottaviano, R., Reddington, J., Sproul, D., Reinhardt, D., Dunican, D., Katz, E., Dixon, J.M., Harrison, D.J., Meehan, R.R., 2012. Tissue type is a major modifier of the 5-hydroxymethylcytosine content of human genes. Genome Res. 22, 467–477.

Nicoletti, C.F., Nonino, C.B., de Oliveira, B.A.P., Pinhel, M.A.d.S., Mansego, M.L., Milagro, F.I., Zulet, M.A., Martinez, J.A., 2016. DNA methylation and hydroxymethylation levels in relation to two weight loss strategies: energy-restricted diet or bariatric surgery. Obes. Surg. 26, 603–611.

Niedzwiecki, M.M., Liu, X., Hall, M.N., Thomas, T., Slavkovich, V., Ilievski, V., Levy, D., Alam, S., Siddique, A.B., Parvez, F., Graziano, J.H., Gamble, M.V., 2015. Sex-specific associations of arsenic exposure with global DNA methylation and hydroxymethylation in leukocytes: results from two studies in Bangladesh. Cancer Epidemiol. Biomark. Prev. 24, 1748–1757.

Nielsen, C.H., Larsen, A., Nielsen, A.L., 2016. DNA methylation alterations in response to prenatal exposure of maternal cigarette smoking: a persistent epigenetic impact on health from maternal lifestyle? Arch. Toxicol. 90, 231–245.

Nilsson, E., Ling, C., 2017. DNA methylation links genetics, fetal environment, and an unhealthy lifestyle to the development of type 2 diabetes. Clin. Epigenetics 9, 105.

Öztürk, N.C., Resendiz, M., Öztürk, H., Zhou, F.C., 2017. DNA methylation program in normal and alcohol-induced thinning cortex. Alcohol 60, 135–147.

Pan, Y., Jing, J., Yeung, L.W.Y., Sheng, N., Zhang, H., Yao, B., Dai, J., 2016. Associations of urinary 5-methyl-2′-deoxycytidine and 5-hydroxymethyl-2′-deoxycytidine with phthalate exposure and semen quality in 562 Chinese adult men. Environ. Int. 94, 583–590.

Papale, L.A., Li, S., Madrid, A., Zhang, Q., Chen, L., Chopra, P., Jin, P., Keleş, S., Alisch, R.S., 2016. Sex-specific hippocampal 5-hydroxymethylcytosine is disrupted in response to acute stress. Neurobiol. Dis. 96, 54–66.

Papale, L.A., Madrid, A., Li, S., Alisch, R.S., 2017. Early-life stress links 5-hydroxymethylcytosine to anxiety-related behaviors. Epigenetics 12, 264–276.

Pastor, W.A., Pape, U.J., Huang, Y., Henderson, H.R., Lister, R., Ko, M., McLoughlin, E.M., Brudno, Y., Mahapatra, S., Kapranov, P., Tahiliani, M., Daley, G.Q., Liu, X.S., Ecker, J.R., Milos, P.M., Agarwal, S., Rao, A., 2011. Genomewide mapping of 5-hydroxymethylcytosine in embryonic stem cells. Nature 473, 394–397.

Pastor, W.A., Aravind, L., Rao, A., 2013. TETonic shift: biological roles of TET proteins in DNA demethylation and transcription. Nat. Rev. Mol. Cell Biol. 14, 341–356.

2. DNA METHYLATION

Pauwels, S., Duca, R.C., Devlieger, R., Freson, K., Straetmans, D., Van Herck, E., Huybrechts, I., Koppen, G., Godderis, L., 2016. Maternal methyl-group donor intake and global DNA (Hydroxy)methylation before and during pregnancy. Nutrients 8, 474.

Pauwels, S., Truijen, I., Ghosh, M., Duca, R.C., Langie, S.A.S., Bekaert, B., Freson, K., Huybrechts, I., Koppen, G., Devlieger, R., Godderis, L., 2017. The effect of paternal methyl-group donor intake on offspring DNA methylation and birth weight. J. Dev. Orig. Health Dis. 8, 311–321.

Penn, N.W., Suwalski, R., O'Riley, C., Bojanowski, K., Yura, R., 1972. The presence of 5-hydroxymethylcytosine in animal deoxyribonucleic acid. Biochem. J. 126, 781–790.

Petterson, A., Chung, T.H., Tan, D., Sun, X., Jia, X.Y., 2014. RRHP: a tag-based approach for 5-hydroxymethylcytosine mapping at single-site resolution. Genome Biol. 15, 456.

Pierce, L.M., Kurata, W.E., Matsumoto, K.W., Clark, M.E., Farmer, D.M., 2016. Long-term epigenetic alterations in a rat model of Gulf War Illness. Neurotoxicology 55, 20–32.

Qian, Y., Tu, J., Tang, N.L.S., Kong, G.W.S., Chung, J.P.W., Chan, W.Y., Lee, T.L., 2015. Dynamic changes of DNA epigenetic marks in mouse oocytes during natural and accelerated aging. Int. J. Biochem. Cell Biol. 67, 121–127.

Razin, A., Riggs, A., 1980. DNA methylation and gene function. Science 210, 604–610.

Riggs, A.D., 1975. X inactivation, differentiation, and DNA methylation. Cytogenet. Genome Res. 14, 9–25.

Robertson, A.B., Dahl, J.A., Vgbø, C.B., Tripathi, P., Krokan, H.E., Klungland, A., 2011. A novel method for the efficient and selective identification of 5-hydroxymethylcytosine in genomic DNA. Nucleic Acids Res. 39, e55.

Robertson, A.B., Dahl, J.A., Ougland, R., Klungland, A., 2012. Pull-down of 5-hydroxymethylcytosine DNA using JBP1-coated magnetic beads. Nat. Protoc. 7, 340–350.

Ross, M.G., Russ, C., Costello, M., Hollinger, A., Lennon, N.J., Hegarty, R., Nusbaum, C., Jaffe, D.B., 2013. Characterizing and measuring bias in sequence data. Genome Biol. 14, R51.

Saldanha, G., Joshi, K., Lawes, K., Bamford, M., Moosa, F., Teo, K.W., Pringle, J.H., 2017. 5-Hydroxymethylcytosine is an independent predictor of survival in malignant melanoma. Mod. Pathol. 30, 60–68.

Sanchez-Guerra, M., Zheng, Y., Osorio-Yanez, C., Zhong, J., Chervona, Y., Wang, S., Chang, D., McCracken, J.P., Díaz, A., Bertazzi, P.A., Koutrakis, P., Kang, C.M., Zhang, X., Zhang, W., Byun, H.M., Schwartz, J., Hou, L., Baccarelli, A.A., 2015. Effects of particulate matter exposure on blood 5-hydroxymethylation: results from the Beijing truck driver air pollution study. Epigenetics 10, 633–642.

Saxonov, S., Berg, P., Brutlag, D.L., 2006. A genome-wide analysis of CpG dinucleotides in the human genome distinguishes two distinct classes of promoters. Proc. Natl. Acad. Sci. 103, 1412–1417.

Scola, G., Kim, H.K., Young, L.T., Salvador, M., Andreazza, A.C., 2014. Lithium reduces the effects of rotenone-induced complex i dysfunction on DNA methylation and hydroxymethylation in rat cortical primary neurons. Psychopharmacology 231, 4189–4198.

Sen, A., Cingolani, P., Senut, M.-C., Land, S., Mercado-Garcia, A., Tellez-Rojo, M.M., Baccarelli, A.A., Wright, R.O., Ruden, D.M., 2015. Lead exposure induces changes in 5-hydroxymethylcytosine clusters in CpG islands in human embryonic stem cells and umbilical cord blood. Epigenetics 10, 607–621.

Shahal, T., Gilat, N., Michaeli, Y., Redy-Keisar, O., Shabat, D., Ebenstein, Y., 2014. Spectroscopic quantification of 5-hydroxymethylcytosine in genomic DNA. Anal. Chem. 86, 8231–8237.

Shang, Y., Das, S., Rabold, R., Sham, J.S.K., Mitzner, W., Tang, W.Y., 2013. Epigenetic alterations by DNA methylation in house dust mite-induced airway hyperresponsiveness. Am. J. Respir. Cell Mol. Biol. 49, 279–287.

Somineni, H.K., Zhang, X., Biagini Myers, J.M., Kovacic, M.B., Ulm, A., Jurcak, N., Ryan, P.H., Khurana Hershey, G.K., Ji, H., 2016. Ten-eleven translocation 1 (TET1) methylation is associated with childhood asthma and traffic-related air pollution. J. Allergy Clin. Immunol. 137, 797–805.e5.

Song, C.X., Clark, T.A., Lu, X.Y., Kislyuk, A., Dai, Q., Turner, S.W., He, C., Korlach, J., 2011a. Sensitive and specific single-molecule sequencing of 5-hydroxymethylcytosine. Nat. Methods 9, 75–77.

Song, C.X., Szulwach, K.E., Fu, Y., Dai, Q., Yi, C., Li, X., Li, Y., Chen, C.H., Zhang, W., Jian, X., Wang, J., Zhang, L., Looney, T.J., Zhang, B., Godley, L.A., Hicks, L.M., Lahn, B.T., Jin, P., He, C., 2011b. Selective chemical labeling reveals the genome-wide distribution of 5-hydroxymethylcytosine. Nat. Biotechnol. 29, 68–75.

Song, C.X., Yu, M., Dai, Q., He, C., 2011c. Detection of 5-hydroxymethylcytosine in a combined glycosylation restriction analysis (CGRA) using restriction enzyme TaqαI. Bioorg. Med. Chem. Lett. 21, 5075–5077.

Song, C.X., Yi, C., He, C., 2012. Mapping recently identified nucleotide variants in the genome and transcriptome. Nat. Biotechnol. 30, 1107–1116.

Strand, S.H., Hoyer, S., Lynnerup, A.-S., Haldrup, C., Storebjerg, T.M., Borre, M., Orntoft, T.F., Sorensen, K.D., 2015. High levels of 5-hydroxymethylcytosine (5hmC) is an adverse predictor of biochemical recurrence after prostatectomy in ERG-negative prostate cancer. Clin. Epigenetics 7, 111.

Stroud, H., Feng, S., Morey Kinney, S., Pradhan, S., Jacobsen, S.E., 2011. 5-Hydroxymethylcytosine is associated with enhancers and gene bodies in human embryonic stem cells. Genome Biol. 12, R54.

Sun, Z., Terragni, J., Borgaro, J.G., Liu, Y., Yu, L., Guan, S., Wang, H., Sun, D., Cheng, X., Zhu, Z., Pradhan, S., Zheng, Y., 2013. High-resolution enzymatic mapping of genomic 5-hydroxymethylcytosine in mouse embryonic stem cells. Cell Rep. 3, 567–576.

Sun, Z., Dai, N., Borgaro, J.G., Quimby, A., Sun, D., Corrêa, I.R., Zheng, Y., Zhu, Z., Guan, S., 2015. A sensitive approach to map genome-wide 5-hydroxymethylcytosine and 5-formylcytosine at single-base resolution. Mol. Cell 57, 750–761.

Sun, X., Chung, T.H., Tan, D., Kim, A., 2016. Practical guidelines and consideration of using RRHP for 5hmC detection. Epigenomics 8, 225–235.

Szulwach, K.E., Li, X., Li, Y., Song, C.X., Wu, H., Dai, Q., Irier, H., Upadhyay, A.K., Gearing, M., Levey, A.I., Vasanthakumar, A., Godley, L.A., Chang, Q., Cheng, X., He, C., Jin, P., 2011. 5-hmC-mediated epigenetic dynamics during postnatal neurodevelopment and aging. Nat. Neurosci. 14, 1607–1616.

Szwagierczak, A., Bultmann, S., Schmidt, C.S., Spada, F., Leonhardt, H., 2010. Sensitive enzymatic quantification of 5-hydroxymethylcytosine in genomic DNA. Nucleic Acids Res. 38, e181.

Szyf, M., 2007. The dynamic epigenome and its implications in toxicology. Toxicol. Sci. 100, 7–23.

Tahiliani, M., Koh, K.P., Shen, Y., Pastor, W.A., Bandukwala, H., Brudno, Y., Agarwal, S., Iyer, L.M., Liu, D.R., Aravind, L., Rao, A., 2009. Conversion of 5-methylcytosine to 5-hydroxymethylcytosine in mammalian DNA by MLL partner TET1. Science 324, 930–935.

Tammen, S., Dolnikowski, G., Ausman, L., Liu, Z., Kim, K., Friso, S., Choi, S., 2014a. Aging alters hepatic DNA hydroxymethylation, as measured by liquid chromatography/mass spectrometry. J. Cancer Prev. 19, 301–308.

Tammen, S., Dolnikowski, G., Ausman, L., Liu, Z., Sauer, J., Friso, S., Choi, S., 2014b. Aging and alcohol interact to alter hepatic DNA hydroxymethylation. Alcohol. Clin. Exp. Res. 38, 2178–2185.

Tammen, S., Park, J., Shin, P., Friso, S., Chung, J., Choi, S., 2016. Iron supplementation reverses reduced global hydroxymethylcytosine associated with chronic alcohol consumption in rats. FASEB J. 21, 264–270.

Tammen, S., Park, L., Dolnikowski, G., Ausman, L., Friso, S., Choi, S., 2017. Hepatic DNA hydroxymethylation is site-specifically altered by chronic alcohol consumption and aging. Eur. J. Nutr. 56, 535–544.

Tan, L., Shi, Y.G., 2012. Tet family proteins and 5-hydroxymethylcytosine in development and disease. Development 139, 1895–1902.

Tan, L., Xiong, L., Xu, W., Wu, F., Huang, N., Xu, Y., Kong, L., Zheng, L., Schwartz, L., Shi, Y., Shi, Y.G., 2013. Genome-wide comparison of DNA hydroxymethylation in mouse embryonic stem cells and neural progenitor cells by a new comparative hMeDIP-seq method. Nucleic Acids Res. 41, e84.

Tang, Y., Chu, J.M., Huang, W., Xiong, J., Xing, X.W., Zhou, X., Feng, Y.Q., Yuan, B.F., 2013. Hydrophilic material for the selective enrichment of 5-hydroxymethylcytosine and its liquid chromatography-tandem mass spectrometry detection. Anal. Chem. 85, 6129–6135.

Tchounwou, P.B., Yedjou, C.G., Patlolla, A.K., Sutton, D.J., 2012. Heavy metal toxicity and the environment. EXS 101, 133–164.

Thomson, J.P., Hunter, J.M., Nestor, C.E., Dunican, D.S., Terranova, R., Moggs, J.G., Meehan, R.R., 2013. Comparative analysis of affinity-based 5-hydroxymethylation enrichment techniques. Nucleic Acids Res. 41, e206.

Tian, X., Sun, B., Chen, C., Gao, C., Zhang, J., Lu, X., Wang, L., Li, X., Xing, Y., Liu, R., Han, X., Qi, Z., Zhang, X., He, C., Han, D., Yang, Y.-G., Kan, Q., 2018. Circulating tumor DNA 5-hydroxymethylcytosine as a novel diagnostic biomarker for esophageal cancer. Cell Res. 28, 597–600.

Truong, T.P., Sakata-Yanagimoto, M., Yamada, M., Nagae, G., Enami, T., Nakamoto-Matsubara, R., Aburatani, H., Chiba, S., 2015. Age-dependent decrease of DNA hydroxymethylation in human T cells. J. Clin. Exp. Hematop. 55, 1–6.

Valentini, E., Zampieri, M., Malavolta, M., Bacalini, M.G., Calabrese, R., Guastafierro, T., Reale, A., Franceschi, C., Hervonen, A., Koller, B., Bernhardt, J., Eline Slagboom, P., Toussaint, O., Sikora, E., Gonos, E.S., Breusing, N., Grune, T., Jansen, E., Dollé, M.E.T., Moreno-Villanueva, M., Sindlinger, T., Bürkle, A., Ciccarone, F., Caiafa, P., 2016. Analysis of the machinery and intermediates of the 5hmC-mediated DNA demethylation pathway in aging on samples from the MARKAGE study. Aging (Albany NY) 8, 1896–1922.

Vincent, J.J., Huang, Y., Chen, P.-Y., Feng, S., Calvopiña, J.H., Nee, K., Lee, S.A., Le, T., Yoon, A.J., Faull, K., Fan, G., Rao, A., Jacobsen, S.E., Pellegrini, M., Clark, A.T., 2013. Stage-specific roles for Tet1 and Tet2 in DNA demethylation in primordial germ cells. Cell Stem Cell 12, 470–478.

Walsh, C.P., Xu, G.L., 2006. Cytosine methylation and DNA repair. Curr. Top. Microbiol. Immunol. 301, 283–315.

2. DNA METHYLATION

Wang, D., Huang, J.H., Zeng, Q.H., Gu, C., Ding, S., Lu, J.Y., Chen, J., Yang, S.B., 2017. Increased 5-hydroxymethylcytosine and ten-eleven translocation protein expression in ultraviolet B-irradiated HaCaT cells. Chin. Med. J. 130, 594–599.

Wei, Y.B., Melas, P.A., Wegener, G., Mathe, A.A., Lavebratt, C., 2015. Antidepressant-like effect of sodium butyrate is associated with an increase in tet1 and in 5-hydroxymethylation levels in the BDNF gene. Int. J. Neuropsychopharmacol. 18, 1–10.

Wen, L., Li, X., Yan, L., Tan, Y., Li, R., Zhao, Y., Wang, Y., Xie, J., Zhang, Y., Song, C., Yu, M., Liu, X., Zhu, P., Li, X., Hou, Y., Guo, H., Wu, X., He, C., Li, R., Tang, F., Qiao, J., 2014. Whole-genome analysis of 5-hydroxymethylcytosine and 5-methylcytosine at base resolution in the human brain. Genome Biol. 15, R49.

Williams, K., Christensen, J., Pedersen, M.T., Johansen, J.V., Cloos, P.A.C., Rappsilber, J., Helin, K., 2011. TET1 and hydroxymethylcytosine in transcription and DNA methylation fidelity. Nature 473, 343–349.

Wossidlo, M., Nakamura, T., Lepikhov, K., Marques, C.J., Zakhartchenko, V., Boiani, M., Arand, J., Nakano, T., Reik, W., Walter, J., 2011. 5-Hydroxymethylcytosine in the mammalian zygote is linked with epigenetic reprogramming. Nat. Commun. 2, 241.

Wu, H., Zhang, Y., 2014. Reversing DNA methylation: mechanisms, genomics, and biological functions. Cell 156, 45–68.

Wu, X., Zhang, Y., 2017. TET-mediated active DNA demethylation: mechanism, function and beyond. Nat. Rev. Genet. 18, 517–534.

Wu, H., D'Alessio, A.C., Ito, S., Wang, Z., Cui, K., Zhao, K., Sun, Y.E., Zhang, Y., 2011. Genome-wide analysis of 5-hydroxymethylcytosine distribution reveals its dual function in transcriptional regulation in mouse embryonic stem cells. Genes Dev. 25, 679–684.

Wyatt, G.R., Cohen, S.S., 1953. The bases of the nucleic acids of some bacterial and animal viruses: the occurrence of 5-hydroxymethylcytosine. Biochem. J. 55, 774–782.

Xiong, J., Liu, X., Cheng, Q.Y., Xiao, S., Xia, L.X., Yuan, B.F., Feng, Y.Q., 2017. Heavy metals induce decline of derivatives of 5-methycytosine in both DNA and RNA of stem cells. ACS Chem. Biol. 12, 1636–1643.

Xu, Y., Wu, F., Tan, L., Kong, L., Xiong, L., Deng, J., Barbera, A.J., Zheng, L., Zhang, H., Huang, S., Min, J., Nicholson, T., Chen, T., Xu, G., Shi, Y., Zhang, K., Shi, Y.G., 2011. Genome-wide regulation of 5hmC, 5mC, and gene expression by Tet1 hydroxylase in mouse embryonic stem cells. Mol. Cell 42, 451–464.

Yang, H., Liu, Y., Bai, F., Zhang, J.-Y., Ma, S.-H., Liu, J., Xu, Z.-D., Zhu, H.-G., Ling, Z.-Q., Ye, D., Guan, K.-L., Xiong, Y., 2012. Tumor development is associated with decrease of TET gene expression and 5-methylcytosine hydroxylation. Oncogene 32, 663–66967.

Yin, R., Mao, S.Q., Zhao, B., Chong, Z., Yang, Y., Zhao, C., Zhang, D., Huang, H., Gao, J., Li, Z., Jiao, Y., Li, C., Liu, S., Wu, D., Gu, W., Yang, Y.G., Xu, G.L., Wang, H., 2013. Ascorbic acid enhances tet-mediated 5-methylcytosine oxidation and promotes DNA demethylation in mammals. J. Am. Chem. Soc. 135, 10396–10403.

Yu, M., Hon, G.C., Szulwach, K.E., Song, C.X., Jin, P., Ren, B., He, C., 2012a. Tet-assisted bisulfite sequencing of 5-hydroxymethylcytosine. Nat. Protoc. 7, 2159–2170.

Yu, M., Hon, G.C., Szulwach, K.E., Song, C.X., Zhang, L., Kim, A., Li, X., Dai, Q., Shen, Y., Park, B., Min, J.H., Jin, P., Ren, B., He, C., 2012b. Base-resolution analysis of 5-hydroxymethylcytosine in the mammalian genome. Cell 149, 1368–1380.

Yuan, C., Zhang, Y., Liu, Y., Wang, S., Wang, Z., 2017. DNA demethylation mediated by down-regulated TETs in the testes of rare minnow Gobiocypris rarus under bisphenol A exposure. Chemosphere 171, 355–361.

Zhang, L., Zhang, L., Zhou, K., Ye, X., Zhang, J., Xie, A., Chen, L., Kang, J.X., Cai, C., 2012. Simultaneous determination of global DNA methylation and hydroxymethylation levels by hydrophilic interaction liquid chromatography-tandem mass spectrometry. J. Biomol. Screen. 17, 877–884.

Zhang, J., Mu, X., Xu, W., Martin, F.L., Alamdar, A., Liu, L., Tian, M., Huang, Q., Shen, H., 2014. Exposure to arsenic via drinking water induces 5-hydroxymethylcytosine alteration in rat. Sci. Total Environ. 497–498, 618–625.

Zhang, L., Li, P., Wang, T., Zhang, X., 2015. Prognostic values of 5-hmC, 5-mC and TET2 in epithelial ovarian cancer. Arch. Gynecol. Obstet. 292, 891–897.

Zhao, C., Wang, H., Zhao, B., Li, C., Yin, R., Song, M., Liu, B., Liu, Z., Jiang, G., 2014. Boronic acid-mediated polymerase chain reaction for gene- and fragment-specific detection of 5-hydroxymethylcytosine. Nucleic Acids Res. 42, e81.

Zheng, H., Zhou, X., Li, D.K., Yang, F., Pan, H., Li, T., Miao, M., Li, R., Yuan, W., 2017. Genome-wide alteration in DNA hydroxymethylation in the sperm from bisphenol A-exposed men. PLoS One. 12e0178535.

2. DNA METHYLATION

NONCODING RNAS

The Role of Noncoding RNAs in Gene Regulation

Emily Woolard*, Brian N. Chorley†

*Oak Ridge Institute for Science and Education at US Environmental Protection Agency, RTP, NC, United States †National Health and Environmental Effects Research Laboratory, US Environmental Protection Agency, RTP, NC, United States

INTRODUCTION

For the greater part of the 20th century, proteins were considered to be the primary class of molecules responsible for carrying out the regulatory functions of the cell, including control of gene expression (Almeida et al., 2011; Sedwick, 2013). With only 1%–2% of the human genome coding for proteins, however, many scientists speculated that other portions of the genome (sometimes referred to as "junk" DNA) might be involved in gene regulation. It is now known that this previously uncharted portion of the genome contains regions of noncoding RNAs (ncRNAs) that have critical regulatory roles for gene transcription and protein translation. In this chapter, we will briefly review the forms of ncRNA with a focus on microRNAs (miRNAs), a type of small ncRNA that has important roles in regulating gene expression and cellular phenotype and may serve as important biomarkers in toxicology. In addition, we highlight current theories of biogenesis and emerging areas of research describing the biological roles of miRNA.

OVERVIEW OF NONCODING RNAS

Noncoding RNAs are roughly divided into two categories based on nucleotide length, either long (> 200) or short (⟨200). While precursors to short RNAs may be longer than 200 nucleotides in length (e.g., pre-miRNAs, precursors to transfer RNAs (tRNAs)), this classification is based on the mature form. Below, both long and short ncRNAs and their biological functions are briefly described.

Transfer and Ribosomal RNA

Transfer RNA (tRNA) and ribosomal RNA (rRNA) were identified in the mid-20th century and were the first ncRNAs described (Holley et al., 1965; Palade, 1955; RajBhandary and Kohrer, 2006; Scherrer, 2003). In 1955, George Palade used electron microscopy to isolate and document spherical granules with diameters of 100–150 Å found in various animal organs and tissues. These granules are now understood to be rRNA-containing ribosomes that have a strong affinity for the endoplasmic reticulum (Palade, 1955; Wells, 2005). Around the same time, tRNA was discovered by Paul Zamecnik and Mahlon Hoagland in protein synthesis studies, where they found an RNA-containing molecule necessary for the transport of amino acids (Kresge et al., 2005). It was later found that both tRNAs and rRNAs interact directly with messenger RNA (mRNA) for protein synthesis. Ribosomal RNAs form large ribosomal complexes with associated proteins and are defined by a small and large subunit. Messenger RNA is decoded by the small ribosomal subunit, and tRNAs deliver amino acids to the large subunit where they are linked by peptide bonds and form newly synthesized proteins (Wohlgemuth et al., 2011). Interestingly, tRNA fragments have also been implicated in translation inhibition and play a role in cell-to-cell communication (Lee and Collins, 2005; Thompson et al., 2008; Uter and Perona, 2004).

Long Noncoding RNA (lncRNA)

Long noncoding RNAs (lncRNAs) are noncoding transcripts >200 nucleotides in length that are now widely recognized for their role in genome regulation (Hung and Chang, 2010; Kung et al., 2013; Rinn and Chang, 2012). lncRNAs are involved in localization of proteins, chromatin remodeling, and organizational scaffolding (Fang and Fullwood, 2016;

Novikova et al., 2013; Wilusz et al., 2009), among other regulatory functions. Many lncRNAs fulfill their regulatory roles by binding chromatin-modifying proteins to specific locations in the genome (Rinn and Chang, 2012). In some cases, lncRNAs regulate local gene expression (Wilusz et al., 2009).

Competitive Endogenous RNA

Some lncRNAs can be classified as competing endogenous RNAs (ceRNAs) and are closely involved with the regulation of miRNAs by sequestering and reducing miRNA activity (Ebert and Sharp, 2010). A special type of ceRNA is known as a circular RNA (circRNA) denoted by a unique circular, covalently bonded large RNA structure of covalent bonds (see "MicroRNA Biogenesis and Regulation" section).

Long Intervening Noncoding RNA

Intergenic in origin, long intervening noncoding RNAs (lincRNAs) function in gene regulation, imprinting, and cell differentiation (Yang et al., 2011; Flynn and Chang, 2014). LincRNA has also been shown to bind chromatin-modifying factors (Khalil et al., 2009; Flynn and Chang, 2014). LincRNAs are highly evolutionarily conserved across species (Guttman et al., 2009) suggesting they have similar roles across species.

Enhancer RNA

Enhancer RNAs (eRNAs) are similar in function to other lncRNAs in that they directly regulate mRNA transcription. ERNAs are produced from gene enhancer regions, function to activate promoter-mediated transcription, and catalyze interactions between enhancer and promoter regions (Kim et al., 2015; Rothschild and Basu, 2017).

Short Noncoding RNA

Short noncoding RNAs (snRNAs) are generally defined as being under 200 nucleotides in length, with most being 20–30 nucleotides long.

Piwi-Interacting RNA

Piwi-interacting RNAs (piRNA) are 26–31 nucleotides in length and regulate gene expression by interacting with P-element induced wimpy testis (piwi) regulatory proteins to form piRNA-induced silencing complexes, which are responsible for silencing transposons in the germ line (Le Thomas et al., 2014; Weick and Miska, 2014). Multiple studies have demonstrated their vital role in both germ line development and carcinogenesis, with specific piRNAs shown to be involved in the onset of colorectal, cervical, breast, and other cancers (Chu et al., 2015a; Wang et al., 2015; Yin et al., 2017; Suzuki et al., 2012).

Small Nuclear RNA

Small nuclear RNAs (snRNAs) process pre-mRNA in the nucleus, maintain telomeres, and aid in the regulation of transcription factors. snRNA proteins form complexes called snRNPs, which include the spliceosome, allowing these short RNA transcripts to directly alter the splicing of introns (Mourao et al., 2010; Valadkhan and Gunawardane, 2013).

Short Nucleolar RNA

Short nucleolar RNAs (snoRNAs) are larger types of short ncRNAs, ranging in length from 60 to 300 nucleotides. They guide covalent modifications of other RNAs via methylation or pseudouridylation (Nogueira Jorge et al., 2017). SnoRNAs are some of the most prevalent short ncRNA among eukaryotes, though they are not present in bacteria or archaea (Omer et al., 2000). Small Cajal body-specific RNAs (scaRNAs) are a class of snoRNA discovered in the early 2000s (Darzacq et al., 2002; Jady and Kiss, 2001). ScaRNAs guide the posttranscriptional methylation and pseudouridylation of spliceosomal snRNAs within nuclear suborganelles (termed Cajal bodies), which is a necessary step for the creation of a functional spliceosome (Patil et al., 2015).

MicroRNA

MicroRNAs (miRNAs) average 22 nucleotides in length and comprise the most widely studied type of small ncRNA. miRNAs posttranscriptionally regulate gene expression by silencing protein expression through cleavage and degradation of the mRNA transcript or inhibiting translation (Valencia-Sanchez et al., 2006). miRNAs can also regulate gene expression by interacting directly or indirectly with regulatory proteins (Eiring et al., 2010). These transcripts are versatile and overlapping in their targeting ability. A single miRNA may bind multiple mRNA targets, while several miRNAs may regulate a single mRNA target (Macfarlane and Murphy, 2010).

Short Interfering RNA

Short (or small) interfering RNAs (siRNAs) are similar to microRNAs in size and function, in that they target and degenerate gene transcripts through a RNA-induced silencing complex (RISC)-mediated process. While miRNAs have the ability to regulate dozens or even hundreds of gene targets through imperfect base pairing, siRNAs bind specifically to a single gene location (Kim, 2005; Lam et al., 2015). This feature has promoted development of molecular tools and therapies using siRNA since the mechanism was shown to promote specific RNA interference (RNAi) in mammalian cells (Elbashir et al., 2001).

MICRORNA DISCOVERY

Beginning in the late 1980s, concurrent efforts by Victor Ambros and Gary Ruvkun led to the discovery of the first miRNA, *lin-4* (Almeida et al., 2011; Lee et al., 1993; Wightman et al., 1993). *Caenorhabditis elegans* deficient of the nonprotein encoding RNA *lin-4* ("lin" stands for "abnormal cell LINeage" in *C. elegans*) displayed severe abnormalities throughout development. Conversely, mutants of the gene that encodes the protein *lin-14* appeared to have developmental timing defects opposite to *lin-4* deficiency. After comparing unpublished findings, Ambros and Ruvkun found that *lin-4* was complementary to a repeated sequence in the 3′ UTR of *lin-14* (Sedwick, 2013). This antisense binding allowed *lin-4* to negatively regulate the expression of the protein *lin-14*, effectively demonstrating that otherwise insignificant RNA fragments could exert a profound role in gene regulation. Ambros and Ruvkun each went on to publish this remarkable finding, establishing the first insights into a completely novel cellular regulatory mechanism (Almeida et al., 2011; Lee et al., 2004a). A few years after the recognition of *lin-4*, *let-7* emerged as the second documented miRNA (Pasquinelli et al., 2000). Since these initial discoveries, thousands of new miRNAs have been identified (Londin et al., 2015; Friedlander et al., 2014).

MICRORNA BIOGENESIS AND REGULATION

The biogenesis of mature miRNA sequences begins with the nuclear transcription of a long (up to 100 kb in length) primary miRNA (pri-miRNA), which is embedded by mono- or polycistronic stem-loop structures (Fig. 1A; Lee et al., 2003). These stem-loop structures, which contain the mature miRNA sequences, are cleaved by the microprocessor complex, forming a stem-loop structure known as precursor miRNA (pre-miRNA; Fig. 1B; Denli et al., 2004; Gregory et al., 2004). The pre-miRNAs are exported from the nucleus to the cytoplasm by the transport karyopherin, exportin-5 (Fig. 1C; Yi et al., 2003). Once in the cytoplasm, a final cleavage step of the processed hairpin structure occurs with the endoribonuclease, Dicer, producing an ~22-nucleotide duplex (Fig. 1D; Hutvagner et al., 2001; Knight and Bass, 2001). The duplex is then loaded into an Argonaute (AGO) protein to form the RNA-induced silencing complex (RISC), the passenger strand of the duplex is removed (Fig. 1E), and guide (mature) strand is then used for mRNA target inhibition or degradation (Fig. 1F; Hammond et al., 2001; Khvorova et al., 2003; Mourelatos et al., 2002).

Pri-miRNA Transcription and Regulation

Transcription of the primary miRNAs (pri-miRNAs), which are located within the intragenic (introns or exons) regions on the same strand of host genes, is mediated by RNA polymerase II (Pol II) and its associated transcriptional machinery (Lee et al., 2004b). Pri-miRNAs can be methylated on the 5' terminal cap (or m7G-cap), polyadenylated, and spliced—all of which are common features of Pol II-transcribed RNA (Cai et al., 2004). Intergenic miRNAs may be derived, however, from either Pol II or Pol III-mediated transcription. For example, miRNA sequences downstream of tRNA, Alu, and mammalian-wide interspersed repeat (MWIR) regions can be derived from Pol III (Borchert et al., 2006). Importantly, these repetitive intergenic regions of the genome may play a role in miRNA biogenesis and gene regulation.

The identification of promoter regions for pri-miRNAs has traditionally been a challenge because of the difficulty in measuring the transient and often large (up to 100 kb in length) pri-miRNA sequences and subsequent mapping of known mature miRNAs to the distal promoter sites of pri-miRNAs (Schanen and Li, 2011). To help overcome this challenge, genome-wide epigenetic measurement techniques that identify open and actively transcribed regions of the genome have assisted in identifying putative transcriptional start sites (TSS) and promoter/enhancer regions of pri-miRNAs. For example, active regulatory regions can be identified by mapping of trimethylated lysine 4 of histone 3 (H3K4me3) and overlapping binding regions for key embryonic stem cell transcription factors (Oct4, Sox2, Nanog, and Tcf3; Marson et al., 2008). It was found that these transcription factors, which mediate the cellular

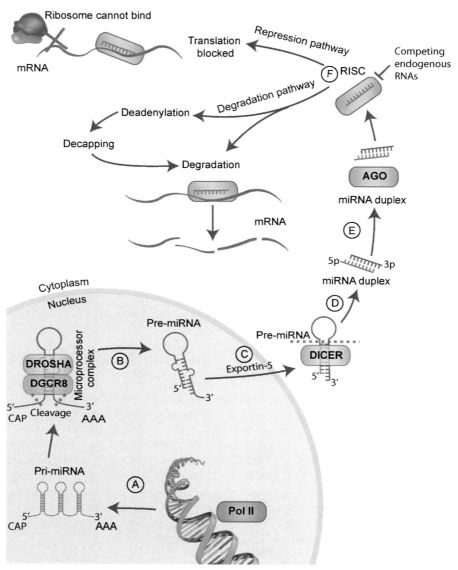

FIG. 1 Steps of microRNA biogenesis. (A) Nuclear transcription by polymerase II of primary miRNA (pri-miRNA) embedded by mono- or polycistronic stem-loop structures. (B) Stem-loop structures, containing the mature miRNA sequences, are cleaved by the microprocessor complex (including proteins DROSHA and DGCR8) and form precursor miRNA (pre-miRNA). (C) Pre-miRNAs are exported to the cytoplasm from the nucleus to the cytoplasm by exportin-5. (D) The final cleavage step of the processed hairpin structure is performed by DICER, producing ~22 nucleotide duplexes. (E) The duplex is then loaded into RNA-induced silencing complex (RISC), and the passenger strand of the duplex is removed. (F) The guide (mature) strand is then utilized for messenger RNA (mRNA) target inhibition or degradation.

pluripotency, occupied putative promoters of several miRNA polycistrons (presumably coexpressed by a common pri-miRNA) implicated in cellular proliferation.

Similar techniques used to identify genome-wide histone-depleted regions, along with other markers of transcriptional initiation, were utilized to identify miRNA promoter and TSS regions in cancer cell lines (Ozsolak et al., 2008). Interestingly, the miRNA promoters identified almost equal amounts of intergenic- and intragenic-derived regions (87 and 88, respectively), with 32 intragenic miRNAs having distinct promoters from the host gene, suggesting separate regulatory regions for the host mRNA and intragenic miRNAs. Recent techniques have also improved the detection of promoter regions for pri-miRNAs, including the use of cap analysis gene expression (CAGE) profiling, which enriches for capped transcripts before deep sequencing. This approach provided promoter information for 1357 human and 804 mouse miRNA (de Rie et al., 2017). In addition, computational methods that integrate many of the aforementioned genome-wide datasets (in particular H3K4me3 patterns) and genomic sequence feature recognition are able to predict miRNA promoters and transcription factors (TFs; Zhao et al., 2017).

Pri-miRNA Processing and Nuclear Export

The transcribed pri-miRNA is typically long (usually >1 kb) and forms one or more localized stem-loop structures that contain the mature miRNA sequence. The upper stem and terminal loop of this duplexed structure is cleaved to form a ~65 nucleotides in length "pre-miRNA" (Lee et al., 2003). In animals, the cleavage step is mediated by the microprocessor complex, which is composed of the endonuclease Drosha and the guiding protein DGCR8 (Denli et al., 2004; Han et al., 2006). DGCR8 measures and determines the cut sites ~11 base pairs from the basal junction (the point at which the single-strand pri-miRNA is beginning to form a duplex with the 3' apical end of the strand; Ha and Kim, 2014) and guides Drosha to cut the pre-miRNA, leaving a two-nucleotide-long 3' overhang (Han et al., 2004). Once processed, the pre-miRNA is transported from the nucleus to the cytoplasm by a protein complex composed of exportin-5 and the RAN-GTP cofactor (Yi et al., 2003). This complex recognizes and binds the double-stranded RNA stem and the 3' overhang of the pre-miRNA, independent of the sequence (Okada et al., 2009).

Maturation of miRNA RISC Formation

After nuclear translocation, the pre-miRNA is further processed by the RNase III type endoribonuclease Dicer that recognizes the 3' overhang of the pre-miRNA and then cleaves the terminal loop from the small RNA duplex (Zhang et al., 2004). It has been suggested by in silico modeling that Dicer may be the target of many environmental mutagens/carcinogens. Surprisingly, the study predicted that these chemicals may have higher affinity for the Dicer endonuclease domain A than double-stranded RNA (e.g., pre-miRNAs), providing a novel, nongenotoxic mechanism by which these mutagens/carcinogens may act to globally downregulate miRNA processing (Ligorio et al., 2011).

Finally, the small RNA duplex is loaded into the AGO protein to form the RNA-induced silencing complex (Mourelatos et al., 2002). In humans, there are four family members of AGO

(1–4) that are capable of loading miRNA. Only AGO2 can mediate slicer activity, whereas all four AGO proteins can inhibit miRNA translation (see "Biological Roles of MicroRNA" section; Huntzinger and Izaurralde, 2011). After loading, the miRNA duplex unwinds and removes the passenger strand and retains the guide strand for gene repression activity (Hammond et al., 2001; Khvorova et al., 2003; Mourelatos et al., 2002). Although miRNAs are not sorted for specific AGO proteins in humans, different family members processing may impact the length of the bound miRNA (Dueck et al., 2012). Strand selection was initially shown to be primarily based on the internal stability of the 5′ end of the duplex (Khvorova et al., 2003; Schwarz et al., 2003) with a reported uracil (U) bias and distinct purine/pyrimidine difference between the guide and passenger stands (Hu et al., 2009), but recent evidence has demonstrated that Dicer and its double-stranded RNA-binding proteins TRBP and PACT may also influence guide strand selection (Wilson et al., 2015). The passenger strand is also sometimes loaded, and there have been reports that such switching can predominantly occur in different tissue types (Chiang et al., 2010).

Regulation of Mature miRNA by ceRNA

Mature miRNAs can be very stable, demonstrating a half-life of a week or more in vivo (van Rooij et al., 2007). Because of this stability, miRNAs may serve as cellular memory after responding to a stressor (Leung and Sharp, 2010), and perhaps not surprisingly, cells have evolved mechanisms to control these stable miRNA levels after maturation. As such, there are a number of RNA "sponges," also known as competing endogenous RNAs (ceRNAs), that work to reduce the miRNA activity, helping fine-tune the levels of miRNA activity (Fig. 1F; Bak and Mikkelsen, 2014; Ebert and Sharp, 2010). Identified types of ceRNAs are pseudogenes and lncRNAs, including circular RNAs (circRNAs). In 2009, Seitz postulated that transcribed pseudotargets are the actual regulators of gene expression because they regulate the miRNAs that, in turn, regulate the target genes (Seitz, 2009). One of the best-studied examples is ceRNA-mediated regulation of the tumor suppressor gene, phosphate and tensin homologue (PTEN). The pseudogene PTENP1 was the first ceRNA identified (Poliseno et al., 2010) and was later found, based on the analysis of several validated PTEN-targeting miRNAs, that several more protein-coding genes (SERINC1, VAPA, and CNOT6L) served as ceRNAs for PTEN in a miRNA-dependent manner (Tay et al., 2011). Moreover, a recent in silico analysis based on experimental miRNA binding and RNA-seq databases predicted an overwhelming 170 putative ceRNAs for PTEN (Zarringhalam et al., 2017).

In vascular endothelial cells, the miRNA let-7e is suppressed by the lncRNA, lnc-MK167IP-3 (Lin et al., 2017). Let-7e acts as pro-inflammatory mediator by increasing activity of a key transcriptional factor for inflammatory genes, NF-κB. Specifically, let-7e targets the NF-κB inhibitory factor, IκBβ, therefore increasing NF-κB activity in this model. Conversely, lnc-MK167IP-3 promotes an antiinflammatory effect by acting as a ceRNA for let-7e. Interestingly, it was also found that let-7e targets and regulates the expression of lnc-MK167IP-3, thereby enhancing the pro-inflammatory effects of let-7e by suppressing its molecular "sponge."

Another type of lncRNA, circular RNAs (circRNAs), can also exhibit ceRNA functions (Hansen et al., 2013; Memczak et al., 2013). Originally thought to be sequencing errors (see review in Greene et al., 2017) or linear transcripts with exon scrambling

(Al-Balool et al., 2011), circRNAs were recently "rediscovered" in RNA-seq data and verified to be abundantly transcribed products (Salzman et al., 2012). CircRNAs are likely derived from either an exon-skipping events, resulting in spliced lariats containing the skipped exons, or 5′–3′ backsplicing events of nascent transcripts (Jeck et al., 2013). Two 2013 papers found that the 1.5 kB circRNA, ciRS-7 (or CDR1as), contained ~70 conserved binding sites for the highly conserved miR-7, and both were colocalized in the mouse brain (Hansen et al., 2013; Memczak et al., 2013). When compared with a perfect anti-miR-7 target site, ciRS-7 demonstrated greater knockdown potential; in vivo, injected plasmid DNA expressing circular ciRS-7 reduced midbrain sizes in zebra fish, which was mitigated by injecting miR-7 precursor. Additionally, CRISPR/Cas9-mediated knockout of ciRS-7 in the mouse brain caused misregulation of miR-7 and another identified miRNA target, miR-671, causing behavior associated with neurological disorders (Piwecka et al., 2017), further implicating the important biological role of circRNAs by regulating miRNA levels.

More recently, toxicological-based studies have begun to focus on alterations of circRNAs and miRNAS in response to specific environmental stressors and the associated regulatory relationships. Many circRNA profiles have been measured in response to xenobiotic and toxicological perturbations. The exposures include the polycyclic aromatic hydrocarbon and potent carcinogen, benzo[a]pyrene (BaP), in a human hepatocellular carcinoma cell line (HepG2; Caiment et al., 2015); free fatty acid in HepG2 cells to induce hepatic steatosis (Guo et al., 2017); irradiation in hepatic stellate cells (Chen et al., 2017); and methionine and choline diet deficiency in a nonalcoholic steatohepatitis (NASH) mouse model (Jin et al., 2016). While the number of differentially expressed circRNAs ranged considerably for these studies (from tens to hundreds of altered circRNAs compared with controls), all studies described integrated mRNA-miRNA-circRNA networks.

BIOLOGICAL ROLES OF MICRORNA

Gene Target Silencing

Two distinct target gene silencing mechanisms are known: slicer-dependent and slicer-independent (Fig. 1F; Valencia-Sanchez et al., 2006). Slicer-dependent activity requires strong base pairing between the miRNA and target mRNA, which is silenced through degradation by Argonaute 2 (AGO2) catalyzed cleavage, which is the exclusive RNAi-dependent gene silencing Argonaute in humans. Slicer-independent silencing occurs when there is limited base pairing between the miRNA and its target, restricting AGO2-mediated cleavage. The miRNA still binds and represses its target, though the inhibition may be reversible if the mRNA transcript is not degraded through other mechanisms (Macfarlane and Murphy, 2010; Meister et al., 2004; Wahid et al., 2010).

Target Degradation Pathway

Target degradation is a slicer-dependent gene regulation mechanism. In this pathway, RISC binds the promoter region and cleaves the mRNA with slicer enzymes. Slicer activity requires that the miRNA contained in RISC have strong complementarity to its target mRNA (Gu and Kay, 2010; Haas et al., 2016; Marzi et al., 2016). Data suggest that mRNA degradation is the most likely outcome following miRNA binding, indicating that reduction of mRNA levels is responsible for $\geq$84% of observable loss in protein output (Guo et al., 2010). A reanalysis of these data, however, by a separate group concluded that this estimate was largely inflated due to mounting evidence that inhibition of mRNA translation precedes mRNA degradation (Larsson and Nadon, 2013); therefore, there is still currently much debate on the primary mechanism by which miRNA suppresses gene targets.

Interestingly, some miRNAs such as miR-125b and let-7 can catalyze and expedite deadenylation in targets bound for degradation (Eulalio et al., 2009; Wu et al., 2006). Deadenylation is the process of having the 3′ poly(A) tail removed from a strand of mRNA during posttranscriptional modification. Following poly(A) tail removal, mRNAs may be either decapped on the 5′ end by Dcp1/Dcp2 enzymes or degraded exonucleolytically by the exosome. When decapped with Dcp1/Dcp2, degradation is catalyzed by the exoribonuclease, Xrn1p (Valencia-Sanchez et al., 2006).

Translational Repression

There is, however, mounting evidence that indicates miRNA binding interferes with translation initiation as its primary mode of action for repressing mRNA output (Wilczynska and Bushell, 2015; Mathonnet et al., 2007; Pillai et al., 2005; Thermann and Hentze, 2007). In the absence of slicer activation, RISC represses translation by binding the promoter region and preventing a ribosome from binding to begin translation. The exact mechanism of this repression is poorly understood. Whether repression and mRNA degradation must take place sequentially is also unknown; however, both in vivo and mammalian in vitro models have demonstrated that repression precedes target decay and that the former mechanism does not require deadenylation to reduce protein production (Bethune et al., 2012; Bazzini et al., 2012; Djuranovic et al., 2012; Fabian et al., 2009). Work has also shown that initial repression of target was a necessary prerequisite for mRNA degradation (Meijer et al., 2013).

miRNA-mediated translational repression is possible with weak base pairing, though stronger complementarity between miRNA and target may allow for enhanced translational repression (Felekkis et al., 2010). Because the mRNA may not necessarily undergo decay, this repression may be reversed with the removal of RISC if no further degradation events are initiated (Bose et al., 2017). Some research supports that target degradation still typically follows such translational repression (Wilczynska and Bushell, 2015).

miRNA-Mediated Gene Target Upregulation

Contrary to the canonical role of miRNAs in gene repression, miRNAs may target gene promoter elements, causing upregulated expression of mRNA. Such upregulation is known as RNA activation (RNAa; Huang et al., 2012). Huang and colleagues examined expression of mouse cyclin B1 (*Ccnb1*) and found three miRNAs to be implicated in the upregulation of *Ccnb1* gene expression in vitro. Removal of one of the identified miRNAs, miR-744, resulted in a *reduction* of Ccnb1 levels. Further analysis demonstrated that miR-744 enhanced translation by increasing enrichment of RNA polymerase II at the target transcription start site and an epigenetic marker of gene activation, trimethylated histone 3 at lysine 4 (H3K4me3; Huang, et al. 2012). The exact mechanism by which miRNAs directly increase levels of target gene transcription of a target gene is still unknown, but indirect activation of a gene or pathway can be mediated through feedback loop circuits.

Enhanced Biological Impact of miRNAs Through Feedback Loops

On average, effects of an individual miRNA have very modest direct effects on gene expression; therefore, it has been speculated (and demonstrated) that miRNAs amplify regulatory effects by self-promoting feedback loops (Fabbri et al., 2011; Tessel et al., 2010). The purposes of such responses are to mediate signaling to tolerate a new, persistent environmental challenge or restore a homeostatic state after perturbation to the cell. Feedback loops amplify a signal and persist even after the initial perturbation has been removed through a positive feedback loop, or they abolish an active signal through a negative feedback loop (Fig. 2A and C). In addition, different types of feedback loops can be utilized to switch biological states (Fig. 2C; Leung and Sharp, 2010).

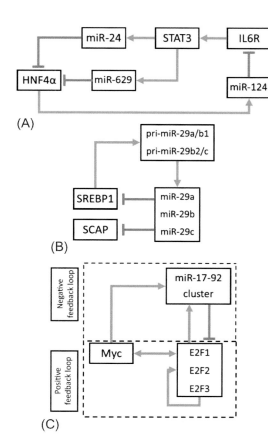

FIG. 2 Examples of feedback loops involving miRNA. (A) The developmental- and cancer-linked liver transcription factor HNF4α is regulated by a positive feedback loop consisting of miR-124, miR-24, and miR-629 and inflammatory modulators interleukin 6 receptor (IL6R) and STAT3. HNF4α upregulates miR-124, which in turn negatively targets the inflammatory modulator IL6R. This protein promotes STAT3, upregulating miR-24 and miR-629, which both downregulate HNF4α, completing the positive feedback loop (Hatziapostolou et al., 2011). (B) SREBP-1, a regulatory element involved with lipid metabolism, is negatively targeted by miR-29 family members miR-29a, miR-29b, and miR-29c. SREBP-1 promotes pri-miRNA expression of miR-29 members, resulting in mature miRNAs that downregulate both SREBP-1 and its associated protein, SCAP, forming a negative feedback loop (Ru and Guo, 2017). (C) The miR-17–92 cluster targets transcription factors Myc and E2F1, E2F2, and E2F3, which in turn regulate the expression of miR-17–92, thus forming a negative feedback loop. A positive feedback loop exists between Myc and E2F members. Quiescent, apoptotic, and proliferative cellular states are ultimately influenced by Myc/E2F levels, which are modulated by the miR-17–92 cluster "tweaking" these feedback mechanisms (Coller et al., 2007; Li et al., 2011). *Solid arrows* represent activation; *capped lines* represent repression in the circuits.

Positive Feedback Loop

An example of a positive feedback loop was demonstrated in an in vitro model of hepatocellular transformation involving three miRNAs (Fig. 2A; (Hatziapostolou et al., 2011). This stable transformation was maintained by a feedback loop consisting of miR-124, miR-24, and miR-629 and inflammatory modulators interleukin 6 receptor (IL6R) and STAT3. It was found that miR-124 was a direct target of the developmental- and cancer-linked liver transcription factor, hepatocyte nuclear factor 4α (HNF4α). The investigators demonstrated that HNF4α levels were reduced and miR-14 levels were also reduced, leading to increased expression and activity of the target protein, IL6R. This subsequently led to phosphorylation and increased activity of the IL6R target, STAT3. To complete the biological circuit, STAT3 activation increased the expression of miR-24 and miR-629, which targeted the expression of HNF4α. This miRNA-linked circuit was shown to be active in human hepatocellular carcinomas (HCC) and served as a molecular indicator of HCC progression (Ning et al., 2014) demonstrating the importance of this feedback loop in clinical settings. In addition, similar positive feedback loops involving miR-124 have also been linked to other cancers including colorectal, prostate, and nonsmall cell lung cancer (Chu et al., 2015b; Liu et al., 2017; Zu et al., 2016) indicating that this miRNA-mediated feedback loop may be ubiquitous mechanism of cancer cell transformation.

Negative Feedback Loop

An example of a negative feedback loop involving a miRNA can be found in signaling pathways of lipid metabolism (Fig. 2B). Sterol regulatory element-binding proteins (SREBPs) consist of three family members (SREBP-1a, SREBP-1c, and SREBP-2), which regulate de novo synthesis of fatty acids, phospholipids, and cholesterol (Ru and Guo, 2017). The miR-29 family members miR-29a, miR-29b, and miR-29c share the same seed sequence and target the 3′ UTR regions of SREBP-1 and complex member SREBP cleavage-activating protein (SCAP; Ru et al., 2016). In addition, SREBP-1 regulates pri-miRNA expression of miR-29 members, which forms a negative feedback loop (Fig. 2B). A potential role of this feedback mechanism is to reduce protein activity after a certain period. Specifically, miR-29 family members need time to mature and increase in numbers before impacting the activity of SREBP, therefore allowing a certain window of activity before miRNA-mediated shutdown of fatty acid synthesis.

Switching Loops

The miRNA cluster miR-17–92 has been implicated in both oncogenic and tumor suppressor functions. This cluster is known to be involved in different regulatory loops that can switch under certain conditions (Fig. 2C), partially explaining the divergent role these miRNAs have in cancer. The miR-17–92 cluster is a polycistronic gene composed of seven mature miRs. These miRNAs have been shown to regulate Myc and E2F1, E2F2, and E2F3 transcription factors that are involved in cellular proliferation. These transcription factors, in turn, inhibit the expression of miR-17–92 (Coller et al., 2007), thus forming a negative feedback loop. In addition, a positive feedback loop exists between Myc and E2F (Li et al., 2011). Computational modeling of these processes demonstrated that the transition of quiescent, apoptotic, and proliferative cellular states was coordinated by the miR-17–92 cluster, which ultimately was regulated by Myc/E2F levels of protein in these coupled feedback loops. Other regulatory loops have been identified in this biological circuit, and as more is learned, computational models will be essential in quantitatively defining the roles of miRNAs in different cellular responses (Lai et al., 2016).

Biofluid-Based Biomarkers and Cell Communication

The biological impact of miRNAs may not be limited to the intracellular environment in which they are produced. Indeed, stable miRNAs are found in almost all biofluids (including blood, urine, sputum, and amniotic fluid) where they are protected through interactions with proteins (such as AGO), lipids, apoptotic bodies, and excretory microvesicles such as exosomes (Arroyo et al., 2011; Valadi et al., 2007; Vickers et al., 2011; Wang et al., 2010). The release of these miRNAs in biofluids can be mediated by tissue and cellular toxicity (i.e., release of highly expressed miRNA due to cellular death), and some of these miRNAs are tissue-specific (see review in Harrill et al., 2016). For these reasons, biofluid-based miRNAs have been extensively studied as putative biomarkers of drug and environmental exposure toxicity. Additionally, biofluid-based miRNAs may be involved in cell-to-cell (paracrine- or endocrine-like) communication (Turchinovich et al., 2013). One of the first published studies of miRNA transfer to donor cells was demonstrated in human mast cells (Valadi et al., 2007). Both mRNA and miRNA were found in exosomes, collectively termed "exosomal shuttle" RNA and altered recipient cell protein production. In later in vitro studies, selectively packaged miR-150 in the microvesicles of cultured human monocyte/macrophage cell line, THP-1, was found to efficiently transfer into human microvascular endothelial cells (HMEC-1; Zhang et al., 2010). Once internalized, miR-150 targeted and reduced levels of c-Myb, which mediated increase endothelial cell migration, thus having potential consequences in disease processes such as atherosclerosis (Li et al., 2013). Not surprisingly, cancer cells utilize this form of communication and may therefore be involved in disease pathogenesis (see review in Salido-Guadarrama et al., 2014). Indeed, cancer cells selectively package miRNA and other related proteins, which has been proposed as a way to pass donor cancer cell traits to recipient cells (Jaiswal et al., 2012). Paracrine signaling mediated by miRNA was also recently found to be important in a mouse model of obesity (Ying et al., 2017). It was discovered that exosomes derived from adipose tissue macrophages of obese mice contained abnormally high levels of miR-155, which through a series of elegant experiments was shown to influence insulin sensitivity and glucose intolerance, both hallmarks of metabolic disease and diabetes.

CONCLUDING REMARKS AND FUTURE DIRECTIONS

RNA from noncoding portions of the genome is now understood to play a critical role in gene regulation and cellular function. Although functions are still being defined for many ncRNAs, evidence of altered cellular profiles following pharmaceutical or chemical exposures and during early disease states and progression suggests important roles in these processes. In this chapter, we focused on introducing miRNA biogenesis and biological function. Since their discovery by Ambros and Ruvkun at the end of the 20th century, miRNAs have quickly been recognized as a critical epigenetic mechanism of gene expression regulation. The role of miRNAs involves several layers of regulation that are applied by the cell during biogenesis and after maturation. In toxicology, knowledge of gene pathway circuits is central to our understanding of cellular homeostasis and perturbation; miRNAs are clear players in these complex processes. By studying miRNAs in conjunction with gene expression and other traditional apical measurements, we may better understand adaptive and adverse response to exogenous exposure. In addition, miRNAs, which exist in both intracellular and extracellular (e.g., biofluids) space, may serve as important early and specific molecular biomarkers of cellular and tissue states. Importantly, these miRNA-based biomarkers that can be noninvasively measured in animals and human subjects are linked to molecular pathways of adversity.

Acknowledgments

The authors wish to thank Dr. Stephanie Padilla and Dr. Charles Wood for their critical reviews of this manuscript. We are also grateful to Molly Windsor and John Havel for their assistance with the figures. This manuscript has been reviewed by the US Environmental Protection Agency and approved for publication. Approval does not signify that the contents reflect the views of the agency, nor mention of trade names or commercial products does not constitute endorsement or recommendation of use.

References

Al-Balool, H.H., Weber, D., Liu, Y., Wade, M., Guleria, K., Nam, P.L., et al., 2011. Post-transcriptional exon shuffling events in humans can be evolutionarily conserved and abundant. Genome Res. 21, 1788–1799.

Almeida, M.I., Reis, R.M., Calin, G.A., 2011. MicroRNA history: discovery, recent applications, and next frontiers. Mutat. Res. 717, 1–8.

Arroyo, J.D., Chevillet, J.R., Kroh, E.M., Ruf, I.K., Pritchard, C.C., Gibson, D.F., et al., 2011. Argonaute2 complexes carry a population of circulating microRNAs independent of vesicles in human plasma. Proc. Natl. Acad. Sci. USA 108, 5003–5008.

Bak, R.O., Mikkelsen, J.G., 2014. miRNA sponges: soaking up miRNAs for regulation of gene expression. Wiley Interdiscip. Rev. RNA 5, 317–333.

Bazzini, A.A., Lee, M.T., Giraldez, A.J., 2012. Ribosome profiling shows that miR-430 reduces translation before causing mRNA decay in zebrafish. Science 336, 233–237.

Bethune, J., Artus-Revel, C.G., Filipowicz, W., 2012. Kinetic analysis reveals successive steps leading to miRNA-mediated silencing in mammalian cells. EMBO Rep. 13, 716–723.

Borchert, G.M., Lanier, W., Davidson, B.L., 2006. RNA polymerase III transcribes human microRNAs. Nat. Struct. Mol. Biol. 13, 1097–1101.

Bose, M., Barman, B., Goswami, A., Bhattacharyya, S.N., 2017. Spatiotemporal uncoupling of microRNA-mediated translational repression and target RNA degradation controls microRNP recycling in mammalian cells. Mol. Cell. Biol. 37, e00464-16.

Cai, X., Hagedorn, C.H., Cullen, B.R., 2004. Human microRNAs are processed from capped, polyadenylated transcripts that can also function as mRNAs. RNA 10, 1957–1966.

Caiment, F., Gaj, S., Claessen, S., Kleinjans, J., 2015. High-throughput data integration of RNA-miRNA-circRNA reveals novel insights into mechanisms of benzo[a]pyrene-induced carcinogenicity. Nucleic Acids Res. 43, 2525–2534.

Chen, Y., Yuan, B., Wu, Z., Dong, Y., Zhang, L., Zeng, Z., 2017. Microarray profiling of circular RNAs and the potential regulatory role of hsa_circ_0071410 in the activated human hepatic stellate cell induced by irradiation. Gene 629, 35–42.

Chiang, H.R., Schoenfeld, L.W., Ruby, J.G., Auyeung, V.C., Spies, N., Baek, D., et al., 2010. Mammalian microRNAs: experimental evaluation of novel and previously annotated genes. Genes Dev. 24, 992–1009.

Chu, H., Xia, L., Qiu, X., Gu, D., Zhu, L., Jin, J., et al., 2015a. Genetic variants in noncoding PIWI-interacting RNA and colorectal cancer risk. Cancer 121, 2044–2052.

Chu, M., Chang, Y., Guo, Y., Wang, N., Cui, J., Gao, W.Q., 2015b. Regulation and methylation of tumor suppressor miR-124 by androgen receptor in prostate cancer cells. PLoS One 10, e0116197.

Coller, H.A., Forman, J.J., Legesse-Miller, A., 2007. "Myc'ed messages": myc induces transcription of E2F1 while inhibiting its translation via a microRNA polycistron. PLoS Genet. 3, e146.

Darzacq, X., Jady, B.E., Verheggen, C., Kiss, A.M., Bertrand, E., Kiss, T., 2002. Cajal body-specific small nuclear RNAs: a novel class of 2′-O-methylation and pseudouridylation guide RNAs. EMBO J. 21, 2746–2756.

de Rie, D., Abugessaisa, I., Alam, T., Arner, E., Arner, P., Ashoor, H., et al., 2017. An integrated expression atlas of miRNAs and their promoters in human and mouse. Nat. Biotechnol. 35, 872–878.

Denli, A.M., Tops, B.B., Plasterk, R.H., Ketting, R.F., Hannon, G.J., 2004. Processing of primary microRNAs by the microprocessor complex. Nature 432, 231–235.

Djuranovic, S., Nahvi, A., Green, R., 2012. miRNA-mediated gene silencing by translational repression followed by mRNA deadenylation and decay. Science 336, 237–240.

Dueck, A., Ziegler, C., Eichner, A., Berezikov, E., Meister, G., 2012. microRNAs associated with the different human Argonaute proteins. Nucleic Acids Res. 40, 9850–9862.

Ebert, M.S., Sharp, P.A., 2010. Emerging roles for natural microRNA sponges. Curr. Biol. 20, R858–R861.

Eiring, A.M., Harb, J.G., Neviani, P., Garton, C., Oaks, J.J., Spizzo, R., et al., 2010. miR-328 functions as an RNA decoy to modulate hnRNP E2 regulation of mRNA translation in leukemic blasts. Cell 140, 652–665.

Elbashir, S.M., Harborth, J., Lendeckel, W., Yalcin, A., Weber, K., Tuschl, T., 2001. Duplexes of 21-nucleotide RNAs mediate RNA interference in cultured mammalian cells. Nature 411, 494–498.

Eulalio, A., Huntzinger, E., Nishihara, T., Rehwinkel, J., Fauser, M., Izaurralde, E., 2009. Deadenylation is a widespread effect of miRNA regulation. RNA 15, 21–32.

Fabbri, M., Bottoni, A., Shimizu, M., Spizzo, R., Nicoloso, M.S., Rossi, S., et al., 2011. Association of a microRNA/TP53 feedback circuitry with pathogenesis and outcome of B-cell chronic lymphocytic leukemia. JAMA 305, 59–67.

Fabian, M.R., Mathonnet, G., Sundermeier, T., Mathys, H., Zipprich, J.T., Svitkin, Y.V., et al., 2009. Mammalian miRNA RISC recruits CAF1 and PABP to affect PABP-dependent deadenylation. Mol. Cell 35, 868–880.

Fang, Y., Fullwood, M.J., 2016. Roles, functions, and mechanisms of long non-coding RNAs in cancer. Genomics Proteomics Bioinformatics 14, 42–54.

Felekkis, K., Touvana, E., Stefanou, C., Deltas, C., 2010. microRNAs: a newly described class of encoded molecules that play a role in health and disease. Hippokratia 14, 236–240.

Flynn, R.A., Chang, H.Y., 2014. Long noncoding RNAs in cell-fate programming and reprogramming. Cell Stem Cell 14, 752–761.

Friedlander, M.R., Lizano, E., Houben, A.J., Bezdan, D., Banez-Coronel, M., Kudla, G., et al., 2014. Evidence for the biogenesis of more than 1,000 novel human microRNAs. Genome Biol. 15, R57.

Greene, J., Baird, A.M., Brady, L., Lim, M., Gray, S.G., McDermott, R., et al., 2017. Circular RNAs: biogenesis, function and role in human diseases. Front. Mol. Biosci. 4, 38.

Gregory, R.I., Yan, K.P., Amuthan, G., Chendrimada, T., Doratotaj, B., Cooch, N., et al., 2004. The microprocessor complex mediates the genesis of microRNAs. Nature 432, 235–240.

Gu, S., Kay, M.A., 2010. How do miRNAs mediate translational repression? Silence 1, 11.

Guo, H., Ingolia, N.T., Weissman, J.S., Bartel, D.P., 2010. Mammalian microRNAs predominantly act to decrease target mRNA levels. Nature 466, 835–840.

Guo, X.Y., He, C.X., Wang, Y.Q., Sun, C., Li, G.M., Su, Q., et al., 2017. Circular RNA profiling and bioinformatic modeling identify its regulatory role in hepatic steatosis. Biomed. Res. Int. 2017, 5936171.

Guttman, M., Amit, I., Garber, M., French, C., Lin, M.F., Feldser, D., et al., 2009. Chromatin signature reveals over a thousand highly conserved large non-coding RNAs in mammals. Nature 458, 223–227.

Ha, M., Kim, V.N., 2014. Regulation of microRNA biogenesis. Nat. Rev. Mol. Cell Biol. 15, 509–524.

Haas, G., Cetin, S., Messmer, M., Chane-Woon-Ming, B., Terenzi, O., Chicher, J., et al., 2016. Identification of factors involved in target RNA-directed microRNA degradation. Nucleic Acids Res. 44, 2873–2887.

Hammond, S.M., Boettcher, S., Caudy, A.A., Kobayashi, R., Hannon, G.J., 2001. Argonaute2, a link between genetic and biochemical analyses of RNAi. Science 293, 1146–1150.

Han, J., Lee, Y., Yeom, K.H., Kim, Y.K., Jin, H., Kim, V.N., 2004. The Drosha-DGCR8 complex in primary microRNA processing. Genes Dev. 18, 3016–3027.

Han, J., Lee, Y., Yeom, K.H., Nam, J.W., Heo, I., Rhee, J.K., et al., 2006. Molecular basis for the recognition of primary microRNAs by the Drosha-DGCR8 complex. Cell 125, 887–901.

Hansen, T.B., Jensen, T.I., Clausen, B.H., Bramsen, J.B., Finsen, B., Damgaard, C.K., et al., 2013. Natural RNA circles function as efficient microRNA sponges. Nature 495, 384–388.

Harrill, A.H., McCullough, S.D., Wood, C.E., Kahle, J.J., Chorley, B.N., 2016. MicroRNA biomarkers of toxicity in biological matrices. Toxicol. Sci. 152, 264–272.

Hatziapostolou, M., Polytarchou, C., Aggelidou, E., Drakaki, A., Poultsides, G.A., Jaeger, S.A., et al., 2011. An HNF4alpha-miRNA inflammatory feedback circuit regulates hepatocellular oncogenesis. Cell 147, 1233–1247.

Holley, R.W., Apgar, J., Everett, G.A., Madison, J.T., Marquisee, M., Merrill, S.H., et al., 1965. Structure of a ribonucleic acid. Science 147, 1462–1465.

Hu, H.Y., Yan, Z., Xu, Y., Hu, H., Menzel, C., Zhou, Y.H., et al., 2009. Sequence features associated with microRNA strand selection in humans and flies. BMC Genomics 10, 413.

Huang, V., Place, R.F., Portnoy, V., Wang, J., Qi, Z., Jia, Z., et al., 2012. Upregulation of cyclin B1 by miRNA and its implications in cancer. Nucleic Acids Res. 40, 1695–1707.

Hung, T., Chang, H.Y., 2010. Long noncoding RNA in genome regulation: prospects and mechanisms. RNA Biol. 7, 582–585.

Huntzinger, E., Izaurralde, E., 2011. Gene silencing by microRNAs: contributions of translational repression and mRNA decay. Nat. Rev. Genet. 12, 99–110.

Hutvagner, G., McLachlan, J., Pasquinelli, A.E., Balint, E., Tuschl, T., Zamore, P.D., 2001. A cellular function for the RNA-interference enzyme Dicer in the maturation of the let-7 small temporal RNA. Science 293, 834–838.

Jady, B.E., Kiss, T., 2001. A small nucleolar guide RNA functions both in 2′-O-ribose methylation and pseudouridylation of the U5 spliceosomal RNA. EMBO J. 20, 541–551.

Jaiswal, R., Luk, F., Gong, J., Mathys, J.M., Grau, G.E., Bebawy, M., 2012. Microparticle conferred microRNA profiles—implications in the transfer and dominance of cancer traits. Mol. Cancer 11, 37.

Jeck, W.R., Sorrentino, J.A., Wang, K., Slevin, M.K., Burd, C.E., Liu, J., et al., 2013. Circular RNAs are abundant, conserved, and associated with ALU repeats. RNA 19, 141–157.

Jin, X., Feng, C.Y., Xiang, Z., Chen, Y.P., Li, Y.M., 2016. CircRNA expression pattern and circRNA-miRNA-mRNA network in the pathogenesis of nonalcoholic steatohepatitis. Oncotarget 7, 66455–66467.

Khalil, A.M., Guttman, M., Huarte, M., Garber, M., Raj, A., Rivea Morales, D., et al., 2009. Many human large intergenic noncoding RNAs associate with chromatin-modifying complexes and affect gene expression. Proc. Natl. Acad. Sci. USA 106, 11667–11672.

Khvorova, A., Reynolds, A., Jayasena, S.D., 2003. Functional siRNAs and miRNAs exhibit strand bias. Cell 115, 209–216.

Kim, V.N., 2005. Small RNAs: classification, biogenesis, and function. Mol. Cell 19, 1–15.

Kim, T.K., Hemberg, M., Gray, J.M., 2015. Enhancer RNAs: a class of long noncoding RNAs synthesized at enhancers. Cold Spring Harb. Perspect. Biol. 7, a018622.

Knight, S.W., Bass, B.L., 2001. A role for the RNase III enzyme DCR-1 in RNA interference and germ line development in Caenorhabditis elegans. Science 293, 2269–2271.

Kresge, N., Simoni, R.D., Hill, R.L., 2005. The discovery of tRNA by Paul C. J. Biol. Chem. 280, e37.

Kung, J.T., Colognori, D., Lee, J.T., 2013. Long noncoding RNAs: past, present, and future. Genetics 193, 651–669.

Lai, X., Wolkenhauer, O., Vera, J., 2016. Understanding microRNA-mediated gene regulatory networks through mathematical modelling. Nucleic Acids Res. 44, 6019–6035.

Lam, J.K., Chow, M.Y., Zhang, Y., Leung, S.W., 2015. siRNA versus miRNA as therapeutics for gene silencing. Mol. Ther. Nucleic Acids 4, e252.

Larsson, O., Nadon, R., 2013. Re-analysis of genome wide data on mammalian microRNA-mediated suppression of gene expression. Translation (Austin) 1, e24557.

Le Thomas, A., Toth, K.F., Aravin, A.A., 2014. To be or not to be a piRNA: genomic origin and processing of piRNAs. Genome Biol. 15, 204.

Lee, S.R., Collins, K., 2005. Starvation-induced cleavage of the tRNA anticodon loop in Tetrahymena thermophila. J. Biol. Chem. 280, 42744–42749.

Lee, R.C., Feinbaum, R.L., Ambros, V., 1993. The C. elegans heterochronic gene lin-4 encodes small RNAs with antisense complementarity to lin-14. Cell 75, 843–854.

Lee, Y., Ahn, C., Han, J., Choi, H., Kim, J., Yim, J., et al., 2003. The nuclear RNase III Drosha initiates microRNA processing. Nature 425, 415–419.

Lee, R., Feinbaum, R., Ambros, V., 2004a. A short history of a short RNA. Cell 116, S89–S92 81 p. following S96.

Lee, Y., Kim, M., Han, J., Yeom, K.H., Lee, S., Baek, S.H., et al., 2004b. MicroRNA genes are transcribed by RNA polymerase II. EMBO J. 23, 4051–4060.

Leung, A.K., Sharp, P.A., 2010. MicroRNA functions in stress responses. Mol. Cell 40, 205–215.

Li, Y., Li, Y., Zhang, H., Chen, Y., 2011. MicroRNA-mediated positive feedback loop and optimized bistable switch in a cancer network involving miR-17-92. PLoS One 6, e26302.

Li, J., Zhang, Y., Liu, Y., Dai, X., Li, W., Cai, X., et al., 2013. Microvesicle-mediated transfer of microRNA-150 from monocytes to endothelial cells promotes angiogenesis. J. Biol. Chem. 288, 23586–23596.

Ligorio, M., Izzotti, A., Pulliero, A., Arrigo, P., 2011. Mutagens interfere with microRNA maturation by inhibiting DICER. An in silico biology analysis. Mutat. Res. 717, 116–128.

Lin, Z., Ge, J., Wang, Z., Ren, J., Wang, X., Xiong, H., et al., 2017. Let-7e modulates the inflammatory response in vascular endothelial cells through ceRNA crosstalk. Sci. Rep. 7, 42498.

Liu, K., Yao, H., Lei, S., Xiong, L., Qi, H., Qian, K., et al., 2017. The miR-124-p63 feedback loop modulates colorectal cancer growth. Oncotarget 8, 29101–29115.

Londin, E., Loher, P., Telonis, A.G., Quann, K., Clark, P., Jing, Y., et al., 2015. Analysis of 13 cell types reveals evidence for the expression of numerous novel primate- and tissue-specific microRNAs. Proc. Natl. Acad. Sci. USA 112, E1106–E1115.

Macfarlane, L.A., Murphy, P.R., 2010. MicroRNA: biogenesis, function and role in cancer. Curr. Genomics 11, 537–561.

Marson, A., Levine, S.S., Cole, M.F., Frampton, G.M., Brambrink, T., Johnstone, S., et al., 2008. Connecting microRNA genes to the core transcriptional regulatory circuitry of embryonic stem cells. Cell 134, 521–533.

Marzi, M.J., Ghini, F., Cerruti, B., de Pretis, S., Bonetti, P., Giacomelli, C., et al., 2016. Degradation dynamics of microRNAs revealed by a novel pulse-chase approach. Genome Res. 26, 554–565.

Mathonnet, G., Fabian, M.R., Svitkin, Y.V., Parsyan, A., Huck, L., Murata, T., et al., 2007. MicroRNA inhibition of translation initiation in vitro by targeting the cap-binding complex eIF4F. Science 317, 1764–1767.

Meijer, H.A., Kong, Y.W., Lu, W.T., Wilczynska, A., Spriggs, R.V., Robinson, S.W., et al., 2013. Translational repression and eIF4A2 activity are critical for microRNA-mediated gene regulation. Science 340, 82–85.

Meister, G., Landthaler, M., Patkaniowska, A., Dorsett, Y., Teng, G., Tuschl, T., 2004. Human Argonaute2 mediates RNA cleavage targeted by miRNAs and siRNAs. Mol. Cell 15, 185–197.

Memczak, S., Jens, M., Elefsinioti, A., Torti, F., Krueger, J., Rybak, A., et al., 2013. Circular RNAs are a large class of animal RNAs with regulatory potency. Nature 495, 333–338.

Mourao, A., Varrot, A., Mackereth, C.D., Cusack, S., Sattler, M., 2010. Structure and RNA recognition by the snRNA and snoRNA transport factor PHAX. RNA 16, 1205–1216.

Mourelatos, Z., Dostie, J., Paushkin, S., Sharma, A., Charroux, B., Abel, L., et al., 2002. miRNPs: a novel class of ribonucleoproteins containing numerous microRNAs. Genes Dev. 16, 720–728.

Ning, B.F., Ding, J., Liu, J., Yin, C., Xu, W.P., Cong, W.M., et al., 2014. Hepatocyte nuclear factor 4alpha-nuclear factor-kappaB feedback circuit modulates liver cancer progression. Hepatology 60, 1607–1619.

Nogueira Jorge, N.A., Wajnberg, G., Ferreira, C.G., de Sa, C.B., Passetti, F., 2017. snoRNA and piRNA expression levels modified by tobacco use in women with lung adenocarcinoma. PLoS One 12, e0183410.

Novikova, I.V., Hennelly, S.P., Sanbonmatsu, K.Y., 2013. Tackling structures of long noncoding RNAs. Int. J. Mol. Sci. 14, 23672–23684.

Okada, C., Yamashita, E., Lee, S.J., Shibata, S., Katahira, J., Nakagawa, A., et al., 2009. A high-resolution structure of the pre-microRNA nuclear export machinery. Science 326, 1275–1279.

Omer, A.D., Lowe, T.M., Russell, A.G., Ebhardt, H., Eddy, S.R., Dennis, P.P., 2000. Homologs of small nucleolar RNAs in archaea. Science 288, 517–522.

Ozsolak, F., Poling, L.L., Wang, Z., Liu, H., Liu, X.S., Roeder, R.G., et al., 2008. Chromatin structure analyses identify miRNA promoters. Genes Dev. 22, 3172–3183.

Palade, G.E., 1955. A small particulate component of the cytoplasm. J. Biophys. Biochem. Cytol. 1, 59–68.

Pasquinelli, A.E., Reinhart, B.J., Slack, F., Martindale, M.Q., Kuroda, M.I., Maller, B., et al., 2000. Conservation of the sequence and temporal expression of let-7 heterochronic regulatory RNA. Nature 408, 86–89.

Patil, P., Kibiryeva, N., Uechi, T., Marshall, J., O'Brien Jr., J.E., Artman, M., et al., 2015. scaRNAs regulate splicing and vertebrate heart development. Biochim. Biophys. Acta 1852, 1619–1629.

Pillai, R.S., Bhattacharyya, S.N., Artus, C.G., Zoller, T., Cougot, N., Basyuk, E., et al., 2005. Inhibition of translational initiation by Let-7 microRNA in human cells. Science 309, 1573–1576.

Piwecka, M., Glazar, P., Hernandez-Miranda, L.R., Memczak, S., Wolf, S.A., Rybak-Wolf, A., et al., 2017. Loss of a mammalian circular RNA locus causes miRNA deregulation and affects brain function. Science 357, 6357–6368.

Poliseno, L., Salmena, L., Zhang, J., Carver, B., Haveman, W.J., Pandolfi, P.P., 2010. A coding-independent function of gene and pseudogene mRNAs regulates tumour biology. Nature 465, 1033–1038.

RajBhandary, U.L., Kohrer, C., 2006. Early days of tRNA research: discovery, function, purification and sequence analysis. J. Biosci. 31, 439–451.

Rinn, J.L., Chang, H.Y., 2012. Genome regulation by long noncoding RNAs. Annu. Rev. Biochem. 81, 145–166.

Rothschild, G., Basu, U., 2017. Lingering questions about enhancer RNA and enhancer transcription-coupled genomic instability. Trends Genet. 33, 143–154.

Ru, P., Guo, D., 2017. microRNA-29 mediates a novel negative feedback loop to regulate SCAP/SREBP-1 and lipid metabolism. RNA Dis. 4, e1525.

Ru, P., Hu, P., Geng, F., Mo, X., Cheng, C., Yoo, J.Y., et al., 2016. Feedback loop regulation of SCAP/SREBP-1 by miR-29 modulates EGFR signaling-driven glioblastoma growth. Cell Rep. 16, 1527–1535.

Salido-Guadarrama, I., Romero-Cordoba, S., Peralta-Zaragoza, O., Hidalgo-Miranda, A., Rodriguez-Dorantes, M., 2014. MicroRNAs transported by exosomes in body fluids as mediators of intercellular communication in cancer. Onco. Targets Ther. 7, 1327–1338.

Salzman, J., Gawad, C., Wang, P.L., Lacayo, N., Brown, P.O., 2012. Circular RNAs are the predominant transcript isoform from hundreds of human genes in diverse cell types. PLoS One 7, e30733.

Schanen, B.C., Li, X., 2011. Transcriptional regulation of mammalian miRNA genes. Genomics 97, 1–6.

Scherrer, K., 2003. Historical review: the discovery of 'giant' RNA and RNA processing: 40 years of enigma. Trends Biochem. Sci. 28, 566–571.

Schwarz, D.S., Hutvagner, G., Du, T., Xu, Z., Aronin, N., Zamore, P.D., 2003. Asymmetry in the assembly of the RNAi enzyme complex. Cell 115, 199–208.

Sedwick, C., 2013. Victor Ambros: the broad scope of microRNAs. Interview by Caitlin Sedwick. J. Cell Biol. 201, 492–493.

Seitz, H., 2009. Redefining microRNA targets. Curr. Biol. 19, 870–873.

Suzuki, R., Honda, S., Kirino, Y., 2012. PIWI expression and function in cancer. Front. Genet. 3, 204.

Tay, Y., Kats, L., Salmena, L., Weiss, D., Tan, S.M., Ala, U., et al., 2011. Coding-independent regulation of the tumor suppressor PTEN by competing endogenous mRNAs. Cell 147, 344–357.

Tessel, M.A., Krett, N.L., Rosen, S.T., 2010. Steroid receptor and microRNA regulation in cancer. Curr. Opin. Oncol. 22, 592–597.

Thermann, R., Hentze, M.W., 2007. Drosophila miR2 induces pseudo-polysomes and inhibits translation initiation. Nature 447, 875–878.

Thompson, D.M., Lu, C., Green, P.J., Parker, R., 2008. tRNA cleavage is a conserved response to oxidative stress in eukaryotes. RNA 14, 2095–2103.

Turchinovich, A., Samatov, T.R., Tonevitsky, A.G., Burwinkel, B., 2013. Circulating miRNAs: cell-cell communication function? Front. Genet. 4, 119.

Uter, N.T., Perona, J.J., 2004. Long-range intramolecular signaling in a tRNA synthetase complex revealed by pre-steady-state kinetics. Proc. Natl. Acad. Sci. USA 101, 14396–14401.

Valadi, H., Ekstrom, K., Bossios, A., Sjostrand, M., Lee, J.J., Lotvall, J.O., 2007. Exosome-mediated transfer of mRNAs and microRNAs is a novel mechanism of genetic exchange between cells. Nat. Cell Biol. 9, 654–659.

Valadkhan, S., Gunawardane, L.S., 2013. Role of small nuclear RNAs in eukaryotic gene expression. Essays Biochem. 54, 79–90.

Valencia-Sanchez, M.A., Liu, J., Hannon, G.J., Parker, R., 2006. Control of translation and mRNA degradation by miRNAs and siRNAs. Genes Dev. 20, 515–524.

van Rooij, E., Sutherland, L.B., Qi, X., Richardson, J.A., Hill, J., Olson, E.N., 2007. Control of stress-dependent cardiac growth and gene expression by a microRNA. Science 316, 575–579.

Vickers, K.C., Palmisano, B.T., Shoucri, B.M., Shamburek, R.D., Remaley, A.T., 2011. MicroRNAs are transported in plasma and delivered to recipient cells by high-density lipoproteins. Nat. Cell Biol. 13, 423–433.

Wahid, F., Shehzad, A., Khan, T., Kim, Y.Y., 2010. MicroRNAs: synthesis, mechanism, function, and recent clinical trials. Biochim. Biophys. Acta 1803, 1231–1243.

Wang, K., Zhang, S., Weber, J., Baxter, D., Galas, D.J., 2010. Export of microRNAs and microRNA-protective protein by mammalian cells. Nucleic Acids Res. 38, 7248–7259.

Wang, J., Song, Y.X., Ma, B., Wang, J.J., Sun, J.X., Chen, X.W., et al., 2015. Regulatory roles of non-coding RNAs in colorectal cancer. Int. J. Mol. Sci. 16, 19886–19919.

Weick, E.M., Miska, E.A., 2014. piRNAs: from biogenesis to function. Development 141, 3458–3471.

Wells, W.A., 2005. Ribosomes, or the particles of Palade. J. Cell Biol. 168, 12.

Wightman, B., Ha, I., Ruvkun, G., 1993. Posttranscriptional regulation of the heterochronic gene lin-14 by lin-4 mediates temporal pattern formation in C. elegans. Cell 75, 855–862.

Wilczynska, A., Bushell, M., 2015. The complexity of miRNA-mediated repression. Cell Death Differ. 22, 22–33.

Wilson, R.C., Tambe, A., Kidwell, M.A., Noland, C.L., Schneider, C.P., Doudna, J.A., 2015. Dicer-TRBP complex formation ensures accurate mammalian microRNA biogenesis. Mol. Cell 57, 397–407.

Wilusz, J.E., Sunwoo, H., Spector, D.L., 2009. Long noncoding RNAs: functional surprises from the RNA world. Genes Dev. 23, 1494–1504.

Wohlgemuth, I., Pohl, C., Mittelstaet, J., Konevega, A.L., Rodnina, M.V., 2011. Evolutionary optimization of speed and accuracy of decoding on the ribosome. Philos. Trans. R. Soc. Lond. Ser. B Biol. Sci. 366, 2979–2986.

Wu, L., Fan, J., Belasco, J.G., 2006. MicroRNAs direct rapid deadenylation of mRNA. Proc. Natl. Acad. Sci. USA 103, 4034–4039.

Yang, L., Lin, C., Rosenfeld, M.G., 2011. A lincRNA switch for embryonic stem cell fate. Cell Res. 21, 1646–1648.

Yi, R., Qin, Y., Macara, I.G., Cullen, B.R., 2003. Exportin-5 mediates the nuclear export of pre-microRNAs and short hairpin RNAs. Genes Dev. 17, 3011–3016.

Yin, J., Jiang, X.Y., Qi, W., Ji, C.G., Xie, X.L., Zhang, D.X., et al., 2017. piR-823 contributes to colorectal tumorigenesis by enhancing the transcriptional activity of HSF1. Cancer Sci. 108, 1746–1756.

Ying, W., Riopel, M., Bandyopadhyay, G., Dong, Y., Birmingham, A., Seo, J.B., et al., 2017. Adipose tissue macrophage-derived exosomal miRNAs can modulate in vivo and in vitro insulin sensitivity. Cell 171, 372–384.

Zarringhalam, K., Tay, Y., Kulkarni, P., Bester, A.C., Pandolfi, P.P., Kulkarni, R.V., 2017. Identification of competing endogenous RNAs of the tumor suppressor gene PTEN: A probabilistic approach. Sci. Rep. 7, 7755.

Zhang, H., Kolb, F.A., Jaskiewicz, L., Westhof, E., Filipowicz, W., 2004. Single processing center models for human Dicer and bacterial RNase III. Cell 118, 57–68.

Zhang, Y., Liu, D., Chen, X., Li, J., Li, L., Bian, Z., et al., 2010. Secreted monocytic miR-150 enhances targeted endothelial cell migration. Mol. Cell 39, 133–144.

Zhao, Y., Wang, F., Chen, S., Wan, J., Wang, G., 2017. Methods of MicroRNA promoter prediction and transcription factor mediated regulatory network. Biomed. Res. Int. 2017, 7049406.

Zu, L., Xue, Y., Wang, J., Fu, Y., Wang, X., Xiao, G., et al., 2016. The feedback loop between miR-124 and TGF-beta pathway plays a significant role in non-small cell lung cancer metastasis. Carcinogenesis 37, 333–343.

miRNAs and lncRNAs as Biomarkers of Toxicant Exposure

Ronit Machtinger*, *Valentina Bollati*[†], *Andrea A. Baccarelli*[‡]

Sheba Medical Center, Ramat-Gan and Tel-Aviv University, Tel Aviv, Israel [†]Center of Molecular and Genetic Epidemiology, Department of Clinical Sciences and Community Health, Università degli Studi di Milano, Milan, Italy [‡]Environmental Precision Biosciences Laboratory, Columbia University, Mailman School of Public Health, New York, NY, United States

INTRODUCTION

Most of the human genome (98%) contains the molecular instruction to produce RNAs that do not code for proteins (noncoding RNAs), but may play important roles in regulating biological processes (Rinn et al., 2007; Rinn, 2014; Brosnan and Voinnet, 2009; Cech and Steitz, 2014; Amaral et al., 2008). Noncoding RNAs are usually divided, based on their number of nucleotides, into long (lncRNA) and short noncoding RNAs.

MicroRNAs (miRNAs) are the most studied class of short noncoding RNAs. These are small, highly conserved noncoding RNA molecules of approximately 22 nucleotides (Bartel, 2004; He and Hannon, 2004). These molecules have well-established roles in regulating gene expression by pairing to mRNAs of protein-coding genes leading to their posttranscriptional repression (Wu and Song, 2011). The genes that encode for miRNAs are first transcribed into long primary transcripts named pri-miRNAs that are processed by an RNase III enzyme (Drosha) into precursor miRNAs (pre-miRNAs) (Lee et al., 2003). These pre-miRNAs are transported into the nucleus where they are further processed by a second RNase III enzyme (Dicer) into mature miRNAs. The mature miRNAs connect with the multiprotein RNA-induced silencing complex (RISC) and guide RISC to silent specific mRNA species by translational inhibition or mRNA degradation (Redfern et al., 2013). In its basic form, RISC is composed of an Argonaute protein bound to a small RNA, and this form has been suggested to be sufficient for target RNA recognition and cleavage (Rivas et al., 2005). However, Argonaute is often associated with many other binding proteins; as such, RISC may present itself in a variety of different ribonucleoproteins, ranging from modest size ($\sim$150 kDa) to an 80S ($\sim$3 MDa) particle. Variants of these complexes include the RNA-induced transcriptional silencing (RITS) complex, isolated from the nuclei of *Schizosaccharomyces pombe*, which has been shown to silence targeted genes through heterochromatin formation, particularly in centromeres. Hence, through RITS, miRNAs may also have a role in controlling chromatin states.

Abnormal miRNA expression has been linked to the pathways of a number of diseases (De Felice et al., 2015; Wu and Song, 2011). Accumulating data indicate that exposure to environmental chemicals and metals can alter miRNA expression. MicroRNAs can not only therefore serve as potential biomarkers of several diseases but also be a key mechanism of action linking environmental exposures to disease development.

lncRNAs are a heterogeneous class of thousands of nonprotein-coding transcripts longer than 200 nucleotides that are emerging as a crucial mechanism of epigenetic control (Fatica and Bozzoni, 2014). lncRNAs, together with transcription factors, are included in a complex regulatory network that controls gene expression in eukaryotes. lncRNAs, in fact, modulate the function of transcription factors in several ways. For example, they can act as coregulators or regulate transcription factor activity. In addition, they have a very strict link with epigenetic regulation, as they can modify transcription through the regulation of histone methylation and chromatin structure and inhibition of miRNA functions (Dempsey and Cui, 2017; Rinn et al., 2007). Although associated with various epigenetic states, lncRNAs share a conserved mechanism of binding to chromatin-modifying and chromatin-remodeling complexes and directing them to specific genomic loci that are critical for proper cellular function. lncRNAs are linked to key biological phenomena such as genomic imprinting, modeling chromosome conformation, and allosterically controlling enzymatic activity (Ponting et al., 2009; Rinn and Chang, 2012; Quinn and Chang, 2016) and the process of X-inactivation by which one of the copies of the X chromosome in females is silenced and unable to undergo transcription. The functions of most of the lncRNAs are still largely unexplored, but the roles and mechanisms of action of some of them are well understood, such as X-inactive specific transcript (XIST; in X chromosome inactivation), HOX transcript antisense RNA (HOTAIR; in positional identity), and telomerase RNA component (TERC; in telomere elongation) [reviewed by Quinn and Chang (2016)]. lncRNAs have been associated with human diseases, including various types of cancer (breast, lung, colorectal, etc.) (Du et al., 2015; Huang et al.,

2014; Khandelwal et al., 2015), neurological disorders such as Alzheimer's disease (Faghihi et al., 2008), metabolic syndrome (Zhao and Lin, 2015), diabetes (Moran et al., 2012), atherosclerosis (Aryal et al., 2014), and autoimmune diseases (Mayama et al., 2016).

miRNAs AND TOXICOLOGICAL RESPONSE TO CHEMICALS

Endocrine Disruptor Chemicals and Altered miRNA Profile

Environmental chemicals may have adverse effects on animal and human health. Some of these exogenous substances can alter the normal function of the endocrine system and therefore are called endocrine-disrupting chemicals (EDC) (Diamanti-Kandarakis et al., 2009). Among the most studied EDCs are bisphenol A (BPA) and phthalates. Several studies have looked at possible correlations between exposure to these EDCs and noncoding RNA profile.

Bisphenol A

BPA is a synthetic chemical and one of the highest-volume industrial chemicals produced worldwide. Potential sources of BPA exposure include ingestion of foods and beverages previously in contact with the lining of cans used for food and beverages, polycarbonate bottles, thermal receipts, dust, water, etc. (Talsness et al., 2009; Vandenberg et al., 2007). BPA was detected in the urine of >90% of participants in the National Health and Nutrition Examination Survey, the United States (Calafat et al., 2008). Animal studies to date showed a possible association between BPA exposure and type II diabetes, cardiovascular disease, cancer, obesity, and reproductive disorders (Diamanti-Kandarakis et al., 2009). Clinical studies reported that higher urinary BPA concentrations were associated with cardiovascular disease, type 2 diabetes, and liver enzyme abnormalities (Groff, 2010; Lang et al., 2008; vom Saal and Myers, 2008).

Avissar-Whiting et al. investigate the effect of BPA on miRNA profile of human placental cells. Using microarray analyses, the authors identified overexpression of miR-146a in BPA-treated placental cell lines (Avissar-Whiting et al., 2010). The capability of BPA to modify miRNA expression in the placenta might reflect a mechanism of BPA toxicity that may lead to harmful effects on the developing fetus. Another study tested miRNA expression in the placenta from pregnant women from a polluted area, exposed to BPA that underwent therapeutic abortion in the second trimester due to fetal anomalies. Using microarray technology, the authors found that miR-146a was significant overexpressed and correlated with BPA accumulation in the placenta among pregnant women with fetal malformations (De Felice et al., 2015).

Increasing evidence of the possible toxic effects of BPA has triggered the industry to use alternative chemicals such as the bisphenol analogues bisphenol S (BPS) and bisphenol F (BPF). Although data to date are scarce, a recent study has shown that in male zebra fish gonad, BPS exposure at doses 5 and 50 µg/L leads to changes in miRNA profile and target gene expression. Pathway analysis revealed that miRNAs significantly altered by BPS exposure were involved in hematopoiesis, lymphoid organ development, steroidogenesis, and immune system development (Lee et al., 2018). Verbanck et al. exposed primary cultures of human adipocytes to both a "low dose" (10 nM) similar to the levels usually detected in human biological fluids and a high dose (10 µM) of BPA, BPS, and BPF. They identified significantly alterations of miRNA and lncRNA profiles after exposure to both low and high doses of these three chemicals. These miRNAs were related to both "cancer" and "organismic injury and abnormality" pathways (Verbanck et al., 2017).

Phthalates

Diester phthalates are man-made chemicals used in everyday products as cosmetics, hair spray, shampoos, deodorants, perfumes, nail polish, body lotions, medication coating, manufacturing of flooring, carpet backings, adhesives, wallpaper, and polyvinyl chloride (PVC) (Just et al., 2010; Meeker et al., 2009; Hauser and Calafat, 2005). Exposure to these substances is ubiquitous and may occur through inhalation, ingestion, or dermal absorption (Lyche et al., 2009). After exposure, diester phthalates are metabolized to monoesters, which are biologically active and exert antiestrogenic, antiandrogenic, or antithyroid activity (Marie et al., 2015; Swan, 2008). La Rocca et al. tested, by qRT-PCR, possible associations between first-trimester urine concentrations of 8 phenols and 11 phthalate metabolites and miRNA expression in the placenta. They detected significant associations between Σphthalates or Σphenols and the expression of miR-142-3p, miR-15a-5p, and miR-185. The target of these miRNAs included the regulation of protein serine/threonine kinase activity (LaRocca et al., 2016).

Exposure to Metals

Experimental data have linked exposure to some metals as aluminum, mercury, cadmium, and lead with altered miRNA expression (Marsit, 2015; Ray et al., 2014; Sanders et al., 2015). Chronic exposure to aluminum is suspected to be one of the mechanisms for Alzheimer's disease (Walton, 2014). Chronic exposure to mercury can affect the nervous system (Evaluating Mercury Exposure, 2009). Lead is a heavy metal that is extensively existing in the environment and toxic to the human health, especially the nervous, hematopoietic, and digestive systems (Nan et al., 2018; Engstrom et al., 2015). Cadmium is a heavy metal commonly present in nickel-cadmium batteries, metal plating, pigments, and plastics. It may accelerate the production of free radicals (Manca et al., 1991), and exposure to cadmium may result in renal and respiratory toxicity.

Recent data have shown that aluminum exposure resulted in altered expression of miRNAs. Treatment of human neural (HN) cells with aluminum sulfate resulted in the upregulation of miR-146a and increased the expression of miR-9, miR-125b, and miR-128. These last three miRNAs were also found to be upregulated in brain cells of Alzheimer patients, suggesting that aluminum exposure may cause toxicity via miRNA-related regulatory elements (Lukiw and Pogue, 2007; Hou et al., 2011; Pogue et al., 2009).

Sanders et al. tested the association between exposure to lead and mercury and miRNA expression in the cervix in women during the second trimester of pregnancy. Seventeen miRNAs were negatively correlated with mercury levels, and lead levels were associated with decreased expression of miR-575 and miR-4286 (Sanders et al., 2015).

Exposure to Particles

Exposure to the particulate component (particulate matter, PM) of ambient air pollution has been linked to increased morbidity and mortality from cardiovascular and respiratory diseases (Baccarelli et al., 2008; Brook et al., 2004; Ciocco and Thompson, 1961; Hou et al., 2011). PM is the sum of liquid and solid particles (including metals as cadmium and lead) suspended in air. These particles, specifically those that are <100 nm, can penetrate through the alveolus in the lung and affect other body organs. Bollati et al. evaluated the effects of exposure to cadmium- and lead-enriched particulate matter (PM) on miRNAs associated with oxidative stress and inflammatory processes in 63 steel workers. MiR-222 and miR-21 expression was increased in samples collected after three workdays, compared with baseline samples before exposure. MiR-222 expression was positively correlated with lead exposure, and the expression of miR-146a was negatively correlated with exposure to lead and cadmium (Bollati et al., 2010). Among the top-ranked pathways, linked with these are miRNAs that regulate genes in pathways related to general functions (e.g., purine metabolism and cell cycle), oxidative stress, and inflammation.

Nanoparticles

Nanoparticles (NPs), a category of fine particles characterized by having all their three dimensions between 1 and 100 nm, are emitted from natural (e.g., volcano) and anthropic (e.g., traffic) sources or produced via nanotechnology.

Since the beginning of the 21st century, the NPs have grown enormously, judging simply by the number of products now on the market and the funds dedicated to research and development. The extremely small size of NPs means that they much more readily gain entry into the human body than larger sized particles. How these NPs behave inside the body is still a major question that needs to be resolved. The behavior of NPs might be a function of their size, shape, and surface reactivity with the surrounding tissue, and it has been hypothesized that NPs are able to modulate miRNA expression.

Bourdon et al. showed that carbon black nanoparticle intratracheal installation modified the expression of miR-135b in the mouse lung (Bourdon et al., 2012). In a very recent study by Sui et al. (2018), exposure of cells to nano-TiO2 in vitro resulted in an altered miRNA pattern. In particular, they observed an increased expression of miR-350, which has an important role in multiple signaling pathways, including MAPK signaling pathway, NF-kappa B signaling pathway, and apoptosis.

miRNAs have been also proposed as sensitive biomarkers of NP toxicity in vivo. In a study conducted by Nagano and colleagues, liver toxicity induced by high intravenous doses of silica nanoparticles of 70 nm diameter was evaluated by serum levels of liver-specific or liver-enriched miRNAs (miR-122, miR-192, and miR-194), comparing them with conventional and well-consolidated hepatic biomarkers [alanine aminotransferase (ALT) and aspartate aminotransferase (AST)]. Surprisingly, the sensitivity of miR-122 for liver damage was at least as good as those of ALT and AST (Nagano et al., 2013).

Cigarette Smoking

According to the most recent WHO data, 1.1 billion people worldwide smoked cigarettes in 2015 (World Health Organization, 2017). Smoking is a significant risk factor for lung cancer and chronic obstructive pulmonary disease (COPD) (Matkovich et al., 2011). Several studies compared miRNA profile among smokers and nonsmokers. Schembri et al. compared miRNA profile of bronchial airway epithelium of smokers and nonsmokers. They identified 28 different miRNAs between the groups, 23 of which were downregulated in smokers. The most significant change was in miR-218, which was downregulated fourfold in smokers (Schembri et al., 2009). Other miRNAs that were significantly expressed between smokers

TABLE 1   miRNAs as Biomarkers of Toxicant Exposure

	Tissue	miRNA	References
Bisphenol A	The human placenta	miR-146a	De Felice et al. (2015) and Avissar-Whiting et al. (2010)
Phthalates and phenols	The human placenta	miR-142-3p miR-15a-5p miR-185	LaRocca et al. (2016)
Aluminum	Human neural (HN) cells	miR-146a miR-9 miR-125b miR-128	Lukiw and Pogue (2007), Hou et al. (2011), and Pogue et al. (2009)
Mercury	Cervical mucous during pregnancy	miR-575 miR-4286	Sanders et al. (2015)
Particulate matter	Plasma	miR-21 miR-146a miR-222	Bollati et al. (2010)
Smoking	Plasma	mir-128b mir-500 mir-181d miR-16 miR-21 miR-146a	Sanders et al. (2015), Schembri et al. (2009), and Maccani et al. (2010)

and nonsmokers in this study included mir-128b, mir-500, and mir-181d. Smoking-induced changes in miRNA profile have previously been linked to altered regulation of oncogenes, tumor suppressor genes, oxidative stress, xenobiotic metabolism, and inflammation (Schembri et al., 2009). Takahashi et al. compared the plasma miRNA profiles of smokers and nonsmokers. The levels of 43 miRNAs were significantly higher in smokers than in nonsmokers, and one miRNA was higher in nonsmokers. Some of these miRNAs were previously reported to be correlated with diseases as nonsmall-cell lung cancer, nasopharyngeal carcinoma, colorectal cancer, diabetes (type 1 and 2), and myocardial infarction. Interestingly, quitting smoking reversed the plasma miRNA profiles resembling those of the nonsmokers (Takahashi et al., 2013). In other studies, miR-16, miR-21, and miR-146a were downregulated in the placentas from women exposed to cigarette smoke compared with unexposed control (Maccani et al., 2010; Maccani and Knopik, 2012; Sanders et al., 2015). In worms, exposure to low doses of nicotine altered global expression profiles of miRNAs not only in F0 but also in subsequent generations (F1 and F2) (Taki et al., 2014). Table 1 summarizes possible correlations between miRNAs and potential toxicants.

lncRNAs AND TOXICOLOGICAL RESPONSE TO CHEMICALS

While miRNAs have established roles, the function of lncRNAs is still being elucidated. In contrast to miRNAs that implicate transcriptional and posttranscriptional gene silencing via

TABLE 2 lncRNAs as Biomarkers of Toxicant Exposure

	Tissue	lncRNA	References
Bisphenol A	Breast cancer cells	HOTAIR	Wang and Chang (2011)
Lead	Neuronal-injury cell model	LncRNAL20992	Engstrom et al. (2015)
Cadmium	Plasma	ENST00000414355	Zhou et al. (2015)
Smoking	Lung tissue	RNA44121 RNA43510	Bi et al. (2015)

specific base pairing with their targets, lncRNAs regulate gene expression by various mechanisms that are not yet fully characterized (Karlsson and Baccarelli, 2016; Wang and Chang, 2011). To date, only a few functional lncRNAs have been well identified. For example, altered expression of lncRNAs as HOTAIR has been linked to diverse cancer types (Bhan et al., 2014; Table 2).

Endocrine Disruptor Chemicals and Altered lncRNA Profile

Bisphenol A

HOTAIR is an lncRNA that is transcriptionally induced by estradiol (E2) and is overexpressed in breast cancer (Bhan et al., 2013). Bhan et al. showed in both in vitro and in vivo model of cultured breast cancer cells, treating rats with BPA, that HOTAIR was upregulated upon exposure to low concentrations of BPA. This study demonstrated that endocrine disruptors can alter the noncoding RNAs (Bhan et al., 2013).

Exposure to Metals and lncRNA Expression

Several studies to date have tested possible associations between exposure to metals as lead and cadmium and altered expression of lncRNAs.

In a neuronal-injury cell model, exposure to lead at 5 or $10\,\mu mol/L$ resulted in increased apoptosis and inhibition in cell proliferation. LncRNAL20992 was significantly upregulated after exposure to lead at these concentrations. The LncRNAL20992 stimulated apoptosis during the process of lead-induced nerve injury (Engstrom et al., 2015).

Among Cd-exposed workers, expression of lncRNA ENST00000414355 in blood was significantly higher in those with higher urinary and blood Cd levels. Moreover, there was a correlation between the expression of ENST00000414355 and DNA damage of blood cells in Cd-exposed workers (Zhou et al., 2015).

Cigarette Smoking and lncRNA Profile

A recent study reported a different lncRNA expression in the lung tissue of smokers and nonsmokers (Bi et al., 2015). The expression of hundreds of lncRNAs was different ($\geq$ twofold change) between smokers and nonsmokers. Specifically, RNA44121 and RNA43510 were the most over- and underexpressed, among smokers and nonsmokers. Pathway analysis showed that these lncRNAs were related to chronic obstructive pulmonary disease (COPD) caused by smoking (Bi et al., 2015). Although the number of participants in this study was small, the results suggest that lncRNAs may be a factor in the pathological changes caused by cigarette smoking.

SUMMARY

Studies on the roles of miRNAs and lncRNAs in toxicology are emerging and provide new insights into the mechanisms of chemical toxicity. Chemicals such as bisphenol A and phthalates; metals including aluminum, mercury, cadmium, and lead; and particles and nanoparticles can alter the expression of miRNAs and lncRNAs and lead to impairment of normal signaling pathways. Further studies are needed to better understand the clinical consequences of the alterations of these pathways and to understand how these substances affect human health.

References

Amaral, P.P., Dinger, M.E., Mercer, T.R., et al., 2008. The eukaryotic genome as an RNA machine. Science 319, 1787–1789.

Aryal, B., Rotllan, N., Fernandez-Hernando, C., 2014. Noncoding RNAs and atherosclerosis. Curr Atheroscler Rep 16, 407.

Avissar-Whiting, M., Veiga, K.R., Uhl, K.M., et al., 2010. Bisphenol A exposure leads to specific microRNA alterations in placental cells. Reprod. Toxicol. 29, 401–406.

Baccarelli, A., Martinelli, I., Zanobetti, A., et al., 2008. Exposure to particulate air pollution and risk of deep vein thrombosis. Arch. Intern. Med. 168, 920–927.

Bartel, D.P., 2004. MicroRNAs: genomics, biogenesis, mechanism, and function. Cell 116, 281–297.

Bhan, A., Hussain, I., Ansari, K.I., et al., 2013. Antisense transcript long noncoding RNA (lncRNA) HOTAIR is transcriptionally induced by estradiol. J. Mol. Biol. 425, 3707–3722.

Bhan, A., Hussain, I., Ansari, K.I., et al., 2014. Bisphenol-A and diethylstilbestrol exposure induces the expression of breast cancer associated long noncoding RNA HOTAIR in vitro and in vivo. J. Steroid Biochem. Mol. Biol. 141, 160–170.

Bi, H., Zhou, J., Wu, D., et al., 2015. Microarray analysis of long non-coding RNAs in COPD lung tissue. Inflamm. Res. 64, 119–126.

Bollati, V., Marinelli, B., Apostoli, P., et al., 2010. Exposure to metal-rich particulate matter modifies the expression of candidate microRNAs in peripheral blood leukocytes. Environ. Health Perspect. 118, 763–768.

Bourdon, J.A., Saber, A.T., Halappanavar, S., et al., 2012. Carbon black nanoparticle intratracheal installation results in large and sustained changes in the expression of miR-135b in mouse lung. Environ. Mol. Mutagen. 53, 462–468.

Brook, R.D., Franklin, B., Cascio, W., et al., 2004. Air pollution and cardiovascular disease: a statement for healthcare professionals from the Expert Panel on Population and Prevention Science of the American Heart Association. Circulation 109, 2655–2671.

Brosnan, C.A., Voinnet, O., 2009. The long and the short of noncoding RNAs. Curr. Opin. Cell Biol. 21, 416–425.

Calafat, A.M., Ye, X., Wong, L.Y., et al., 2008. Exposure of the U.S. population to bisphenol A and 4-tertiary-octylphenol: 2003-2004. Environ. Health Perspect. 116, 39–44.

Cech, T.R., Steitz, J.A., 2014. The noncoding RNA revolution-trashing old rules to forge new ones. Cell 157, 77–94.

Ciocco, A., Thompson, D.J., 1961. A follow-up of Donora ten years after: methodology and findings. Am. J. Public Health Nations Health 51, 155–164.

De Felice, B., Manfellotto, F., Palumbo, A., et al., 2015. Genome-wide microRNA expression profiling in placentas from pregnant women exposed to BPA. BMC Med. Genet. 8, 56.

Dempsey, J.L., Cui, J.Y., 2017. Long non-coding RNAs: a novel paradigm for toxicology. Toxicol. Sci. 155, 3–21.

Diamanti-Kandarakis, E., Bourguignon, J.P., Giudice, L.C., et al., 2009. Endocrine-disrupting chemicals: an Endocrine Society scientific statement. Endocr. Rev. 30, 293–342.

Du, M., Wang, W., Jin, H., et al., 2015. The association analysis of lncRNA HOTAIR genetic variants and gastric cancer risk in a Chinese population. Oncotarget 6, 31255–31262.

Engstrom, A., Wang, H., Xia, Z., 2015. Lead decreases cell survival, proliferation, and neuronal differentiation of primary cultured adult neural precursor cells through activation of the JNK and p38 MAP kinases. Toxicol. In Vitro 29, 1146–1155.

Evaluating Mercury Exposure, 2009. Information for Health Care Providers. www.atsdr.cdc.gov/mercury/docs/physician_hg_flier.pdf.

Faghihi, M.A., Modarresi, F., Khalil, A.M., et al., 2008. Expression of a noncoding RNA is elevated in Alzheimer's disease and drives rapid feed-forward regulation of beta-secretase. Nat. Med. 14, 723–730.

Fatica, A., Bozzoni, I., 2014. Long non-coding RNAs: new players in cell differentiation and development. Nat. Rev. Genet. 15, 7–21.

Groff, T., 2010. Bisphenol A: invisible pollution. Curr. Opin. Pediatr. 22, 524–529.

Hauser, R., Calafat, A.M., 2005. Phthalates and human health. Occup. Environ. Med. 62, 806–818.

He, L., Hannon, G.J., 2004. MicroRNAs: small RNAs with a big role in gene regulation. Nat. Rev. Genet. 5, 522–531.

Hou, L., Wang, D., Baccarelli, A., 2011. Environmental chemicals and microRNAs. Mutat. Res. 714, 105–112.

Huang, J., Zhou, N., Watabe, K., et al., 2014. Long non-coding RNA UCA1 promotes breast tumor growth by suppression of p27 (Kip1). Cell Death Dis. 5, e1008.

Just, A.C., Adibi, J.J., Rundle, A.G., et al., 2010. Urinary and air phthalate concentrations and self-reported use of personal care products among minority pregnant women in New York city. J. Expo. Sci. Environ. Epidemiol. 20, 625–633.

Karlsson, O., Baccarelli, A.A., 2016. Environmental health and long non-coding RNAs. Curr. Environ. Health Rep. 3, 178–187.

Khandelwal, A., Bacolla, A., Vasquez, K.M., et al., 2015. Long non-coding RNA: a new paradigm for lung cancer. Mol. Carcinog. 54, 1235–1251.

Lang, I.A., Galloway, T.S., Scarlett, A., et al., 2008. Association of urinary bisphenol A concentration with medical disorders and laboratory abnormalities in adults. JAMA 300, 1303–1310.

LaRocca, J., Binder, A.M., McElrath, T.F., et al., 2016. First-trimester urine concentrations of phthalate metabolites and phenols and placenta miRNA expression in a cohort of U.S. women. Environ. Health Perspect. 124, 380–387.

Lee, Y., Ahn, C., Han, J., Choi, H., et al., 2003. The nuclear RNase III Drosha initiates microRNA processing. Nature 425 (6956), 415–419.

Lee, J., Kho, Y., Kim, P.G., et al., 2018. Exposure to bisphenol S alters the expression of microRNA in male zebrafish. Toxicol. Appl. Pharmacol. 338, 191–196.

Lukiw, W.J., Pogue, A.I., 2007. Induction of specific micro RNA (miRNA) species by ROS-generating metal sulfates in primary human brain cells. J. Inorg. Biochem. 101, 1265–1269.

Lyche, J.L., Gutleb, A.C., Bergman, A., et al., 2009. Reproductive and developmental toxicity of phthalates. J. Toxicol. Environ. Health B Crit. Rev. 12, 225–249.

Maccani, M.A., Knopik, V.S., 2012. Cigarette smoke exposure-associated alterations to non-coding RNA. Front. Genet. 3, 53.

Maccani, M.A., Avissar-Whiting, M., Banister, C.E., et al., 2010. Maternal cigarette smoking during pregnancy is associated with downregulation of miR-16, miR-21, and miR-146a in the placenta. Epigenetics 5, 583–589.

Manca, D., Ricard, A.C., Trottier, B., et al., 1991. Studies on lipid peroxidation in rat tissues following administration of low and moderate doses of cadmium chloride. Toxicology 67, 303–323.

Marie, C., Vendittelli, F., Sauvant-Rochat, M.P., 2015. Obstetrical outcomes and biomarkers to assess exposure to phthalates: a review. Environ. Int. 83, 116–136.

Marsit, C.J., 2015. Influence of environmental exposure on human epigenetic regulation. J. Exp. Biol. 218, 71–79.

Matkovich, S.J., Van Booven, D.J., Eschenbacher, W.H., et al., 2011. RISC RNA sequencing for context-specific identification of in vivo microRNA targets. Circ. Res. 108, 18–26.

Mayama, T., Marr, A.K., Kino, T., 2016. Differential expression of glucocorticoid receptor noncoding RNA repressor Gas5 in autoimmune and inflammatory diseases. Horm. Metab. Res. 48, 550–557.

Meeker, J.D., Sathyanarayana, S., Swan, S.H., 2009. Phthalates and other additives in plastics: human exposure and associated health outcomes. Philos. Trans. R. Soc. Lond. Ser. B Biol. Sci. 364, 2097–2113.

Moran, I., Akerman, I., van de Bunt, M., et al., 2012. Human beta cell transcriptome analysis uncovers lncRNAs that are tissue-specific, dynamically regulated, and abnormally expressed in type 2 diabetes. Cell Metab. 16, 435–448.

Nagano, T., Higashisaka, K., Kunieda, A., et al., 2013. Liver-specific microRNAs as biomarkers of nanomaterial-induced liver damage. Nanotechnology 24, 405102.

Nan, A., Jia, Y., Li, X., et al., 2018. LncRNAL20992 regulates apoptotic proteins to promote lead-induced neuronal apoptosis. Toxicol. Sci. 161 (1), 115–124.

Pogue, A.I., Li, Y.Y., Cui, J.G., et al., 2009. Characterization of an NF-kappaB-regulated, miRNA-146a-mediated down-regulation of complement factor H (CFH) in metal-sulfate-stressed human brain cells. J. Inorg. Biochem. 103, 1591–1595.

Ponting, C.P., Oliver, P.L., Reik, W., 2009. Evolution and functions of long noncoding RNAs. Cell 136, 629–641.

Quinn, J.J., Chang, H.Y., 2016. Unique features of long non-coding RNA biogenesis and function. Nat. Rev. Genet. 17 (1), 47–62. https://dx.doi.org/10.1038/nrg.2015.10.

Ray, P.D., Yosim, A., Fry, R.C., 2014. Incorporating epigenetic data into the risk assessment process for the toxic metals arsenic, cadmium, chromium, lead, and mercury: strategies and challenges. Front. Genet. 5, 201. PubMed PMID: 25076963. Pubmed Central PMCID: 4100550.

Redfern, A.D., Colley, S.M., Beveridge, D.J., et al., 2013. RNA-induced silencing complex (RISC) Proteins PACT, TRBP, and Dicer are SRA binding nuclear receptor coregulators. Proc. Natl. Acad. Sci. USA 110 (16), 6536–6541. https://dx.doi.org/10.1073/pnas.1301620110.

Rinn, J.L., 2014. lncRNAs: linking RNA to chromatin. Cold Spring Harb. Perspect. Biol. 6(8). https://dx.doi.org/10.1101/cshperspect.a018614.

Rinn, J.L., Chang, H.Y., 2012. Genome regulation by long noncoding RNAs. Annu. Rev. Biochem. 81, 145–166.

Rinn, J.L., Kertesz, M., Wang, J.K., et al., 2007. Functional demarcation of active and silent chromatin domains in human HOX loci by noncoding RNAs. Cell 129, 1311–1323.

Rivas, F.V., Tolia, N.H., Song, J.J., et al., 2005. Purified Argonaute2 and an siRNA form recombinant human RISC. Nat. Struct. Mol. Biol. 12, 340–349.

Sanders, A.P., Burris, H.H., Just, A.C., et al., 2015. Altered miRNA expression in the cervix during pregnancy associated with lead and mercury exposure. Epigenomics 7, 885–896.

Schembri, F., Sridhar, S., Perdomo, C., et al., 2009. MicroRNAs as modulators of smoking-induced gene expression changes in human airway epithelium. Proc. Natl. Acad. Sci. USA 106, 2319–2324.

Sui, J., Fu, Y., Zhang, Y., Ma, S., Yin, L., Pu, Y., Liang, G., 2018. Molecular mechanism for miR-350 in regulating of titanium dioxide nanoparticles in macrophage RAW264.7 cells. Chem. Biol. Interact. 280, 77–85.

Swan, S.H., 2008. Environmental phthalate exposure in relation to reproductive outcomes and other health endpoints in humans. Environ. Res. 108, 177–184.

Takahashi, K., Yokota, S., Tatsumi, N., et al., 2013. Cigarette smoking substantially alters plasma microRNA profiles in healthy subjects. Toxicol. Appl. Pharmacol. 272, 154–160.

Taki, F.A., Pan, X., Lee, M.H., et al., 2014. Nicotine exposure and transgenerational impact: a prospective study on small regulatory microRNAs. Sci. Rep. 4, 7513.

Talsness, C.E., Andrade, A.J., Kuriyama, S.N., et al., 2009. Components of plastic: experimental studies in animals and relevance for human health. Philos. Trans. R. Soc. Lond. Ser. B Biol. Sci. 364, 2079–2096.

Vandenberg, L.N., Hauser, R., Marcus, M., et al., 2007. Human exposure to bisphenol A (BPA). Reprod. Toxicol. 24, 139–177.

Verbanck, M., Canouil, M., Leloire, A., et al., 2017. Low-dose exposure to bisphenols A, F and S of human primary adipocyte impacts coding and non-coding RNA profiles. PLoS One 12, e0179583.

vom Saal, F.S., Myers, J.P., 2008. Bisphenol A and risk of metabolic disorders. JAMA 300, 1353–1355.

Walton, J.R., 2014. Chronic aluminum intake causes Alzheimer's disease: applying Sir Austin Bradford Hill's causality criteria. J. Alzheimers Dis. 40, 765–838.

Wang, K.C., Chang, H.Y., 2011. Molecular mechanisms of long noncoding RNAs. Mol. Cell 43, 904–914.

World Health Organization, 2017. http://www.who.int/gho/tobacco/use/en/.

Wu, X., Song, Y., 2011. Preferential regulation of miRNA targets by environmental chemicals in the human genome. BMC Genomics 12, 244. www.atsdr.cdc.gov/mercury/docs/physician_hg_flier.pdf.

Zhao, X.Y., Lin, J.D., 2015. Long noncoding RNAs: a new regulatory code in metabolic control. Trends Biochem. Sci. 40, 586–596.

Zhou, Z., Liu, H., Wang, C., et al., 2015. Long non-coding RNAs as novel expression signatures modulate DNA damage and repair in cadmium toxicology. Sci. Rep. 5, 15293.

Further Reading

Pratt, A.J., MacRae, I.J., 2009. The RNA-induced silencing complex: a versatile gene-silencing machine. J. Biol. Chem. 284 (27), 17897–17901.

Verdel, A., Jia, S., Gerber, S., et al., 2004. RNAi-mediated targeting of heterochromatin by the RITS complex. Science 303, 672–676.

Yosim, A., Fry, R.C., 2014. Incorporating epigenetic data into the risk assessment process for the toxic metals arsenic, cadmium, chromium, lead, and mercury: strategies and challenges. Front. Genet. 5, 201.

SPECIAL CONSIDERATIONS IN TOXICOEPIGENETICS RESEARCH

Germline and Transgenerational Impacts of Toxicant Exposures

Jessica A. Camacho*, Patrick Allard[†]

*Molecular Toxicology Interdepartmental Program, University of California, Los Angeles, Los Angeles, CA, United States [†]Institute for Society and Genetics, University of California, Los Angeles, Los Angeles, CA, United States

INTRODUCTION

When a pregnant woman is exposed to an epigenotoxicant, it may directly impact not only her epigenome but also the epigenome of her offspring and grand offspring, commonly referred to as inter- or multigenerational effects. Much attention has been given to G0 exposure and F1 effects. Much less attention, however, has been given to direct effects of exposures on the germ line, the eventual F2 (grand offspring) generation. This may be due to the intense focus over the last decade on the potential for exposures to influence transgenerational effects (F3 and beyond). In this chapter, we will explore the impact of environmental exposures on future generations, mainly focusing on germ-cell-mediated effects. Thus, to explain the mechanisms of inheritance stemming from various environmental cues and some of the discussions in the field, we first need to understand the developmental odyssey that germ cells are subjected to.

THE GERMLINE

Germ cells are the bridge between generations. They are responsible for passing down information from one generation to the next and consequently are often referred to as "immortal cells." The specification of germ cells occurs early during embryogenesis, yet their growth and development can span multiple years in mammals. A comprehensive description of germ-line development and across species has been reviewed elsewhere (e.g., Robert et al., 2015) and is beyond the scope of this chapter. Below, we will examine several critical periods of germ-cell development, how they may offer windows of susceptibility, and how they relate to the question of transgenerational inheritance.

GERMLINE SPECIFICATION AND THE IMPORTANCE OF THE EPIGENETIC REPRESSION OF SOMATIC FATES

In most animals, primordial germ cells (PGCs), the precursors to gametes, are formed during embryogenesis. There are two well-understood modes of PGC specification. One is the "preformation" mode, in which PGCs are specified by a specialized maternal cytoplasm or germ plasm that is asymmetrically divided during oogenesis or after fertilization to specify the cells to enter the germ-line lineage. Preformation is common among model organisms like *Drosophila*, *Caenorhabditis elegans* (*C. elegans*), *Xenopus*, and zebra fish. The second mechanism is termed epigenesis, where PGCs are induced during early embryogenesis by extracellular signals promoting pluripotent progenitor differentiation into germ cells (Seydoux and Braun, 2006). This mode was first observed in mice (Tam and Zhou, 1996) and appears to be the most widespread mechanism of germ-cell specification in metazoans (Extavour and Akam, 2003). These two modes of germ-cell specification have obvious distinct implications for transgenerational inheritance as the cytoplasmic continuity offered by preformation and the germplasm could act as a vector of information across generations. However, despite these differences, both mechanisms rely on the inhibition of the expression of somatic genes (Seydoux and Braun, 2006). Indeed, germ-line formation is guided by the inhibition of transcription and the use of repressive chromatin modifications in both modes of specification (Seydoux and Braun, 2006). For example, in mice, *Blimp1* (also known as PR domain zinc finger protein 1 (*Prdm1*)) is expressed during PGC specification and has been hypothesized to promote the repression of the somatic program, consistent with its known activity as a transcriptional repressor (Keller and Maniatis, 1991). *Tcfap2c*, a putative Blimp1 target, is expressed in PGCs from E7.25 and functions downstream to suppress mesodermal differentiation (Ohinata et al., 2005). In *Drosophila* and *C. elegans*, PGCs (preformation models), inhibition of RNA polymerase II is observed, alongside a decrease in zygotic mRNAs until hours post fertilization at the onset of gastrulation. This transcriptional silencing is mediated by germplasm components, PIE-1 in *C. elegans* (Batchelder et al., 1999), and *germ cell-less* (*gcl*) and *polar granule component* (*pgc*) in *Drosophila* (Leatherman et al., 2002; Martinho et al., 2004; Schaner and Kelly 2006). In addition to germplasm components, chromatin-based mechanisms play a large part in specification of germ cells. In *Drosophila*, there is a decrease in the activating histone mark H3K4me2 (Rudolph et al., 2007), whereas in *C. elegans*, there is also a decrease in H3K4me2 alongside an increase in the repressive mark H3K27me3 (Katz et al., 2009; Schaner and Kelly, 2006). As detailed in a later section, the deregulation of histone marks in the germ line following environmental exposure has recently been shown to serve as a potent transgenerational signal in species such as *C. elegans*.

PGC'S EPIGENETIC REPROGRAMMING: A CRITICAL PERIOD

In mammals, soon after their specification, PGCs initiate a dramatic remodeling of their chromatin, a stage referred to as the period of epigenetic reprogramming. The complex kinetics of epigenetic modifications unfold over the course of several days to several weeks, depending on the species, and include a dramatic loss of global DNA methylation and changes to the levels of various histone marks (reviewed in Tang et al. (2016)). It is that period of epigenetic reprogramming that conceptually forms the biggest barrier to epigenetic inheritance, as epigenetic marks such as DNA methylation are broadly erased to reach the lowest level of DNA methylation of any mammalian cell type (Gkountela et al., 2015; Hackett et al., 2013; Seisenberger et al., 2012) (Fig. 1). However, the identification of DNA demethylation-resistant loci and other epigenetic marks that are not erased may provide a mechanistic link bridging generations.

The process of DNA demethylation in PGCs is remarkably extensive, leading to the levels measured to near or below 10 % average CpG methylation across the genome in E13.5 mouse PGCs (Hackett et al., 2013; Seisenberger et al., 2012) and is comparable with the levels reached

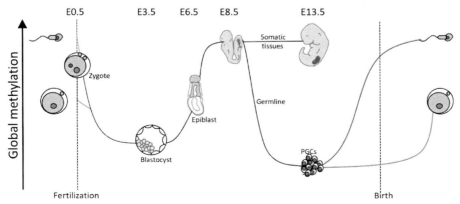

FIG. 1 Changes in global methylation levels during early embryonic development.

in human PGCs (Gkountela et al., 2015; Tang et al., 2015). The study of the dynamics and requirements for DNA demethylation in mouse primordial germ cells revealed that it is achieved in two phases. The first phase of DNA demethylation initiated in the mouse around E7.25 is both global and passive, that is, it is seen genome-wide and caused by the combination of active cellular replication and the simultaneous downregulation of expression of the *de novo* methyltransferases DNMT3A and DNMT3B, thus leading to the dilution of methylated CpGs over the course of several cell divisions (Grabole et al., 2013). In contrast, the second phase, which starts at E9.5, is dependent on the ten-eleven translocation (TET) enzymes, TET1 and TET2, and is more specific in its action (Hackett et al., 2013; Vincent et al., 2013). The methylation of imprinted control regions (ICRs), promoters of genes necessary for germ-cell formation and meiosis, and CpG islands of the inactive X chromosome in females survive the first wave of demethylation and appear to only reach full unmethylated state after the second wave of demethylation (Hackett et al., 2013; Hackett and Surani, 2013a,b; Tang et al., 2015). The supporting evidence includes the fact that *Tet1* and *Tet2* double knockout mice retain methylation at some ICRs and that *Tet1* is required for the demethylation of germ-cell- and meiosis-specific genes (Vincent et al., 2013; Yamaguchi et al., 2012). As mentioned above, while DNA methylation in PGCs is remarkably extensive, it is not complete. Notably, whole-genome bisulfite sequencing (WGBS) has revealed a total of 4730 loci that escape DNA demethylation in mouse PGCs, the vast majority of them being repeat associated (Hackett et al., 2013). These escapees are mainly evolutionarily young retrotransposons such as IAP elements in mice, which are known in other contexts to be sensitive to environmental exposures (see below) and LINE-1 L1HS elements in humans (Gkountela et al., 2015). Other regions also include pericentromeric satellite repeats (Tang et al., 2015) and subtelomeric regions (Guibert et al., 2012). There is clear interest in elucidating the consequence of perturbing the methylation at these loci, especially through the means of environmental exposures, and understanding how other epigenetic mechanisms, such as histone modifications, may be involved in the transcriptional repression of evolutionary older demethylated transposons.

However, in addition to DNA methylation erasure, histone modifications are also extensively reprogrammed during PGC differentiation. Notably, in mouse germ cells at around E8, there is a marked reduction in H3K9me2 that dovetails with an elevation of H3K27me3 (Kurimoto et al., 2015; Mansour et al., 2012; Seki et al., 2005, 2007; Shirane et al., 2016). The repressive mark H3K9me3 appears maintained throughout the reprogramming phase in the mouse and shows a "spotted" pattern in PGC nuclei corresponding to pericentromeric heterochromatin (Kim et al., 2014; Seki et al., 2007). One crucial factor for the regulation of these marks in PGCs is the N-methyltransferase SETDB1 that regulates H3K9me3 levels. Interestingly, the loss of SETDB1 leads to a reduction in H3K9me3 and H3K27me3 at IAPs and a reduction in DNA methylation at these loci (Leung et al., 2014; Liu et al., 2014) highlighting the importance of a concerted action between the different marks for transcriptional repression and the cross talk between these marks in PGCs. In other model organisms, such as *C. elegans*, where 5mC is not found, histone modifications are also remodeled during the germ-line cycle (Schaner and Kelly, 2006). However, as seen further below, the alteration of various histone marks, namely, H3K4me3, H3K9me3, and H3K27me3, has been associated with transgenerational inheritance of environmentally induced effects.

In summary, the transfer of epigenetic information across the period of PGC differentiation is a tightly regulated process that sees a global erasure of methylated CG dinucleotides and a remodeling of histone modifications. While it has been argued that this heavy remodeling may prevent the inheritance of environmentally induced epimutations (heritable changes in gene activity not directly associated with the presence DNA mutations), the identification of demethylation-resistant young transposable elements among other loci and the stable expression of some histone marks, such as H3K9me3, open the possibility for a transfer of epimutations across that period.

TRANSGENERATIONAL EFFECTS STEMMING FROM ENVIRONMENTAL EXPOSURES

The possibility that organisms may pass down traits elicited by toxicant exposures has been controversial for a variety of reasons (Heard and Martienssen, 2014; Hughes, 2014), chief among them being an absence of clear mechanisms of transgenerational inheritance. However, recent developments from multiple laboratories, working with various model organisms, have reversed this situation such that there are now multiple lines of evidence that environmental exposures can lead to transgenerational effects and several proposed mechanisms for their inheritance. We will now highlight some of the critical findings in the field that were foundational to some recent exciting mechanistic discoveries.

FROM MULTI- TO TRANS-GENERATIONAL

An increasingly accepted nomenclature is used to distinguish between direct and indirect environmental effects on given generations: multigenerational versus transgenerational effects, respectively (Skinner, 2008). This nomenclature addresses the need to separate these effects as they would be born from different exposure windows and distinct mechanisms. In the multigenerational model of exposure, the F0 (sometimes more logically called "P0" for parental generation 0) is exposed to the environmental cue, thereby also exposing its germ cells. However, if the P0 parent is a pregnant mother, the exposure may also affect the fetus and the fetal germ cells, which represent the precursors to the filial generations F1 and F2, respectively. Thus, any health effects displayed by the P0, F1, and/or F2 may have been caused by the direct exposure to the environmental cue and may not be inherited. However, any effect shown by the F3 generation and beyond cannot be caused by direct exposure and therefore requires the transmission of a memory of the environmental exposure. This phenomenon is referred to as transgenerational exposure. The scenario is different if P0 male mice are used for exposure as only the P0 and the F1 germ cells are represented during the exposure; thus, any effects shown at the F2 and beyond represent a heritable transgenerational effect (Fig. 2).

The prior lack of agreement on a nomenclature may be leading to some confusion with regard to the ability of various environmental exposures to elicit transgenerational effects as defined above. The famous case of the Dutch famine is one such example (Ravelli et al., 1976). As described by Susser and others, pregnant women (P0) exposed to low calorie intake during pregnancy at the end of World War II gave birth to boys (F1) who showed a significant increase in their BMI at 19 years of age (Ravelli et al., 1976). A subsequent publication examined the F2 generation for various health end points and concluded that there was increased

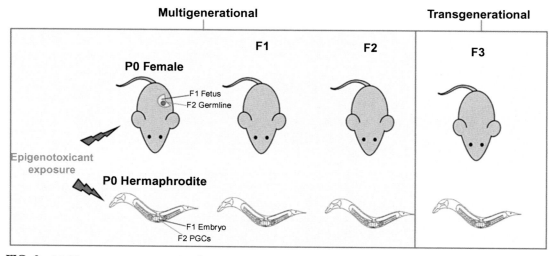

FIG. 2   Multi- versus transgenerational exposures in rodents and in *C. elegans*.

adiposity in that generation and thus transgenerational effect (Painter et al., 2008). In this exposure scenario, the F2 was represented as germ cells within the exposed fetus (F1). Therefore, one cannot exclude the possibility that the effects observed in the F2 are caused by direct exposure to low nutrient conditions as opposed to a heritable effect. Such studies however can be considered as foundational to transgenerational studies as they establish that the exposure in question has a direct effect on the fetus and that its germ cells and their epigenome may therefore also be impacted. The ability of *in utero* environmental exposures to alter health end points later in life led to the concept of the developmental origin of health and disease or DOHaD (Wadhwa et al., 2009). Importantly, it has now been clearly established that the type of environmental exposures able to act on the fetus is limited to not only "natural" environmental cues, such as diet, but also man-made ones, such as exposure to lead or plastics. In an important study carried by Dolinoy and colleagues, perinatal exposure to environmentally and physiologically relevant doses of the plastic manufacturing chemical bisphenol A led to an alteration of the DNA methylation levels of the $A(vy)$ metastable epiallele involving an IAP element and leading to a change in its expression (Dolinoy et al., 2007). Together, these findings demonstrate a clear environmental impact on the fetus and extend the narrative related to early-life exposures to artificial environmental cues.

The ability of man-made environmental exposures to cause "true" transgenerational effects was established in 2005 by Skinner and colleagues (Anway et al., 2005). In their seminal study, gestational exposure (E8–E15) of rats by IP injection of high doses of the fungicide vinclozolin led to a strong effect on the reproductive function of males down to the F4 generation. These defects, which were transmitted through the male lineage, included a dramatic alteration of the testis morphology and a decrease in sperm count and quality (Anway et al., 2005). In line with the male lineage mode of inheritance, further analysis revealed that the F3 sperm carried a DNA methylation signature of ancestral vinclozolin exposure that consisted of 52 differentially methylated regions (DMRs) identified by MeDIP-chip (Guerrero-Bosagna et al., 2010). This work was later extended to other compounds such as a mixtures of plastic compounds (BPA and phthalates) with similar conclusions (Manikkam et al., 2013). Some of these studies have since come under scrutiny for the chemical doses used (reviewed in (Hughes, 2014)), and it remains to be determined whether transgenerational effects of these compounds could be detected at environmentally relevant levels and routes of exposure. In those carefully defined and relevant conditions, it will also be important to test whether the identified epimutations caused by environmental exposure resist the period of DNA demethylation observed in PGCs and in early embryos, thus contributing to potential transgenerational inheritance. Nonetheless, it is undeniable that the studies by Skinner and colleagues ignited a vivid interest to examine toxicological end points over several generations and to investigate the role of the epigenome as a mechanism of transgenerational effects stemming from environmental chemicals.

TOWARD A MECHANISM OF INHERITANCE: IT'S NOT JUST ABOUT DNA METHYLATION

While not focused on environmental exposures, a recent study nonetheless demonstrated that overexpression of a histone lysine demethylase is sufficient to elicit a transmissible epigenetic impact on development (Siklenka et al., 2015). In this study, overexpression of the H3K4 demethylase KDM1A in transgenic male mice from only one generation was sufficient to result in transgenerational (F2 and beyond in this case) effects including a variety of severe developmental defects and a decrease in offspring survival. Interestingly, while there was a multigenerational decrease in H3K4me2 levels at 2300 loci corresponding to developmental genes in transgenic animals, this was not observed in nontransgenic animals and therefore, by inference, transgenerationally. Furthermore, DNA methylation was not altered in the sperm of transgenic or nontransgenic animals at loci where H3K4me2 levels were changed or genome-wide. However, gene expression in both sperm and embryos generated from sperm with ancestral KDM1A overexpression altered transgenerationally indicating that while the ancestral initiating event may be the change in H3K4me2, it is transmitted and exerts its developmental effects through another mechanism. This work is important because it suggests that an exposure-mediated change in expression levels of the epigenetic machinery might be sufficient to elicit transgenerational effects. It also emphasizes the fact that changes in DNA methylation are not necessary for the manifestation of transgenerational effects of environmental exposures.

A series of elegant and paradigm-shifting studies from several laboratories have also highlighted the importance of small tRNA-like RNAs in the transfer of information related to paternal diet to the next generation where they regulate metabolism. Rando and colleagues demonstrated that a low-protein paternal diet is sufficient to induce a metabolic reprogramming in the next generation, including an alteration in hepatic cholesterol production (Carone et al., 2010). However, sperm methylome profiling under various dietary condition revealed that diet alone had little influence on sperm methylation patterns (Shea et al., 2015). Instead, the low-protein diet led to the accumulation of small ~28–34 nt tRNA fragments that appear to be derived from the 5′ ends of tRNAs (Sharma et al.,

2016). A significant amount of those tRNA fragments is provided to the sperm during its passage through the epididymis through vesicles called epididymosomes that originate from the epididymal epithelium (Caballero et al., 2013). Importantly, injection of the sperm-derived small RNAs from low- versus high-protein males in zygotes was sufficient to differentially alter gene expression in preimplantation embryos, suggesting that later metabolic defects may be caused by the initial transcriptional landscape established by the presence of the small RNAs (Sharma et al., 2016). The effects of paternal high-fat diet (HFD) were also transmissible to the offspring, causing impaired glucose tolerance and insulin resistance (Chen et al., 2016). Injection of the sperm head from HFD males directly into the oocyte of female fed with a normal diet was sufficient to induce these changes (Chen et al., 2016). Furthermore, the injection of small 30–40 nucleotide RNAs, but not smaller or larger RNAs, isolated from these sperm into a zygote generated from normal diet mice was also sufficient to induce these metabolic phenotypes in the offspring (Chen et al., 2016). Finally, the authors were able to monitor the expression changes of genes in early embryos and in islets from mice obtained from injected embryos and showed that metabolic genes are perturbed in both cases, suggesting an alteration of these genes throughout the differentiation history of the islet cells.

Together, these studies highlight the role of small RNAs in the inheritance of paternal diet exposures; however, it is not clear whether any of these effects are transmissible to the subsequent generation (i.e., beyond the F1) or whether other paternal environmental conditions may also lead to the transfer of phenotypes to the next generation *via* small RNAs. Interestingly, a study on trauma and stress in mice has also shown that injection of sperm RNA was sufficient to induce behavioral and metabolic changes in the offspring (Gapp et al., 2014), suggesting it is possible that the sperm RNA content may transmit a wide range of paternal life experience.

NONMAMMALIAN MODELS AND THE CENTRAL ROLE OF HISTONE MODIFICATIONS

Transgenerational effects have also been identified in a variety of nonmammalian models (Baker et al., 2014; Brookheart and Duncan, 2016; Rankin, 2015). Crucially, some of these "alternative" models have been particularly conducive to the dissection of the mechanisms underlying environmental inheritance. This is particularly true of the nematode *C. elegans*, which offers the significant advantages of a short generation time (3–4 days) and a high level of genetic tractability.

The short generation time of the nematode and its utility for transgenerational studies are highlighted by a recent study on the effect of high temperature on the epigenome of the worm (Klosin et al., 2017). The authors examined the alteration of chromatin repression *via* desilencing of a heterochromatin-like repetitive transgene reporter following a five-generation high-temperature exposure. The study showed that such high-temperature exposure caused a derepression of the transgene expression lasting for 14 generations before returning to basal levels. The authors then examined the genetic requirement underlying this inheritance and showed that the transgene desilencing correlated with a decrease of the repressive mark H3K9me3 in the germ line. SET-25, the methyltransferase responsible for

H3K9 trimethylation, is required for the silencing of the reporter, and its deletion suppresses the difference between ancestrally exposed worms and nonexposed controls.

A variety of histone marks have been implicated in the inheritance of environmental effects that might be a reflection of the different environmental cues used in the studies. For example, exposure to arsenite leads to reproductive health defects, including reduction in brood size in *C. elegans*, in P0 generation and in subsequent generations (F1–F5) (Yu and Liao, 2016). In this case, the effects of arsenite are dependent on the levels of H3K4me2, a transcriptionally activating histone modification, as there is a decrease in the expression of the H3K4me2 demethylase, spr-5, together with an increase in H3K4me2 levels in the arsenite-exposed P0 generation and subsequent F1–F3 generations (Yu and Liao, 2016). H3K4 methylation has also been implicated as a mediator of the effects of various environmental exposures and stressors such as arsenite exposure, hyperosmosis, and starvation (Kishimoto et al., 2017). Exposure to these stressors during development resulted in an increased resistance to proteotoxicity and to the normally lethal oxidative stressor hydrogen peroxide that lasted until the F3 generation; however, worms with inactivating mutations in H3K4 methyltransferase complex components (*wdr-5.1* and *set-2*) failed to inherit resistance, reinforcing the role of H3K4 methylation in the transmission of the effect (Kishimoto et al., 2017). By comparing different exposures, this study suggests that the transgenerational effects that result from independent exposures may be transmitted through similar mechanisms. By contrast, a recent study identified the regulation of both H3K9me3 and H3K27me3 as mediators of memory of BPA exposure (Camacho et al., 2018) raising several hypotheses discussed below.

It is important to note that beside histone modifications, small RNAs have also been implicated as a mechanism of transgenerational inheritance in *C. elegans*. Starvation of worms, larvae, for example, induces the expression of several small RNAs, termed STGs for "small RNAs targeting a given gene." These STGs regulate nutrient reserves and have been detected in the both starved generation (P0) and their descendants through the F3 generation. Inheritance of the STGs is dependent on the presence of the argonaute proteins RDE-4 and HRDE-1 (Rechavi et al., 2014). Such small RNA mechanisms may work in concert with histone marks to regulate transgenerational inheritance as recently demonstrated in another study (Lev et al., 2017). That study examined the progressive mortal germ-line phenotype of the *met-2* mutants defective in H3K9 mono and dimethylation. In these worms, there was a progressive reduction in fertility that unfolds over 10–30 generations. Interestingly, the argonaute factor *hdre-1*, associated with small RNAs, is required for the progressive sterility phenotype of *met-2* mutant. From these results and others, the authors propose a model of inheritance where MET-2 functions to suppress the transgenerational transfer of small RNAs via the regulation of H3K9me. These findings directly link repressive histone marks and small RNAs; however, whether a similar connection can be established for environmentally induced phenotype remains to be explored.

FINAL CONSIDERATIONS

As illustrated above, there is currently not a single mechanism responsible for the transfer of environmental exposure effects from one generation to the next. The potential cross talk between epigenetic mechanisms, as observed in *C. elegans*, may explain why various epigenetic marks have independently been shown to be important for the inheritance of exposure effects. Distinct environmental exposures may also act through different individual, or combinations of, epigenetic modifications. Efforts to distinguish between these possibilities would benefit greatly from a concerted and comprehensive effort between investigators in which standardized approaches are used to examine DNA methylation, histone modifications, and small RNAs simultaneously. As discussed earlier, it is also important for the community to establish a common nomenclature (e.g., definition of multi-, inter-, or transgenerational) and corresponding guidelines to examine these new models of inheritance. A movement in that direction has already been proposed by several key actors in the field (Bohacek and Mansuy, 2017; Skinner, 2008). Future studies must also determine whether a connection exists between specific environmental exposures and alterations in specific epigenetic factors. Similarly, there is also a need to examine how environmentally altered levels or distribution of epigenetic marks in germ cells may modify the function of adult organs. In conclusion, it is now well established that organisms have the ability to transmit various environmental cues across generations. This provides a fascinating reshaping of Darwinist evolutionary views to leave room to a certain degree of environmental responsiveness of animals over several generations. Despite these advances, we have yet to fully understand how environmental exposures alter the epigenome, how epimutations persist through developmental epigenetic reprogramming, and how epigenetic alterations in germ cells influence organ structure and function.

References

Anway, M.D., Cupp, A.S., Uzumcu, M., Skinner, M.K., 2005. Epigenetic transgenerational actions of endocrine disruptors and male fertility. Science 308, 1466–1469.

Baker, T.R., King-Heiden, T.C., Peterson, R.E., Heideman, W., 2014. Dioxin induction of transgenerational inheritance of disease in zebrafish. Mol. Cell. Endocrinol. 398, 36–41.

Batchelder, C., Dunn, M.A., Choy, B., Suh, Y., Cassie, C., Shim, E.Y., et al., 1999. Transcriptional repression by the caenorhabditis elegans germ-line protein pie-1. Genes Dev. 13, 202–212.

Bohacek, J., Mansuy, I.M., 2017. A guide to designing germline-dependent epigenetic inheritance experiments in mammals. Nat. Methods 14, 243–249.

Brookheart, R.T., Duncan, J.G., 2016. Drosophila melanogaster: an emerging model of transgenerational effects of maternal obesity. Mol. Cell. Endocrinol. 435, 20–28.

Caballero, J.N., Frenette, G., Belleannee, C., Sullivan, R., 2013. Cd9-positive microvesicles mediate the transfer of molecules to bovine spermatozoa during epididymal maturation. PLoS One 8, e65364.

Camacho, J., Truong, L., Kurt, Z., Chen, Y.W., Morselli, M., Gutierrez, G., Pellegrini, M., Yang, X., Allard, P., 2018. The memory of environmental chemical exposure in C. elegans is dependent on the Jumonji demethylases jmjd-2 and jmjd-3/utx-1. Cell Rep. 23 (8), 2392–2404.

Carone, B.R., Fauquier, L., Habib, N., Shea, J.M., Hart, C.E., Li, R., et al., 2010. Paternally induced transgenerational environmental reprogramming of metabolic gene expression in mammals. Cell 143, 1084–1096.

Chen, Q., Yan, M., Cao, Z., Li, X., Zhang, Y., Shi, J., et al., 2016. Sperm tsrnas contribute to intergenerational inheritance of an acquired metabolic disorder. Science 351, 397–400.

Dolinoy, D.C., Huang, D., Jirtle, R.L., 2007. Maternal nutrient supplementation counteracts bisphenol a-induced DNA hypomethylation in early development. Proc. Natl. Acad. Sci. U. S. A. 104, 13056–13061.

Extavour, C.G., Akam, M., 2003. Mechanisms of germ cell specification across the metazoans: Epigenesis and preformation. Development 130, 5869–5884.

Gapp, K., Jawaid, A., Sarkies, P., Bohacek, J., Pelczar, P., Prados, J., et al., 2014. Implication of sperm rnas in transgenerational inheritance of the effects of early trauma in mice. Nat. Neurosci. 17, 667–669.

Gkountela, S., Zhang, K.X., Shafiq, T., Liao, W.-W., Hargan-Calvopina, J., Chen, P.-Y., Clark, A.T., 2015. DNA demethylation dynamics in the human prenatal germline. Cell 161, 1425–1436.

Grabole, N., Tischler, J., Hackett, J.A., Kim, S., Tang, F., Leitch, H.G., et al., 2013. Prdm14 promotes germline fate and naive pluripotency by repressing fgf signalling and DNA methylation. EMBO Rep. 14, 629–637.

Guerrero-Bosagna, C., Settles, M., Lucker, B., Skinner, M.K., 2010. Epigenetic transgenerational actions of vinclozolin on promoter regions of the sperm epigenome. PLoS One. 5.

Guibert, S., Forne, T., Weber, M., 2012. Global profiling of DNA methylation erasure in mouse primordial germ cells. Genome Res. 22, 633–641.

Hackett, J.A., Sengupta, R., Zylicz, J.J., Murakami, K., Lee, C., Down, T.A., Surani, M.A., 2013. Germline DNA demethylation dynamics and imprint erasure through 5-hydroxymethylcytosine. Science 339 (6118), 448–452.

Hackett, J.A., Surani, M.A., 2013b. DNA methylation dynamics during the mammalian life cycle. Philos. Trans. R. Soc. Lond. Ser. B: Biol. Sci. 368 (1609), 20110328.

Hackett, J.A., Surani, M.A., 2013a. Beyond DNA: programming and inheritance of parental methylomes. Cell 153, 737–739.

Heard, E., Martienssen, R.A., 2014. Transgenerational epigenetic inheritance: myths and mechanisms. Cell 157, 95–109.

Hughes, V., 2014. Epigenetics: the sins of the father. Nature 507, 22–24.

Katz, D.J., Edwards, T.M., Reinke, V., Kelly, W.G., 2009. A *C. elegans* lsd1 demethylase contributes to germline immortality by reprogramming epigenetic memory. Cell 137, 308–320.

Keller, A.D., Maniatis, T., 1991. Identification and characterization of a novel repressor of beta- interferon gene expression. Genes Dev. 5, 868–879.

Kim, S., Gunesdogan, U., Zylicz, J.J., Hackett, J.A., Cougot, D., Bao, S., et al., 2014. Prmt5 protects genomic integrity during global DNA demethylation in primordial germ cells and preimplantation embryos. Mol. Cell 56, 564–579.

Kishimoto, S., Uno, M., Okabe, E., Nono, M., Nishida, E., 2017. Environmental stresses induce transgenerationally inheritable survival advantages via germline-to-soma communication in *Caenorhabditis elegans*. Nat. Commun. 8:14031, https://dx.doi.org/10.1038/ncomms14031.

Klosin, A., Casas, E., Hidalgo-Carcedo, C., Vavouri, T., Lehner, B., 2017. Transgenerational transmission of environmental information in *C. elegans*. Science 356, 320–323.

Kurimoto, K., Yabuta, Y., Hayashi, K., Ohta, H., Kiyonari, H., Mitani, T., et al., 2015. Quantitative dynamics of chromatin remodeling during germ cell specification from mouse embryonic stem cells. Cell Stem Cell 16, 517–532.

Leatherman, J.L., Levin, L., Boero, J., Jongens, T.A., 2002. Germ cell-less acts to repress transcription during the establishment of the drosophila germ cell lineage. Curr. Biol. 12, 1681–1685.

Leung, D., Du, T., Wagner, U., Xie, W., Lee, A.Y., Goyal, P., et al., 2014. Regulation of DNA methylation turnover at ltr retrotransposons and imprinted loci by the histone methyltransferase setdb1. Proc. Natl. Acad. Sci. U. S. A. 111, 6690–6695.

Lev, I., Seroussi, U., Gingold, H., Bril, R., Anava, S., Rechavi, O., 2017. Met-2-dependent h3k9 methylation suppresses transgenerational small rna inheritance. Curr. Biol. 27, 1138–1147.

Liu, S., Brind'Amour, J., Karimi, M.M., Shirane, K., Bogutz, A., Lefebvre, L., et al., 2014. Setdb1 is required for germline development and silencing of h3k9me3-marked endogenous retroviruses in primordial germ cells. Genes Dev. 28, 2041–2055.

Manikkam, M., Tracey, R., Guerrero-Bosagna, C., Skinner, M.K., 2013. Plastics derived endocrine disruptors (BPA, DEHP and DBP) induce epigenetic transgenerational inheritance of obesity, reproductive disease and sperm epimutations. PLoS One 8, e55387.

Mansour, A.A., Gafni, O., Weinberger, L., Zviran, A., Ayyash, M., Rais, Y., et al., 2012. The h3k27 demethylase utx regulates somatic and germ cell epigenetic reprogramming. Nature 488, 409–413.

Martinho, R.G., Kunwar, P.S., Casanova, J., Lehmann, R., 2004. A noncoding rna is required for the repression of rnapolii-dependent transcription in primordial germ cells. Curr. Biol. 14, 159–165.

Ohinata, Y., Payer, B., O'carroll, D., Ancelin, K., Ono, Y., Sano, M., et al., 2005. Blimp1 is a critical determinant of the germ cell lineage in mice. Nature 436, 207–213.

Painter, R.C., Osmond, C., Gluckman, P., Hanson, M., Phillips, D.I., Roseboom, T.J., 2008. Transgenerational effects of prenatal exposure to the dutch famine on neonatal adiposity and health in later life. BJOG: Int. J. Obstet. Gynaecol. 115, 1243–1249.

Rankin, C.H., 2015. A review of transgenerational epigenetics for rnai, longevity, germline maintenance and olfactory imprinting in *Caenorhabditis elegans*. J. Exp. Biol. 218, 41–49.

Ravelli, G.P., Stein, Z.A., Susser, M.W., 1976. Obesity in young men after famine exposure in utero and early infancy. N. Engl. J. Med. 295, 349–353.

Rechavi, O., Houri-Ze'evi, L., Anava, S., Goh, W.S., Kerk, S.Y., Hannon, G.J., et al., 2014. Starvation-induced transgenerational inheritance of small rnas in *C. elegans*. Cell 158, 277–287.

Robert, V.J., Garvis, S., Palladino, F., 2015. Repression of somatic cell fate in the germline. Cell. Mol. Life Sci. 72, 3599–3620.

Rudolph, T., Yonezawa, M., Lein, S., Heidrich, K., Kubicek, S., Schafer, C., et al., 2007. Heterochromatin formation in drosophila is initiated through active removal of h3k4 methylation by the lsd1 homolog su(var)3-3. Mol. Cell 26, 103–115.

Schaner, C.E., Kelly, W.G., 2006. Germline chromatin. In: WormBook: The Online Review of *C. elegans* Biology. WormBook, Pasadena, CA, pp. 1–14.

Seisenberger, S., Andrews, S., Krueger, F., Arand, J., Walter, J., Santos, F., et al., 2012. The dynamics of genome-wide DNA methylation reprogramming in mouse primordial germ cells. Mol. Cell 48, 849–862.

Seki, Y., Hayashi, K., Itoh, K., Mizugaki, M., Saitou, M., Matsui, Y., 2005. Extensive and orderly reprogramming of genome-wide chromatin modifications associated with specification and early development of germ cells in mice. Dev. Biol. 278, 440–458.

Seki, Y., Yamaji, M., Yabuta, Y., Sano, M., Shigeta, M., Matsui, Y., et al., 2007. Cellular dynamics associated with the genome-wide epigenetic reprogramming in migrating primordial germ cells in mice. Development 134, 2627–2638.

Seydoux, G., Braun, R.E., 2006. Pathway to totipotency: lessons from germ cells. Cell 127, 891–904.

Sharma, U., Conine, C.C., Shea, J.M., Boskovic, A., Derr, A.G., Bing, X.Y., et al., 2016. Biogenesis and function of trna fragments during sperm maturation and fertilization in mammals. Science 351, 391–396.

Shea, J.M., Serra, R.W., Carone, B.R., Shulha, H.P., Kucukural, A., Ziller, M.J., et al., 2015. Genetic and epigenetic variation, but not diet, shape the sperm methylome. Dev. Cell 35, 750–758.

Shirane, K., Kurimoto, K., Yabuta, Y., Yamaji, M., Satoh, J., Ito, S., et al., 2016. Global landscape and regulatory principles of DNA methylation reprogramming for germ cell specification by mouse pluripotent stem cells. Dev. Cell 39, 87–103.

Siklenka, K., Erkek, S., Godmann, M., Lambrot, R., McGraw, S., Lafleur, C., et al., 2015. Disruption of histone methylation in developing sperm impairs offspring health transgenerationally. Science 350, aab2006.

Skinner, M.K., 2008. What is an epigenetic transgenerational phenotype? F3 or f2. Reprod. Toxicol. 25, 2–6.

Tam, P.P., Zhou, S.X., 1996. The allocation of epiblast cells to ectodermal and germ-line lineages is influenced by the position of the cells in the gastrulating mouse embryo. Dev. Biol. 178, 124–132.

Tang, W.W., Dietmann, S., Irie, N., Leitch, H.G., Floros, V.I., Bradshaw, C.R., et al., 2015. A unique gene regulatory network resets the human germline epigenome for development. Cell 161, 1453–1467.

Tang, W.W., Kobayashi, T., Irie, N., Dietmann, S., Surani, M.A., 2016. Specification and epigenetic programming of the human germ line. Nat. Rev. Genet. 17, 585–600.

Vincent, J.J., Huang, Y., Chen, P.Y., Feng, S., Calvopina, J.H., Nee, K., et al., 2013. Stage-specific roles for tet1 and tet2 in DNA demethylation in primordial germ cells. Cell Stem Cell 12, 470–478.

Wadhwa, P.D., Buss, C., Entringer, S., Swanson, J.M., 2009. Developmental origins of health and disease: brief history of the approach and current focus on epigenetic mechanisms. Semin. Reprod. Med. 27, 358–368.

Yamaguchi, S., Hong, K., Liu, R., Shen, L., Inoue, A., Diep, D., et al., 2012. Tet1 controls meiosis by regulating meiotic gene expression. Nature 492, 443–447.

Yu, C.W., Liao, V.H., 2016. Transgenerational reproductive effects of arsenite are associated with h3k4 dimethylation and spr-5 downregulation in *Caenorhabditis elegans*. Environ. Sci. Technol. 50, 10673–10681.

Further Reading

Allergrucci, C., 2005. Reproduction.

Blackwell, T.K., 2004. Germ cells: finding programs of mass repression. Curr. Biol. 14, R229–R230.

Chiquoine, A.D., 1954. The identification, origin, and migration of the primordial germ cells in the mouse embryo. Anat. Rec. 118 (2), 135–146. https://dx.doi.org/10.1002/ar.1091180202. PMID 13138919.

Chuva De Sousa Lopes, S.M., Hayashi, K., Shovlin, T.C., Mifsud, W., Surani, M.A., Mclaren, A., 2008. X chromosome activity in mouse XX primordial germ cells. PLoS Genetics 4, e30.

Copeland, N.G., Gilbert, D.J., Cho, B.C., Donovan, P.J., Jenkins, N.A., Cosman, D., et al., 1990. Mast cell growth factor maps near the steel locus on mouse chromosome 10 and is deleted in a number of steel alleles. Cell 63, 175–183.

Fullston, T., et al., 2013. FASEB J. 27, 4226–4243.

Ginsburg, M., Snow, M.H., McLaren, A., 1990. Primordial germ cells in the mouse embryo during gastrulation. Development 110 (2), 521–528.

Greer, E.L., et al., 2011. Nature 479, 365–371.

Hajkova, P., Ancelin, K., Waldmann, T., Lacoste, N., Lange, U.C., Cesari, F., Lee, C., Almouzni, G., Schneider, R., Surani, M.A., 2008. Chromatin dynamics during epigenetic reprogramming in the mouse germ line. Nature 452 (7189), 877–881.

Hajkova, P., Jeffries, S.J., Lee, C., Miller, N., Jackson, S.P., Surani, M.A., 2010. Genome-wide reprogramming in the mouse germ line entails the base excision repair pathway. Science 329 (5987), 78–82.

Huang, E., Nocka, K., Beier, D.R., Chu, T.Y., Buck, J., Lahm, H.W., et al., 1990. The hematopoietic growth factor KL is encoded by the Sl locus and is the ligand of the c-kit receptor, the gene product of the W locus. Cell 63, 225–233.

Kohli, R.M., Zhang, Y., 2013. TET enzymes, TDG and the dynamics of DNA demethylation. Nature 502 (7472), 472–479.

Lambrot, R., et al., 2013. Nat. Commun. 4, 2889.

Lawson, K.A., Hage, W.J., 1994. In: Clonal analysis of the origin of primordial germ cells in the mouse. Ciba Foundation Symposium. Novartis Foundation Symposia, 182, pp. 68–84. discussion 84–91.

Maatouk, D.M., Resnick, J.L., 2006. DNA methylation is a primary mechanism for silencing postmigratory primordial germ cell genes in both germ cell and somatic cell lineages. Development 133, 3411–3418.

Marczylo, E.L., Amoako, A.A., Konje, J.C., Gant, T.W., Marczylo, T.H., 2012. Epigenetics 7, 432–439.

Messerschmidt, D.M., Knowles, B.B., Solter, D., 2014. DNA methylation dynamics during epigenetic reprogramming in the germline and preimplantation embryos. Genes Dev. 28, 812–828.

Ng, S.-F., et al., 2010. Nature 467, 963–966.

Ohinata, Y., Ohta, H., Shigeta, M., Yamanaka, K., Wakayama, T., Saitou, M., 2009. A signaling principle for the specification of the germ cell lineage in mice. Cell 137, 571–584.

Pembrey, M.E., et al., 2006. Eur. J. Hum. Genet. 14, 159–166.

Saitou, M., Barton, S.C., Surani, M.A., 2002. A molecular programme for the specification of germ cell fate in mice. Nature 418, 293–300.

Saitou, M., Kagiwada, S., Kurimoto, K., 2012. Epigenetic reprogramming in mouse pre-implantation development and primordial germ cells. Development 139, 15–31.

Wei, Y., et al., 2014. Proc. Natl. Acad. Sci. U. S. A. 111, 1873–1878.

Yamaji, M., Seki, Y., Kurimoto, K., Yabuta, Y., Yuasa, M., Shigeta, M., et al., 2008. Crit- ical function of Prdm14 for the establishment of the germ cell lineage in mice. Nature Genetics 40, 1016–1022.

Yoshimizu, T., Obinata, M., Matsui, Y., 2001. Stage-specific tissue and cell interactions play key roles in mouse germ cell specification. Development 128, 481–490.

Zwaka, T.P., Thomson, J.A., 2005. A germ cell origin of embryonic stem cells? Development 132, 227–233.

Novel Bioinformatics Methods for Toxicoepigenetics

Raymond G. Cavalcante*,a, Tingting Qin†,a, Maureen A. Sartor*,†,‡

*Epigenomics Core, Biomedical Research Core Facility, University of Michigan, Ann Arbor, MI, United States †Department of Computational Medicine and Bioinformatics, University of Michigan, Ann Arbor, MI, United States ‡Biostatistics Department, University of Michigan, Ann Arbor, MI, United States

a Equal contribution.

INTRODUCTION

The rapid increase in genome-wide assays, especially high-throughput sequencing-based assays, available for assessing epigenetic marks has led to an equally rapid expansion in the number of bioinformatics approaches used to analyze the resulting data. The most recent expansion has occurred in a few main areas: an improved experimental approaches for an epigenetic assay (e.g., ATAC-seq instead of DNAse-seq or MNase-seq) (Buenrostro et al., 2013); an increase in the number of sites covered (e.g., the Illumina EPIC BeadChip instead of the HumanMethylation450K BeadChip) (Pidsley et al., 2016); an increase in experiments integrating multiple epigenomic assays to obtain a more comprehensive signature of the cell's regulatory state, especially for complex diseases such as cancers (Meldi and Figueroa, 2015; Messier et al., 2016; Lin et al., 2017) and type 2 diabetes (Barres and Zierath, 2016); and the introduction of single-cell genome-wide epigenomic technologies (Clark et al., 2016). Each of these advancements has led to opportunities for new analysis methods (Stricker et al., 2017; Zhou et al., 2017; Koenecke et al., 2016). For toxicoepigenetics experiments in particular, there are often fewer observed changes and with smaller magnitude differences than observed in complex diseases, such as cancers (Breton et al., 2017). Thus, bioinformatics approaches should be chosen that are capable of detecting these changes and that can facilitate the interpretation of them.

In addition to the new assays and analysis methods, new annotation resources are becoming available to predict the regulatory effect of an epigenetic change. Genome-wide human enhancer regions are now known for many specific tissues and cell types, and various methods are being used to identify the regulatory target genes from each enhancer (He et al., 2014; Mora et al., 2016; O'Connor et al., 2017). For example, genome-wide chromosome conformation capture (Hi-C) and chromatin interaction analysis with paired-end tag (ChIA-PET) experiments can identify three-dimensional loops from an enhancer region to a proximal promoter region or transcription start site, which has been shown to yield high-quality target predictions (Belton et al., 2012; Denker and de Laat, 2016; Li et al., 2014). RegulomeDB is one key resource for identifying the regulatory effect from an SNP or epigenetic change at a particular genomic region (Boyle et al., 2012). For a broader perspective, observed genome-wide epigenomic changes can be input into novel bioinformatics tools for gene set enrichment testing to determine the affected cellular pathways or to predict associated diseases.

Together, the rapid advances in assays and in bioinformatics methods and resources for epigenomic data make it difficult for even advanced bioinformatics users to stay abreast of the field. In this chapter, we hope to introduce the reader to some of the novel bioinformatics approaches, summary measures, and integrative techniques used to arrive at biologically meaningful conclusions from epigenomic studies, especially those studying cellular responses to environmental exposures and toxins.

THE CHALLENGE AND COMPLEXITY OF EPIGENOMICS DATA ANALYSIS MOTIVATES THE NEED FOR NOVEL BIOINFORMATICS METHODS

The importance of studying epigenetics as a mediator between environmental exposures and later life disease has long been appreciated. Epigenetic modifications are often more stable than gene expression levels and can provide a blueprint not only of the current regulatory state of a cell type but also of the past and a prediction of the future. Furthermore, certain epigenetic marks are more stable than others, such that integrating data from multiple epigenetic marks can provide complementary results for interpreting different timescales. For example, whereas aberrant DNA methylation marks may remain present throughout a life span, the presence of DNA hydroxymethylation often suggests more recently activated genes (Greco et al., 2016; Serandour et al., 2012). Unlike genomics, which is the same for all tissues in an individual, epigenomic states differ dramatically across tissues (Wang et al., 2018) and may differ in a nonmonotonic fashion across exposure levels/doses (Gore et al., 2015; Kim et al., 2014). These complex differences across tissues, time, doses, and epigenetic marks motivate the need for correspondingly complex bioinformatics approaches to study toxicoepigenomics.

The genome-wide epigenomic marks and most popular technologies being used to interrogate them are (1) DNA methylation (enhanced reduced representation bisulfite sequencing (ERRBS), whole-genome bisulfite sequencing (WGBS), Illumina EPIC BeadChip, and MeDIP-seq), (2) histone modifications (ChIP-seq), (3) open-chromatin regions (ATAC-seq), and (4) DNA hydroxymethylation (hMeDIP-seq). Recent years have seen the introduction of methods to test for differential DNA methylation or hydroxymethylation and histone modifications or open-chromatin regions that allow for more complex experimental designs so that appropriate covariates and interactions can be modeled. New smoothing and summary measures that more accurately capture subtle DNA methylation differences in or near CpG islands and shores or promoter regions have also been developed. A new approach, informME, applies information theory to model the epigenomic landscape using WGBS data and has been shown to be a powerful tool to study the stochastic nature of the epigenome (Jenkinson et al., 2017, 2018). Novel tools are now being introduced to help researchers integrate the results from these experiments to find biologically meaningful results, such as ELMER (Yao et al., 2015a), TENET (Rhie et al., 2016), and RegNetDriver (Dhingra et al., 2017). Finally, downstream analysis tools can help interpret results at the level of pathways and biological processes and in relation to other publicly available datasets, such as ChIP-Enrich (which now includes gene sets from the Comparative Toxicogenomics Database (CTD)) (Welch et al., 2014), WashU Epigenome Browser (Zhou and Wang, 2012), FEM (Jiao et al., 2014), and SMITE Bioconductor package (Wijetunga et al., 2017).

BIOINFORMATICS APPROACHES FOR DNA METHYLATION

Introduction

In addition to the well-studied DNA methylation (5mC) epigenetic mark, other variants of DNA methylation exist such as 5-hydroxymethylcytosine (5hmC), 5-formylcytosine (5fC), and 5-carboxycytosine (5caC) that could be assessed genome wide (Plongthongkum et al., 2014). Each occurs in the active demethylation pathway (Wu and Zhang, 2017), but whereas measurement of 5fC and 5caC has not gained traction, studies of 5hmC are increasing and indicate it is a stable mark (Bachman et al., 2014). Moreover, a unique binding protein (Mbd3) that recognizes 5hmC suggests it has its own epigenetic function (Yildirim et al., 2011). Expanding on this idea, there is evidence indicating that 5hmC may be maintained across DNA replication via a complex of the DNMT1, TET, and UHRF1 proteins (Shen and Zhang, 2013). Thus, we limit this section to the detection and analysis of 5mC and 5hmC.

Bisulfite Sequencing Approaches

Bisulfite treatment of DNA converts cytosine to uracil unless the cytosine is protected by a methyl (5mC) or hydroxymethyl (5hmC) group. This enables the detection of DNA methylation but introduces ambiguity as to whether 5mC or 5hmC is the source of the signal. The gold-standard approach for assessing DNA methylation is whole-genome bisulfite sequencing (WGBS). It enables absolute quantification of methylation levels at base-pair resolution at over 90 % of CpGs in a genome. However, unbiased coverage requires deep sequencing, incurring a higher cost (Ziller et al., 2015). An alternative approach is reduced representation bisulfite sequencing (RRBS), which uses restriction enzymes to cut the DNA (biased to CpG-rich regions). This approach covers 10%–20% of CpGs, reducing the required depth of sequencing and alleviating costs (Table 1). While neither WGBS nor RRBS can distinguish between 5mC and 5hmC, two recently developed protocols reportedly can. Oxidative bisulfite sequencing (oxBS-seq) uses an oxidative step prior to bisulfite treatment, enabling the detection of 5mC alone (Booth et al., 2013). Conversely, Tet-assisted bisulfite sequencing (TAB-seq) uses a glucosylation step that enables the detection of 5hmC alone (Yu et al., 2012). Both approaches provide absolute methylation quantification at base-pair resolution, and both can be modified to use the RRBS protocol after their initial steps.

A common set of software packages can be used to analyze WGBS, RRBS, oxBS-seq, and TAB-seq data. Assessing read quality with FastQC is essential. Next, adapter sequence must be trimmed. For RRBS, 2 bp from the 5′ end of the read is removed because the end-repair step results in a noninformative CpG site. The *Bismark* aligner and *methylation extractor* are commonly used tools for aligning reads to an *in silico* bisulfite-treated reference genome and quantifying absolute methylation levels (Krueger and Andrews, 2011). Bismark can be used for WGBS, RRBS, oxBS-seq, and TAB-seq, where the meaning of "methylation" is dependent on the experiment. The MultiQC tool (Ewels et al., 2016) is a useful aggregator of reports from these software, enabling the assessment of quality, alignment, and methylation levels for all samples.

Toxicoepigenomics studies often test for effects from an exposure (e.g., lead and BPA) that requires testing for differential methylation between groups of samples. Several tools are effective in testing for differential methylation at a CpG resolution and/or at a region

TABLE 1 DNA Methylation Assays and Their Characteristics

Category	Method	Mark	CpG Coverage	Pros	Cons
Bisulfite conversion + sequencing	WGBS	5mC + 5hmC	>90%	• Unbiased coverage • Absolute quantification at CpG resolution	• Expensive to sequence at depth required for DM tests
	RRBS	5mC + 5hmC	10%–20%	• Less expensive than WGBS while targeting many CpGs • Absolute quantification at CpG resolution	• Biased coverage of CpG-rich locations
	oxBS-seq	5mC	>90%	• Unbiased coverage • Differentiates between 5mC and 5hmC • Absolute quantification at CpG resolution	• Difficult protocol owing to high levels of DNA degradation
	TAB-seq	5hmC	>90%		
Antibody pulldown	MeDIP-seq	5mC	60%–90%	• Relatively inexpensive • 5mC/5hmC-specific antibody differentiates the two marks	• Relative quantification at region resolution
	hMeDIP-seq	5hmC	60%–90%		
Bisulfite conversion + array	450K	5mC +5hmC	2%	• Inexpensive, allows for large-scale population studies	• Replaced by EPIC array • Was available for humans only
	EPIC	5mC +5hmC	~3%	• Inexpensive, allows for large-scale population studies • Increased coverage over 450K includes FANTOM5 enhancers and ENCODE open chromatin • Covers >90% of 450K probes	• Lower coverage of CpGs compared with sequencing-based assays • Available for humans only

resolution. The methylSig R package uses a beta-binomial model to test for differences in group methylation levels (Park et al., 2014). MethylSig filters sites by coverage thresholds and C > T SNPs and can take into account local information from neighboring CpGs. It can test for differential methylation at CpGs, in genomic windows, or in user-defined regions. Moreover, methylSig can now test for differential methylation under general experimental designs with covariates by implementing some of the functionality of the DSS R/ Bioconductor package (Park and Wu, 2016), which is increasingly necessary with the reduction of sequencing cost and corresponding increase in experimental design complexity due to larger cohorts.

Antibody Pulldown Followed by Sequencing

Methylated DNA immunoprecipitation (IP) sequencing (MeDIP-seq) and the hydroxymethylation variety (hMeDIP-seq) are IP-based approaches providing region-resolution assessment of DNA methylation and hydroxymethylation using antibodies specific to 5mC and 5hmC, respectively. Such approaches are useful if absolute quantification of methylation is not necessary or if cost is an issue. The analysis of (h)MeDIP-seq is similar to that of ChIP-seq experiments, including the need for control samples (e.g., input DNA or nonspecific antibodies). Sequence quality assessed with FastQC and adapter trimming are standard initial steps. Aligning to the genome with Bowtie 2 is typical, followed by either sample-wise quantification of peaks or group-wise tests for regions of differential methylation.

Sample-wise peak calling for regions of methylation is often performed with MACS2 (Feng et al., 2012) and uses the paired control experiment to determine enrichment against background. Group-wise testing for differentially methylated regions can be performed with PePr (Zhang et al., 2014), which models read counts across replicates and between groups using a negative binomial distribution. If a more general experimental design is called for, the R package methylAction (Bhasin et al., 2016) or csaw (Lun and Smyth, 2016) may be more appropriate.

Array-Based Approaches

For studies containing many samples (enough to make sequencing prohibitively expensive) and where unbiased coverage of the entire genome is less important, array-based technologies such as the Illumina EPIC BeadChIP are viable options (Table 1). An advantage is that it guarantees coverage of the same CpGs, whereas RRBS coverage of sites can vary substantially among samples. R/Bioconductor packages developed for the 450K array have been modified for the new >850,000 CpG site EPIC array. In particular, minfi, ENmix, and ChAMP (Fortin et al., 2017; Morris et al., 2014; Xu et al., 2016) all have functions to display QC measures and perform initial sample and probe filtering. Preprocessing and normalization are the next steps of array-based analysis. New dye bias corrections specific to the EPIC array are implemented in the ENmix and minfi packages. Each has an implementation of quantile normalization (localized differences, small effect) and functional normalization (global differences, large effect). Probe filtering for cross-reactivity (Pidsley et al., 2016) and/or the presence of SNPs (Fortin et al., 2017) is also recommended.

Model building and testing for differentiation with the limma R/Bioconductor package (Ritchie et al., 2015) enable explicit correction for confounders. The ChAMP package has a useful function and visualization (champ.SVD) that correlates any covariates provided with principal components. The ComBat function within the sva R/Bioconductor package (Leek et al., 2017) is also able to adjust for known batches using an empirical Bayesian framework. If users want to determine differential methylation over regions instead of probes, R packages such as DMRcate (Peters et al., 2015) perform tests based on the grouping of probes.

Cell-Type Heterogeneity and Deconvolution

Blood is a widely used surrogate tissue to assess epigenetic changes in toxicoepigenetic studies. However, the cell-type composition of blood may be a confounder in the search for differentially methylated cytosines (Houseman et al., 2012; Liu et al., 2013). There are a variety of algorithms for cell-type deconvolution that can be divided into two classes: reference-based and reference-free (reviewed in (Teschendorff and Relton, 2018)). Most reference-based methods have been developed for array-based technologies such as the 450K arrays, with more methods being updated to work with the EPIC BeadChip (reviewed in (Teschendorff and Zheng, 2017)).

Integration and Contextualization

As discussed, widely used platforms (e.g., WGBS and RRBS) capture a mixture of 5mC and 5hmC. Orthogonal experiments (such as oxBS-seq and TAB-seq) are needed to differentiate the two marks. However, neither oxBS-seq nor TAB-seq has been widely adopted, so some studies have proceeded with hMeDIP-seq as a way to determine regions of 5mC from 5hmC. A recently published pipeline named mint (Cavalcante et al., 2017) enables joint or separate analysis of bisulfite-treated and pull-down methods. In the joint case, an additional integration step gives users lists of genomic regions categorized as 5mC, 5hmC, or a combination.

Common to all genomic investigations is a need to assign genomic regions to genomic annotations. Various tools in the R/Bioconductor ecosystem, such as annotatr (Cavalcante and Sartor, 2017), Goldmine (Bhasin and Ting, 2016), and LOLA (Sheffield and Bock, 2016), offer similar capabilities to annotate genomic regions, while some provide additional summary visualizations (annotatr), and others provide statistical tests for enrichment (Goldmine and LOLA).

BIOINFORMATICS APPROACHES FOR HISTONE MODIFICATIONS

Introduction

The nucleosome, with its core histone octamer and polypeptide tails, is a crucial player in higher-order chromatin organization and gene regulation. Measuring posttranslational modifications (PTMs) to the histone tails, also known as histone modifications (HMs), has turned out to be remarkably fruitful in understanding their role in gene regulation. The primary assay for measuring PTMs is chromatin immunoprecipitation followed by sequencing (ChIP-seq).

Experimental Considerations

Important to the success of a ChIP-seq experiment are the use of control samples, the need for replicates, and a consideration of the sequencing depth. Input DNA (DNA removed before IP), mock IP DNA (DNA from IP without antibodies), or nonspecific IP (such as IgG) are all possibilities for control samples (Park, 2009) to account for background. At least two biological replicates are recommended to ensure that the IP signal is reproducible. The depth of sequencing is also important. As discussed in Park (2009), sequencing more deeply can lead to calling a peak when the underlying enrichment of IP over input is the same but the number of supporting reads is greater. Therefore, there is a point in ChIP-seq experiments where reads saturate. Generally, 20 million uniquely mapped reads are recommended for narrow peaks, while 40 million are recommended for broader peaks (Furey, 2012).

Analysis and Interpretation

As with all sequencing experiments, assessment of read quality with FastQC is an important first step. After reads are assessed to be of sufficient quality, alignment to the genome with a short-read aligner (e.g., BWA or Bowtie 2) allows for downstream analysis. MACS2 (Feng et al., 2012) is a widely used tool to call sample-wise peaks, enrichment of IP signal over background signal within individual samples. However, differential histone modification between groups is often of interest when comparing exposures in toxicoepigenomics (as with differential methylation). As discussed previously, PePr (Zhang et al., 2014) can be used to test for group-wise differences, while csaw (Lun and Smyth, 2016) can be used for more general experimental designs, including with covariates.

Annotation to genomic features after differential peak calling can lend biological insight into the effect of the condition being studied. Tools such as annotatr (Cavalcante and Sartor, 2017) can easily, with minimal programming, give summary visualizations of genomic annotations for each direction of differential binding. Alternate tools include Goldmine (Bhasin and Ting, 2016) and LOLA (Sheffield and Bock, 2016). Moreover, detailed annotation tables enable the investigation of particular genes that may, a priori, be of known biological importance. Gene set enrichment is another functional analysis tool that can lend biological insight into differential peak calling. Tools such as Broad-Enrich (Cavalcante et al., 2014) were designed specifically for functional enrichment of broad genomic regions (e.g., histone PTMs). Such analyses enable investigators to explore beyond known genes and determine higher-level biological effects of differential peaks.

CHROMATIN ACCESSIBILITY ASSAYS

DNA regulation is tightly linked to chromatin accessibility, and open-chromatin sites have been regarded as the primary regulatory elements (John et al., 2011). The positioning of nucleosomes on a genome determines the in vivo availability of DNA binding sites to transcriptional factors (TFs) and transcriptional machinery, controlling DNA-dependent biological processes such as gene transcription and DNA repair, replication, and recombination (Radman-Livaja and Rando, 2010). In the advent of next-generation sequencing (NGS), a massive amount of genome-wide chromatin accessibility data has been generated. The open or protected genomic regions are generally isolated by an enzymatic or chemical method, followed by parallel sequencing. Compared with MNase-seq, which indirectly assays chromatin accessibility by isolating nucleosomes of a genome (Schones et al., 2008), DNase-seq (Song and Crawford, 2010; Thurman et al., 2012) and ATAC-seq (Buenrostro et al., 2013, 2015) are two currently widely used approaches to directly evaluate the accessible genomic locations. ATAC-seq, the most current method, has significantly simplified library preparation allowing low cell input and the ability to study multiple aspects of chromatin architecture simultaneously at high resolution. In the following, we discuss the popular bioinformatics pipelines used for DNase-seq and ATAC-seq data analysis.

Data Quality Control

Most data preprocessing and QC steps are the same for all chromatin accessibility assays, including demultiplexing, alignment to a reference genome, read filtering, and sequencing quality control (QC). The goal is to determine if the sequencing was done with the required depth of coverage and to prepare BAM files for downstream assay-specific analysis. BWA (Li and Durbin, 2009) and Bowtie (Langmead et al., 2009) are two most common tools for alignment. Specifically for ATAC-seq data, special QC steps need to be taken after read alignment: (i) Since the minimum transposition spacing is 38 bp, the mapped fragments below that length are removed (Adey et al., 2010); (ii) the reads mapped to the mitochondrial genome are also discarded as unrelated to the scope of ATAC-seq. The mapped reads are then visualized by constructing composite plots to determine if the experiment is successful. For example, since TSSs are known to be chromatin accessible on average in eukaryotic genomes, the composite plot signal intensity is expected to increase at this feature in both DNase-seq and ATAC-seq data. DeepTools is one of the most comprehensive and user-friendly tools available to generate these kinds of QC plots (Ramirez et al., 2016). ATAC-seq QC can be further performed by estimating the percentage of contaminated mitochondrial reads and by generating "insert size metric plots" using Picard tools (http://broadinstitute.github.io/picard). High-quality ATAC-seq data will have a low percentage of mitochondrial reads, low background, high promoter signal, and a periodic distribution of insert size representing 5–6 nucleosomal arrays along with 10 bp patterns.

Detection of Open Chromatin Regions

Researchers generally apply algorithms developed for ChIP-seq without an input control to detect the enriched DNase hypersensitive sites (DHSs) for both DNase-seq and ATAC-seq data. However, careful adjustment of default settings needs to be taken (Madrigal and Krajewski, 2012). MACS2 (Liu, 2014; Liu et al., 2008), which is a model-based algorithm with wide application in ChIP-seq data analysis, has been integrated as the peak caller in the ENCODE ATAC-seq analysis pipeline (ENCODE, n.d.). PePr is a differential peak caller applicable to both DNase-seq and ATAC-seq data (Zhang et al., 2014) and can also take biological variance into account when calling peaks on replicate samples. Recently, an R package for downstream workflow of defining ALTered Regulatory Elements (ALTRE) using chromatin accessibility data was developed to streamline differential analysis of regulatory elements genome wide (Baskin et al., 2017). ALTRE takes alignment BAM files and peak BED files as input and performs consensus peak filtering, peak annotation, differential peak identification by DESeq2 (Love et al., 2014), visualization track generation, and gene set enrichment analysis (GSEA) by GREAT (McLean et al., 2010).

Estimation of Genomic Footprinting and Transcription Factor Occupancy

Regulatory proteins, such as transcription factors (TFs), protect their binding DNA sequences from nuclease cleavage, resulting in the markedly increased accessibility surrounding their binding sites and over neighboring chromatin (Hesselberth et al., 2009). Therefore, TF binding sites can be inferred within open-chromatin sites by the regions with a steep drop in accessibility. Both high-depth DNase-seq and paired-end ATAC-seq data can be used to systematically reveal the genome-wide cis-regulatory elements, that is, digital genomic footprinting (DGF) (Hesselberth et al., 2009). BinDNase is a supervised method to detect TF footprinting by automatically extracting features from the DNase-seq (or ATAC-seq) data for each TF that maximally discriminate bound and unbound genomic locations (Kahara and Lahdesmaki, 2015). Bivariate genomic footprinting (BaGFoot) is another approach that effectively detects TF activity by capturing two TF-dependent effects on chromatin accessibility, footprinting and motif-flanking accessibility (Baek et al., 2017). BaGFoot is robust to different accessibility assays (DNase-seq and ATAC-seq), all examined peak-calling programs, and to a variety of cut bias correction approaches.

CHROMOSOMAL INTERACTIONS

Increasing evidence has shown that the three-dimensional (3-D) organization of chromosomes has an important role of determining gene expression patterns and biological functions (Bulger and Groudine, 2011; Sanyal et al., 2012). Although the location of potential regulatory elements can easily be mapped genome wide using various chromatin accessibility and ChIP-seq assays (Buenrostro et al., 2013; Hesselberth et al., 2009; Mikkelsen et al., 2007), it is challenging to determine the targeted genes of the regulatory elements because they can reside at large distances from their targets, such as distal enhancers that can bind to gene promoters through chromatin looping (Sanyal et al., 2012).

Overview of 3C-Based Technologies

Chromosome conformation capture (3C) technology is essential to understand the physical interactions between distal regulatory elements and gene promoters (Tolhuis et al., 2002; Stamatoyannopoulos, 2016; Dixon et al., 2012; Nora et al., 2012; Hsieh et al., 2015, 2016). Multiple 3C-derived assays with various scopes and resolutions have been developed, including circular chromosome conformation capture (4C) (Splinter et al., 2012), chromosome conformation capture carbon copy (5C) (van Berkum and Dekker, 2009), Hi-C (Belton et al., 2012), and chromatin interaction analysis with paired-end tag (ChIA-PET) sequencing (Fullwood and Ruan, 2009). All of the approaches share the first basic three steps in library preparation (cross-linking, digestion, and ligation) but vary in the downstream protocols leading to different scales of chromosomal link data: 4C detects contacts between a selected genomic site ("viewpoint") with all other possible genomic fragments (one vs all); 5C concurrently determines interactions among any sequences (many vs many), however suffering lower resolution than 4C, Hi-C, or ChIA-PET due to the limitation of its primer design; Hi-C is the first "all-to-all" method allowing identification of all possible genomic interactions genome wide; ChIA-PET combines chromatin immunoprecipitation (ChIP) with a 3C-based assay to detect all-to-all chromatin interactions formed between sites bound by a chosen DNA-binding protein (Denker and de Laat, 2016; Li et al., 2014; Davies et al., 2017; Schmitt et al., 2016). Appropriate applications of the different methods depend on the specific study aims: 3C and 4C are suitable for the confirmation of small numbers of interactions, while 5C and Hi-C are good for deciphering chromatin 3-D structure, and ChIA-PET offers the highest resolution (from 100 bp to 1 kb) of all-to-all interactions bound by the same protein of interest (Davies et al., 2017; Dekker et al., 2013). Currently, ChIA-PET and Hi-C are frequently used methods to investigate genome-wide chromatin arrangement.

Data Analysis of High-Throughput 3C-Based Chromosomal Interaction Assays

The general data analysis pipeline of 3C-based assays includes adapter trimming, splitting reads at the ligation junction into their compartment fragments, read alignment, filtering out spurious reads (e.g., PCR duplicates, undigested reads, and off-target capture), windowing/binning of data, and statistical analysis (Davies et al., 2017). A characteristic bias in 3C-based approaches is that very strong interactions commonly occur between neighboring fragments due to their close spatial proximity, resulting in background interactions reversely related to the distance between the linked genomic loci. Several analysis methods use distance modeling to determine contacts that are significantly different from background interaction profiles. Mango is the first complete ChIA-PET data analysis pipeline that addresses the distance bias by modeling the likelihood of interactions between genomic loci as a function of both distance and peak depth (Phanstiel et al., 2015). It takes raw fastq files as input and completes all analysis steps to output interactions associated with statistical confidence. More recently, an improved analysis pipeline, ChIA-PET2, has been published (Li et al., 2017). Besides the common analysis steps in Mango, ChIA-PET2 provides comprehensive quality control and allele-specific analysis modules. In contrast to Mango, ChIA-PET2 is applicable to the datasets generated by the new "bridge linker" ChIA-PET protocol for long reads, which requires a special linker trimming algorithm. It outperforms the existing pipelines in terms of sensitivity and reproducibility. Likewise, HiC-Pro is a versatile and flexible pipeline for processing Hi-C data from raw reads to normalized contact maps (symmetrical heat map matrices showing strengths of contact), which can also use phased genotype data to build allele-specific contact maps (Servant et al., 2015). In addition to building genome-wide chromatin interactions, various computational methods have been developed to identify topologically associating domains (TADs) using Hi-C data, such as InsulationScore (Crane et al., 2015), TADbit (Serra et al., 2017), and TADtree (Weinreb and Raphael, 2016). However, due to the lack of a clear biological or operational definition of TAD, different TAD prediction tools were developed based on varied assumptions about TADs such as size distribution, type of Hi-C signal detected, and the presence/absence of overlap and/or nesting, resulting in discordant results (Dali and Blanchette, 2017). Researchers need to take data resolution and specific study goals into consideration when choosing an appropriate tool and parameters.

INTERPRETATION OF EPIGENOMIC DATA

Inferring Biological Functions of Epigenetic Changes

Observed epigenetic alterations of interest can be associated with the deregulation of genes or biological pathways after adjusting for confounders, and thus, gene set enrichment (GSE) testing is often applied to test for significantly enriched gene ontology (GO) terms or pathways in the epigenetic signals. One such tool is missMethyl, an R/Bioconductor package performing GSE for Illumina's HumanMethylation450 or EPIC data (Phipson et al., 2016). This method adjusts for the different number of CpGs associated with each gene. Because other epigenetic features of interest such as DMRs and ChIP-seq peaks are often genomic regions and different genes can have vastly different length regions associated with them, it is critically important to adjust for the gene locus length to avoid bias when applying GSE to ChIP-seq or similar data (Geeleher et al., 2013). GSE approaches specifically tailored for genomic regions with different characteristics include ChIP-Enrich (Welch et al., 2014), Poly-Enrich (http://chip-enrich.med. umich.edu), and Broad-Enrich (Cavalcante et al., 2014) (Table 2). If only epigenetic changes in proximal promoter regions are of interest, then, in many cases, a GSE method based on Fisher's exact test, such as DAVID, is sufficient (da Huang et al., 2009).

Integrative Analysis With Other Multi-Omics Data

Analyzing different types of epigenetic data in conjunction with transcriptomic data allows the construction of epigenetic landscapes that can help elucidate system biological principles underlying diverse phenomena such as cell-type heterogeneity and cancer and therefore improve our understanding of "system epigenomics" (Teschendorff and Relton, 2018). Much effort has been devoted to the integration of epigenome and transcriptome data, which usually requires the preassignment of an epigenetic value to a gene by focusing on a specific predictive region (e.g., nearest TSS, promoter, or first exon) (Jones, 2012; Yang et al., 2014; Schlosberg et al., 2017) (Table 2). A fully Bayesian latent variable model called iClusterBayes can jointly model different types of omics data to identify sample subtypes and relevant omics features (Mo et al., 2018), which is applicable to infer subgroups of individuals exposed to environmental toxins with respect to different biological and/or clinical effects. By using the spatial distribution of the methylation profile over a gene and beyond as a predictor of gene expression, Methylation-based Gene Expression Classification (ME-Class) has been shown to improve the prediction of gene expression by using methylation spatial profiles as features (Schlosberg et al., 2017).

A more powerful system-level integrative approach is to infer regulatory patterns by incorporating the inverse correlation between methylation regulation and TF-binding activity. For example, enhancer linking by methylation/expression relationships (ELMER) is able to identify cancer-specific enhancer-gene networks (Yao et al., 2015b); tracing enhancer networks using epigenetic traits (TENET) (Rhie et al., 2016) further refined the method by identifying tissue-specific enhancer-gene links; and RegNetDriver (Dhingra et al., 2017) constructs tissue-specific regulatory networks by integrating cell-type-specific open-chromatin data with regulatory elements from ENCODE (Consortium et al., 2012) and RMEC (Roadmap Epigenomics et al., 2015). By integrating epigenetic and gene expression data in the context of a gene function network (e.g., protein-protein interaction (PPI) network), the functional epigenetic module (FEM) approach can identify differentially expressed gene modules under epigenetic control in relation to a phenotype of interest (Jiao et al., 2014). Significance-based modules integrating the transcriptome and epigenome (SMITE) further

TABLE 2 Resources and Tools for Interpretation of Epigenomic Data

Name	Description	Web Links	Refs.
GSEA			
missMethyl	Gene ontology and gene set enrichment analysis for methylation BeadChip data	https://www.bioconductor.org/packages/release/bioc/html/missMethyl.html	Phipson et al. (2016)
ChIP-Enrich/Poly-Enrich	Gene ontology and gene set enrichment analysis for ChIP-seq data with narrow peaks (e.g., TF binding sites and H3K4me3)	http://chip-enrich.med.umich.edu	Welch et al. (2014)
Broad-Enrich	Gene ontology and gene set enrichment analysis for broad genomic regions (e.g., ChIP-seq data for histone modifications or copy-number variations)	http://broad-enrich.med.umich.edu	Cavalcante et al. (2014)
Integrative analysis			
ELMER	Enhancer linking by methylation/expression relationships	https://bioconductor.org/packages/release/bioc/html/ELMER.html	Yao et al. (2015a)
TENET	Tracing enhancer networks using epigenetic traits	http://farnhamlab.com/software	Rhie et al. (2016)
RegNetDriver	A framework for integration of genetic and epigenetic alterations with tissue-specific regulatory network	Linux executable and R	Dhingra et al. (2017)
SMITE	Significance-based modules integrating the transcriptome and epigenome	https://bioconductor.org/packages/release/bioc/html/SMITE.html	Wijetunga et al. (2017)
FEM	Functional epigenetic modules (integrative analysis of DNA methylation and gene expression)	https://bioconductor.org/packages/release/bioc/html/FEM.html	Jiao et al. (2014)
MultiassayExperiment	Software for the integration of multiomics experiments in Bioconductor	http://bioconductor.org/packages/release/bioc/html/MultiAssayExperiment.html	
Data visualization			
UCSC Genome Browser	University of California at Santa Cruz (UCSC) Genome Browser	http://genome.ucsc.edu	Kent et al. (2002)

Continued

4. SPECIAL CONSIDERATIONS IN TOXICOEPIGENETICS RESEARCH

TABLE 2   Resources and Tools for Interpretation ofEpigenomic Data—cont'd

Name	Description	Web Links	Refs.
WashU EpiGenome Browser	Washington University Epigenome Browser	https://epigenomegateway.wustl.edu	Zhou et al. (2013)
Juicebox	Cloud-based visualization software for Hi-C data	https://www.aidenlab.org/juicebox/	Durand et al. (2016)
Three-dimensional Genome Browser	Web-based browser that allows users to smoothly explore both published and their own chromatin interaction data	http://3dgenome.org	Wang et al. (2017b)

Supplementary datasets

ENCODE	Public resource of regulatory elements in the human genome that were generated from cell lines by the Encyclopedia of DNA Elements (ENCODE) Consortium	https://www.encodeproject.org	Consortium et al. (2012)
Roadmap Epigenome	Public resource of epigenomic data across hundreds of human cell types and tissues that were generated by US National Institutes of Health (NIH) Roadmap Epigenomics Mapping Consortium	http://www.roadmapepigenomics.org/data/	Roadmap Epigenomics et al. (2015)
TaRGET	Toxicant Exposures and Responses by Genomic and Epigenomic Regulators of Transcription (TaRGET)	https://target.wustl.edu/about.html	Wang et al. (2018)

extend the FEM algorithm without direct integration with a PPI network (Wijetunga et al., 2017). To streamline integrative analyses in the same environment, a Bioconductor package, MultiassayExperiment, was recently developed that simplifies data representation, statistical analysis, and visualization (Ramos et al., 2017).

Data Visualization

Data visualization is an indispensable step in data analysis, especially in multiomics integrative analysis, by making huge amounts of data more accessible and understandable. However, visualizing many different epigenetic marks simultaneously is not trivial due to the complex 2-D and 3-D organization of the marks that can interact with many intermediate genes/proteins along the genome. A number of genome browser tools are publicly available (Nielsen et al., 2010), and the University of California at Santa Cruz (UCSC) Genome Browser (Kent et al., 2002) and the Integrative Genomics Viewer (IGV) (Robinson et al., 2011; Thorvaldsdottir et al., 2013) are two options (Table 2). UCSC Genome Browser provides a spectrum of information on different types of NGS-based OMICS data from ENCODE (Consortium et al., 2012) and other research projects and supports the incorporation of personally generated data. IGV represents another efficient genomic visualization and exploration tool, allowing users to handle large and diverse datasets on a local computer. Users can also compare a variety of personally generated data to publicly available data from ENCODE, The Cancer Genome Atlas (Cancer Genome Atlas Research Network et al., 2013), 1000 Genomes (Genomes Project et al., 2015), etc. In contrast, the WashU EpiGenome Browser displays epigenomic data (Zhou and Wang, 2012; Zhou et al., 2013), allowing users to browse the epigenomes of over 100 cells and tissues that were recently generated by the US National Institutes of Health (NIH) Roadmap Epigenomics Mapping Consortium (Roadmap Epigenomics et al., 2015) and will be soon generated by the Toxicant Exposures and Responses by Genomic and Epigenomic Regulators of Transcription (TaRGET II) Consortium (Wang et al., 2018).

In the context of higher-order chromosome structure, multiple visualization modes are required to view the data where the linear visualization tools cannot meet the need. A variety of robust and informative methods have been developed recently for the interpretation of chromosomal interaction data. For example, Juicebox (Durand et al., 2016; Robinson et al., 2018) employs heat maps to display both intra- and interchromosomal interactions, in which users can zoom into a region of the full contact matrix, and by the embedded loop-finding algorithm, it can show the loop annotations directly on a contact map. The 3-D Genome Browser extends the browser framework to visualize Hi-C data at a particular region by using a rotated local heat map and local arc track, which also allows viewing other types of epigenomic data including virtual 4C, ChIA-PET, and cross cell-type correlation of proximal and distal DHSs (Wang et al., 2017a). The tools of choice for data visualization depend on the nature of the inquiry regarding data type and scale.

FUTURE DIRECTIONS AND CONCLUSIONS

Given the extent of evidence for the important roles epigenetics plays in toxicology and environmental health research, we can expect toxicoepigenomics to develop and expand in the coming years. Researchers in the field can learn much from the cutting-edge bioinformatics approaches being developed in the cancer epigenetic field. Similar to epigenomic studies in cancer, toxicoepigenomics research must deal with heterogeneity in cell-type mixtures, genotypes, and demographic/phenotypic characteristics. However, unlike epigenomic studies in cancer, toxicoepigenomics experimental design and bioinformatics approaches should carefully consider each study's power to detect small effect sizes in relatively few target genes.

These complex issues of experimental design and bioinformatics strategies are currently being considered by researchers in the TaRGET II Consortium, which consists of researchers from seven US institutions and is funded by the National Institute of Environmental Health Sciences (NIEHS). The goal of the TaRGET II Consortium is to determine when and to what extent surrogate tissues (e.g., blood) can be used to detect the effects of environmental exposures in place of more difficult to access target tissues (e.g., lung, liver, or brain). TaRGET II will address these questions using mouse models, with the goal of translating the findings to humans in the future TaRGET III Consortium. The current participating institutions and their exposures are Johns Hopkins (the metal arsenic (As) and fine particulate matter (PM2.5)), Case Western (PM2.5), North Carolina State (the dioxin TCDD), Baylor College of Medicine (the obesogen tributyltin), the University of Pennsylvania (the endocrine-active compound bisphenol A), the University of Chicago (PM2.5), and the University of Michigan (the metal lead (Pb) and the phthalate plasticizer DEHP). Together, in coordination with the Data Coordinating Center (Washington University), they are integrating data from DNA methylation (whole-genome bisulfite sequencing), gene expression (RNA-seq), and chromatin accessibility (ATAC-seq), in addition to epigenomic assays being used by a subset of the institutions (ChIP-seq to measure histone modifications and hMeDIP-seq to assay 5-hydroxymethylation) (Wang et al., 2018).

Moving forward, researchers in the toxicoepigenomics field will be adopting newer techniques, such as single-cell RNA-seq, which allows the measurement of gene expression for individual cells in a complex tissue sample; identification of novel biomarkers for individual cell types; and detection of expression changes for individual cell types. Several analysis pipelines, preprocessing methods, and clustering approaches have already been introduced to aid in single-cell RNA-seq analysis (Bacher and Kendziorski, 2016). Single-cell assays have also been developed for epigenomic assays, including single-cell bisulfite sequencing and single-cell Hi-C (Nagano et al., 2015). These single-cell molecular techniques hold great promise in a field like toxicoepigenomics, where the targeted molecular changes may only occur in a rare cell type, resulting in small, difficult-to-detect changes in bulk samples. Time will tell the extent to which these technologies hold the key to understanding complex molecular effects from environmental exposures; novel bioinformatics methods will play a large role in unlocking that mystery.

References

Adey, A., Morrison, H.G., Asan, Xun, X., Kitzman, J.O., Turner, E.H., Stackhouse, B., AP, M.K., Caruccio, N.C., Zhang, X., Shendure, J., 2010. Rapid, low-input, low-bias construction of shotgun fragment libraries by high-density in vitro transposition. Genome Biol. 11 (12), R119. https://dx.doi.org/10.1186/gb-2010-11-12-r119. 21143862.

Bacher, R., Kendziorski, C., 2016. Design and computational analysis of single-cell RNA-sequencing experiments. Genome Biol. 17, 63. https://dx.doi.org/10.1186/s13059-016-0927-y. 27052890.

Bachman, M., Uribe-Lewis, S., Yang, X., Williams, M., Murrell, A., Balasubramanian, S., 2014. 5-Hydroxymethylcytosine is a predominantly stable DNA modification. Nat. Chem. 6 (12), 1049–1055. https://dx.doi.org/10.1038/nchem.2064. 25411882.

Baek, S., Goldstein, I., Hager, G.L., 2017. Bivariate genomic footprinting detects changes in transcription factor activity. Cell Rep. 19 (8), 1710–1722. https://dx.doi.org/10.1016/j.celrep.2017.05.003. 28538187.

Barres, R., Zierath, J.R., 2016. The role of diet and exercise in the transgenerational epigenetic landscape of T2DM. Nat. Rev. Endocrinol. 12 (8), 441–451. https://dx.doi.org/10.1038/nrendo.2016.87. 27312865.

Baskin, E., Farouni, R., Mathe, E.A., 2017. ALTRE: workflow for defining ALTered regulatory elements using chromatin accessibility data. Bioinformatics 33 (16), 2609. https://dx.doi.org/10.1093/bioinformatics/btx386. 28666331.

Belton, J.M., McCord, R.P., Gibcus, J.H., Naumova, N., Zhan, Y., Dekker, J., 2012. Hi-C: a comprehensive technique to capture the conformation of genomes. Methods 58 (3), 268–276. https://dx.doi.org/10.1016/j.ymeth.2012.05.001. 22652625.

Bhasin, J.M., Hu, B., Ting, A.H., 2016. MethylAction: detecting differentially methylated regions that distinguish biological subtypes. Nucleic Acids Res. 44 (1), 106–116. https://dx.doi.org/10.1093/nar/gkv1461. 26673711.

Bhasin, J.M., Ting, A.H., 2016. Goldmine integrates information placing genomic ranges into meaningful biological contexts. Nucleic Acids Res. 44 (12), 5550–5556. https://dx.doi.org/10.1093/nar/gkw477. 27257071.

Booth, M.J., Ost, T.W.B., Beraldi, D., Bell, N.M., Branco, M.R., Reik, W., Balasubramanian, S., 2013. Oxidative bisulfite sequencing of 5-methylcytosine and 5-hydroxymethylcytosine. Nat. Protocols 8 (10), 1841–1851. https://dx.doi.org/10.1038/nprot.2013.115. 24008380.

Boyle, A.P., Hong, E.L., Hariharan, M., Cheng, Y., Schaub, M.A., Kasowski, M., Karczewski, K.J., Park, J., Hitz, B.C., Weng, S., Cherry, J.M., Snyder, M., 2012. Annotation of functional variation in personal genomes using RegulomeDB. Genome Res. 22 (9), 1790–1797. https://dx.doi.org/10.1101/gr.137323.112. 22955989.

Breton, C.V., Marsit, C.J., Faustman, E., Nadeau, K., Goodrich, J.M., Dolinoy, D.C., Herbstman, J., Holland, N., JM, L.S., Schmidt, R., Yousefi, P., Perera, F., Joubert, B.R., Wiemels, J., Taylor, M., Yang, I.V., Chen, R., Hew, K.M., Freeland, D.M., Miller, R., Murphy, S.K., 2017. Small-Magnitude Effect Sizes in Epigenetic End Points are Important in Children's Environmental Health Studies: The Children's Environmental Health and Disease Prevention Research Center's Epigenetics Working Group. Environ. Health Perspect. 125 (4), 511–526. https://dx.doi.org/10.1289/EHP595. 28362264.

Buenrostro, J.D., Giresi, P.G., Zaba, L.C., Chang, H.Y., Greenleaf, W.J., 2013. Transposition of native chromatin for fast and sensitive epigenomic profiling of open chromatin, DNA-binding proteins and nucleosome position. Nat. Methods 10 (12), 1213–1218. https://dx.doi.org/10.1038/nmeth.2688. 24097267.

Buenrostro, J.D., Wu, B., Chang, H.Y., Greenleaf, W.J., 2015. ATAC-seq: a method for assaying chromatin accessibility genome-wide. Curr Protoc Mol Biol. 109, 21.91-9, https://doi.org/10.1002/0471142727.mb2129s10925559105.

Bulger, M., Groudine, M., 2011. Functional and mechanistic diversity of distal transcription enhancers. Cell 144 (3), 327–339. https://dx.doi.org/10.1016/j.cell.2011.01.024. 21295696.

Cancer Genome Atlas Research Network, Weinstein, J.N., Collisson, E.A., Mills, G.B., Shaw, K.R., Ozenberger, B.A., Ellrott, K., Shmulevich, I., Sander, C., Stuart, J.M., 2013. The Cancer Genome Atlas Pan-Cancer analysis project. Nat Genet. 45 (10), 1113–1120. https://dx.doi.org/10.1038/ng.2764. 24071849.

Cavalcante, R.G., Lee, C., Welch, R.P., Patil, S., Weymouth, T., Scott, L.J., Sartor, M.A., 2014. Broad-Enrich: functional interpretation of large sets of broad genomic regions. Bioinformatics 30 (17), i393–i400. https://dx.doi.org/10.1093/bioinformatics/btu444. 25161225.

Cavalcante, R.G., Patil, S., Park, Y., Rozek, L.S., Sartor, M.A., 2017. Integrating DNA methylation and hydroxymethylation data with the mint pipeline. Cancer Res. 77 (21), e27–e30. https://dx.doi.org/10.1158/0008-5472.CAN-17-0330. 29092933.

Cavalcante, R.G., Sartor, M.A., 2017. annotatr: genomic regions in context. Bioinformatics 33 (15), 2381–2383. https://dx.doi.org/10.1093/bioinformatics/btx183. 28369316.

Clark, S.J., Lee, H.J., Smallwood, S.A., Kelsey, G., Reik, W., 2016. Single-cell epigenomics: powerful new methods for understanding gene regulation and cell identity. Genome Biol. 17, 72. https://dx.doi.org/10.1186/s13059-016-0944-x. 27091476.

Consortium, E.P., Dunham, I., Kundaje, A., Aldred, S.F., Collins, P.J., Davis, C.A., Doyle, F., Epstein, C.B., Frietze, S., Harrow, J., Kaul, R., Khatun, J., Lajoie, B.R., Landt, S.G., Lee, B.K., Pauli, F., Rosenbloom, K.R., Sabo, P., Safi, A., Sanyal, A., Shoresh, N., Simon, J.M., Song, L., Trinklein, N.D., Altshuler, R.C., Birney, E., Brown, J.B., Cheng, C., Djebali, S., Dong, X., Dunham, I., Ernst, J., Furey, T.S., Gerstein, M., Giardine, B., Greven, M., Hardison, R.C., Harris, R.S., Herrero, J., Hoffman, M.M., Iyer, S., Kelllis, M., Khatun, J., Kheradpour, P., Kundaje, A., Lassmann, T., Li, Q., Lin, X., Marinov, G.K., Merkel, A., Mortazavi, A., Parker, S.C., Reddy, T.E., Rozowsky, J., Schlesinger, F., Thurman, R.E., Wang, J., Ward, L.D., Whitfield, T.W., Wilder, S.P., Wu, W., Xi, H.S., Yip, K.Y., Zhuang, J., Bernstein, B.E., Birney, E., Dunham, I., Green, E.D., Gunter, C., Snyder, M., Pazin, M.J., Lowdon, R.F., Dillon, L.A., Adams, L.B., Kelly, C.J., Zhang, J., Wexler, J.R., Green, E.D., Good, P.J., Feingold, E.A., Bernstein, B.E., Birney, E., Crawford, G.E., Dekker, J., Elinitski, L., Farnham, P.J., Gerstein, M., Giddings, M.C., Gingeras, T.R., Green, E.D., Guigo, R., Hardison, R.C., Hubbard, T.J., Kellis, M., Kent, W.J., Lieb, J.D., Margulies, E.H., Myers, R.M., Snyder, M., Starnatoyannopoulos, J.A., Tennebaum, S.A., Weng, Z.,

White, K.P., Wold, B., Khatun, J., Yu, Y., Wrobel, J., Risk, B.A., Gunawardena, H.P., Kuiper, H.C., Maier, C.W., Xie, L., Chen, X., Giddings, M.C., Bernstein, B.E., Epstein, C.B., Shoresh, N., Ernst, J., Kheradpour, P., Mikkelsen, T.S., Gillespie, S., Goren, A., Ram, O., Zhang, X., Wang, L., Issner, R., Coyne, M.J., Durham, T., Ku, M., Truong, T., Ward, L.D., Altshuler, R.C., Eaton, M.L., Kellis, M., Djebali, S., Davis, C.A., Merkel, A., Dobin, A., Lassmann, T., Mortazavi, A., Tanzer, A., Lagarde, J., Lin, W., Schlesinger, F., Xue, C., Marinov, G.K., Khatun, J., Williams, B.A., Zaleski, C., Rozowsky, J., Roder, M., Kokocinski, F., Abdelhamid, R.F., Alioto, T., Antoshechkin, I., Baer, M.T., Batut, P., Bell, I., Bell, K., Chakrabortty, S., Chen, X., Chrast, J., Curado, J., Derrien, T., Drenkow, J., Dumais, E., Dumais, J., Duttagupta, R., Fastuca, M., Fejes-Toth, K., Ferreira, P., Foissac, S., Fullwood, M.J., Gao, H., Gonzalez, D., Gordon, A., Gunawardena, H.P., Howald, C., Jha, S., Johnson, R., Kapranov, P., King, B., Kingswood, C., Li, G., Luo, O.J., Park, E., Preall, J.B., Presaud, K., Ribeca, P., Risk, B.A., Robyr, D., Ruan, X., Sammeth, M., Sandu, K.S., Schaeffer, L., See, L.H., Shahab, A., Skancke, J., Suzuki, A.M., Takahashi, H., Tilgner, H., Trout, D., Walters, N., Wang, H., Wrobel, J., Yu, Y., Hayashizaki, Y., Harrow, J., Gerstein, M., Hubbard, T.J., Reymond, A., Antonarakis, S.E., Hannon, G.J., Giddings, M.C., Ruan, Y., Wold, B., Carninci, P., Guigo, R., Gingeras, T.R., Rosenbloom, K.R., Sloan, C.A., Learned, K., Malladi, V.S., Wong, M.C., Barber, G.P., Cline, M.S., Dreszer, T.R., Heitner, S.G., Karolchik, D., Kent, W.J., Kirkup, V.M., Meyer, L.R., Long, J.C., Maddren, M., Raney, B.J., Furey, T.S., Song, L., Grasfeder, L.L., Giresi, P.G., Lee, B.K., Battenhouse, A., Sheffield, N.C., Simon, J.M., Showers, K.A., Safi, A., London, D., Bhinge, A.A., Shestak, C., Schaner, M.R., Kim, S.K., Zhang, Z.Z., Mieczkowski, P.A., Mieczkowska, J.O., Liu, Z., RM, M.D., Ni, Y., Rashid, N.U., Kim, M.J., Adar, S., Zhang, Z., Wang, T., Winter, D., Keefe, D., Birney, E., Iyer, V.R., Lieb, J.D., Crawford, G.E., Li, G., Sandhu, K.S., Zheng, M., Wang, P., Luo, O.J., Shahab, A., Fullwood, M.J., Ruan, X., Ruan, Y., Myers, R.M., Pauli, F., Williams, B.A., Gertz, J., Marinov, G.K., Reddy, T.E., Vielmetter, J., Partridge, E.C., Trout, D., Varley, K.E., Gasper, C., Bansal, A., Pepke, S., Jain, P., Amrhein, H., Bowling, K.M., Anaya, M., Cross, M.K., King, B., Muratet, M.A., Antoshechkin, I., Newberry, K.M., McCue, K., Nesmith, A.S., Fisher-Aylor, K.I., Pusey, B., De Salvo, G., Parker, S.L., Balasubramanian, S., Davis, N.S., Meadows, S.K., Eggleston, T., Gunter, C., Newberry, J.S., Levy, S.E., Absher, D.M., Mortazavi, A., Wong, W.H., Wold, B., Blow, M.J., Visel, A., Pennachio, L.A., Elnitski, L., Margulies, E.H., Parker, S.C., Petrykowska, H.M., Abyzov, A., Aken, B., Barrell, D., Barson, G., Berry, A., Bignell, A., Boychenko, V., Bussotti, G., Chrast, J., Davidson, C., Derrien, T., Despacio-Reyes, G., Diekhans, M., Ezkurdia, I., Frankish, A., Gilbert, J., Gonzalez, J.M., Griffiths, E., Harte, R., Hendrix, D.A., Howald, C., Hunt, T., Jungreis, I., Kay, M., Khurana, E., Kokocinski, F., Leng, J., Lin, M.F., Loveland, J., Lu, Z., Manthravadi, D., Mariotti, M., Mudge, J., Mukherjee, G., Notredame, C., Pei, B., Rodriguez, J.M., Saunders, G., Sboner, A., Searle, S., Sisu, C., Snow, C., Steward, C., Tanzer, A., Tapanari, E., Tress, M.L., van Baren, M.J., Walters, N., Washieti, S., Wilming, L., Zadissa, A., Zhengdong, Z., Brent, M., Haussler, D., Kellis, M., Valencia, A., Gerstein, M., Raymond, A., Guigo, R., Harrow, J., Hubbard, T.J., Landt, S.G., Frietze, S., Abyzov, A., Addleman, N., Alexander, R.P., Auerbach, R.K., Balasubramanian, S., Bettinger, K., Bhardwaj, N., Boyle, A.P., Cao, A.R., Cayting, P., Charos, A., Cheng, Y., Cheng, C., Eastman, C., Euskirchen, G., Fleming, J.D., Grubert, F., Habegger, L., Hariharan, M., Harmanci, A., Iyenger, S., Jin, V.X., Karczewski, K.J., Kasowski, M., Lacroute, P., Lam, H., Larnarre-Vincent, N., Leng, J., Lian, J., Lindahl-Allen, M., Min, R., Miotto, B., Monahan, H., Moqtaderi, Z., Mu, X.J., O'Geen, H., Ouyang, Z., Patacsil, D., Pei, B., Raha, D., Ramirez, L., Reed, B., Rozowsky, J., Sboner, A., Shi, M., Sisu, C., Slifer, T., Witt, H., Wu, L., Xu, X., Yan, K.K., Yang, X., Yip, K.Y., Zhang, Z., Struhl, K., Weissman, S.M., Gerstein, M., Farnham, P.J., Snyder, M., Tenebaum, S.A., Penalva, L.O., Doyle, F., Karmakar, S., Landt, S.G., Bhanvadia, R.R., Choudhury, A., Domanus, M., Ma, L., Moran, J., Patacsil, D., Slifer, T., Victorsen, A., Yang, X., Snyder, M., White, K.P., Auer, T., Centarin, L., Eichenlaub, M., Gruhl, F., Heerman, S., Hoeckendorf, B., Inoue, D., Kellner, T., Kirchmaier, S., Mueller, C., Reinhardt, R., Schertel, L., Schneider, S., Sinn, R., Wittbrodt, B., Wittbrodt, J., Weng, Z., Whitfield, T.W., Wang, J., Collins, P.J., Aldred, S.F., Trinklein, N.D., Partridge, E.C., Myers, R.M., Dekker, J., Jain, G., Lajoie, B.R., Sanyal, A., Balasundaram, G., Bates, D.L., Byron, R., Canfield, T.K., Diegel, M.J., Dunn, D., Ebersol, A.K., Ebersol, A.K., Frum, T., Garg, K., Gist, E., Hansen, R.S., Boatman, L., Haugen, E., Humbert, R., Jain, G., Johnson, A.K., Johnson, E.M., Kutyavin, T.M., Lajoie, B.R., Lee, K., Lotakis, D., Maurano, M.T., Neph, S.J., Neri, F.V., Nguyen, E.D., Qu, H., Reynolds, A.P., Roach, V., Rynes, E., Sabo, P., Sanchez, M.E., Sandstrom, R.S., Sanyal, A., Shafer, A.O., Stergachis, A.B., Thomas, S., Thurman, R.E., Vernot, B., Vierstra, J., Vong, S., Wang, H., Weaver, M.A., Yan, Y., Zhang, M., Akey, J.A., Bender, M., Dorschner, M.O., Groudine, M., MJ, M.C., Navas, P., Stamatoyannopoulos, G., Kaul, R., Dekker, J., Stamatoyannopoulos, J.A., Dunham, I., Beal, K., Brazma, A., Flicek, P., Herrero, J., Johnson, N., Keefe, D., Lukk, M., Luscombe, N.M., Sobral, D., Vaquerizas, J.M., Wilder, S.P., Batzoglou, S., Sidow, A., Hussami, N.,

Kyriazopoulou-Panagiotopoulou, S., Libbrecht, M.W., Schaub, M.A., Kundaje, A., Hardison, R.C., Miller, W., Giardine, B., Harris, R.S., Wu, W., Bickel, P.J., Banfai, B., Boley, N.P., Brown, J.B., Huang, H., Li, Q., Li, J.J., Noble, W.S., Bilmes, J.A., Buske, O.J., Hoffman, M.M., Sahu, A.O., Kharchenko, P.V., Park, P.J., Baker, D., Taylor, J., Weng, Z., Iyer, S., Dong, X., Greven, M., Lin, X., Wang, J., Xi, H.S., Zhuang, J., Gerstein, M., Alexander, R.P., Balasubramanian, S., Cheng, C., Harmanci, A., Lochovsky, L., Min, R., Mu, X.J., Rozowsky, J., Yan, K.K., Yip, K.Y., Birney, E., 2012. An integrated encyclopedia of DNA elements in the human genome. Nature 489 (7414), 57–74. https://dx.doi.org/10.1038/nature11247. 22955616.

Crane, E., Bian, Q., McCord, R.P., Lajoie, B.R., Wheeler, B.S., Ralston, E.J., Uzawa, S., Dekker, J., Meyer, B.J., 2015. Condensin-driven remodelling of X chromosome topology during dosage compensation. Nature 523 (7559), 240–244. https://dx.doi.org/10.1038/nature14450. 26030525.

da Huang, W., Sherman, B.T., Lempicki, R.A., 2009. Systematic and integrative analysis of large gene lists using DAVID bioinformatics resources. Nat. Protocols. 4 (1), 44–57. https://dx.doi.org/10.1038/nprot.2008.211. 19131956.

Dali, R., Blanchette, M., 2017. A critical assessment of topologically associating domain prediction tools. Nucleic Acids Res. 45 (6), 2994–3005. https://dx.doi.org/10.1093/nar/gkx145. 28334773.

Davies, J.O., Oudelaar, A.M., Higgs, D.R., Hughes, J.R., 2017. How best to identify chromosomal interactions: a comparison of approaches. Nat. Methods 14 (2), 125–134. https://dx.doi.org/10.1038/nmeth.4146. 28139673.

Dekker, J., Marti-Renom, M.A., Mirny, L.A., 2013. Exploring the three-dimensional organization of genomes: interpreting chromatin interaction data. Nat. Rev. Genet. 14 (6), 390–403. https://dx.doi.org/10.1038/nrg3454. 23657480.

Denker, A., de Laat, W., 2016. The second decade of 3C technologies: detailed insights into nuclear organization. Genes Dev. 30 (12), 1357–1382. https://dx.doi.org/10.1101/gad.281964.116. 27340173.

Dhingra, P., Martinez-Fundichely, A., Berger, A., Huang, F.W., Forbes, A.N., Liu, E.M., Liu, D., Sboner, A., Tamayo, P., Rickman, D.S., Rubin, M.A., Khurana, E., 2017. Identification of novel prostate cancer drivers using RegNetDriver: a framework for integration of genetic and epigenetic alterations with tissue-specific regulatory network. Genome Biol. 18 (1), 141. https://dx.doi.org/10.1186/s13059-017-1266-3. 28750683.

Dixon, J.R., Selvaraj, S., Yue, F., Kim, A., Li, Y., Shen, Y., Hu, M., Liu, J.S., Ren, B., 2012. Topological domains in mammalian genomes identified by analysis of chromatin interactions. Nature 485 (7398), 376–380. https://dx.doi.org/10.1038/nature11082. 22495300.

Durand, N.C., Robinson, J.T., Shamim, M.S., Machol, I., Mesirov, J.P., Lander, E.S., Aiden, E.L., 2016. Juicebox provides a visualization system for Hi-C contact maps with unlimited zoom. Cell Syst. 3 (1), 99–101. https://dx.doi.org/10.1016/j.cels.2015.07.012. 27467250.

ENCODE. n.d. ATAC-seq Data Standards and Prototype Processing Pipeline. Available from: https://www.encodeproject.org/atac-seq/.

Ewels, P., Magnusson, M., Lundin, S., Käller, M., 2016. MultiQC: summarize analysis results for multiple tools and samples in a single report. Bioinformatics 32 (19), 3047–3048. https://dx.doi.org/10.1093/bioinformatics/btw354. 27312411.

Feng, J., Liu, T., Qin, B., Zhang, Y., Liu, X.S., 2012. Identifying ChIP-seq enrichment using MACS. Nat. Protocols 7 (9), 1728–1740. https://dx.doi.org/10.1038/nprot.2012.101. 22936215.

Fortin, J.-P., Triche, T.J., Hansen, K.D., 2017. Preprocessing, normalization and integration of the Illumina HumanMethylationEPIC array with minfi. Bioinformatics 33 (4), 558–560. https://dx.doi.org/10.1093/bioinformatics/btw691. 28035024.

Fullwood, M.J., Ruan, Y., 2009. ChIP-based methods for the identification of long-range chromatin interactions. J. Cell Biochem. 107 (1), 30–39. https://dx.doi.org/10.1002/jcb.22116. 19247990.

Furey, T.S., 2012. ChIP–seq and beyond: new and improved methodologies to detect and characterize protein–DNA interactions. Nat. Rev. Genet. 13 (12), 1–13. https://dx.doi.org/10.1038/nrg3306.

Geeleher, P., Hartnett, L., Egan, L.J., Golden, A., Raja Ali, R.A., Seoighe, C., 2013. Gene-set analysis is severely biased when applied to genome-wide methylation data. Bioinformatics 29 (15), 1851–1857. https://dx.doi.org/10.1093/bioinformatics/btt311. 23732277.

Genomes Project, C., Auton, A., Brooks, L.D., Durbin, R.M., Garrison, E.P., Kang, H.M., Korbel, J.O., Marchini, J.L., McCarthy, S., McVean, G.A., Abecasis, G.R., 2015. A global reference for human genetic variation. Nature 526 (7571), 68–74. https://dx.doi.org/10.1038/nature15393. 26432245.

Gore, A.C., Chappell, V.A., Fenton, S.E., Flaws, J.A., Nadal, A., Prins, G.S., Toppari, J., Zoeller, R.T., 2015. EDC-2: the endocrine society's second scientific statement on endocrine-disrupting chemicals. Endocrine Rev. 36 (6), E1–E150. https://dx.doi.org/10.1210/er.2015-1010. 26544531.

Greco, C.M., Kunderfranco, P., Rubino, M., Larcher, V., Carullo, P., Anselmo, A., Kurz, K., Carell, T., Angius, A., Latronico, M.V., Papait, R., Condorelli, G., 2016. DNA hydroxymethylation controls cardiomyocyte gene expression in development and hypertrophy. Nat. Commun. 7, 12418. https://dx.doi.org/10.1038/ncomms12418. 27489048.

He, B., Chen, C., Teng, L., Tan, K., 2014. Global view of enhancer-promoter interactome in human cells. Proc. Natl. Acad. Sci. U. S. A. 111 (21), E2191–E2199. https://dx.doi.org/10.1073/pnas.1320308111. 24821768.

Hesselberth, J.R., Chen, X., Zhang, Z., Sabo, P.J., Sandstrom, R., Reynolds, A.P., Thurman, R.E., Neph, S., Kuehn, M.S., Noble, W.S., Fields, S., Stamatoyannopoulos, J.A., 2009. Global mapping of protein-DNA interactions in vivo by digital genomic footprinting. Nat. Methods 6 (4), 283–289. https://dx.doi.org/10.1038/nmeth.1313. 19305407.

Houseman, E.A., Accomando, W.P., Koestler, D.C., Christensen, B.C., Marsit, C.J., Nelson, H.H., Wiencke, J.K., Kelsey, K.T., 2012. DNA methylation arrays as surrogate measures of cell mixture distribution. BMC Bioinform. 13 (1), 86. https://dx.doi.org/10.1186/1471-2105-13-86. 22568884.

Hsieh, T.H., Weiner, A., Lajoie, B., Dekker, J., Friedman, N., Rando, O.J., 2015. Mapping nucleosome resolution chromosome folding in yeast by micro-C. Cell 162 (1), 108–119. https://dx.doi.org/10.1016/j.cell.2015.05.048. 26119342.

Hsieh, T.S., Fudenberg, G., Goloborodko, A., Rando, O.J., 2016. Micro-C XL: assaying chromosome conformation from the nucleosome to the entire genome. Nat. Methods 13 (12), 1009–1011. https://dx.doi.org/10.1038/nmeth.4025. 27723753.

Jenkinson, G., Abante, J., Feinberg, A.P., Goutsias, J., 2018. An information-theoretic approach to the modeling and analysis of whole-genome bisulfite sequencing data. BMC Bioinform. 19 (1), 87. https://dx.doi.org/10.1186/s12859-018-2086-5. 29514626.

Jenkinson, G., Pujadas, E., Goutsias, J., Feinberg, A.P., 2017. Potential energy landscapes identify the information-theoretic nature of the epigenome. Nat. Genet. 49 (5), 719–729. https://dx.doi.org/10.1038/ng.3811. 28346445.

Jiao, Y., Widschwendter, M., Teschendorff, A.E., 2014. A systems-level integrative framework for genome-wide DNA methylation and gene expression data identifies differential gene expression modules under epigenetic control. Bioinformatics 30 (16), 2360–2366. Epub 2014/05/06, https://doi.org/10.1093/bioinformatics/btu31624794928.

John, S., Sabo, P.J., Thurman, R.E., Sung, M.H., Biddie, S.C., Johnson, T.A., Hager, G.L., Stamatoyannopoulos, J.A., 2011. Chromatin accessibility pre-determines glucocorticoid receptor binding patterns. Nat Genet. 43 (3), 264–268. https://dx.doi.org/10.1038/ng.759. 21258342.

Jones, P.A., 2012. Functions of DNA methylation: islands, start sites, gene bodies and beyond. Nat. Rev. Genet. 13 (7), 484–492. Epub 2012/05/30https://doi.org/10.1038/nrg323022641018.

Kahara, J., Lahdesmaki, H., 2015. BinDNase: a discriminatory approach for transcription factor binding prediction using DNase I hypersensitivity data. Bioinformatics 31 (17), 2852–2859. https://dx.doi.org/10.1093/bioinformatics/btv294. 25957350.

Kent, W.J., Sugnet, C.W., Furey, T.S., Roskin, K.M., Pringle, T.H., Zahler, A.M., Haussler, D., 2002. The human genome browser at UCSC. Genome Res. 12 (6), 996–1006. https://dx.doi.org/10.1101/gr.229102. 12045153.

Kim, J.H., Sartor, M.A., Rozek, L.S., Faulk, C., Anderson, O.S., Jones, T.R., Nahar, M.S., Dolinoy, D.C., 2014. Perinatal bisphenol A exposure promotes dose-dependent alterations of the mouse methylome. BMC Genomics 15, 30. https://dx.doi.org/10.1186/1471-2164-15-30. 24433282.

Koenecke, N., Johnston, J., Gaertner, B., Natarajan, M., Zeitlinger, J., 2016. Genome-wide identification of Drosophila dorso-ventral enhancers by differential histone acetylation analysis. Genome Biol. 17 (1), 196. https://dx.doi.org/10.1186/s13059-016-1057-2. 27678375.

Krueger, F., Andrews, S.R., 2011. Bismark: a flexible aligner and methylation caller for Bisulfite-Seq applications. Bioinformatics 27 (11), 1571–1572. https://dx.doi.org/10.1093/bioinformatics/btr167. 21493656.

Langmead, B., Trapnell, C., Pop, M., Salzberg, S.L., 2009. Ultrafast and memory-efficient alignment of short DNA sequences to the human genome. Genome Biol. 10 (3), R25. https://dx.doi.org/10.1186/gb-2009-10-3-r25. 19261174.

Leek, J.T., Johnson, W.E., Parker, H.S., Fertig, E.J., Jaffe, A.E., Storey, J.D., Zhang, Y., Torres, L.C., 2017. svg: Surrogate Variable Analysis. R package. version 3.26.0.

Li, G., Cai, L., Chang, H., Hong, P., Zhou, Q., Kulakova, E.V., Kolchanov, N.A., Ruan, Y., 2014. Chromatin Interaction Analysis with Paired-End Tag (ChIA-PET) sequencing technology and application. BMC Genomics 15 (Suppl 12), S11. https://dx.doi.org/10.1186/1471-2164-15-S12-S11. 25563301.

Li, G., Chen, Y., Snyder, M.P., Zhang, M.Q., 2017. ChIA-PET2: a versatile and flexible pipeline for ChIA-PET data analysis. Nucleic Acids Res. 45 (1), e4. https://dx.doi.org/10.1093/nar/gkw809. 27625391.

Li, H., Durbin, R., 2009. Fast and accurate short read alignment with Burrows-Wheeler transform. Bioinformatics 25 (14), 1754–1760. https://dx.doi.org/10.1093/bioinformatics/btp324. 19451168.

Lin, D.C., Dinh, H.Q., Xie, J.J., Mayakonda, A., Silva, T.C., Jiang, Y.Y., Ding, L.W., He, J.Z., Xu, X.E., Hao, J.J., Wang, M.R., Li, C., Xu, L.Y., Li, E.M., Berman, B.P., Phillip, K.H., 2017. Identification of distinct mutational patterns and new driver genes in oesophageal squamous cell carcinomas and adenocarcinomas. Gut.. https://dx.doi.org/10.1136/gutjnl-2017-31460728860350.

Liu, T., 2014. Use model-based Analysis of ChIP-Seq (MACS) to analyze short reads generated by sequencing protein-DNA interactions in embryonic stem cells. Methods Mol. Biol. 1150, 81–95. https://dx.doi.org/10.1007/978-1-4939-0512-6_4. 24743991.

Liu, Y., Aryee, M.J., Padyukov, L., Fallin, M.D., Hesselberg, E., Runarsson, A., Reinius, L., Acevedo, N., Taub, M., Ronninger, M., Shchetynsky, K., Scheynius, A., Kere, J., Alfredsson, L., Klareskog, L., Ekström, T.J., Feinberg, A.P., 2013. Epigenome-wide association data implicate DNA methylation as an intermediary of genetic risk in rheumatoid arthritis. Nat. Biotechnol. 31 (2), 142–147. https://dx.doi.org/10.1038/nbt.2487. 23334450.

Love, M.I., Huber, W., Anders, S., 2014. Moderated estimation of fold change and dispersion for RNA-seq data with DESeq2. Genome Biol. 15 (12), 550. https://dx.doi.org/10.1186/s13059-014-0550-8. 25516281.

Lun, A.T.L., Smyth, G.K., 2016. csaw: a Bioconductor package for differential binding analysis of ChIP-seq data using sliding windows. Nucleic Acids Res.. 44 (5) e45-e, https://doi.org/10.1093/nar/gkv119126578583.

Madrigal, P., Krajewski, P., 2012. Current bioinformatic approaches to identify DNase I hypersensitive sites and genomic footprints from DNase-seq data. Front Genet. 3, 230. https://dx.doi.org/10.3389/fgene.2012.00230. 23118738.

McLean, C.Y., Bristor, D., Hiller, M., Clarke, S.L., Schaar, B.T., Lowe, C.B., Wenger, A.M., Bejerano, G., 2010. GREAT improves functional interpretation of cis-regulatory regions. Nat. Biotechnol. 28 (5), 495–501. https://dx.doi.org/10.1038/nbt.1630. 20436461.

Meldi, K.M., Figueroa, M.E., 2015. Cytosine modifications in myeloid malignancies. Pharmacol. Ther. 152, 42–53. https://dx.doi.org/10.1016/j.pharmthera.2015.05.002. 25956466.

Messier, T.L., Gordon, J.A., Boyd, J.R., Tye, C.E., Browne, G., Stein, J.L., Lian, J.B., Stein, G.S., 2016. Histone H3 lysine 4 acetylation and methylation dynamics define breast cancer subtypes. Oncotarget 7 (5), 5094–5109. https://dx.doi.org/10.18632/oncotarget.6922. 26783963.

Mikkelsen, T.S., Ku, M., Jaffe, D.B., Issac, B., Lieberman, E., Giannoukos, G., Alvarez, P., Brockman, W., Kim, T.K., Koche, R.P., Lee, W., Mendenhall, E., O'Donovan, A., Presser, A., Russ, C., Xie, X., Meissner, A., Wernig, M., Jaenisch, R., Nusbaum, C., Lander, E.S., Bernstein, B.E., 2007. Genome-wide maps of chromatin state in pluripotent and lineage-committed cells. Nature 448 (7153), 553–560. https://dx.doi.org/10.1038/nature06008.

Mo, Q., Shen, R., Guo, C., Vannucci, M., Chan, K.S., Hilsenbeck, S.G., 2018. A fully Bayesian latent variable model for integrative clustering analysis of multi-type omics data. Biostatistics 19 (1), 71–86. https://dx.doi.org/10.1093/biostatistics/kxx017. 28541380.

Mora, A., Sandve, G.K., Gabrielsen, O.S., Eskeland, R., 2016. In the loop: promoter-enhancer interactions and bioinformatics. Briefings Bioinform. 17 (6), 980–995. https://dx.doi.org/10.1093/bib/bbv097. 26586731.

Morris, T.J., Butcher, L.M., Feber, A., Teschendorff, A.E., Chakravarthy, A.R., Wojdacz, T.K., Beck, S., 2014. ChAMP: 450k Chip analysis methylation pipeline. Bioinformatics 30 (3), 428–430. https://dx.doi.org/10.1093/bioinformatics/btt684. 24336642.

Nagano, T., Lubling, Y., Yaffe, E., Wingett, S.W., Dean, W., Tanay, A., Fraser, P., 2015. Single-cell Hi-C for genome-wide detection of chromatin interactions that occur simultaneously in a single cell. Nat. Protocols 10 (12), 1986–2003. https://dx.doi.org/10.1038/nprot.2015.127. 26540590.

Nielsen, C.B., Cantor, M., Dubchak, I., Gordon, D., Wang, T., 2010. Visualizing genomes: techniques and challenges. Nat. Methods 7 (3 Suppl), S5–S15. https://dx.doi.org/10.1038/nmeth.1422. 20195257.

Nora, E.P., Lajoie, B.R., Schulz, E.G., Giorgetti, L., Okamoto, I., Servant, N., Piolot, T., van Berkum, N.L., Meisig, J., Sedat, J., Gribnau, J., Barillot, E., Bluthgen, N., Dekker, J., Heard, E., 2012. Spatial partitioning of the regulatory landscape of the X-inactivation centre. Nature 485 (7398), 381–385. https://dx.doi.org/10.1038/nature11049. 22495304.

O'Connor, T., Boden, M., Bailey, T.L., 2017. CisMapper: predicting regulatory interactions from transcription factor ChIP-seq data. Nucleic Acids Res.. 45(4)e19https://dx.doi.org/10.1093/nar/gkw95628204599.

Park, P.J., 2009. ChIP–seq: advantages and challenges of a maturing technology. Nat. Rev. Genet. 10 (10), 669–680. https://dx.doi.org/10.1038/nrg2641.

Park, Y., Figueroa, M.E., Rozek, L.S., Sartor, M.A., 2014. MethylSig: a whole genome DNA methylation analysis pipeline. Bioinformatics. 30 (17), 2414–2422. https://dx.doi.org/10.1093/bioinformatics/btu339. 24836530.

Park, Y., Wu, H., 2016. Differential methylation analysis for BS-seq data under general experimental design. Bioinformatics 32 (10), 1446–1453. https://dx.doi.org/10.1093/bioinformatics/btw026. 26819470.

Peters, T.J., Buckley, M.J., Statham, A.L., Pidsley, R., Samaras, K., Lord, R.V., Clark, S.J., Molloy, P.L., 2015. De novo identification of differentially methylated regions in the human genome. Epigenet. Chromatin 8, 6. https://dx.doi.org/10.1186/1756-8935-8-6. 25972926.

Phanstiel, D.H., Boyle, A.P., Heidari, N., Snyder, M.P., 2015. Mango: a bias-correcting ChIA-PET analysis pipeline. Bioinformatics 31 (19), 3092–3098. https://dx.doi.org/10.1093/bioinformatics/btv336. 26034063.

Phipson, B., Maksimovic, J., Oshlack, A., 2016. missMethyl: an R package for analyzing data from Illumina's HumanMethylation450 platform. Bioinformatics 32 (2), 286–288. https://dx.doi.org/10.1093/bioinformatics/btv560. 26424855.

Pidsley, R., Zotenko, E., Peters, T.J., Lawrence, M.G., Risbridger, G.P., Molloy, P., Van Djik, S., Muhlhausler, B., Stirzaker, C., Clark, S.J., 2016. Critical evaluation of the Illumina MethylationEPIC BeadChip microarray for whole-genome DNA methylation profiling. Genome Biol. 17 (1), 208. https://dx.doi.org/10.1186/s13059-016-1066-1. 27717381.

Plongthongkum, N., Diep, D.H., Zhang, K., 2014. Advances in the profiling of DNA modifications: cytosine methylation and beyond. Nat. Rev. Genet. 15 (10), 647–661. https://dx.doi.org/10.1038/nrg3772. 25159599.

Radman-Livaja, M., Rando, O.J., 2010. Nucleosome positioning: how is it established, and why does it matter? Dev. Biol. 339 (2), 258–266. https://dx.doi.org/10.1016/j.ydbio.2009.06.012. 19527704.

Ramirez, F., Ryan, D.P., Gruning, B., Bhardwaj, V., Kilpert, F., Richter, A.S., Heyne, S., Dundar, F., Manke, T., 2016. deepTools2: a next generation web server for deep-sequencing data analysis. Nucleic Acids Res. 44 (W1), W160–W165. https://dx.doi.org/10.1093/nar/gkw257. 27079975.

Ramos, M., Schiffer, L., Re, A., Azhar, R., Basunia, A., Rodriguez, C., Chan, T., Chapman, P., Davis, S.R., Gomez-Cabrero, D., Culhane, A.C., Haibe-Kains, B., Hansen, K.D., Kodali, H., Louis, M.S., Mer, A.S., Riester, M., Morgan, M., Carey, V., Waldron, L., 2017. Software for the integration of multiomics experiments in bioconductor. Cancer Res. 77 (21), e39–e42. https://dx.doi.org/10.1158/0008-5472.CAN-17-0344. 29092936.

Rhie, S.K., Guo, Y., Tak, Y.G., Yao, L., Shen, H., Coetzee, G.A., Laird, P.W., Farnham, P.J., 2016. Identification of activated enhancers and linked transcription factors in breast, prostate, and kidney tumors by tracing enhancer networks using epigenetic traits. Epigenetics Chromatin 9, 50. https://dx.doi.org/10.1186/s13072-016-0102-4. 27833659.

Ritchie, M.E., Phipson, B., Wu, D., Hu, Y., Law, C.W., Shi, W., Smyth, G.K., 2015. limma powers differential expression analyses for RNA-sequencing and microarray studies. Nucleic Acids Res. 43 (7). e47-e, https://doi.org/10.1093/nar/gkv00725605792.

Roadmap Epigenomics, C., Kundaje, A., Meuleman, W., Ernst, J., Bilenky, M., Yen, A., Heravi-Moussavi, A., Kheradpour, P., Zhang, Z., Wang, J., Ziller, M.J., Amin, V., Whitaker, J.W., Schultz, M.D., Ward, L.D., Sarkar, A., Quon, G., Sandstrom, R.S., Eaton, M.L., Wu, Y.C., Pfenning, A.R., Wang, X., Claussnitzer, M., Liu, Y., Coarfa, C., Harris, R.A., Shoresh, N., Epstein, C.B., Gjoneska, E., Leung, D., Xie, W., Hawkins, R.D., Lister, R., Hong, C., Gascard, P., Mungall, A.J., Moore, R., Chuah, E., Tam, A., Canfield, T.K., Hansen, R.S., Kaul, R., Sabo, P.J., Bansal, M.S., Carles, A., Dixon, J.R., Farh, K.H., Feizi, S., Karlic, R., Kim, A.R., Kulkarni, A., Li, D., Lowdon, R., Elliott, G., Mercer, T.R., Neph, S.J., Onuchic, V., Polak, P., Rajagopal, N., Ray, P., Sallari, R.C., Siebenthall, K.T., Sinnott-Armstrong, N.A., Stevens, M., Thurman, R.E., Wu, J., Zhang, B., Zhou, X., Beaudet, A.E., Boyer, L.A., De Jager, P.L., Farnham, P.J., Fisher, S.J., Haussler, D., Jones, S.J., Li, W., Marra, M.A., MT, M.M., Sunyaev, S., Thomson, J.A., Tlsty, T.D., Tsai, L.H., Wang, W., Waterland, R.A., Zhang, M.Q., Chadwick, L.H., Bernstein, B.E., Costello, J.F., Ecker, J.R., Hirst, M., Meissner, A., Milosavljevic, A., Ren, B., Stamatoyannopoulos, J.A., Wang, T., Kellis, M., 2015. Integrative analysis of 111 reference human epigenomes. Nature 518 (7539), 317–330. https://dx.doi.org/10.1038/nature14248. 25693563;.

Robinson, J.T., Thorvaldsdottir, H., Winckler, W., Guttman, M., Lander, E.S., Getz, G., Mesirov, J.P., 2011. Integrative genomics viewer. Nat. Biotechnol. 29 (1), 24–26. https://dx.doi.org/10.1038/nbt.1754. 21221095.

Robinson, J.T., Turner, D., Durand, N.C., Thorvaldsdottir, H., Mesirov, J.P., Aiden, E.L., 2018. Juicebox.js provides a cloud-based visualization system for Hi-C data. Cell Syst. 6 (2), 256–258. e1, https://doi.org/10.1016/j.cels.2018.01.00129428417.

Sanyal, A., Lajoie, B.R., Jain, G., Dekker, J., 2012. The long-range interaction landscape of gene promoters. Nature 489 (7414), 109–113. https://dx.doi.org/10.1038/nature11279. 22955621.

Schlosberg, C.E., VanderKraats, N.D., Edwards, J.R., 2017. Modeling complex patterns of differential DNA methylation that associate with gene expression changes. Nucleic Acids Res. 45 (9), 5100–5111. https://dx.doi.org/10.1093/nar/gkx078. 28168293.

Schmitt, A.D., Hu, M., Jung, I., Xu, Z., Qiu, Y., Tan, C.L., Li, Y., Lin, S., Lin, Y., Barr, C.L., Ren, B., 2016. A compendium of chromatin contact maps reveals spatially active regions in the human genome. Cell Rep. 17 (8), 2042–2059. https://dx.doi.org/10.1016/j.celrep.2016.10.061. 27851967.

Schones, D.E., Cui, K., Cuddapah, S., Roh, T.Y., Barski, A., Wang, Z., Wei, G., Zhao, K., 2008. Dynamic regulation of nucleosome positioning in the human genome. Cell 132 (5), 887–898. https://dx.doi.org/10.1016/j.cell.2008.02.022. 18329373.

Serandour, A.A., Avner, S., Oger, F., Bizot, M., Percevault, F., Lucchetti-Miganeh, C., Palierne, G., Gheeraert, C., Barloy-Hubler, F., Peron, C.L., Madigou, T., Durand, E., Froguel, P., Staels, B., Lefebvre, P., Metivier, R., Eeckhoute, J., Salbert, G., 2012. Dynamic hydroxymethylation of deoxyribonucleic acid marks differentiation-associated enhancers. Nucleic Acids Res. 40 (17), 8255–8265. 22730288.

Serra, F., Bau, D., Goodstadt, M., Castillo, D., Filion, G.J., Marti-Renom, M.A., 2017. Automatic analysis and 3D-modelling of Hi-C data using TADbit reveals structural features of the fly chromatin colors. PLoS Comput. Biol. 13 (7), e1005665. https://dx.doi.org/10.1371/journal.pcbi.1005665. 28723903.

Servant, N., Varoquaux, N., Lajoie, B.R., Viara, E., Chen, C.J., Vert, J.P., Heard, E., Dekker, J., Barillot, E., 2015. HiC-Pro: an optimized and flexible pipeline for Hi-C data processing. Genome Biol. 16, 259. https://dx.doi.org/10.1186/s13059-015-0831-x. 26619908.

Sheffield, N.C., Bock, C., 2016. LOLA: enrichment analysis for genomic region sets and regulatory elements in R and Bioconductor. Bioinformatics 32 (4), 587–589. https://dx.doi.org/10.1093/bioinformatics/btv612. 26508757.

Shen, L., Zhang, Y., 2013. 5-Hydroxymethylcytosine: generation, fate, and genomic distribution. Curr. Opin. Cell Biol. 25 (3), 289–296. https://dx.doi.org/10.1016/j.ceb.2013.02.017. 23498661.

Song, L., Crawford, G.E., 2010. DNase-seq: a high-resolution technique for mapping active gene regulatory elements across the genome from mammalian cells. Cold Spring Harb Protoc. 2010 (2). pdb prot5384, https://doi.org/10.1101/pdb.prot538420150147.

Splinter, E., de Wit, E., van de Werken, H.J., Klous, P., de Laat, W., 2012. Determining long-range chromatin interactions for selected genomic sites using 4C-seq technology: from fixation to computation. Methods 58 (3), 221–230. https://dx.doi.org/10.1016/j.ymeth.2012.04.009. 22609568.

Stamatoyannopoulos, J., 2016. Connecting the regulatory genome. Nat. Genet. 48 (5), 479–480. https://dx.doi.org/10.1038/ng.3553. 27120444.

Stricker, S.H., Koferle, A., Beck, S., 2017. From profiles to function in epigenomics. Nat. Rev. Genet. 18 (1), 51–66. https://dx.doi.org/10.1038/nrg.2016.138. 27867193.

Teschendorff, A.E., Relton, C.L., 2018. Statistical and integrative system-level analysis of DNA methylation data. Nat. Rev. Genet. 19 (3), 129–147. https://dx.doi.org/10.1038/nrg.2017.86. 29129922.

Teschendorff, A.E., Zheng, S.C., 2017. Cell-type deconvolution in epigenome-wide association studies: a review and recommendations. Epigenomics 9 (5), 757–768. https://dx.doi.org/10.2217/epi-2016-0153. 28517979.

Thorvaldsdottir, H., Robinson, J.T., Mesirov, J.P., 2013. Integrative Genomics Viewer (IGV): high-performance genomics data visualization and exploration. Briefings Bioinform. 14 (2), 178–192. https://dx.doi.org/10.1093/bib/bbs017. 22517427.

Thurman, R.E., Rynes, E., Humbert, R., Vierstra, J., Maurano, M.T., Haugen, E., Sheffield, N.C., Stergachis, A.B., Wang, H., Vernot, B., Garg, K., John, S., Sandstrom, R., Bates, D., Boatman, L., Canfield, T.K., Diegel, M., Dunn, D., Ebersol, A.K., Frum, T., Giste, E., Johnson, A.K., Johnson, E.M., Kutyavin, T., Lajoie, B., Lee, B.K., Lee, K., London, D., Lotakis, D., Neph, S., Neri, F., Nguyen, E.D., Qu, H., Reynolds, A.P., Roach, V., Safi, A., Sanchez, M.E., Sanyal, A., Shafer, A., Simon, J.M., Song, L., Vong, S., Weaver, M., Yan, Y., Zhang, Z., Zhang, Z., Lenhard, B., Tewari, M., Dorschner, M.O., Hansen, R.S., Navas, P.A., Stamatoyannopoulos, G., Iyer, V.R., Lieb, J.D., Sunyaev, S.R., Akey, J.M., Sabo, P.J., Kaul, R., Furey, T.S., Dekker, J., Crawford, G.E., Stamatoyannopoulos, J.A., 2012. The accessible chromatin landscape of the human genome. Nature 489 (7414), 75–82. https://dx.doi.org/10.1038/nature11232. 22955617.

Tolhuis, B., Palstra, R.J., Splinter, E., Grosveld, F., de Laat, W., 2002. Looping and interaction between hypersensitive sites in the active beta-globin locus. Mol. Cell 10 (6), 1453–1465. 12504019.

van Berkum, N.L., Dekker, J., 2009. Determining spatial chromatin organization of large genomic regions using 5C technology. Methods Mol Biol. 567, 189–213. https://dx.doi.org/10.1007/978-1-60327-414-2_13. 19588094.

Wang, T., Pehrsson, E.C., Purushotham, D., Li, D., Zhuo, X., Zhang, B., Lawson, H.A., Province, M.A., Krapp, C., Lan, Y., Coarfa, C., Katz, T.A., Tang, W.Y., Wang, Z., Biswal, S., Rajagopalan, S., Colacino, J.A., Tsai, Z.T.,

Sartor, M.A., Neier, K., Dolinoy, D.C., Pinto, J., Hamanaka, R.B., Mutlu, G.M., Patisaul, H.B., Aylor, D.L., Crawford, G.E., Wiltshire, T., Chadwick, L.H., Duncan, C.G., Garton, A.E., KA, M.A., RIIC, T., Bartolomei, M.S., Walker, C.L., Tyson, F.L., 2018. The NIEHS TaRGET II Consortium and environmental epigenomics. Nat. Biotechnol. 36 (3), 225–227. https://dx.doi.org/10.1038/nbt.4099. 29509741.

Wang, Y., Zhang, B., Zhang, L., An, L., Xu, J., Li, D., Choudhary, M.N.K., Li, Y., Hu, M., Hardison, R., Wang, T., Yue, F., 2017b. The 3D Genome Browser: a web-based browser for visualizing 3D genome organization and long-range chromatin interactions. bioRxiv. 112268 https://dx.doi.org/10.1101/112268.

Wang, Y., Zhang, B., Zhang, L., An, L., Xu, J., Li, D., Choudhary, M.N.K., Li, Y., Hu, M., Hardison, R., Wang, T., Yue, F., 2017a. The 3D Genome Browser: a web-based browser for visualizing 3D genome organization and long-range chromatin interactions. bioRxiv. 112268 https://dx.doi.org/10.1101/112268.

Weinreb, C., Raphael, B.J., 2016. Identification of hierarchical chromatin domains. Bioinformatics 32 (11), 1601–1609. https://dx.doi.org/10.1093/bioinformatics/btv485. 26315910.

Welch, R.P., Lee, C., Imbriano, P.M., Patil, S., Weymouth, T.E., Smith, R.A., Scott, L.J., Sartor, M.A., 2014. ChIP-Enrich: gene set enrichment testing for ChIP-seq data. Nucleic Acids Res. 42 (13), e105. https://dx.doi.org/10.1093/nar/gku463. 24878920.

Wijetunga, N.A., Johnston, A.D., Maekawa, R., Delahaye, F., Ulahannan, N., Kim, K., Greally, J.M., 2017. SMITE: an R/Bioconductor package that identifies network modules by integrating genomic and epigenomic information. BMC Bioinformatics 18 (1), 41. https://dx.doi.org/10.1186/s12859-017-1477-3. 28100166.

Wu, X., Zhang, Y., 2017. TET-mediated active DNA demethylation: mechanism, function and beyond. Nat. Rev. Genet. 1–18. https://dx.doi.org/10.1038/nrg.2017.33.

Xu, Z., Niu, L., Li, L., Taylor, J.A., 2016. ENmix: a novel background correction method for Illumina HumanMethylation450 BeadChip. Nucleic Acids Res.. 44 (3) e20-e, https://doi.org/10.1093/nar/gkv90726384415.

Yang, X., Han, H., De Carvalho, D.D., Lay, F.D., Jones, P.A., Liang, G., 2014. Gene body methylation can alter gene expression and is a therapeutic target in cancer. Cancer Cell 26 (4), 577–590. https://dx.doi.org/10.1016/j.ccr.2014.07.028. 25263941.

Yao, L., Berman, B.P., Farnham, P.J., 2015b. Demystifying the secret mission of enhancers: linking distal regulatory elements to target genes. Crit. Rev. Biochem. Mol. Biol. 50 (6), 550–573. https://dx.doi.org/10.3109/10409238.2015.1087961. 26446758.

Yao, L., Shen, H., Laird, P.W., Farnham, P.J., Berman, B.P., 2015a. Inferring regulatory element landscapes and transcription factor networks from cancer methylomes. Genome Biol. 16, 105. https://dx.doi.org/10.1186/s13059-015-0668-3. 25994056.

Yildirim, O., Li, R., Hung, J.H., Chen, P.B., Dong, X., Ee, L.S., Weng, Z., Rando, O.J., Fazzio, T.G., 2011. Mbd3/NURD complex regulates expression of 5-hydroxymethylcytosine marked genes in embryonic stem cells. Cell 147 (7), 1498–1510. https://dx.doi.org/10.1016/j.cell.2011.11.054. 22196727.

Yu, M., Hon, G.C., Szulwach, K.E., Song, C.X., Zhang, L., Kim, A., Li, X., Dai, Q., Shen, Y., Park, B., Min, J.H., Jin, P., Ren, B., He, C., 2012. Base-resolution analysis of 5-hydroxymethylcytosine in the mammalian genome. Cell 149 (6), 1368–1380. https://dx.doi.org/10.1016/j.cell.2012.04.027. 22608086.

Zhang, Y., Lin, Y.H., Johnson, T.D., Rozek, L.S., Sartor, M.A., 2014. PePr: a peak-calling prioritization pipeline to identify consistent or differential peaks from replicated ChIP-Seq data. Bioinformatics 30 (18), 2568–2575. https://dx.doi.org/10.1093/bioinformatics/btu372. 24894502.

Zhang, Y., Liu, T., Meyer, C.A., Eeckhoute, J., Johnson, D.S., Bernstein, B.E., Nusbaum, C., Myers, R.M., Brown, M., Li, W., Liu, X.S., 2008. Model-based analysis of ChIP-Seq (MACS). Genome Biol. 9 (9), R137. https://dx.doi.org/10.1186/gb-2008-9-9-r137. 18798982.

Zhou, W., Laird, P.W., Shen, H., 2017. Comprehensive characterization, annotation and innovative use of Infinium DNA methylation BeadChip probes. Nucleic Acids Res. 45(4):e22. https://dx.doi.org/10.1093/nar/gkw96727924034.

Zhou, X., Lowdon, R.F., Li, D., Lawson, H.A., Madden, P.A., Costello, J.F., Wang, T., 2013. Exploring long-range genome interactions using the WashU Epigenome Browser. Nat. Methods 10 (5), 375–376. https://dx.doi.org/10.1038/nmeth.2440. 23629413.

Zhou, X., Wang, T., 2012. Using the Wash U Epigenome Browser to examine genome-wide sequencing data. Curr. Protoc. Bioinform. Chapter 10:Unit10, https://doi.org/10.1002/0471250953.bi1010s4023255151.

Ziller, M.J., Hansen, K.D., Meissner, A., Aryee, M.J., 2015. Coverage recommendations for methylation analysis by whole-genome bisulfite sequencing. Nat. Methods 12 (3), 230–232. 1 p following 2. https://doi.org/10.1038/nmeth.315225362363.

Incorporating Epigenetics Into a Risk Assessment Framework

Ila Cote*, John J. Vandenberg†, Ingrid L. Druwe†, Michelle M. Angrish†

*Cote and Associates, Boulder, Colorado, United States †U.S. Environmental Protection Agency, National Center for Environmental Assessment, Research Triangle Park, NC, United States

INTRODUCTION

The Environmental Protection Agency (EPA) is charged with protecting human health and the environment from the adverse effects of exposure to environmental pollution. The evidence that epigenomic modification by environmental factors can pose significant public

Toxicoepigenetics
https://doi.org/10.1016/B978-0-12-812433-8.00013-7

health risks is substantial.[1,2] Despite these concerns, there is limited application of epigenetic information in risk assessments. Consequently, health risks from epigenomic modifications may go unaddressed. In this chapter, we provide some examples of epigenomic-related environmental risks, review the risk assessment process, and propose a risk assessment framework to enhance characterization of epigenomic risks that extends traditional risk assessment. Such improvements in risk assessment are anticipated to be of interest to risk managers, risk assessors, and the public. While epigenomics and risk assessment are equally relevant to both human and nonhuman species, this chapter will focus on human health risk assessment.

Epigenomic toxicants include a variety of environmental contaminants, for example, air pollution, heavy metals, pesticides, aflatoxin B1, tobacco smoke, bisphenol A, polycyclic aromatic hydrocarbons, persistent organic pollutants, and endocrine disruptors (Martin and Fry, 2016; Hou et al., 2012). It is known that epigenomic changes have a major role in the incidence of complex diseases, including cancer, cardiovascular and respiratory diseases, metabolic syndrome (e.g., diabetes and obesity), neurological impairments, and developmental effects (Laubach et al., 2018; Helsley and Zhou, 2017; Han and He, 2016; McCullough et al., 2016; Sales et al., 2017; Janesick and Blumberg, 2016). Alterations of chromatin structure, gene expression, and genome structure are major features of epigenomic modifications (Angrish et al., 2018). Consequently, implications of chemical exposures resulting in epigenomic modifications can include induction of an unusually wide variety of health effects, early-life-stage sensitivities (i.e., fetus, neonates, and young children) to both early-life and later-in-life diseases, and potential multigenerational effects (Chapter 4-1; Bowers and McCullough, 2017; Cote et al., 2017; Sales et al., 2017; Han and He, 2016; Martin and Fry, 2016; Green and Marsit, 2015; Ladd-Acosta and Fallin, 2016; Marsit, 2015; Olden et al., 2015, 2016; Smith et al., 2016; Xin et al., 2015; Janesick et al., 2014; Boekelheide et al., 2012, Baccarelli and Bollati, 2009; Jirtle and Skinner, 2007). Moreover, epigenomic modification can accumulate with age and is thought to underlie the overall effects of aging and aging-related diseases, like neurodegenerative diseases (Kochmanski et al., 2017; Han and He, 2016). Assessment of chemical epigenomic toxicants, however, is complicated by interactions among a wide variety of not only epigenetic but also genetic and/or biological (intrinsic) and environmental (extrinsic) factors.

BRIEF DESCRIPTION OF HUMAN HEALTH RISK ASSESSMENT

Human health risk assessment is the characterization of the nature, magnitude, and probability of adverse health effects resulting from human exposure to environmental hazards.

---

[1] As noted in previous chapters, epigenomics is the study of changes in gene expression that do not involve changes to the DNA sequence. Essentially, epigenomics can change an organism's phenotype without changing the genotype.

[2] In this context, human population risk is being considered; risk is commonly defined as the probability (or likelihood) of disease or harm. For environmental risk assessment, it is usually calculated as the difference in probability between those who are exposed and those who are not. Generally, risk of a disease or disorder exists in the unexposed and the exposed population. Chemical exposure both shifts the risk distribution toward increased risk and increases the severity of health effects. The risk distribution is composed of all the individual risks in the population that is due to heterogeneity in individual responses.

This is a multistep process in which scientific data are synthesized to estimate risk(s). Risk assessment informs risk management decisions that impact public health, but risk management also considers legal, economic, social, technological, and public policy and other factors. Due to the consideration of risk assessment in support of risk management decisions, it is of upmost importance that assessments are made with the most informative and sound science available. Data to inform risk assessments are typically gathered from peer-reviewed studies and/or Good Laboratory Practice studies. A range of data types—epidemiology; clinical studies; animal studies; and, increasingly, in vitro and in silico studies—can be used to inform an assessment. If available, exposure data are evaluated with pharmacokinetic and pharmacodynamic models to improve our understanding between the relationship with dose, target tissues, and mechanisms underlying toxicity (Angrish et al., 2018; Cote et al., 2017; Wetmore, 2015; US EPA, 2012a, 2014; IPCS, 2010). Taken together, these data are used to derive estimates of risk.[3]

In 1983, the National Academy of Sciences detailed the risk assessment process used by the US federal government and established four basic components of risk assessment that are still used today: (1) *hazard identification*, (2) *dose-response assessment*, (3) *exposure assessment*, and (4) *risk characterization* (NRC, 1983). In 2009, the process was further elaborated with the development of a framework for risk assessment in which these four basic components are embedded (NRC, 2009). The framework consists of three phases: (I) problem formulation and scoping, in which the problem is described and available risk management options are identified (for the environmental problem of interest); (II) planning and assessment, in which the type of risk assessment needed to address the problem is determined and, then, the risk assessment tools described above are used to determine risks under existing conditions and under potential risk management options; and (III) risk management, in which risk and nonrisk information is integrated to inform choices among options (NRC, 2009; Fig. 1).

The following discussion focuses only on phase II portion of the framework, that is, the risk assessment process itself (NRC, 1983, 2009):

- *Risk assessment planning consists of formulating the questions and key science issues that the assessment will address.* Issues are likely to include the following: What is the environmental problem of concern? Who might be at risk and under what conditions of exposure and location? The following risk assessment components strive to address these questions. It is important to note that the purpose of risk assessment planning is to help target the type of risk assessment conducted to the problem of concern (NRC, 2009). Such target assessments are often referred to as "fit for purpose" leading to a portfolio of assessment options depending on need. In this manner, risk assessment can be flexible and efficient.
- *Hazard identification (HI). What adverse health effects are associated with the agents of concern?* This component examines whether an environmental agent has the potential to cause harm to humans (e.g., adverse developmental effects, cancer, asthma, and cognitive dysfunction) and, if so, under what circumstances. Additionally, HI evaluates underlying

---

[3]For those interested in further details of traditional risk assessment, much guidance is freely available from EPA (https://www.epa.gov/risk) and the World Health Organization and the United Nations International Program on Chemical Safety (http://www.inchem.org/documents/harmproj/harmproj/harmproj8.pdf), as well as other organizations.

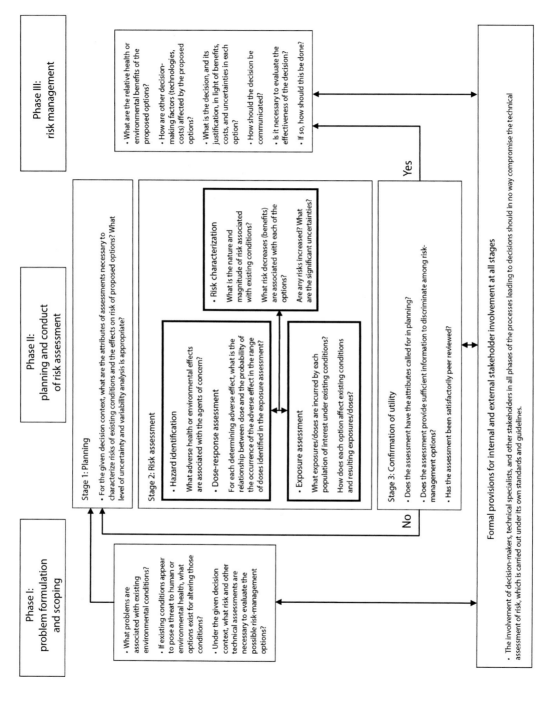

FIG. 1 A framework for risk-based decision-making that maximizes the utility of risk assessment. *Reproduced with permission from NRC (National Research Council), 2009. Science and Decisions: Advancing Risk Assessment. The National Academies Press, Washington, DC. https://doi.org/10.17226/12209, page 11.*

mechanisms of toxicity to inform biological plausibility, if feasible. This information is integrated, and evidence for a causal relationship between exposure and effects is weighed. The decision to consider an effect as adverse, or not, is a key decision in risk assessment.

This decision may be complicated by context-dependent effects, such as life stage where susceptibility to an environmental agent may differ between an adult population and a juvenile population due to differences in sensitivity and/or metabolism at these different life stages. Adverse versus adaptive responses to environmental agents are discussed under "Challenges to Incorporating Epigenomics into Risk Assessment" section.

- *Dose-response (DR)[4] assessment. For each determining adverse effect, what is the quantitative relationship between the dose and the probability of occurrence of the effect in the range of doses identified in the exposure assessment?* This component also explores how the relationship may change with different levels or durations of exposures. Moreover, DR assessment identifies and quantifies uncertainty and variability, as feasible. How dose-response assessment is combined with exposure assessment is described in risk characterization.

Adjustments to the data generally need to be made to facilitate comparison of dose-response information to exposure. Pharmacokinetic models are used to convert either exposure to dose or vis versa, as needed (US EPA, 2006, 2012a, 2014). Additionally, adjustment factors can be applied to convert among interspecific and intraspecies differences in the absence of pharmacokinetic models (US EPA, 2012a, 2014, 2017). Notably, clinical or epidemiological studies of environmental media concentrations directly predict risks, although adjustments for portions of the population may need to be made (e.g., estimated risks from a study of adults would need to be adjusted by body weight, respiration, and metabolic rate when extrapolating to children).

There are three common approaches to estimating dose-response relationships used by all regulatory agencies worldwide. These three methods are discussed here in some detail as they also are applicable to epigenomic data. They are arranged from the simplest and least data intensive to most complex and more data intensive:

1. *NOAEL/LOAEL approach.* A point of departure[5] (POD) for an assessment can be identified and calculated from a single "best" study (ideally from multiple studies) evaluating the most sensitive end point (or critical effect within the data set). The POD is then divided by an uncertainty factor to estimate a value assumed to be without appreciable risk to the human population, including sensitive populations and life stages (US EPA, 2002, 2012a, 2014). In no. 3, metaanalysis techniques for combining this type of information from multiple studies are discussed. These uncertainty factors (UFs) are used to account for uncertainty in animal to human extrapolation, variability within the human population, duration of exposure, database uncertainty, or use of a LOAEL versus the preferred NOAEL. Pharmacokinetic models can usually help refine

---

[4]In place of dose response, the terms exposure response when considering human data or concentration response when considering air exposures also are used.

[5]Definition: Point of departure (POD) is the dose-response point that marks the beginning of a low-dose extrapolation. This point can be the lower bound on dose for an estimated incidence or a change in response level from a dose-response model (BMD_L) or a lowest-observed-adverse-effect level (LOAEL) or no-observed-adverse-effect level (NOAEL) for an observed incidence or change in level of response (US EPA, 2012b).

and reduce uncertainty. At EPA, resulting values are termed a reference dose (RfD) for oral exposure or reference concentration (RfC) for inhalation exposure (US EPA, 2002, 2012a, 2014). This method is generally applied to noncancer end points. Advantages of the uncertainty factor approach are that it has a long history of use and is less data intensive than the subsequently discussed approaches. The major disadvantage is that risk estimates are not developed for exposure levels between the POD and the reference value.

2. *Dose-response modeling.* Mathematical curve fitting approaches are used to quantitatively model the dose-response relationship. For most chemicals, this modeling approach also includes low-dose extrapolation in order to evaluate environmental levels of exposure and acceptable levels of risk. For example, a common scenario for estimating risks from occupational or animal bioassay exposures is that the exposure- or dose-response data are collected at a higher concentration than environmental exposures and have considerable higher response rates than is acceptable for the general public (e.g., 20%). Hence, using DR modeling, available data are used to extrapolate to lower exposure levels and lower response rates (e.g., increased incidence of an adverse effect over background of 1/1000–1/1,000,000 people exposed). Importantly, the assumed shape of the low-dose or low-exposure relationship may differ (e.g., linear, linear quadratic, and u-shaped) and potentially can be informed by mechanistic information. For some chemicals, the databases are robust enough to obviate the need for low-dose extrapolation (e.g., particulate matter, ozone, and lead), but this is relatively rare. Two approaches for low-dose extrapolation can be used. One approach called benchmark dose (BMD) modeling (https://www.epa.gov/bmds) is used to calculate the lower 95% confidence limit (the benchmark dose lower bound (BMDL)) for a 10%, 5%, or 1% response rate as a surrogate for a point of departure and, then, apply the UFs described above (US EPA, 2012b). This approach is considered an improvement to the UF approach described above because it uses all of the data in the dose-response curve and includes the variance, to estimate the point of departure (POD) versus a single measured data point, such as the LOAEL or NOAEL. A second approach is to use mathematical modeling to extrapolate to low-dose or low-exposure levels. Numerous mathematical models exist that are useful for DR assessment and give very similar estimates of risk within the range of the data available. The different models, however, make different assumptions about the shape of the dose-response curve outside the range of data (i.e., in the low dose-response range), and consequently, modeled risk estimates can diverge noticeably when extrapolated to lower doses. Variation on multihit models may be most appropriate to consider the multiple events that are often characteristics of epigenomic data. Thus, there is uncertainty associated with model choice. The advantage of the DR modeling extrapolation approach is that it provides an estimate of the magnitude of risk; the disadvantage is that it is more data and resource (e.g., time) intensive.

3. *Metaanalyses and probabilistic methods.* Recently, there also has been a concerted effort to utilize metaanalyses and probabilistic techniques, such as Monte Carlo sampling, to better define DR, and the accompanying uncertainties and variability (Shao et al., 2017; Chiu and Slob, 2015; IPCS, 2014). The advantages of these approaches are that they combine data sets, allow for utilization of more data from more studies, and better quantitatively describe DR; the disadvantages are that they are more resource intensive and more

complicated to describe to risk managers and the public. In the era of "big data," metaanalyses and probabilistic evaluations of data are increasingly prevalent as they allow the integration and analyses of large amounts of data intended to increase the confidence and precision of the assessment.

- *Exposure assessment.* What exposures/doses does each population of interest under existing conditions incur? This component examines the frequency, timing, and levels of contact with the environmental agent. As this paper focuses on the health-related components of risk assessment, exposure assessment will not be discussed further. Interestingly, specific epigenomic modifications have been proposed as biomarkers of exposures (Olden et al., 2016).
- *Risk characterization. What is the nature and magnitude of the risk associated with existing conditions or with risk management options.* This component combines the information from the previous steps: HI, DR, and exposure assessment. For identified hazards (usually those considered known or likely hazards to humans), DR information and exposure are combined to describe risks along with information on data variability and significant uncertainties that were identified throughout the process, to derive a scientifically based conclusion about the potential risk(s) of the agent in question. How well the data support the conclusions about risks also are conveyed.

Table 1 shows the types of questions that human health risk assessment may address.

CHALLENGES TO INCORPORATING EPIGENOMICS INTO RISK ASSESSMENT

The field of epigenomics has substantially progressed leading to significant advances and a deeper understanding of epigenomic mechanisms and tools useful for evaluating diverse pathways connected to those mechanisms. Epigenomic marks are known to be important for early-life development (e.g., epigenetic imprinting that are erased and reestablished during germ cell development and again after fertilization), phenotype, and disease (Han

TABLE 1   Types of Questions Human Health Assessments Address

Human health risk assessment addresses questions such as the following:

- What types of health problems may be caused by environmental stressors such as chemicals or radiation?
- What is the chance that people will experience health problems when exposed to different levels of environmental stressors?
- Is there a level below which some chemicals don't pose a human health risk?
- What environmental stressors are people exposed to and at what levels and for how long?
- Are some people more likely to be susceptible to environmental stressors because of factors such as age, genetics, preexisting health conditions, cultural practices, and gender?
- Are some people more likely to be exposed to environmental stressors because of factors such as where they work, where they play, what they like to eat, etc.?

The answers to these types of questions helps decision makers, whether they are parents or public officials, understand the possible human health risks from environmental media.

Source: EPA, 2018. https://www.epa.gov/risk/human-health-risk-assessment.

and He, 2016). Yet, the interaction between epigenomic modifications, gene regulation, and intrinsic (e.g., age, sex, and genetics) and extrinsic (e.g., diet, socioeconomic status, and environment) factors is inherently complex and difficult to untangle (McCullough et al., 2016; Mirbahai and Chipman, 2014; Saban et al., 2014; Tammen et al., 2013; Bailey et al., 2013).

As discussed in more detail in other chapters, epigenomic alterations encompass modifications to DNA, histone proteins, and histone variants; nucleosome positioning; and noncoding RNAs that collectively regulate the expression of genes without changing DNA sequence (Chapters 1-2, 2-1, and 3-1; Angrish et al., 2018; Han and He, 2016). Therefore, interpreting these various types of epigenomic evidence is compounded by not only mechanisms that regulate epigenomic state but also polymorphic DNA sequence, complex gene regulatory networks that include a host of various transcription factors and regulatory elements, and biological context (e.g., developmental age and disease status). Another layer of complexity is added by epigenomic marks (i.e., DNA methylation and histone modification) that might have allelic imbalance because of imprinting and/or abnormal epigenomic reprogramming. Collectively, these genetic, epigenomic, and trans-acting transcriptional layers make up a complex gene regulatory network that may interact with or respond to environmental factors to drive functional (and possibly) adverse health outcomes.

The understanding of the intrinsic factors described above is further complicated by experimental model (e.g., in vitro vs in vivo), species, sex, tissue type, developmental stage, and exposure considerations (Martin and Fry, 2016; Bowers and McCullough, 2017; Cote et al., 2017; Chappell et al., 2014). Ultimately, integrated analysis of multilayer epigenomic, genomic, transcriptomic, and proteomic data will be essential to understanding how epigenomic information contributes to controlling complex regulatory processes (Bowers and McCullough, 2017; Han and He, 2016). In the interim, the major challenge of risk assessment is to consider pieces of the puzzle in the absence of the whole picture, identify biologically meaningful adverse changes, and interpret these changes in terms of the nature and magnitude of public health risks.

Interpreting the complex regulatory landscape in the context of the risk assessment process described above is an integrative challenge. Epigenomic evaluation techniques are fundamentally resource intensive. Other basic limitations include identifying the backdrop for comparisons that distinguish signal from noise. Specifically, setting the "normal" epigenome as a comparator circles back to the human heterogeneity problem; like the genome, the epigenome also has interindividual and population level variability. In addition, the epigenome exhibits plasticity; thus, relative interindividual and total population variability will likely change as a function of time.

Furthermore, the epigenome contributes to response variance across species, strain, and subpopulation, as well as tissue, cell types, and their phenotypic differences. As an example, a study by Israel et al. (2018) evaluated tissue and strain response differences after exposure to the DNA-damaging carcinogen butadiene. They found that "tissue effects dominate differences in both gene expression and chromatin states, followed by strain effects," and concluded that "variation in the basal states of epigenome and transcriptome may be useful indicators for individuals or tissues susceptible to genotoxic environmental chemicals." Similarly, Bowers and McCullough (2017) illustrated how differences in susceptible human populations can be linked to epigenomic differences. These epigenome-associated variations can pose problems for extrapolation across tissues, strains, and subpopulations. As an example, many epigenomic studies use blood leukocytes as a sample source. Although blood is technically easier to sample than a solid organ, it may not reflect key epigenomic events and responses of target tissue(s) of interest. Moreover, both blood and solid organs have mixed cell populations. Each cell type has the same genome yet a unique epigenome. Therefore, the relative proportion of each cell type will drive the overall distribution of epigenomic marks. Analyses rarely include locus-specific epigenomic details, like that of Cheung et al. (2017) where epigenomic profiling of the methylome and histone code identified allele and noncoding region-specific epigenomic patterns. Therefore, it is more realistic that a variety of data from the public realm will need to be integrated into frameworks that not only consider the evidence but also study validity and methodological variation, sensitivity, precision, specificity, and accuracy (for a further review of relevant epigenomic assay technologies, readers are directed elsewhere) (DeAngelis et al., 2008; Yan et al., 2016).

Notably, in traditional risk assessment, chemical-specific exposures are generally associated with a specific health outcome, and mechanisms are often viewed as linear sets of events linking exposure to effects. Given what is known about the complexities of epigenomic modification and downstream events, a different approach is needed. A network approach to characterizing interacting events is more likely to robustly represent mechanisms of disease (Cote et al., 2016; US EPA, 2014).

As part of this complexity, four different effect types can result from epigenomic modification (Angrish et al., 2018):

(1) Adverse—an exposure-induced epigenomic change that clearly results in an adverse effect.

(2) Null—exposure-induced epigenomic change with no effect.

(3) Adaptive—an exposure-induced change in the epigenome that renders the cell, tissue, organism, or population better suited to its environment.

(4) Emergent—a change in epigenomic state that results in an adverse outcome at a later life stage.

While adaptive effects[6] may pose no immediate health risk and in fact may improve health or survival status, they can propagate across generations and become emergent later in life, as thought to be the case for fetuses gestating during the Dutch Hunger Winter of 1944–45 and with developmental origins of disease in general (Chapter 2-2; Marsit, 2015; Wallack and Thornburg, 2016; Heindel and Vandenberg, 2015). Obesity, diabetes, cardiovascular morbidity, and neuropsychiatric diseases are some of the disorders considered to be, at least in part, of developmental origin (Heindel and Vandenberg, 2015). Understanding the effect type resulting from a chemically related epigenomic change is an important factor in risk assessment. Notably, many epigenetic alterations are reversible; thus, alterations that are longer lasting may be more important for risk assessment.

[6]Due to evolutionary pressure, it is reasonable to assume that most epigenomic effects, occurring in response to the natural environment, will tend to fall into either adaptive or emergent effect types. An epigenomic effect can also be both simultaneously adaptive and adverse or emergent. As biological systems, however, did not evolve to respond to industrial chemical exposures, this assumption may not apply to chemical exposures (Laubach et al., 2018).

EVALUATION AND INTERPRETATION OF EVIDENCE

In general, relative measures, but not absolute omics measurements, agree well across laboratories. Interindividual variability in response, different statistical approaches, use of different experimental protocols, and numerous other factors can introduce variability and uncertainty in results. As suggested in a European Center for Ecotoxicology and Toxicology of Chemicals workshop report, using fold changes in omics changes (as compared with the levels observed in control groups) together with P-value thresholds has enhanced reproducibility while still optimally balancing sensitivity and specificity. The application of multiple testing adjustment factors may enable more a stringent statistical filtering than the use of fold change thresholds (Buesen et al., 2017; Chu and Huang, 2017). The European Center for Ecotoxicology and Toxicology of Chemicals workshop report also has helped to reduce variability among studies and laboratories by establishing best practices for collecting, storing, and curating omics data; processing of omics data, and weight-of-evidence approaches for integrating omics data (Buesen et al., 2017).

The development of a risk assessment begins with a systematic review of the available literature. Such a review can be narrowly focused on the problem identified in the planning stage (e.g., chemical X and mammalian DNA methylation and atherosclerosis) or broadly focused (e.g., chemicals and diseases). Studies selected for detailed review from the literature search are generally well designed, carefully conducted, and transparently reported. Study quality criteria originally developed for genomic data are available and can be considered as resources to evaluate the quality of epigenomic data (NCBI/GEO, 2018; Roadmap Epigenomics Consortium et al., 2015; McConnell et al., 2014). Epigenomic study quality criteria developed for research and clinical purposes also can be considered but are evolving as new information becomes available (NCBI/Roadmap, 2018; Sosnowski et al., 2018; Pollock et al., 2017; Han and He, 2016; Bock, 2012). Also see Chapter 2-1 of this book for additional information on specific protocols.

Evidence from the most informative studies is evaluated and integrated across data types and studies to inform hazard identification and dose response. Importantly, only studies that directly inform components of the risk assessment paradigm (discussed above) are considered. Confidence in the conclusions of the assessment can be bolstered by a number of factors, for example, more studies with consistent results; use of vertebrate, ideally human, derived data, tissue-type or cell-type-specific data, primary versus immortalized cells; and coherence across evidence streams (e.g., agreement between epidemiological and clinical data and animal toxicological and/or mechanistic data). Notably, in traditional risk assessment based on specific health outcomes, consistency of findings across species, sexes, and tissues adds to the weight of evidence; for epigenomics, it is recognized that there are important epigenomic differences that may not be preserved across these variables.

Importantly, the data linking epigenomic change and the downstream adverse outcomes need not be chemical related. In recent years, mechanistic models have drawn information from multiple chemicals, normal biology, and pathophysiology not directly associated with chemical exposures. In this fashion, more robust and complete conceptual mechanisms have been developed that can be used to explore events induced or modified by chemical exposures and other environmental or genomic factors related to disease.

Ideally, exposure-dose-response relationships in humans can be quantified; however, this type of data is relatively uncommon for toxicants. Consequently, in the absence of human data, default assumptions or pharmacokinetic models are used to address interspecific and intraspecies extrapolation and in vitro to in vivo extrapolation of dose-response data (US EPA, 2002, 2006, 2012a, 2014). Quantitative interpretations of exposure-dose relationships particularly from in vitro studies or dam to fetus are particularly uncertain, although pharmacokinetic models can be developed for these purposes (Jaroch et al., 2018; Brinkmann et al., 2017; Pearce et al., 2017; El-Masri et al., 2016; Jamei, 2016; Martin et al., 2015). Once the exposure dose is estimated, the three quantitative approaches (discussed above) can be applied to the data (i.e., the NOAEL/LOAEL approach, dose-response modeling, and metaanalyses and probabilistic methods).

Fig. 2 illustrates a common method, for displaying metadata (called data arrays) across multiple studies and study types to help integrate and evaluate data. Omics data can be integrated in two ways: (1) horizontal integration, where data from different studies of the same type (e.g., DNA methylation) are pooled, generally across labs and platforms and (2) vertical

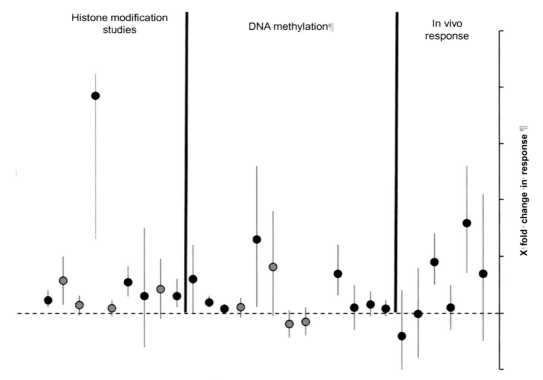

FIG. 2 This figure illustrates how multiple studies of the same chemical can be presented together as a means of integrating and evaluating information. Each bar is the result of a different study. Several types of epigenomic studies are shown. *Dotted line* represents no response. *Dots* are median responses, and *lines* are error bars. *Black dots* are in vivo studies. *Red dots* are in vitro studies.

integration that combines different types of omics data (e.g., DNA methylation and gene expression) from the same subjects or groups of subjects (Chu and Huang, 2017). See Chiu et al. (2017, 2018), Williams et al. (2017), Attene-Ramos et al. (2013a,b, 2015), Tice et al. (2013), and Judson et al. (2011) for illustrations of other ways to analyze and display omics data.

FRAMEWORK FOR EPIGENOMIC RISK ASSESSMENT

Consideration of the type of environmental problem being addressed and how the type of risk assessment supports risk management is a key factor in the proposed conceptual framework for epigenomic risk assessment. This, in turn, will heavily influence the types of epigenomic data needed and its application in risk assessment (Cote et al., 2016, 2017; US EPA, 2014; Thomson et al., 2014). The framework uses three illustrative "fit-for-purpose" assessment types as examples. The three assessment types are listed in order of increasing complexity, increasing reliance on in vivo data, and the ability to support progressively more costly risk management decisions. These generalized decision contexts do not, and are not meant to, capture all decisions or situational nuances that risk manager face:

- *Chemical screening and prioritization assessments* are generally used to rank a large number of chemicals and/or evaluate data-limited chemicals for further testing and research or for emergency response situations. These assessments generally (and often heavily) rely on in vitro, in silico, and structure-activity data. Evaluations of these data types are anchored to in vivo data as feasible but may also be tied to mechanistic events upstream from apical outcomes such as epigenomic modification and gene expression. It is proposed for screening and prioritization assessments that any significant modification of the epigenome associated with chemical exposure can be considered *potentially* adverse. Fairly limited data could be used for this purpose. As noted above, a key challenge in the assessment of epigenomic change is to determine whether the modifications are adverse, including emergent, or not. This approach assumes that significant modification of the epigenome is concerning and that the greater the fold change, the higher the priority for additional testing and research. Future advances in research may well allow us to differentiate or better understand epigenomic changes associated with adaptive versus adverse events, and at that point, this assumption may no longer be valid.

Additional testing and research would investigate the relationship between epigenomic changes and possible outcomes and the nature of the outcomes. Such an evaluation can be based on a variety of medium- and high-quality data types (structure activity, in vitro, in silico, and in vivo). Read-across[7] techniques can also be useful. The evidence supporting such a decision could be, for example, from well-conducted high-throughput in vitro or in silico studies (Knudsen et al., 2015, 2017; Parfett and Desaulniers, 2017) or alternative

---

[7]Read across is a technique for predicting end point information for one substance (target substance), by using data from the same end point from (an)other substances, (source substance(s)) (European Chemical Agency, 2014, https://echa.europa.eu/documents/10162/13628/09_read_across_webinar_en.pdf/4dbb2e64-408c-4d12-a605-e9f9b75615d8). Also see the Organization for Economic Co-operation and Development guidance on grouping and read across via the document "Guidance on Grouping of Chemicals, Second Edition," available at http://www.oecd.org/chemicalsafety/risk-assessment/groupingofchemicalschemicalcategoriesandread-across.htm.

species studies (Aluru, 2017; Cavalieri and Spinelli, 2017), or data mined results from data warehouses such as the Gene Expression Omnibus or the Roadmap Epigenomics Consortium et al. (2015) (Fingerman et al., 2013). Additionally, chemicals could be ranked based on their potency to disrupt the epigenome. Chemicals could be ranked by comparing dose-response curves (e.g., Attene-Ramos, 2013a,b, 2015) or NOAELs, LOAELs, or BMD(L)s (e.g., modeled NOAELs) (Chiu et al., 2018; US EPA, 2012b; Judson et al., 2011). Such a ranking is most easily done among similar studies. Care is needed when comparing across variables such as specific types of modifications, platforms, and experimental paradigms. Potency estimates may need to be normalized across dissimilar studies, in certain cases. This type of assessment generally focuses on higher throughput methods capable of evaluating many chemicals in a relatively short time period and cost-effective manner. All other data types (e.g., epidemiological data), however, could still be used to flag a chemical for more testing or research. US EPA's ToxCast program has developed approaches and tools that facilitate chemical screening and prioritization in this manner (https://www.epa.gov/chemical-research/toxicity-forecasting). Similarly, the multiagency Tox21 program continues to develop methods and data for high-throughput epigenomic evaluations (Parfett and Desaulniers, 2017). As our knowledge of epigenomics progresses, epigenomic modifications, coupled with understanding of the biological implications for disease, might eventually be sufficient to support limited and major scope assessments in the absence of chemical-specific in vivo data.

- *Limited scope assessments* generally are used to evaluate somewhat limited exposure, hazard, or data situations and often support nonregulatory decision-making. For example, Superfund site cleanup decisions may use this type of assessment to determine what chemicals will be remediated at a site and to what concentrations. These assessments generally rely on a range of limited data types (structure-activity data and read-across, in vitro, in silico, and in vivo data) and are generally more resource intensive than chemical screening and prioritization. For limited scope assessments, the association of a significant modification of the epigenome with chemical exposure alone could be considered potentially adverse for some limited scope actions (emergency response and Superfund clean-up decisions). Supporting mechanistic data strengthen the weight of evidence and may warrant additional actions. The evidence supporting such a decision is unlikely to be the result of a single experiment but rather be a synthesis across a number of well-conducted experiments and/or observations. Depending on the type and quality of dose-response data available, epigenomic data could be used to develop quantitative comparisons with exposure levels, that is, dose-response assessment.
- *Major scope assessments* usually support regulatory decision-making and are aimed at national exposures wherein many people are potentially affected or less-widespread exposures associated with potentially high individual risks. At this point in time, major scope assessments are almost invariably based on strong epidemiological, clinical, or in vivo animal bioassay data linking exposure to adverse outcome(s). They are always the most in-depth type of assessment and the most data and resource intensive. The required certainty in the assessment generally is directly linked to the potential cost of regulation and the expected extent of public health impacts. Proposed major roles for epigenomic data to support major scope assessments are (1) to elucidate the mechanistic links between exposure and adverse outcome data derived from in vivo

studies in which epigenomics were investigated, thus strengthening the evidence for a causal determination and confidence in the assessment; (2) to help unravel the complex quantitative links between molecular initiating events and adverse outcomes; (3) to develop insight into the basis for human variability—an important issue for all assessments; and (4) to provide insight into response differences across experimental paradigms, cell types, tissues, species, and/or sexes experiencing similar exposures.

In the example provided below, epigenomic data are helping to increase confidence in the causal nature of arsenic-associated health effects beyond cancer, elucidating the hazards of early-life arsenic exposures, and providing insights into biomarkers of exposure and effects.

Table 2 identifies information that advances application of epigenomics to risk assessment. It is intended that this table may help research scientists target their research in a manner that facilitates risk assessment.

For inclusion of hazard identification in an assessment, the evidence for causality is increased by consistency of the data across multiple, independent studies; adherence to commonly accepted study quality criteria; and the coherence of the data across different streams of evidence (e.g., human, animal, and mechanistic). Chance, bias, and confounding should be ruled out or minimized with reasonable confidence to infer a causal or likely causal relationship between exposure and effect (US EPA, 2015). When chance, bias, or confounding cannot be minimized, data are generally considered suggestive or insufficient (US EPA, 2015). Importantly, confidence in the assessment's ability to predict human health effects reflects the overall confidence in the body of evidence rather than confidence in individual studies.

TABLE 2   Information That Advances Application of Epigenomics to Risk Assessment

---

Evaluation of single chemicals

- Identify epigenomic modification(s) associated with chemical exposures
- Explore mechanisms of action
- Generate hypotheses regarding compound action(s) and apical adverse outcomes
- Determine the no or lowest observed effect levels observed in studies or dose-response modeled estimates of response rates (e.g., 1%–10% response rates). Apply uncertainty factors or modeled low-dose extrapolation to estimate risks in the range of environmental exposures

Evaluation of multiple chemicals

- Categorize chemicals by epigenomic signature(s), downstream mechanism(s), and apical outcomes to the extent known
- Relatively rank chemicals by their potency to induce epigenomic modifications using lowest observed effect in "best" study or studies
  - o. In screening, rank any chemicals based on any epigenomic modification
  - o. For limited or major assessments, compare similar epigenomic outcomes measured in generally similar protocols
- Consider interactions among chemicals relative to apical outcome

Biomarkers of exposure and response

- Discover biomarkers of exposure and toxicity
- Verify and quantify biomarker signatures
- Estimate dose-response relationships at near environmental exposure levels, if feasible

---

Examples of the types of studies, or combinations of studies, that could increase the weight of evidence for causal or likely causal relationships among exposure, an epigenomic event, and adverse outcome include the following:

- Metaanalyses of multiple well-conducted observational or experimental studies that provide consistent findings of significant associations among exposures, intermediate effects, and outcomes. When these data are available, mechanistic details are less important.
- Consistent and coherent observations across multiple studies. Traditional exposure outcome data (e.g., whole animal bioassay data) augmented by epigenomic mechanistic information.
- Pharmacological interventions that block exposure-dependent epigenomic alterations or linked epigenomic and genomic expression and, concomitantly, block or mitigate adverse outcomes.

PRACTICAL APPLICATION

Arsenic is an example of a toxicant acting, at least partially, via an epigenomic mechanism. Associations have been observed in humans and animals between arsenic exposure, epigenomic modifications, transcriptional changes, and adverse effects (Martin and Fry, 2016; Rager et al., 2017; Argos, 2015; Cardenas et al., 2015; Muenyi et al., 2015; Broberg et al., 2014; Koestler et al., 2013; Hou et al., 2012; Kile et al., 2014; Pilsner et al., 2012; Treas et al., 2012; Tsang et al., 2012). Arsenic has been reported to induce genome-wide global DNA hypomethylation, specific gene promoter methylation changes, epigenomic regulatory-gene expression changes, and histone modification (Martin and Fry, 2016; Mauro et al., 2016; NRC, 2013; Treas et al., 2012; Reichard and Puga, 2010). Data, however, appear somewhat inconsistent, likely due to variability introduced by tissue, species, and sex differences and methodological approaches (Martin and Fry, 2016; Cote et al., 2017; Argos, 2015). Elucidating this epigenomic mechanism has strengthened the causal connection between a variety of noncancer end points and arsenic exposure (Martin et al., 2017), partially explained observed human variation in response to chemical exposures (Martin and Fry, 2016), provided biomarkers of effect and arsenic exposure, and highlighted risks associated with early-life exposure (Farzan et al., 2013). Moreover, epidemiological studies of early-life exposures and adverse effects and the multitude of tissues affected by arsenic exposures suggest that significant modification may occur early in the developmental process before tissue differentiation (Smeester and Fry, 2018; Rahman et al., 2017; Farzan et al., 2013; Smith et al., 2012). Unfortunately, the complexities of the epigenomic mechanism and other operative mechanisms make the use of mechanistic data for dose-response assessment difficult. To date, efforts to quantitatively correlate epigenomic changes with epidemiological outcomes have been limited. Rager et al. (2017) estimated arsenic doses that correspond to changes in "transcriptomic, proteomic, epigenomic, and integrated multiomic signatures in human cord blood through benchmark dose modeling" and found similarities in the doses at which epigenomic modification occurred and epidemiological outcomes were observed. Further exploration of such approaches is needed.

SUMMARY AND CONCLUSIONS

Human health risk assessment is a complex, multistep process where the available scientific evidence is combined to estimate the potential risks to public health from exposure to agents of concern, such as air and water pollutants. The products of human health risk assessment are used to support a wide variety of decisions, ranging from setting regulatory standards, to estimating risks to support programmatic priority setting, to identification of critical research needs. The rapid expansion of scientific evidence indicating epigenetic modification result from exposure to environmental agents provides an opportunity for new, innovative research to inform health risk assessment. To date, however, there has been little incorporation of epigenetic information into health risk assessments.

There is sufficient scientific basis to justify developing approaches to incorporate epigenomic modifications into risk assessment. Without this action, important public health risks may not be fully addressed. Numerous examples exist and demonstrate that exposure to environmental chemicals can induce epigenomic changes that result in adverse health effects. This chapter evaluates what types of information are most valuable in informing health assessments and proposes a framework with some decision points that can guide the use of epigenomic data in risk assessments. The proposed framework is anticipated to be of interest to researchers generating epigenomic data, risk assessors using epigenomic data, and risk managers basing decisions with consideration of epigenomic information.

Challenges to incorporating advanced molecular data into risk assessment include both normal and adverse perturbations of the epigenome that are very complex and at times difficult to interpret; furthermore, tools that facilitate their incorporation into risk assessment are limited. Lastly, risk assessors and risk managers, in general, are relatively unfamiliar with epigenomic data. Progress in these areas will necessitate the development of epigenomic risk assessment examples to address these issues.

Acknowledgments

The authors would like to thank Dr. Ronald Hines and Dr. Marie Fortin for their contributions to our consideration of epigenomic information in health risk assessment and their constructive comments on a draft of this chapter.

References

Aluru, N., 2017. Epigenomic effects of environmental chemicals: insights from zebrafish. Curr. Opin. Toxicol. 6 (October), 26–33. https://doi.org/10.1016/j.cotox.2017.07.004.

Angrish, M.M., Allard, P., McCullough, S.D., Druwe, I.L., Helbling Chadwick, L., Hines, E., Chorley, B.N., 2018. Epigenetic applications in adverse outcome pathways and environmental risk evaluation. Environ. Health Perspect. 126 (4), 045001. https://doi.org/10.1289/EHP2322.

Argos, M., 2015. Arsenic exposure and epigenetic alterations: recent findings based on the illumina 450K DNA methylation array. Curr. Environ. Health Rep. 2 (2), 137–144. https://doi.org/10.1007/s40572-015-0052-1. F.

Attene-Ramos, M.S., Huang, R., Sakamuru, S., Witt, K.L., Beeson, G.C., Shou, L., Schnellmann, R.G., Beeson, C.C., Tice, R.R., Austin, C.P., Xia, M., 2013a. Systematic study of mitochondrial toxicity of environmental chemicals using quantitative high throughput screening. Chem. Res. Toxicol. 26 (9), 1323–1332. https://doi.org/10.1021/tx4001754.

Attene-Ramos, M.S., Miller, N., Huang, R., Michael, S., Itkin, M., Kavlock, R.J., Austin, C.P., Shinn, P., Simeonov, A., Tice, R.R., Xia, M., 2013b. The Tox21 robotic platform for the assessment of environmental chemicals—from vision to reality. Drug Discov. Today. 18 (15–16), 716–723. https://doi.org/10.1016/j.drudis.2013.05.015.

Attene-Ramos, M.S., Huang, R., Michael, S., Witt, K.L., Richard, A., Tice, R.R., Simeonov, A., Austin, C.P., Xia, M., 2015. Profiling of the Tox21 chemical collection for mitochondrial function to identify compounds that acutely decrease mitochondrial membrane potential. Environ. Health Perspect. 123 (1), 49–56. https://doi.org/10.1289/ehp.1408642.

Baccarelli, A., Bollati, V., 2009. Epigenomics and environmental chemicals. Curr. Opin. Pediatr. 21 (2), 243–251.

Bailey, K.A., Wu, M.C., Ward, W.O., Smeester, L., Rager, J.E., García-Vargas, G., Del Razo, L.M., Drobná, Z., Stýblo, M., Fry, R.C., 2013. Arsenic and the epigenome: interindividual differences in arsenic metabolism related to distinct patterns of DNA methylation. J. Biochem. Mol. Toxicol. 27 (2), 106–115. https://doi.org/10.1002/jbt.21462.

Bock, C., 2012. Analysing and interpreting DNA methylation data. Nat. Rev. Genet. 13 (10), 705–719. https://doi.org/10.1038/nrg3273.

Boekelheide, K., Blumberg, B., Chapin, R.E., Cote, I., Graziano, J.H., Janesick, A., Lane, R., Lillycrop, K., Myatt, L., States, J.C., Thayer, K.A., Waalkes, M.P., Rogers, J.M., 2012. Predicting later-life outcomes of early-life exposures. Environ. Health Perspect. 120 (10), 1353–1361. https://doi.org/10.1289/ehp.1204934.

Bowers, E.C., McCullough, S.D., 2017. Linking the epigenome with exposure effects and susceptibility: the epigenomic seed and soil model. Toxicol. Sci. 155 (2), 302–314. https://doi.org/10.1093/toxsci/kfw215.

Brinkmann, M., Preuss, T.G., Hollert, H., 2017. Advancing in vitro-in vivo extrapolations of mechanism-specific toxicity data through toxicokinetic modeling. Adv. Biochem. Eng. Biotechnol. 157, 293–317. https://doi.org/10.1007/10_2015_5015.

Broberg, K., Ahmed, S., Engstrom, K., et al., 2014. Arsenic exposure in early pregnancy alters genome-wide DNA methylation in cord blood, particularly in boys. J. Dev. Orig. Health Dis. 5, 288–298.

Buesen, R., Chorley, B.N., da Silva, L.B., Daston, G., Deferme, L., Ebbels, T., Gant, T.W., Goetz, A., Greally, J., Gribaldo, L., Hackermüller, J., Hubesch, B., Jennen, D., Johnson, K., Kanno, J., Kauffmann, H.M., Laffont, M., McMullen, P., Meehan, R., Pemberton, M., Perdichizzi, S., Piersma, A.H., Sauer, U.G., Schmidt, K., Seitz, H., Sumida, K., Tollefsen, K.E., Tong, W., Tralau, T., van Ravenzwaay, B., Weber, R.J.M., Worth, A., Yauk, C., Poole, A., 2017. Applying 'omics technologies in chemicals risk assessment: report of an ECETOC workshop. Regul. Toxicol. Pharmacol. 91 (Suppl. 1), S3–S13. https://doi.org/10.1016/j.yrtph.2017.09.002 Epub 2017 Sep 25.

Cardenas, A., Houseman, E.A., Baccarelli, A.A., Quamruzzaman, Q., Rahman, M., Mostofa, G., Wright, R.O., Christiani, D.C., Kile, M.L., 2015. In utero arsenic exposure and epigenome-wide associations in placenta, umbilical artery, and human umbilical vein endothelial cells. Epigenetics 10 (11), 1054–1063. https://doi.org/10.1080/15592294.2015.1105424.

Cavalieri, V., Spinelli, G., 2017. Environmental epigenomics in zebrafish. Epigenomics Chromatin 10 (1), 46. https://doi.org/10.1186/s13072-017-0154-0.

Chappell, G., Kobets, T., O'Brien, B., Tretyakova, N., Sangaraju, D., Kosyk, O., Sexton, K.G., Bodnar, W., Pogribny, I.P., Rusyn, I., 2014. Epigenetic events determine tissue-specific toxicity of inhalational exposure to the genotoxic chemical 1,3-butadiene in male C57BL/6J mice. Toxicol. Sci. 142 (2), 375–384. https://doi.org/10.1093/toxsci/kfu191.

Cheung, N.K.M., Nakamura, R., Uno, A., Kumagai, M., Fukushima, H.S., Morishita, S., Takeda, H., 2017. Unlinking the methylome pattern from nucleotide sequence, revealed by large-scale in vivo genome engineering and methylome editing in medaka fish. PLoS Genet. 13(12). e1007123. https://doi.org/10.1371/journal.pgen.1007123.

Chiu, W.A., Slob, W., 2015. A unified probabilistic framework for dose-response assessment of human health effects. Environ. Health Perspect. 123 (12), 1241–1254. https://doi.org/10.1289/ehp.1409385 (Epub 2015 May 22).

Chiu, W.A., Wright, F.A., Rusyn, I., 2017. A tiered, Bayesian approach to estimating of population variability for regulatory decision-making. ALTEX 34 (3), 377–388. https://doi.org/10.14573/altex.1608251.

Chiu, W.A., Guyton, K.Z., Martin, M.T., Reif, D.M., Rusyn, I., 2018. Use of high-throughput in vitro toxicity screening data in cancer hazard evaluations by IARC Monograph Working Groups. ALTEX 35 (1), 51–64. https://doi.org/10.14573/altex.1703231.

Chu, S.H., Huang, Y.T., 2017. Integrated genomic analysis of biological gene sets with applications in lung cancer prognosis. BMC Bioinf. 18 (1), 336. https://doi.org/10.1186/s12859-017-1737-2.

Cote, I., Andersen, M.E., Ankley, G.T., Barone, S., Birnbaum, L.S., Boekelheide, K., Bois, F.Y., Burgoon, L.D., Chiu, W.A., Crawford-Brown, D., Crofton, K.M., DeVito, M., Devlin, R.B., Edwards, S.W., Guyton, K.Z., Hattis, D., Judson, R.S., Knight, D., Krewski, D., Lambert, J., Maull, E.A., Mendrick, D., Paoli, G.M., Patel, C.J., Perkins, E.J., Poje, G., Portier, C.J., Rusyn, I., Schulte, P.A., Simeonov, A., Smith, M.T., Thayer, K.A., Thomas, R.S., Thomas, R., Tice, R.R., Vandenberg, J.J., Villeneuve, D.L., Wesselkamper, S., Whelan, M.,

Whittaker, C., White, R., Xia, M., Yauk, C., Zeise, L., Zhao, J., DeWoskin, R.S., 2016. The next generation of risk assessment multi-year study-highlights of findings, applications to risk assessment, and future directions. Environ. Health Perspect. 124 (11), 1671–1682.

Cote, I., McCullough, S.D., Hines, R.N., Vandenberg, J.J., 2017. Application of epigenomic data in human health risk assessment. Curr. Opin. Toxicol. 6, 71–78. https://doi.org/10.1016/j.cotox.2017.09.002.

DeAngelis, J.T., Farrington, W.J., Tollefsbol, T.O., 2008. An overview of epigenetic assays. Mol. Biotechnol. 38, 179–183.

El-Masri, H., Kleinstreuer, N., Hines, R.N., Adams, L., Tal, T., Isaacs, K., Wetmore, B.A., Tan, Y.M., 2016. Integration of life-stage physiologically based pharmacokinetic models with adverse outcome pathways and environmental exposure models to screen for environmental hazards. Toxicol. Sci. 152 (1), 230–243. https://doi.org/10.1093/toxsci/kfw082.

Farzan, S.F., Karagas, M.R., Chen, Y., 2013. In utero and early life arsenic exposure in relation to long-term health and disease. Toxicol. Appl. Pharmacol. 272 (2), 384–390. https://doi.org/10.1016/j.taap.2013.06.030.

Fingerman IM, Zhang X, Ratzat W, Husain N, Cohen RF, Schuler GD. NCBI epigenomics: what's new for 2013, Nucleic Acids Res., Volume 41, Issue D1, 2013, D221–D225. (accessed February 6, 2018). Available at https://doi.org/10.1093/nar/gks1171.

Green, B.B., Marsit, C.J., 2015. Select prenatal environmental exposures and subsequent alterations of gene-specific and repetitive element DNA methylation in fetal tissues. Curr. Environ. Health Rep. 2 (2), 126–136. https://doi.org/10.1007/s40572-015-0045-0.

Han, Y., He, X., 2016. Integrating epigenomics into the understanding of biomedical insight. Bioinform. Biol. Insights 10, 267–289. https://doi.org/10.4137/BBI.S38427.

Heindel, J.J., Vandenberg, L.N., 2015. Developmental origins of health and disease: a paradigm for understanding disease cause and prevention. Curr. Opin. Pediatr. 27 (2), 248–253. https://doi.org/10.1097/MOP.0000000000000191.

Helsley, R.N., Zhou, C., 2017. Epigenomic impact of endocrine disrupting chemicals on lipid homeostasis and atherosclerosis: a pregnane X receptor-centric view. Environ. Epigenet. 3 (4) https://doi.org/10.1093/eep/dvx017 pii:dvx017.

Hou, L., Zhang, X., Wang, D., Baccarelli, A., 2012. Environmental chemical exposures and human epigenomics. Int. J. Epidemiol. 41, 79–105.

IPCS (International Programme on Chemical Safety) 2010 Characterization and application of physiologically based pharmacokinetic models in risk assessment, (Harmonization Project Document No. 9). World Health Organization Geneva. (accessed February 6, 2018). Available at http://www.inchem.org/documents/harmproj/harmproj/harmproj9.pdf.

IPCS (International Programme on Chemical Safety) 2014. Guidance Document on Evaluating and Expressing uncertainty in Hazard Characterization. (Harmonization Project Document 11). Geneva, World Health Organization. (accessed February 6, 2018). Available at http://www.inchem.org/documents/harmproj/harmproj/harmproj11.pdf.

Israel, J.W., Chappell, G.A., Simon, J.M., Pott, S., Safi, A., Lewis, L., Cotney, P., Boulos, H.S., Bodnar, W., Lieb, J.D., Crawford, G.E., Furey, T.S., Rusyn, I., 2018. Tissue- and strain-specific effects of a genotoxic carcinogen 1,3-butadiene on chromatin and transcription. Mamm. Genome 29 (1–2), 153–167. https://doi.org/10.1007/s00335-018-9739-6.

Jamei, M., 2016. Recent advances in development and application of physiologically-based pharmacokinetic (PBPK) models: a transition from academic curiosity to regulatory acceptance. Curr. Pharmacol. Rep. 2, 161–169.

Janesick, A.S., Blumberg, B., 2016. Obesogens: an emerging threat to public health. Am. J. Obstet. Gynecol. 214 (5), 559–565. https://doi.org/10.1016/j.ajog.2016.01.182.

Janesick, A.S., Shioda, T., Blumberg, B., 2014 Dec. Transgenerational inheritance of prenatal obesogen exposure. Mol. Cell. Endocrinol. 398 (1–2), 31–35. https://doi.org/10.1016/j.mce.2014.09.002.

Jaroch, K., Jaroch, A., Bojko, B., 2018. Cell cultures in drug discovery and development: the need of reliable in vitro-in vivo extrapolation for pharmacodynamics and pharmacokinetics assessment. J. Pharm. Biomed. Anal. 147, 297–312. https://doi.org/10.1016/j.jpba.2017.07.023.

Jirtle, R.L., Skinner, M.K., 2007 Apr. Environmental epigenomics and disease susceptibility. Nat. Rev. Genet. 8 (4), 253–262.

Judson, R.S., Kavlock, R.J., Setzer, R.W., Hubal, E.A., Martin, M.T., Knudsen, T.B., Houck, K.A., Thomas, R.S., Wetmore, B.A., Dix, D.J., 2011. Estimating toxicity-related biological pathway altering doses for high-throughput chemical risk assessment. Chem. Res. Toxicol. 24 (4), 451–462. https://doi.org/10.1021/tx100428e.

4. SPECIAL CONSIDERATIONS IN TOXICOEPIGENETICS RESEARCH

Kile, M.L., Houseman, E.A., Baccarelli, A.A., et al., 2014. Effect of prenatal arsenic exposure on DNA methylation and leukocyte subpopulations in cord blood. Epigenomics 9, 774–782.

Knudsen, T.B., Keller, D.A., Sander, M., Carney, E.W., Doerrer, N.G., Eaton, D.L., Fitzpatrick, S.C., Hastings, K.L., Mendrick, D.L., Tice, R.R., Watkins, P.B., Whelan, M., 2015. FutureTox II: in vitro data and in silico models for predictive toxicology. Toxicol. Sci. 143 (2), 256–267. https://doi.org/10.1093/toxsci/kfu234.

Knudsen, T.B., Klieforth, B., Slikker Jr., W., 2017. Programming microphysiological systems for children's health protection. Exp. Biol. Med. (Maywood) 242 (16), 1586–1592. https://doi.org/10.1177/1535370217717697 (Epub 2017 June 28).

Kochmanski, J., Montrose, L., Goodrich, J.M., Dolinoy, D.C., 2017. Environmental deflection: the impact of toxicant exposures on the aging epigenome. Toxicol. Sci. 156 (2), 325–335. https://doi.org/10.1093/toxsci/kfx005.

Koestler, D.C., Avissar-Whiting, M., Houseman, E.A., et al., 2013. Differential DNA methylation in umbilical cord blood of infants exposed to low levels of arsenic in utero. Environ. Health Perspect. 121, 971–977.

Ladd-Acosta, C., Fallin, M.D., 2016. The role of epigenomics in genetic and environmental epidemiology. Epigenomics 8 (2), 271–283. https://doi.org/10.2217/epi.15.102.

Laubach, Z.M., Perng, W., Dolinoy, D.C., Faulk, C.D., Holekamp, K.E., Getty, T., 2018. Epigenetics and the maintenance of developmental plasticity: extending the signalling theory framework. Biol. Rev. Camb. Philos. Soc. (Jan 21). https://doi.org/10.1111/brv.12396.

Marsit, C.J., 2015. Influence of environmental exposure on human epigenomic regulation. J. Exp. Biol. 218 (1), 71–79. https://doi.org/10.1242/jeb.106971.

Martin, E.M., Fry, R.C., 2016. A cross-study analysis of prenatal exposures to environmental contaminants and the epigenome: support for stress-responsive transcription factor occupancy as a mediator of gene-specific CpG methylation patterning. Environ. Epigenet. 2 (1), dvv011. https://doi.org/10.1093/eep/dvv011.

Martin, S.A., McLanahan, E.D., Bushnell, P.J., Hunter 3rd, E.S., El-Masri, H., 2015. Species extrapolation of life-stage physiologically-based pharmacokinetic (PBPK) models to investigate the developmental toxicology of ethanol using in vitro to in vivo (IVIVE) methods. Toxicol. Sci. 143 (2), 512–535. https://doi.org/10.1093/toxsci/kfu246.

Martin, E.M., Stýblo, M., Fry, R.C., 2017. Genetic and epigenomic mechanisms underlying arsenic-associated diabetes mellitus: a perspective of the current evidence. Epigenomics 9 (5), 701–710. https://doi.org/10.2217/epi-2016-0097.

Mauro, M., Caradonna, F., Klein, C.B., 2016. Dysregulation of DNA methylation induced by past arsenic treatment causes persistent genomic instability in mammalian cells. Environ. Mol. Mutagen. 57 (2), 137–150. https://doi.org/10.1002/em.21987.

McConnell, E.R., Bell, S.M., Cote, I., Wang, R.L., Perkins, E.J., Garcia-Reyero, N., Gong, P., Burgoon, L.D., 2014. Systematic omics analysis review (SOAR) tool to support risk assessment. PLoS One 9 (12), e110379. https://doi.org/10.1371/journal.pone.0110379.

McCullough, S.D., Bowers, E.C., On, D.M., Morgan, D.S., Dailey, L.A., Hines, R.N., Devlin, R.B., Diaz-Sanchez, D., 2016. Baseline chromatin modification levels may predict interindividual variability in ozone-induced gene expression. Toxicol. Sci. 150 (1), 216–224. https://doi.org/10.1093/toxsci/kfv324.

Mirbahai, L., Chipman, J.K., 2014. Epigenomic memory of environmental organisms: a reflection of lifetime stressor exposures. Mutat. Res. Genet. Toxicol. Environ. Mutagen. 764–765 (April), 10–17. https://doi.org/10.1016/j.mrgentox.2013.10.003.

Muenyi, C.S., Ljungman, M., States, J.C., 2015. Arsenic disruption of DNA damage responses-potential role in carcinogenesis and chemotherapy. Biomol. Ther. 5 (4), 2184–2193. https://doi.org/10.3390/biom5042184 (Review).

NCBI (National Center for Biotechnology Information), 2018. Epigenome Roadmap. Available at http://www.roadmapepigenomics.org/data/. [(Accessed 6 February 2018)].

NCBI GEO (National Center for Biotechnology Information Gene Expression Omnibus), 2018. Submitting high-throughput sequence data to GEO. Available at https://www.ncbi.nlm.nih.gov/geo/info/seq.html-intro. [(Accessed 6 February 2018)].

NRC (National Research Council), 1983. Risk Assessment in the Federal Government: Managing the Process. The National Academies Press, Washington, DC. https://doi.org/10.17226/366.

NRC (National Research Council), 2009. Science and Decisions: Advancing Risk Assessment. The National Academies Press, Washington, DC. https://doi.org/10.17226/12209.

NRC (National Research Council), 2013. Critical Aspects of EPA's IRIS Assessment of Inorganic Arsenic: Interim Report. The National Academies Press, Washington, DC. https://doi.org/10.17226/18594.

Olden, K., Olden, H.A., Lin, Y.S., 2015. The role of the epigenome in translating neighborhood disadvantage into health disparities. Curr. Environ. Health Rep. 2 (2), 163–170. https://doi.org/10.1007/s40572-015-0048-x.

Olden, K., Lin, Y., Bussard, D., 2016. Epigenome: a biomarker or screening tool to evaluate health impact of cumulative exposure to chemical and non-chemical stressors. Biosensors (Basel) 6 (2), 12.

Parfett, C.L., Desaulniers, D., 2017. A Tox21 approach to altered epigenomic landscapes: assessing epigenomic toxicity pathways leading to altered gene expression and oncogenic transformation in vitro. Int. J. Mol. Sci. 18 (6) https://doi.org/10.3390/ijms18061179 pii:E1179.

Pearce, R., Strope, C., Setzer, W., Sipes, N., Wambaugh, J., 2017. HTTK: R package for high-throughput toxicokinetics. J. Stat. Softw. 79 (4), 1–26.

Pilsner, J.R., Hall, M.N., Liu, X., Ilievski, V., Slavkovich, V., Levy, D., Factor-Litvak, P., Yunus, M., Rahman, M., Graziano, J.H., Gamble, M.V., 2012. Influence of prenatal arsenic exposure and newborn sex on global methylation of cord blood DNA. PLoS ONE. 7(5), e37147. https://doi.org/10.1371/journal.pone.0037147.

Pollock, R.A., Abji, F., Gladman, D.D., 2017. Epigenomics of psoriatic disease: a systematic review and critical appraisal. J. Autoimmun. 78, 29–38. https://doi.org/10.1016/j.jaut.2016.12.002.

Rager, J.E., Auerbach, S.S., Chappell, G.A., Martin, E., Thompson, C.M., Fry, R.C., 2017. Benchmark dose modeling estimates of the concentrations of inorganic arsenic that induce changes to the neonatal transcriptome, proteome, and epigenome in a pregnancy cohort. Chem. Res. Toxicol. 30 (10), 1911–1920. https://doi.org/10.1021/acs.chemrestox.7b00221.

Rahman, A., Granberg, C., Persson, L.Å., 2017. Early life arsenic exposure, infant and child growth, and morbidity: a systematic review. Arch. Toxicol. 91 (11), 3459–3467. https://doi.org/10.1007/s00204-017-2061-3.

Reichard, J.F., Puga, A., 2010 Feb. Effects of arsenic exposure on DNA methylation and epigenetic gene regulation. Epigenomics 2 (1), 87–104. https://doi.org/10.2217/epi.09.45.

Roadmap Epigenomics Consortium, Kundaje, A., Meuleman, W., Ernst, J., Bilenky, M., Yen, A., et al., 2015. Integrative analysis of 111 reference human epigenomes. Nature 518 (7539), 317–330. https://doi.org/10.1038/nature14248.

Saban, K.L., Mathews, H.L., DeVon, H.A., Janusek, L.W., 2014. Epigenomics and social context: implications for disparity in cardiovascular disease. Aging Dis. 5 (5), 346–355. https://doi.org/10.14336/AD.2014.0500346.

Sales, V.M., Ferguson-Smith, A.C., Patti, M.E., 2017. Epigenomic mechanisms of transmission of metabolic disease across generations. Cell Metab. 25 (3), 559–571. https://doi.org/10.1016/j.cmet.2017.02.016.

Shao, K., Allen, B.C., Wheeler, M.W., 2017. Bayesian hierarchical structure for quantifying population variability to inform probabilistic health risk assessments. Risk Anal. 37 (10), 1865–1878. https://doi.org/10.1111/risa.12751.

Smeester, L., Fry, R.C., 2018. Long-term health effects and underlying biological mechanisms of developmental exposure to arsenic. Curr. Environ. Health Rep. https://doi.org/10.1007/s40572-018-0184-1.

Smith, A.H., Marshall, G., Liaw, J., Yuan, Y., Ferreccio, C., Steinmaus, C., 2012. Mortality in young adults following in utero and childhood exposure to arsenic in drinking water. Environ. Health Perspect. 120 (11), 1527–1531. https://doi.org/10.1289/ehp.1104867.

Smith, M.T., Guyton, K.Z., Gibbons, C.F., Fritz, J.M., Portier, C.J., Rusyn, I., DeMarini, D.M., Caldwell, J.C., Kavlock, R.J., Lambert, P.F., Hecht, S.S., Bucher, J.R., Stewart, B.W., Baan, R.A., Cogliano, V.J., Straif, K., 2016. Key characteristics of carcinogens as a basis for organizing data on mechanisms of carcinogenesis. Environ. Health Perspect. 124 (6), 713–721. https://doi.org/10.1289/ehp.1509912.

Sosnowski, D.W., Booth, C., York, T.P., Amstadter, A.B., Kliewer, W., 2018. Maternal prenatal stress and infant DNA methylation: a systematic review. Dev. Psychobiol. https://doi.org/10.1002/dev.21604.

Tammen, S.A., Friso, S., Choi, S.W., 2013. Epigenomics: the link between nature and nurture. Mol. Asp. Med. 34 (4), 753–764. https://doi.org/10.1016/j.mam.2012.07.018.

Thomson, J.P., Moggs, J.G., Wolf, C.R., Meehan, R.R., 2014. Epigenetic profiles as defined signatures of xenobiotic exposure. Mutat. Res. Genet. Toxicol. Environ. Mutagen. 764–765, 3–9. https://doi.org/10.1016/j.mrgentox.2013.08.007.

Tice, R.R., Austin, C.P., Kavlock, R.J., Bucher, J.R., 2013. Improving the human hazard characterization of chemicals: a Tox21 update. Environ. Health Perspect. 121 (7), 756–765. https://doi.org/10.1289/ehp.1205784.

Treas, J.N., Tyagi, T., Singh, K.P., 2012. Effects of chronic exposure to arsenic and estrogen on epigenetic regulatory genes expression and epigenetic code in human prostate epithelial cells. PLoS One 7 (8), e43880. https://doi.org/10.1371/journal.pone.0043880.

Tsang, V., Fry, R.C., Niculescu, M.D., Rager, J.E., Saunders, J., Paul, D.S., Zeisel, S.H., Waalkes, M.P., Stýblo, M., Drobná, Z., 2012. The epigenetic effects of a high prenatal folate intake in male mouse fetuses exposed in utero to arsenic. Toxicol. Appl. Pharmacol. 264 (3), 439–450. https://doi.org/10.1016/j.taap.2012.08.022.

US EPA (Environmental Protection Agency). 2002. A review of the reference dose and reference concentration processes. (accessed February 6, 2018). EPA/630/P-02/002F, Washington, DC. Available at https://www.epa.gov/sites/production/files/2014-12/documents/rfd-final.pdf.

US EPA (Environmental Protection Agency). 2006. Approaches for the application of physiologically based pharmacokinetic (PBPK) models and supporting data in risk assessment (final report). (accessed February 6, 2018). EPA/600/R-05/043F, Washington, DC. Available at http://cfpub.epa.gov/ncea/cfm/recordisplay.cfm?deid=157668.

US EPA (Environmental Protection Agency), 2012a. Advances in inhalation gas dosimetry for derivation of a reference concentration (RfC) and use in risk assessment. EPA/600/R-12/044, Research Triangle Park, NC.

US EPA (Environmental Protection Agency), 2012b. Benchmark Dose Technical Guidance. Research Triangle Park, NC, Available at https://www.epa.gov/bmds. [(Accessed 6 February 2018)].

US EPA (Environmental Protection Agency), 2014. Guidance for applying quantitative data to develop data-derived extrapolation factors for interspecies and intraspecies extrapolation. EPA/100/R-14/002, Washington, DC.

US EPA (Environmental Protection Agency), 2015. Preamble to the Integrated Science Assessments. Research Triangle Park, NC, Available at https://cfpub.epa.gov/ncea/isa/recordisplay.cfm?deid=310244. [(Accessed 6 February 2018)].

US EPA (Environmental Protection Agency), 2017. Exposure Factor Handbook. Washington, DC, Available at https://www.epa.gov/expobox/about-exposure-factors-handbook. [(Accessed 6 February 2018)].

Wallack, L., Thornburg, K., 2016. Developmental origins, epigenomics, and equity: moving upstream. Matern. Child Health J. 20 (5), 935–940. https://doi.org/10.1007/s10995-016-1970-8.

Wetmore, B.A., 2015. Quantitative in vitro-to-in vivo extrapolation in a high-throughput environment. Toxicology 332, 94–101. https://doi.org/10.1016/j.tox.2014.05.012.

Williams, A.J., Grulke, C.M., Edwards, J., McEachran, A.D., Mansouri, K., Baker, N.C., Patlewicz, G., Shah, I., Wambaugh, J.F., Judson, R.S., Richard, A.M., 2017. The CompTox Chemistry Dashboard: a community data resource for environmental chemistry. J. Cheminform. 9 (1), 61. https://doi.org/10.1186/s13321-017-0247-6.

Xin, F., Susiarjo, M., Bartolomei, M.S., 2015. Multigenerational and transgenerational effects of endocrine disrupting chemicals: a role for altered epigenomic regulation? Semin. Cell. Dev. Biol. 43, 66–75. https://doi.org/10.1016/j.semcdb.2015.05.008.

Yan, H., Tian, S., Slager, S.L., Sun, Z., Ordog, T., 2016. Genome-wide epigenetic studies in human disease: a primer on -omic technologies. Am. J. Epidemiol. 183 (2), 96–109. https://doi.org/10.1093/aje/kwv187.

Further Reading

Angrish, M.M., McQueen, C.A., Cohen-Hubal, E., Bruno, M., Ge, Y., Chorley, B.N., 2017. Editor's highlight: mechanistic toxicity tests based on an adverse outcome pathway network for hepatic steatosis. Toxicol. Sci. 159 (1), 159–169. https://doi.org/10.1093/toxsci/kfx121.

Best, L.G., García-Esquinas, E., Yeh, J.L., Yeh, F., Zhang, Y., Lee, E.T., Howard, B.V., Farley, J.H., Welty, T.K., Rhoades, D.A., Rhoades, E.R., Umans, J.G., Navas-Acien, A., 2015. Association of diabetes and cancer mortality in American Indians: the Strong Heart Study. Cancer Causes Control 26 (11), 1551–1560. https://doi.org/10.1007/s10552-015-0648-7.

Faulk, C., in press. Implications of DNA methylation in toxicology. In: Epigenetic Toxicology: Core Principles and Applications.

IARC (International Agency for Research on Cancer), 2012. A review of human carcinogens: arsenic, metals, fibres, and dusts. IARC monographs on the evaluation of carcinogenic risks to humans, vol. 100C, Lyon, France. Available at http://monographs.iarc.fr/ENG/Monographs/vol100C/mono100C.pdf. [(Accessed 6 February 2018)].

NTP (National Toxicology Program). 2016. Report on Carcinogens. 14th ed.; Research Triangle Park, NC: U.S. Department of Health and Human Services, Public Health Service. (accessed February 6, 2018). Available at http://ntp. niehs.nih.gov/go/roc14.

Smeester, L., Rager, J.E., Bailey, K.A., Guan, X., Smith, N., Garcia-Vargas, G., Del Razo, L.M., Drobna, Z., Kelkar, H., Styblo, M., Fry, R.C., 2011. Epigenomic changes in individuals with arsenicosis. Chem. Res. Toxicol. 24 (2), 165–167.

WHO (World Health Organization), 2011. Arsenic in drinking-water. WHO/SDE/WSH/03.04/75/rev1, Geneva, Switzerland. Available at http://www.who.int/water_sanitation_health/publications/arsenic/en/. [(Accessed 6 February 2018)].

Woolard, E., Chorley, B.N., in press. The role of non-coding RNAs in gene regulation. In: Epigenetic Toxicology: Core Principles and Applications.

PROTOCOLS FOR TOXICOEPIGENETICS RESEARCH

Chromatin Immunoprecipitation: An Introduction, Overview, and Protocol ☆

Elizabeth M. Martin*,†, *Doan M. On**,‡, *Emma C. Bowers*†, *Shaun D. McCullough**,1

*National Health and Environmental Effects Research Laboratory, US Environmental Protection Agency, Research Triangle Park, NC, United States †Curriculum in Toxicology, University of North Carolina—Chapel Hill, Chapel Hill, NC, United States ‡Department of Pharmacology and Toxicology, Medical College of Virginia, Richmond, VA, United States

OUTLINE

☆ Most of the material (text, figures, and tables) presented here has been previously published in *Current Protocols in Toxicology* (Wiley) and used in accordance with the original publisher's copyright policies. Reference: McCullough SD, On DM, Bowers EC (2017). Using Chromatin Immunoprecipitation in Toxicology: A Step-by-Step Guide to Increasing Efficiency, Reducing Variability, and Expanding Applications. *Curr. Protoc. Toxicol.* 72: 3.14.1–3.14.28.

[1] The contributions made by Shaun McCullough is in Public domain

INTRODUCTION

The genomic DNA of every eukaryotic cell is organized on nucleosomes, repeating heterooctameric structures containing two copies of histone 2A (H2A), H2B, H3, and H4, collectively referred to as chromatin. While the genome contains the blueprint for cellular structure and function, the accessibility and the use of this information are regulated by covalent modifications to the DNA and its histone protein scaffolding known as the epigenome. In contrast to DNA, histones are subject to a broad range of posttranslational modifications, including methylation, acetylation, phosphorylation, ubiquitination, and SUMOylation (Kouzarides, 2007). To date, more than 130 discrete histone modifications have been identified (Tan et al., 2011), and along with DNA methylation, these modifications are analogous to individual letters in a complex alphabet that encodes the instruction manual for the regulation of gene expression. This epigenetic code is "read" by specific binding domains in chromatin-associated proteins, which control chromatin accessibility and the recruitment of the transcriptional machinery (Strahl and Allis, 2000; Jenuwein and Allis, 2001; Cedar and Bergman, 2009). The epigenome is dynamic and responsive to both chemical and nonchemical aspects of an individual's environment (Baccarelli and Bollati, 2009; Burris and Baccarelli, 2014). While environmentally induced changes in some epigenetic modifications are relatively transient, others are persistent and have the potential to impact cellular function and organismal health throughout an organism's lifetime and across generations. Further, pre-exposure levels of certain histone modifications have been shown to correlate with the magnitude of postexposure gene induction (McCullough et al., 2016), suggesting that baseline histone modification levels have potential as predictors of exposure outcomes. Furthering our understanding of interactions between the environment and the epigenome will provide novel insight into interindividual variability in susceptibility, mechanisms underlying exposure-related disease, multi- and transgenerational health effects, and modifiable factors that can be leveraged to mitigate exposure effects.

The NIH Roadmap Epigenomics Consortium recently highlighted the variation in genome-wide patterns of histone modifications across gene regulatory regions and between cell and tissue types (Roadmap Epigenomics Consortium). Chromatin immunoprecipitation (ChIP) allows the researcher to explore the distribution and abundance of specific histone modifications either globally (via ChIP-seq) or at specific loci of interest (via ChIP-qPCR). While Basic Protocol 1 describes the protocol for ChIP-qPCR, this protocol can be easily adapted for ChIP-seq experiments. The ChIP protocol presented here is generally applicable to cells grown in culture and peripheral leukocytes isolated from whole blood (Support Protocol 2) and provides guidance on optimizing critical parameters of the protocol for use with other cell lines/types and target histone modifications/genomic loci of interest. This protocol can also be adapted for use with tissue or biopsy specimens. Specific details are presented here for conducting ChIP in primary bronchial epithelial cells grown in air-liquid interface (ALI) culture and the monocytic leukemia cell line THP-1 for specific histone modifications within the regulatory regions of four genes (*IL-8*, *IL-6*, *COX2*, and *HMOX1*) that are commonly considered in toxicological studies.

STRATEGIC PLANNING

It is important to optimize the ChIP protocol for the cell type/line of interest and carefully plan studies that will involve ChIP experiments. Optimization of critical aspects of the protocol, such as the use of a sufficient number of cells per sample for the histone modification and genomic locus of interest, sample fixation conditions, chromatin fragmentation parameters, and choice of antibody, facilitates assay success and consistency. After conditions have been optimized, the number of ChIP assays for each antibody target should be estimated and used to calculate the total amount of antibody required for all the technical and biological replicates in the study. The required amount of antibody should be obtained, combined, and separated into single-use aliquots. Doing so will increase reproducibility across experiments within the study (explained further in critical parameters below). Care should be taken at all steps to limit technical variability thus increasing reproducibility.

BASIC PROTOCOL 1

Quantifying the Abundance of Epigenetic Modifications Within the Regulatory Regions of Target Genes

This protocol describes the use of chromatin immunoprecipitation (ChIP) in cultured cells to quantify the relative abundance of specific epigenetic modifications, transcription factors, and other chromatin-associated proteins within target areas of the genome. The protocol can be used for either adherent cells (cell lines or primary cells), cells grown in suspension, or peripheral leukocytes isolated from whole blood. Chromatin within a given genomic region typically contains a variety of epigenetic modifications; however, it is important to note that not all epigenetic modifications exist within the regulatory regions of all genes and thus may not be universally detectable. The ChIP protocol described below has been optimized for the detection of histone H3 lysine 4 trimethylation (H3K4me3), H3K27me3, H3K27ac, pan-acetylated histone H4 (H4ac), total H3, and 5-hydroxymethylcytosine (5-hmC) within the promoter regions of interleukin-8 (*IL-8*), *IL-6*, cyclooxygenase 2 (*COX2* also known as *PTGS2*), and hemeoxygenase 1 (*HMOX1*); however, it can serve as a starting point for the quantification of other histone modifications within additional regulatory regions of interest. An overview of the ChIP workflow is shown in Fig. 1. The protocol described here represents a traditional approach to ChIP that is suitable for analyzing interactions between target histone modifications and/or chromatin-associated proteins (e.g., histone-modifying enzymes,

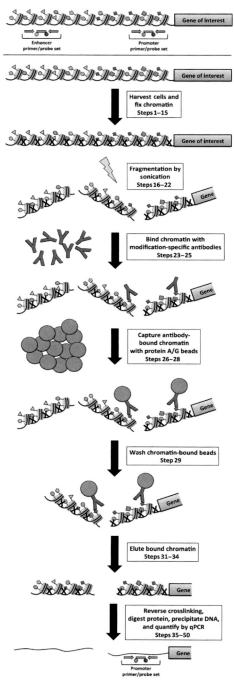

FIG. 1 Workflow for chromatin immunoprecipitation. Chromatin immunoprecipitation allows for the relative quantification of histone modifications within the regulatory regions of target genomic loci. Protein-DNA interactions are chemically cross-linked together by treatment with formaldehyde prior to being fragmented by sonication. Fragmented chromatin containing target histone modifications is bound by target-specific antibodies, and the antibody-chromatin complexes are captured by protein A-bound beads. After washing, bound chromatin is eluted from the beads, the cross-links are reversed, proteins are digested with proteinase K, and the purified DNA is precipitated. Isolated DNA is quantified by qPCR.

transcription factors, and coactivators/corepressors) and genomic loci of interest; however, there are alternatives such as native ChIP (O'Neill and Turner, 2003) and microchip (Dahl and Collas, 2008) that have more specialized applications, which will not be discussed here.

Materials

Cells in culture
ChIP-grade antibodies (see Table 1 for antibodies that we have used
successfully with this protocol)
Dulbecco's phosphate-buffered saline (DPBS, room temperature)
0.25% Trypsin (for adherent cultures)
DPBS-TI (for adherent cultures)
Cell strainer (70 μm pore)
16% Paraformaldehyde (in single-use ampules)
2.5 M Glycine

TABLE 1   Antibodies Successfully Used in ChIP as Described in Basic Protocol 1

ChIP Target	Mono/Polyclonal	Antibody Manufacturer	Product Number
H2BK5ac	Polyclonal	Active motif	39123
H2BK120ub	Monoclonal	Active motif	39623
H2BK120ub	Monoclonal	Millipore	17-650
Total H3	Polyclonal	Active motif	39163
Total H3	Monoclonal	Active motif	61475
H3K4me1	Polyclonal	Active motif	61633
H3K4me3	Polyclonal	Active motif	39915
H3K4me3	Monoclonal	Active motif	61379
H3K9ac	Monoclonal	Active motif	61251
H3K27me3	Monoclonal	Active motif	61017
H3K27me2/3	Polyclonal	Active motif	39535
H3K27ac	Polyclonal	Active motif	39133
H3K27ac	Monoclonal	Active motif	39685
H3K79me1	Polyclonal	Active motif	39921
H4ac	Polyclonal	Active motif	39925
H4K20me1	Monoclonal	Active motif	39727
5-hmC	Monoclonal	Active motif	39999
RNA PolII	Monoclonal	Active motif	39097
RNA PolII	Monoclonal	Millipore	17-620
RNA PolII phosphor-Ser5	Polyclonal	Active motif	39749

Protease inhibitor solution (PIS, ice cold; see recipe)
Hemocytometer
Liquid nitrogen
Sonicator with microprobe tip (Fisher Scientific Sonic Dismembrator Model 500)
End-over-end rotator
Digital temperature-controlled dry bath
ChIP lysis buffer (see recipe)
ChIP dilution buffer (see recipe)
Low-salt wash buffer (see recipe)
High-salt wash buffer (see recipe)
LiCl wash buffer (see recipe)
TE buffer (see recipe)
Elution buffer (see recipe)
Elution addition buffer (see recipe)
3 M Sodium acetate and pH 5.2 (see recipe)
100% Ethanol
Protein A/G polyacrylamide beads
10 mg/mL Proteinase K (see recipe)
Phenol/chloroform/isoamyl alcohol (25:24:1)
Chloroform/isoamyl alcohol (1:1)
10 mg/mL Glycogen (see recipe)

Protocol Steps

Cell Collection

Adherent cells in submerged culture

1. Following the completion of the desired exposure, aspirate the medium (for submerged cells) and wash with DPBS.
2. Add 0.25% trypsin and incubate at 37°C to detach cells from the culture dish:
   (a) *Optimization*: the duration of trypsinization required to remove cells from the culture dish will vary between cell types. Determine the optimal trypsinization time for your cell type(s) of interest as undertrypsinization will reduce the cell yield and overtrypsinization will result in cell clumping. Depending on the cell type, a less stringent digestion, for example, using Accutase (Sigma), may be preferred.
   (b) *Reproducibility*: the activity of enzyme solutions changes with each warming and cooling cycle. Prepare single-use aliquots of the trypsin solution to decrease variability due to differential trypsinization.
   (c) *Variation*: if using cells in air-liquid interface cultures, then add trypsin to the apical and basolateral compartments of the Transwell insert. Incubate at 37°C for 5–10 min or until cells detach:
       (i) For 24 mm inserts, use 500 and 1500 μL, respectively.
       (ii) For 12 mm inserts, use 250 and 750 μL, respectively.
       (iii) For 6.5 mm inserts, use 100 and 500 μL, respectively.
       (iv) *Note*: cell detachment should be monitored closely to prevent overtrypsinization, which will reduce reproducibility between replicates.

(d) *Variation*: if using THP-1, then trypsinization is not required. Instead, transfer the medium containing cells to a suitably sized conical tube, wash the plate with fresh growth medium, and add the wash medium to the same conical tube. Proceed to step 5.

3. Immediately after cells become detached, add an equal volume of DPBS-TI.

4. Collect the trypsinized cells by adding ice-cold DPBS-TI and triturating to dissociate clumped cells:

   (a) *Reproducibility*: if multiple culture dishes are being used to generate sufficient starting material for each treatment condition, then combine them into a single conical tube.

5. Pellet cells by centrifugation at $1000 \times g$ for 4 min at 4°C.

6. Carefully aspirate the supernatant, resuspend the cell pellet in 1 mL of room-temperature DPBS, and add DPBS to give a final cell density of 1×10^6 cells/mL:

   (a) *Reproducibility*: the cell suspension does not need to be quantified at this point since the added time may augment the epigenetic modifications of interest; however, an average number of cells per culture dish for each treatment condition should be determined prior to conducting ChIP experiments. Consistent cell density during fixation will increase the reproducibility of the procedure.

7. Remove cell clumps by passing the cell suspension through a 70 μm pore cell strainer.

8. Cross-link the chromatin by adding 16% paraformaldehyde to the strained cell suspension to a final concentration of 1% and mixing by gentle inversion. Incubate at room temperature for 5 min without agitation:

   (a) *Reproducibility*: paraformaldehyde ampules are preferable, and paraformaldehyde should only be used within 24 h of the ampule being opened. The use of larger volume commercial preparations of formaldehyde solution is discouraged since it will reduce chromatin yield when "older" fixative solutions are used.

9. Quench the fixation reaction by adding 2.5 M glycine to a final concentration of 125 mM and mixing by gentle inversion. Incubate for 5 min at room temperature without agitation.

10. Pellet cross-linked cells at $1000 \times g$ for 4 min at room temperature.

11. Aspirate the supernatant, and wash the fixed cells by resuspending in 1 mL of ice-cold PIS then adding 9 mL of additional ice-cold PIS and mixing by gentle inversion.

12. Take an aliquot (dilute if desired), and enumerate the cells with a hemocytometer.

13. Pellet the cells at $1000 \times g$ for 4 min at room temperature.

14. Aspirate the supernatant, and resuspend the cells at a concentration that is twice the desired number of cells per immunoprecipitation (IP) per milliliter (e.g., if 1×10^6 cells per IP is desired, then prepare a cell suspension of 2×10^6 cells/mL):

   (a) *Note*: the number of cells that should be used for each IP may vary depending on the target and genomic locus of interest and should be empirically determined.

15. Aliquot 1 mL of the cell suspension into individual Eppendorf tubes, pellet the cells at $1000 \times g$ for 4 min at 4°C, and aspirate the supernatant:

   (a) *Stopping point*: if desired, the cell pellets can be flash frozen in liquid nitrogen and stored at −80°C. Freezing in liquid nitrogen is optimal for preserving protein modifications.

   (b) *Reproducibility*: if pellets are frozen prior to continuing, then ensure that all replicates/samples are similarly frozen before proceeding to the fragmentation step.

Chromatin fragmentation and immunoprecipitation

16. If cell pellets were frozen after step 15, then thaw cell pellets on ice.

17. Resuspend the cell pellet in 300 μL of room-temperature ChIP lysis buffer and incubate on ice for 10 min:

    (a) *Note*: ChIP lysis buffer has a high concentration of SDS, which will precipitate if the buffer gets too cold. To prevent precipitation, lyse cells by placing tubes transversely on top of ice.

18. Fragment the chromatin:

    (a) *Note*: this protocol uses a probe-tip sonicator, but enzymatic techniques are available, as are other sonication instruments (e.g., Bioruptor).

    (b) *Note*: place each tube on ice for at least 90 s between each sonication cycle. Excess heating will compromise the samples and reduce IP efficiency.

    (c) Sonication conditions for primary bronchial epithelial cells when using the Fisher Scientific Sonic Dismembrator Model 500.

        (i) Sonicate the fixed cell pellets for seven cycles, each consisting of 10 pulses (0.9 s on and 0.1 s off) at 20% output.

    (d) Sonication conditions for THP-1 cells when using the Fisher Scientific Sonic Dismembrator Model 500:

        (ii) Sonicate the fixed cell pellets for six cycles, each consisting of 10 pulses (0.9 s on and 0.1 s off) at 20% output.

        (iii) Fig. 2 demonstrates the effect of sonication cycle number on chromatin fragmentation in fixed THP-1 cells.

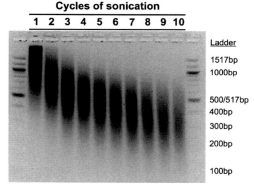

FIG. 2 Optimization of chromatin fragmentation in THP-1 cells. The chromatin fragmentation conditions should be optimized for each cell type/line and fixation condition used for ChIP. These results demonstrate the effect of successive cycles of sonication on THP-1 chromatin fragmentation. Samples were generated by fixing THP-1 cells with formaldehyde for 5 min prior to preparing pellets containing 2.0×10^6 cells. Pellets were thawed, resuspended in 600 μL ChIP lysis buffer, and divided into equal fractions, so each sonication sample contained 1×10^6 cells in 300 μL. The 300 μL samples were then subjected to a range of sonication cycles where each cycle consists of 10 pulses (0.9 s on and 0.1 s off) at 20% output on a Fisherbrand Sonic Dismembrator 500. DNA was isolated from fragmented chromatin as described in Support Protocol 1 and separated on a 3% agarose-TBE gel containing 0.4 μg/mL ethidium bromide. After running, the gel was destained overnight in TBE and imaged while being subjected to transillumination with UV light. These results demonstrate that six cycles of sonication yield chromatin fragments with an optimal size for ChIP (average length of 500 bp with a range of 250–750 bp).

(e) *Optimize*: the duration, power, and number of sonication cycles must be empirically determined for each cell type that will be used in the ChIP protocol. Further, the sonication conditions will depend on the wattage of the sonicator being used. The optimal number of cycles can be determined using Support Protocol 1. Using the data shown in Fig. 2, optimal fragmentation was accomplished with six cycles of sonication.

(f) *Reproducibility*: if multiple pellets from the same sample are being used to achieve the required amount of starting material, then combine the fragmented chromatin into a single prechilled Eppendorf tube prior to proceeding.

19. Clarify the fragmented chromatin by centrifugation at $\geq 20{,}000 \times g$ for 10 min at 10°C.

    (a) *Note*: centrifugation results in the collection of any insoluble material at the bottom of the tube.

20. Transfer the supernatant (clarified chromatin) to a prechilled conical tube on ice.

21. Dilute the clarified chromatin with nine volumes of ice-cold ChIP dilution buffer.

22. Take a sample of the diluted chromatin to be used as the "input" sample and store at −80°C until step 30:

    (a) The input sample is typically either 1% or 5% of the diluted chromatin volume. Make note of the percentage of the diluted chromatin taken for the input sample(s) as it will be needed during data analysis. If the chromatin is pooled prior to beginning the IPs, then only one input sample for each treatment condition (instead of each sample) is required.

23. Divide the diluted chromatin evenly into individual snap-lock Eppendorf tubes (one tube for each IP):

    (a) *Note*: snap-lock Eppendorf tubes are used to prevent the caps from opening during the subsequent antibody and bead-binding steps.

24. Add the antibody targeting the epigenetic modification of interest to each tube:

    (a) *Optimize*: the amount of antibody used for each IP must be determined empirically for each antibody and cell type. Typically, an IP from 1×10^6 cells will require between 1 and 15 µg of antibody.

    (b) *Reproducibility*: there is considerable lot-to-lot variability in antibody IP efficiency, especially in polyclonal antibodies. Avoid directly comparing data from IPs conducted with different antibodies and antibody preparations from different production lots. It is recommended that the user either acquire a sufficient amount of antibody preparation to complete the study from a single lot or pool different lots of the same antibody and prepare single-use aliquots before beginning the project. When possible, use monoclonal antibodies to reduce the variability in avidity that can occur between lots of polyclonal antibodies. When possible, use purified antibody preparations to reduce interactions between nonspecific antibodies and the IP reaction.

25. Incubate overnight at 4°C on an end-over-end rotator.

Capture antibody-chromatin complexes

26. Prepare protein A/G polyacrylamide beads:

    (a) For *n* immunoprecipitations, transfer $(n + 2) \times 50$ µL of protein A/G polyacrylamide bead slurry into a prechilled conical tube:

       (i) *Note*: determine the bead volume of the slurry. Many agarose and polyacrylamide bead preparations are supplied as a 50% slurry (1:1 volume of beads and buffer).

  (b) Pellet beads at $3000 \times g$ for 2 min at 4°C:

       (i) *Note*: if using agarose beads instead of polyacrylamide beads, spin at $1000 \times g$ for all subsequent bead collection steps.

  (c) Carefully aspirate the supernatant, leaving a small volume to prevent bead loss, and resuspend the beads in 10 bead volumes of ice-cold ChIP dilution buffer:

       (i) *Note*: use a 25 ga needle attached to a vacuum aspirator.

  (d) Wash the beads by repeating steps 25b and c.

  (e) Pellet beads at $3000 \times g$ for 2 min at 4°C.

  (f) Resuspend the beads in $(n+2) \times 24\,\mu L$ of ice-cold ChIP dilution buffer.

27. Add 50 μL of the bead slurry to each IP tube and incubate on an end-over-end rotator for 1 h at 4°C:

  (i) *Note*: vortex the bead slurry briefly before transferring to each IP tube.

28. Pellet the bead-antibody-chromatin complexes at $3000 \times g$ for 2 min at 4°C.

29. Aspirate the supernatant with a 25 ga needle, and wash the complexes with 1 mL of each of the following buffers for 2 min on an end-over-end rotator at room temperature. Resuspend the beads in each wash by pipetting. Do not vortex:

  (a) Low-salt wash buffer (room temperature)

  (b) High-salt wash buffer (room temperature)

  (c) LiCl wash buffer (room temperature)

  (d) TE buffer (room temperature)

       (i) After resuspending in TE buffer, transfer the bead slurry into a new, prechilled Eppendorf tube before pelleting the beads by centrifugation. This added step reduces carryover of contaminants bound to the walls of the original Eppendorf tube used in the IP step.

30. Thaw input samples from step 22. Digest RNA by adding 20 μg of RNaseA to both the input and IP samples and incubate at 37°C for 30 min.

31. Pellet beads at $3000 \times g$ for 2 min at room temperature, and carefully aspirate the supernatant.

32. Elute immunoprecipitated material from the beads by adding 265 μL of SDS elution buffer, vortexing, and incubating at 65°C for 5 min.

33. Pellet beads at $3000 \times g$ for 2 min at room temperature, and transfer 250 μL of the supernatant to a new Eppendorf tube.

34. Repeat steps 32 and 33, and combine the two elution volumes (now referred to as the "elution") in a single Eppendorf tube.

Crosslink reversal, proteinase digestion, and DNA purification

35. Add SDS elution buffer to input samples (thawed in step 30) to a total volume of 500 μL.

36. Add 50 μL of elution addition buffer to the diluted input (500 μL) and pooled elution (500 μL) from each IP and vortex.

37. Add 2 μL of 10 mg/mL proteinase K to each sample and incubate at 65°C for 4 h:

  (a) *Note*: if desired, this incubation step can be extended to overnight; however, to ensure reproducibility between samples, the same incubation duration should be used for the entire study.

38. Extract immunoprecipitated DNA by adding an equal volume (550 μL) of phenol/chloroform/isoamyl alcohol (PCI; 25:24:1 v/v ratio), vortexing thoroughly, and centrifuging at 13,000 × *g* for 3 min at room temperature:

(a) *Note*: after vortexing, the solution will have an opaque white appearance, which will be separated into aqueous (upper) and organic (lower) phases during centrifugation.

39. Transfer 500 μL of the aqueous (top) phase to a new Eppendorf tube:

(a) *Note*: the aqueous phase contains the immunoprecipitated DNA.

(b) *Note*: the use of filter tips will reduce the dripping of the aqueous phase during transfer.

(c) *Note*: do not attempt to collect the entire aqueous phase as doing so will result in the carryover of phenol, which will interfere with PCR reactions and will impair accurate quantification of immunoprecipitated DNA.

40. Back extract the aqueous phase by adding an equal volume (500 μL) of 1:1 (v:v) chloroform/isoamyl alcohol solution, vortexing thoroughly, and centrifuging at 13000 × *g* for 3 min at room temperature:

(a) *Note*: ensure that the 1:1 chloroform/isoamyl alcohol solution is prepared fresh before use.

41. Transfer 450 μL of the aqueous (top) phase to a new Eppendorf tube.

42. Precipitate the immunoprecipitated DNA from the back-extracted aqueous phase by adding 44 μL of 3 M sodium acetate, pH 5.2 (see recipe), 2 μL of 10 mg/mL glycogen, and 1 mL of 100% ethanol; vortexing; and incubating at −80°C overnight:

(a) *Note*: the addition of glycogen increases the precipitation of small amounts of low-molecular-weight DNA.

(b) *Stopping point*: the precipitation can be left at −80°C for up to several days if necessary.

43. Pellet precipitated DNA at ≥20,000 × *g* for 15 min at 4°C.

44. Carefully aspirate the supernatant with a 25 ga needle:

(a) *Note*: the DNA pellet is not always visible following centrifugation. To prevent unintended sample loss, place all of the Eppendorf tubes in the same orientation in the centrifuge rotor so that the pellet will form in a predictable location in all tubes.

45. Wash the pellets by adding 1 mL of 70% ethanol (chilled to −20°C) to each tube and vortexing.

46. Repellet the washed DNA by centrifugation at ≥20,000 × *g* for 10 min at 4°C.

47. Carefully aspirate the supernatant with a 25 ga needle, leaving approximately 10–15 μL volume to prevent the accidental aspiration of DNA.

48. Dry the DNA pellets in a speed-vac, fume hood, or laminar flow hood until all of the residual ethanol has evaporated:

(a) *Note*: ensure that all of the residual ethanol has evaporated before continuing. Residual ethanol will interfere with PCR reactions and will impair accurate quantification of immunoprecipitated DNA.

49. Resuspend the DNA in 50 μL of TE buffer and store at −20°C:

(a) *Note*: depending on the number of target loci to be evaluated by qPCR, it may be necessary to dilute the input DNA. If so, then dilute the input DNA in TE buffer and account for the dilution factor in the calculations used to determine % input values for immunoprecipitated DNA.

Quantify immunoprecipitated DNA by qPCR

50. Use 2 μL of the purified DNA from step 49 per 20 μL qPCR reaction:

(a) *Note*: if desired, the purified DNA can be diluted prior to use in qPCR reactions; however, this must be determined empirically for each genomic locus to be assessed as the distribution and abundance of epigenetic modifications varies between genomic regions. If targeting a low-abundance epigenetic modification, then a greater volume of purified DNA may be required in each qPCR reaction for reproducible quantification.

(b) *Note*: recommendations on ChIP-qPCR data analysis are given in the *expected results* section below.

SUPPORT PROTOCOL 1

Determine Optimal Sonication Conditions for ChIP

The target locus specificity and reproducibility of ChIP relies on the use of fragmentation conditions that consistently generate chromatin fragments that average ~500 bp with a range of ~250–750 bp (Fig. 2). The use of chromatin fragments within this size range allows for the reliable detection of association between target histone modifications and discrete genomic loci by ChIP-qPCR. Larger chromatin fragments reduce the specificity of detection (target histone modifications could be occurring on more distant nucleosomes that remain linked to target genomic loci), and smaller fragments can hinder the detection of target genomic loci by qPCR. The sonication conditions required to achieve this level of fragmentation will vary based on several parameters, including cell type/cell line, sonicator wattage, sonicator probe tip, and buffer conditions. This protocol describes the optimization of chromatin fragmentation by sonication. During early protocol development, prepare a large number of chromatin pellets to determine the sonication/fragmentation conditions. Then, test the conditions across several chromatin preparations to ensure that the sonication/fragmentation is reproducible across replicates.

Materials

Cells in culture
Dulbecco's phosphate-buffered saline (DPBS, room temperature)
0.25% Trypsin (for adherent cultures)
Soybean trypsin inhibitor (for adherent cultures)
Cell strainer (70 μm pore)
16% Paraformaldehyde (in single-use ampules)
2.5 M Glycine (see recipe)
Protease inhibitor solution (PIS, ice cold; see recipe)
Hemocytometer
Liquid nitrogen
Sonicator with microprobe tip (Fisher Scientific Sonic Dismembrator Model 500)

Digital temperature-controlled dry bath
ChIP lysis buffer (see recipe)
ChIP dilution buffer (see recipe)
TE buffer (see recipe)
3 M Sodium acetate and pH 5.2 (see recipe)
100% Ethanol
10 mg/mL Proteinase K (see recipe)
RNaseA (20 mg/mL)
Phenol/chloroform/isoamyl alcohol (25:24:1)
Chloroform/isoamyl alcohol (1:1, v:v)
10 mg/mL Glycogen (see recipe)
3% Agarose/TBE gel (see recipe)
TBE buffer (see recipe)
10 mg/mL Ethidium bromide
100 bp DNA ladder
6X Purple gel loading dye
Ultraviolet light box

Protocol Steps

Determine Optimal Chromatin Fragmentation Parameters

1. Complete steps 1–15 of Basic Protocol 1 to prepare fixed chromatin.
2. Fragment the chromatin using an increasing duration of sonication:
   (a) *Note*: start with four cycles at 10 pulses (0.9 s on and 0.1 s off), and increase the number of cycles. Be sure to place each tube on ice for at least 90 s between each cycle. This is important because excessive heat generated by sonication may compromise sample integrity, decreased IP efficiency, and increased variability.
   (b) *Note*: before beginning the sonication time course, determine the most rigorous sonication conditions for your particular sonicator and sample. The goal of this step is to identify the highest output that can be used consistently without causing foaming of the chromatin samples. Foaming should be avoided because the bubbles reduce the transmission of energy from the sonicator probe tip to the sample, thus limiting sonication efficiency. If using a 500 W sonicator (as described here), begin by sonicating in cycles of 10 pulses (0.9 s on and 0.1 s off) at 20% output. Using 30% output on a 500 W sonicator can reduce the total number of pulses/cycles required to achieve the desired chromatin fragmentation; however, using a higher output increases the likelihood of foaming in chromatin samples of a relatively low volume, such as those described here. The likelihood of foaming at higher output can be reduced in samples of larger volume or when the microprobe tip is able to project into the sample (e.g., sonication of chromatin in a volume of 300 μL is more successfully executed in a 0.5 mL microcentrifuge tube than a 1.5 mL Eppendorf tube). If using a sonicator with a lower power output, then a greater percentage of output may be required to achieve the desired level of chromatin fragmentation.

Consult the user's manual provided with your sonicator to avoid exceeding the recommended maximum output for a microprobe tip.

3. Clarify the fragmented chromatin by centrifugation at $\geq 20{,}000 \times g$ for 10 min at 4°C.

4. Transfer the clarified chromatin to a new Eppendorf tube.

5. Add SDS elution buffer to each sample to give a total volume of 500 µL, and then, add 50 µL of elution addition buffer and vortex.

6. Add 2 µL of 10 mg/mL proteinase K to each sample and incubate at 65°C for 4 h:
   (c) *Note*: if desired, this incubation step can be extended to overnight; however, to ensure reproducibility between samples, the same incubation duration should be used for the entire study.

7. Extract DNA from the fragmented chromatin samples by adding an equal volume (550 µL) of phenol/chloroform/isoamyl alcohol (PCI, 25:24:1 v/v ratio), vortexing thoroughly, and centrifuging at $13000 \times g$ for 3 min at room temperature.

8. Transfer 500 µL of the aqueous (top) phase to a new Eppendorf tube:
   (d) *Note*: the use of filter tips will reduce the dripping of the aqueous phase during transfer.

9. Back extract the aqueous phase by adding an equal volume (500 µL) of 1:1 (v:v) chloroform/isoamyl alcohol solution, vortexing thoroughly, and centrifuging at $13000 \times g$ for 3 min at room temperature:
   (e) *Note*: ensure that the 1:1 chloroform/isoamyl alcohol solution is prepared fresh before use.

10. Transfer 450 µL of the aqueous (top) phase to a new Eppendorf tube.

11. Precipitate the immunoprecipitated DNA from the back-extracted aqueous phase by adding 44 µL of 3 M sodium acetate, pH 5.2 (see recipe), 2 µL of 20 mg/mL glycogen, and 1 mL of 100% ethanol; vortexing; and incubating at −80°C overnight:
    (f) *Stopping point*: the precipitation can be left at −80°C for up to several days if necessary.

12. Pellet precipitated DNA at $\geq 20{,}000 \times g$ for 15 min at 4°C.

13. Carefully aspirate the supernatant with a 25 ga needle:
    (g) *Note*: the DNA pellet is not always visible following centrifugation. To prevent unintended sample loss, place all of the Eppendorf tubes in the same orientation in the centrifuge rotor, so the pellet will form in a predictable location in all tubes.

14. Wash the pellets by adding 1 mL of 70% ethanol (chilled to −20°C) to each tube and vortexing.

15. Repellet the washed DNA by centrifugation at $\geq 20{,}000 \times g$ for 10 min at 4°C.

16. Carefully aspirate the supernatant with a 25 ga needle.

17. Air dry the DNA pellets in   flow hood until all of the residual ethanol has evaporated:
    (h) *Note*: ensure that all of the residual ethanol has evaporated before continuing. Residual ethanol will interfere with PCR reactions and will impair accurate quantification of immunoprecipitated DNA.

18. Resuspend the DNA in 50 µL of TE buffer.

19. Add 20 µg of RNaseA and incubate at 37°C for 30 min:
    (a) *Stopping point*: samples can be stored at −20°C until use.

20. Separate the fragmented DNA on a 3% agarose/TBE gel:
    (a) Add ethidium bromide to a final concentration of 0.4 µg/mL to the molten agarose prior to casting the gel.

(b) If starting with 1×10^6 cells/condition, then load 24 µL per well of a mixture containing 20 µL of the resuspended DNA with 4 µL of a suitable 6X loading buffer.

(c) Use a 100 bp ladder as a basis for the assessment of average chromatin fragment size.

21. After running the gel for the desired amount of time, destain the gel in TBE buffer in 15–20 min intervals (use fresh TBE buffer for each interval) to remove unbound ethidium bromide from the gel and image on a UV light source.

SUPPORT PROTOCOL 2

Isolation of Leukocytes From Peripheral Whole Blood

Peripheral blood samples can be readily collected during toxicology studies; however, whole blood is not suitable for ChIP. This support protocol describes a rapid method for the isolation of leukocytes from peripheral whole blood for use in ChIP experiments.

Materials

ACK lysis buffer (see recipe)
Cell resuspension solution (see recipe)
Hemocytometer

Protocol Steps

1. Transfer fresh whole blood from each collection tube into individual 50 mL conical tubes on ice:
    (a) *Note*: blood should be used immediately after collection in tubes with either EDTA or sodium citrate as an anticoagulant.
2. Add five volumes of ACK buffer and mix gently but thoroughly prior to incubating on ice for 5 min:
    (a) *Note*: incubation with ACK lysis buffer lyses erythrocytes, but not leukocytes.
3. Pellet intact cells by centrifugation at $500 \times g$ for 4 min at 4°C.
4. Carefully aspirate the supernatant:
    (a) *Note*: the aspirate should be treated as potentially biohazardous material until decontaminated with bleach (final concentration of $\geq 10\%$) or other suitable decontaminating agent.
5. Resuspend the cell pellet in one volume (original blood volume transferred in step 1) of ACK lysis buffer, and transfer it to a new 15 mL conical tube and incubate on ice for 2 min:
    (a) *Note*: a single ACK lysis step is often insufficient to lyse all of the erythrocytes in the blood sample. This second ACK lysis step ensures thorough erythrocyte lysis.
6. Centrifuge at $500 \times g$ for 4 min at 4°C and aspirate the supernatant.
7. Wash the pellet by resuspending in 1 mL of cell resuspension buffer, and transfer the cell suspension to a 15 mL conical tube. Add cell resuspension buffer to a final volume of 10 mL.
8. Enumerate cells with a hemocytometer.

9. Pellet cells at $500 \times g$ for 4 min at 4°C, and aspirate the supernatant.
10. Resuspend the pellet *with room-temperature* cell resuspension buffer to the desired cell density.
11. Continue ChIP by starting at step 8 of Basic Protocol 1:

   (a) *Note*: the relative abundance of different cell populations should be considered when using total leukocytes from whole blood as starting material. Ideally, a white blood cell differential count should be conducted to determine the composition of the leukocyte population in the sample (or blood collected at the same time as the sample).

REAGENTS AND SOLUTIONS

Use distilled and deionized water (or equivalent) for all reagents and buffers. For common stock solutions, see Appendix; for suppliers, see Supplier Appendix.

ACK lysis buffer

150 mM Ammonium chloride
10 mM Potassium bicarbonate
500 mM EDTA
Prepare immediately before use, and keep on ice

Cell resuspension solution

Hank's balanced salt solution
1X cOmplete protease inhibitor cocktail (Roche)
20 mM Sodium butyrate
25 mM Sodium fluoride
1 mM Sodium pyrophosphate
0.1% Bovine serum albumin
Prepare immediately before use, and keep on ice

DPBS-TI

Dulbecco's phosphate-buffered saline
30 mg/mL Soybean trypsin inhibitor

Protease inhibitor solution (PIS)

Dulbecco's phosphate-buffered saline (Gibco)
1X cOmplete protease inhibitor cocktail (Roche)
25 mM Sodium fluoride
1 mM Sodium pyrophosphate
20 mM Sodium butyrate
Prepare at the day of use, and keep on ice

ChIP lysis buffer

50 mM Tris and pH 8.1 (see recipe)
10 mM EDTA (see recipe)

1% SDS (w/v)
1 mM Phenylmethylsulfonyl fluoride (PMSF; see recipe)
1X cOmplete protease inhibitor cocktail (Roche)
Prepare at the day of use, and keep at room temperature

ChIP dilution buffer

16.7 mM Tris and pH 8.1 (see recipe)
167 mM NaCl
1.2 mM EDTA (see recipe)
1.1% Triton X-100
0.01% SDS (w/v)
1 mM Phenylmethylsulfonyl fluoride (PMSF; see recipe)
1X cOmplete protease inhibitor cocktail (Roche)

Low salt wash buffer

20 mM Tris and pH 8.1
150 mM NaCl
2 mM EDTA
1% Triton X-100
0.1% SDS

High salt wash buffer

20 mM Tris and pH 8.1
500 mM NaCl
2 mM EDTA
1% Triton X-100
0.1% SDS

Lithium chloride wash buffer

10 mM Tris and pH 8.1
1 mM EDTA
1% IGEPAL CA-630
250 mM LiCl
1% Sodium deoxycholate

TE buffer

10 mM Tris and pH 8.0
1 mM EDTA

Elution buffer

100 mM NaHCO$_3$
1% SDS

Elution addition buffer

400 mM Tris and pH 6.5
2.0 M NaCl
100 mM EDTA

Proteinase K (10 mg/mL)

20 mM Tris and pH 8.0
40% Glycerol
1 mM $CaCl_2$

Use this buffer to resuspend powdered proteinase K in the manufacturer's container to a concentration of 10 mg/mL, and then, prepare 50 μL aliquots and store at −20°C.

Tris borate electrophoresis (TBE) buffer

8.9 mM Tris base
8.9 mM Sodium borate
2.0 mM EDTA and pH 8.0

COMMENTARY

Background Information

Chromatin immunoprecipitation is a powerful technique for exploring the role of the epigenome in environmental exposure-mediated disease and susceptibility, which has been adapted and improved over the last three decades (reviewed in Carey et al., 2009). An early precursor of the procedure was pioneered by Gilmour and Lis (1984) to examine interactions between RNA polymerase and specific regions within the *Escherichia coli* and *Salmonella* spp genomes. While these bacteria lack the chromatin structure found in eukaryotes that would ultimately be the namesake of the technique, by identifying specific RNA polymerase binding sites within the bacterial genome, Gilmour and Lis (1984) demonstrated that protein-DNA interactions could be fixed by covalent cross-linking, purified by immunoprecipitation, and mapped by hybridization. They transitioned these principles to a eukaryotic model by mapping the distribution of RNA polymerase II (RNAPII) in *Drosophila melanogaster* (Gilmour and Lis, 1985, 1986). These early ChIP protocols relied on UV light to cross-link DNA and closely associated proteins (Gilmour and Lis, 1984, 1985, 1986); however, the utility and efficacy of UV cross-linking are limited. The ability of formaldehyde to form DNA-protein and protein-protein cross-links had been previously characterized (Brutlag et al., 1969; Ilyin and Georgiev, 1969; Van Lente et al., 1975; Jackson, 1978), applied to whole-cell fixation to study chromatin dynamics (Jackson and Chalkley, 1981; Solomon and Varshavsky, 1985), and replaced UV light as a cross-linking agent in ChIP (Solomon et al., 1988). Nearly a decade later, the traditional probe-based blotting methods used for quantifying immunoprecipitated DNA was replaced with a PCR-based detection method (Hecht et al., 1996). The introduction of formaldehyde and PCR facilitated the emergence and popularization of ChIP (reviewed in Carey et al., 2009). ChIP has continued to evolve with subsequent improvements in discrete aspects of the procedure to become more sensitive, accurate, and reproducible. In more recent years, the quantification of immunoprecipitated DNA by semi-quantitative PCR was replaced with SYBR green-based quantitative PCR (qPCR) or the more target-specific TaqMan qPCR. Specificity and reproducibility have also been increased by the increased availability of monoclonal antibodies and the emerging use of recombinant antibodies as a replacement for polyclonal antibodies to target proteins and histone modifications of interest.

While ChIP is a complex and potentially intimidating procedure, with careful optimization of key steps, it can easily become a reliable way to readily explore the role of the epigenetic modifications in toxicology studies. Given the central role that the epigenome plays in regulating gene expression and DNA replication and damage repair among other functions, understanding the relationship between environmental exposures and histone modifications will provide critical insight into the mechanisms underlying exposure effects and provides novel biomarkers of susceptibility to exposure-related disease. Here, we have described a streamlined protocol for quantifying the relative abundance of specific histone modifications within regulatory regions of four genes (*IL-8*, *IL-6*, *COX2*, and *HMOX1*) that are commonly studied in the field of toxicology. This protocol is directed toward examining the relationship between histone modifications and the regulatory regions of four toxicity-related genes; however, using the principles outlined here, the protocol can be readily adapted to examine the relative abundance of target histone modifications, transcription factors, and other chromatin-associated proteins within the regulatory regions of any gene(s) of interest.

Critical Parameters and Troubleshooting

The ChIP procedure relies on several critical parameters that play an important role in both the success of the assay and reproducibility between replicates. The intrinsic variability in this complex procedure can be minimized by thoughtful attention to detail and careful optimization of several key steps. Further, it is important to consider that nucleosomal density (Fig. 4A) and the distribution of histone modifications vary between target genomic loci.

Experimental Treatment Conditions

As with all experiments, it is critically important to carefully control the conditions under which the cells being used are grown, exposed, and harvested. We recommend plating cells at a calculated and consistent cell density during regular passage and in preparation for experimental treatments. Further, we recommend plating cells for experiments in a consistent manner across biological replicates (e.g., plating cells for an experiment with the same number of days after being passaged). When possible, the experimental treatment agent should be prepared in batch and frozen in single-use aliquots prior to use.

Buffer Preparation

Buffers can be prepared ahead of conducting the ChIP protocol; however, aliquots should be removed and treated with the indicated protease (PMSF and cOmplete protease inhibitor cocktail), phosphatase (sodium fluoride and sodium pyrophosphate), and histone deacetylase (butyric acid) inhibitors immediately prior to use. Protease inhibitors are intentionally omitted from all of the wash buffers because residual carryover can hinder the subsequent proteinase K digestion step. Further, if histone deacetylase inhibitors are desired in the binding and washing buffers, then butyrate should be used in place of butyric acid in the

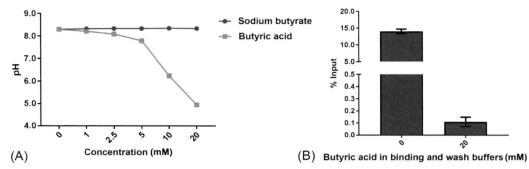

FIG. 3 The histone deacetylase inhibitor butyric acid lowers buffer pH and inhibits chromatin immunoprecipitation. Butyric acid and its conjugate base butyrate are commonly used to inhibit deacetylation of histone proteins during ChIP. (A) Butyric acid is effective at inhibiting histone deacetylation; however, its inclusion in buffers used in the ChIP procedure (low-salt wash buffer is shown, which is identical to the buffer condition after the dilution of the fragmented chromatin in step 21 of Basic Protocol 1) results in a concentration-dependent reduction in buffer pH. Despite being used at similar concentrations, its conjugate base butyrate does not alter ChIP buffers. (B) The inclusion of butyric acid in associated binding and wash buffers inhibits ChIP efficiency. The effect observed in (B) is likely the result of poor antibody-chromatin binding due to the effect of decreased pH in buffers containing 20 mM butyric acid on antibody folding. To facilitate optimal immunoprecipitation efficiency, sodium butyrate should be used as a replacement for butyric acid in the ChIP protocol.

PIS because the use of butyric acid alters the buffers' pH sufficiently to hinder antibody-antigen binding and thus immunoprecipitation efficiency (Fig. 3).

Fixation

Treatment of cells with formaldehyde/paraformaldehyde results in the formation of inter- and intramolecular cross-links between DNA and proximal proteins (discussed in detail by Hoffman et al., 2015). The optimal fixation time for ChIP differs among cell types/lines and can vary from 1 to 30 min. While overfixation can impair chromatin fragmentation and interfere with antibody binding by deforming epitopes on the coordinate antigen, underfixation can reduce ChIP efficiency by failing to covalently link DNA with its associated proteins. Thus, the length of fixation should be determined empirically for each application by testing chromatin fragmentation and ChIP efficiency across a range of fixation times. The fixation described in this protocol is conducted at room temperature, but fixation can also be conducted at 37°C or on ice. Once optimal conditions are determined, the fixation conditions should be closely followed to limit the intrinsic variability in fixation between samples and replicates (Fig. 4B).

Sonication

Chromatin fragmentation is a critical aspect of successful ChIP experiments. Reproducible sonication conditions should be determined empirically as described here in Support Protocol 1. Underfragmented chromatin has limited solubility and will reduce ChIP efficiency and the resolution of detection for target genomic loci. Excessively, fragmented chromatin will

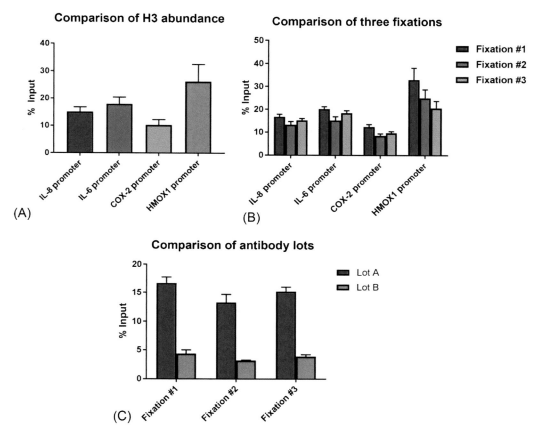

FIG. 4 Variability in ChIP. (A) Nucleosomal density, as indicated by the abundance of histone H3, varies between promoters of genes that are commonly examined in toxicology studies. The data shown in (A) are the mean ± SD of three independent fixations that were subjected to αH3 ChIP in triplicate, with target quantification of each genomic target by qPCR in triplicate. (B) Once conditions are carefully established, independent fixation replicates should exhibit limited variability; however, variability can increase in samples with a relatively high % input, such as those observed for total H3 abundance at the HMOX1 promoter. The data shown in (B) are the mean ± SD of three αH3 ChIP assays using THP-1 cells that were fixed on three different days, frozen as fixed cell pellets (Basic Protocol 1 steps 1–15), and subjected to αH3 ChIP simultaneously with the quantification of each genomic locus by qPCR in triplicate. (C) Two different lots of the same manufacturer product number were used in side-by-side ChIP assays from pooled starting material. Data shown in (A), (B), and (C) represent αH3 ChIP conducted in THP-1 cells according to the conditions described in Basic Protocol 1 and IL-8, IL-6, COX-2, and HMOX1 promoter primer/probe sets.

reduce qPCR detection of target genomic loci following immunoprecipitation. While not discussed here, it should be noted that chromatin can also be fragmented by micrococcal nuclease (MNase) digestion. While MNase digestion can be used to fragment fixed chromatin, it is more commonly used in the fragmentation of native (unfixed) chromatin as discussed by O'Neill and Turner (2003).

IP Antibody

Choice of immunoprecipitation antibody is instrumental to the success of the ChIP protocol. While many antibodies are recommended for use in other immunodetection assays (e.g., Western blotting, ELISA, and immunofluorescence staining), this does not ensure that they will work well in ChIP. Check datasheets provided by the antibody manufacturer to ensure that a specific antibody preparation has been validated for use in ChIP. Further, conduct an antibody titration to determine that a sufficient quantity of antibody is being used for the immunoprecipitation. Polyclonal antibodies are readily available for most histone modifications; however, while their sensitivity of detection is lower, monoclonal antibody preparations are more consistent, and their use can increase the reproducibility of results across replicates in ChIP experiments. Experiments should be carefully planned such that a sufficient amount of each antibody for all replicates in a study can be obtained prior to beginning. Ideally, a single lot should be used as interlot variability can introduce variability in the assay results (Fig. 4C); however, if not practical, several lots of the same antibody can be pooled before beginning the study. Regardless of the number of lots used, the tubes of antibody preparation should be pooled, distributed into single-use aliquots, and stored according to the manufacturer's recommendation before beginning the study. Antibodies that we have used successfully with the protocol described here are listed in Table 1.

Capture Resin

The choice of capture depends on the desired downstream application of ChIP material and can have a substantial impact on the immunoprecipitation efficiency. The antibody-chromatin capture step in ChIP is typically accomplished through the use of either agarose/Sepharose, magnetic, or polyacrylamide beads conjugated to the IgG-binding factor protein A, protein G, or both proteins A and G. Care should be taken to determine the efficiency with which the binding protein (e.g., protein A or protein G) binds the immunoprecipitation antibody. The binding affinity of protein A and G vary depending on antibody host and isotype. These different capture resin formats have individual advantages and disadvantages as shown in Fig. 5 and summarized in Table 2. Agarose and Sepharose beads have been commonly used in ChIP given their relatively low cost and antibody binding capacity; however, the porous nature of agarose often results in a high background signal. To reduce the background signal, agarose beads are typically preadsorbed with salmon sperm DNA to reduce the nonspecific binding of DNA from ChIP samples during the capture step. While salmon sperm DNA effectively reduces nonspecific DNA binding to the beads, it is also carried over into the eluted immunoprecipitated DNA. While salmon sperm DNA is sufficiently divergent from human DNA to prevent it from being amplified with primers/probes targeting human loci used in ChIP-qPCR, it can contaminate downstream sequencing in ChIP-seq experiments. To avoid salmon sperm DNA contamination, ChIP-seq protocols utilize magnetic beads, which are nonporous and do not readily bind nonspecific sample DNA. Magnetic beads also require the use of a magnetic tube rack to collect the beads for binding and wash steps, which precludes the need to collect beads by centrifugation during each of these steps. Despite these benefits, the use of magnetic beads in ChIP-qPCR is often limited by higher cost. To overcome the limitations of both agarose/Sepharose and magnetic beads, we

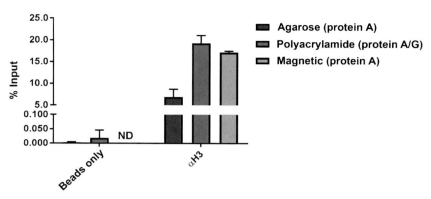

FIG. 5 Comparison of capture bead materials. The use of different capture beads can impact the outcome of ChIP experiments. ChIP experiments were conducted side by side using fragmented chromatin from 1.0×10^6 THP-1 cells, αH3 monoclonal antibody (isotype IgG3), and three different capture bead materials according to Basic Protocol 1. Each IP sample contained the equivalent of 50 μL of resuspended bead slurry as provided by the manufacturer. The immunoprecipitated material was subjected to qPCR with an IL-8 promoter primer/probe set. The "beads only" data are shown to demonstrate the nonspecific binding capacity of each capture bead material. The data shown represent the mean ± SD of three independent immunoprecipitations, which were quantified by qPCR in technical triplicate. ND indicates no detectable qPCR signal.

TABLE 2 Comparison of Capture Beads Commonly Used in ChIP

	Protein A Agarose w/Salmon Sperm DNA	Protein A Magnetic	Protein A/G Polyacrylamide
Antibody binding capacity	+	+++	+++
Cost	$	$$$	$
Background	Low/medium	Low	Low
ChIP-qPCR compatible	Yes	Yes	Yes
ChIP-seq compatible	No	Yes	Yes
Bead collection method	Centrifugation	Magnetic rack	Centrifugation

use polyacrylamide beads because they are low cost, do not require preadsorption with salmon sperm DNA, and have an equal/greater binding capacity as magnetic beads.

Controls

Once determined, ChIP conditions should be validated with the use of suitable positive and negative controls. Either a "no antibody" (beads only) or nonspecific IgG antibody can be used as a negative control to test the stringency of the ChIP conditions. In contrast, a total H3 antibody should be used, in conjunction with qPCR primers/probes for a genomic locus that is known to contain nucleosomes, as a positive control.

Troubleshooting

Sub-optimal chromatin fragmentation

- Under- or overfragmentation
  - Increase or decrease the number of sonication cycles or the number/duration of pulses in each sonication cycle.
  - Underfragmentation can also result from overfixing the sample. Conduct a time course to determine the optimal duration of fixation for the cell type/line being used.
  - Sonication should be optimized as described in Support Protocol 1 for each cell type/line and set of fixation conditions being used for ChIP.
- Foaming during sonication

  - Improper or variable probe placement during sonication. Ensure that the probe is in a consistent position near the bottom of the sample volume during sonication; however, avoid touching the sides of the tube with the probe. If sonicating in a volume is <500 μL, use a 0.5 mL microcentrifuge tube instead of a 1.5 mL Eppendorf tube to increase the depth of the sample.
  - Sonicator output is too high. Reduce the sonicator output and/or the number/duration of pulses in each sonication cycle.

Low or no detectable PCR signal

- Poor or no PCR amplification of purified DNA. Perform a controlled test with known material to ensure that the qPCR reaction for each target genomic locus is working efficiently. Include a standard curve for each qPCR primer/probe set on each PCR plate to evaluate PCR function and efficiency.
- Insufficient or no immunoprecipitation of target DNA. Increase the amount of chromatin used for the immunoprecipitation by increasing the number of cells used to prepare the cell pellets in step 15. The distribution of histone modifications varies greatly across the genome, and some histone modifications are sparse or absent within target genomic loci.
- Low immunoprecipitation efficiency. Ensure that ChIP-validated antibodies are being used. Many antibodies that are recommended for use in other immunodetection assays (e.g., Western blotting, ELISA, and immunofluorescence staining) do not work well in ChIP. Check the datasheets provided by the antibody manufacturer to determine whether a specific antibody preparation has been validated for use in ChIP. Further, conduct an antibody titration to determine that a sufficient quantity of antibody is being used for the immunoprecipitation.
- Check reagents and buffers. Ensure that reagents and buffers were made properly and that stock solutions of buffering solutions (e.g., Tris) are at the correct pH. Ensure that inhibitors (other than those recommended here) have not altered the pH of buffers used in the protocol. For example, the use of butyric acid as a replacement for its conjugate base sodium butyrate will significantly augment buffer pH and impair the immunoprecipitation (Fig. 3).
- Ensure that all of the residual ethanol has evaporated in step 48 prior to the resuspension of the precipitated DNA in step 50.

High non-specific background signal

- Prolonged incubation with protein A-acrylamide beads. Limit the incubation with protein A-acrylamide beads in step 27 to 1 h or less.
- Insufficient wash stringency. Increase the number of washes or the stringency of the washes by increasing the NaCl or Triton X-100 concentration in wash buffers.
- Ensure that the beads are transferred to a new tube in step 29d(i). Transferring the beads to a new tube in this step reduces the carryover of unbound DNA adhered to the wall of the original tube.

High variability between replicates

- Variability between lots of antibody used for immunoprecipitation. Obtain a sufficient amount of each antibody for all ChIP experiments in a given study from a single lot. After arrival, pool the antibody, and prepare single-use aliquots prior to storage under the conditions recommended by the antibody supplier. Alternatively, if it is not possible to obtain sufficient antibody for the study from a single lot, then obtain the total amount of antibody required from the necessary number of lots and pool and aliquot as described earlier.
- Utilize fresh single-use ampules of formaldehyde for fixation. The yield of the ChIP protocol decreases with increasing age of the formaldehyde used for fixation in step 8.
- Ensure the consistency of all indicated incubation times across replicates.
- Ensure that consistent numbers of cells are used in each ChIP sample. Carefully count the cells in each sample or treatment group in step 12.
- Ensure that fresh phenol/chloroform/isoamyl alcohol is used, and prepare the chloroform/isoamyl alcohol solution immediately prior to each use.

Anticipated Results

Data Analysis

The data collected from ChIP-qPCR experiments do not reflect an absolute quantification of epigenetic modification or chromatin-associated protein levels; rather, they are relative measures that are influenced by the strength of antibody-epitope interactions, the accessibility of the epitope, and the absolute amount of epitope present at the target genomic locus (Milne et al., 2009). While ChIP-qPCR data can be analyzed by a simple ΔC_t comparison of input and immunoprecipitated (IP) values, this method fails to account for the differences in PCR efficiency between primer/probe sets. Further, the PCR reaction efficiency of each primer/probe set varies between PCR runs, and accounting for this intrinsic variability is crucial to obtain the most accurate data from ChIP-qPCR reactions. PCR reaction efficiency can be calculated from the slope of the line of a standard curve for each primer/probe set, which should be included for each primer/probe set on every qPCR plate:

$$\text{Efficiency } (E) = 10^{(-1/\text{slope})}$$

For example, if the line resulting from standard curve values for a specific primer/probe set had a slope of −3.425, then the efficiency of the PCR reaction would be determined as follows:

$$E = 10^{(-1/-3.525)}$$

$$E = 10^{(0.283)}$$

$$E = 1.92$$

The efficiency of the PCR reaction is a measure of the number of copies of each template molecule generated during each PCR cycle. A PCR reaction that results in two copies of each template molecule being produced during each amplification cycle (100% efficient reaction) has a slope of −3.3 and an E value of 2.0. In the example described earlier, an E value of 1.92 indicates that 1.92 copies are made from each template molecule during each PCR amplification cycle, meaning that the PCR reaction is 92% efficient.

Chromatin immunoprecipitation data are most frequently expressed as the amount of immunoprecipitated genomic locus relative to the total amount of that genomic locus in the input sample. The values obtained from immunoprecipitated samples are expressed as "% input," which requires the calculation of the total amount of target material in the chromatin sample prior to beginning the immunoprecipitation. This value can be obtained by correcting the raw C_t values obtained for the input samples according to their relative fraction to the original sample (noted in step 22 of Basic Protocol 1). The input fraction noted in step 21 is converted to a decimal and used to calculate a "correction value," which accounts for the $\log_2$ nature of C_t values:

$$\text{Correction value} = -\log_2(\text{input fraction})$$

For example, if 1% of the diluted chromatin was taken as the input sample, then convert 1% into decimal form (0.01) and determine $-\log_2(0.01)$:

$$-\log_2(0.01) = 6.644$$

The correction value is then used to correct the raw C_t values of the input samples by subtracting the correction value from the raw C_t value for each input sample. The corrected input C_t values reflect the total amount of quantifiable target in the original chromatin sample prior to beginning the immunoprecipitation:

$$\text{Corrected } C_t = [(\text{raw } C_t) - (\text{correction value})]$$

For example, if a 1% input sample had a raw C_t value of 29.5, then the corrected C_t value would be calculated as follows:

$$\text{Corrected } C_t = [(29.5) - (6.644)]$$

$$\text{Corrected } C_t = 22.856$$

The corrected input values are then used as the basis for determining the relative quantity of the target genomic locus in the immunoprecipitated samples using a comparison of C_t values relative to the efficiency of the qPCR reaction:

$$\text{Immunoprecipitated quantity relative to input} = E^{(C_t \text{ input} - C_t \text{ IP})}$$

For example, if an IP sample from the same chromatin sample used in the examples above had a raw C_t value of 28.0 measured in a PCR reaction with an E value of 1.92, then the fold enrichment would be calculated as follows:

$$\text{Immunoprecipitated quantity relative to input} = 1.92^{(22.856-28.0)}$$

$$\text{Immunoprecipitated quantity relative to input} = 0.0349$$

The immunoprecipitated quantity relative to input can then be converted from a decimal to a percentage to give the % input value:

$$\text{Immunoprecipitated quantity relative to input} \times 100 = \%\text{input}$$

Using the example above,

$$0.0349 \times 100 = 3.49\%\text{input}$$

The importance of using observed efficiency values for each qPCR run can be highlighted using the sample data given in the above examples by comparing the result as calculated with an observed efficiency value and as calculated with the efficiency assumed ($E=2.0$, 100% efficient PCR reaction) in the simple ΔC_t method:

$$\text{Calculated with observed efficiency} : 1.92^{(22.856-28.0)} = 0.0349 = 3.49\%\text{input}$$

$$\text{Calculated with assumed efficiency of } 100\% : 2.0^{(22.856-28.0)} = 0.0283 = 2.83\%\text{input}$$

The baseline relative abundance of histone modifications at target genomic loci can be expressed as % input, and treatment-induced changes can be expressed as a fold change (treatment/control). It is important to remember that the distribution of histone modifications can vary considerably between different genomic loci and experimental treatment conditions, and thus, the % input values can vary dramatically between immunoprecipitation targets and genomic loci of interest. Typically, total H3 can range from <1% to >10% input, while values obtained following immunoprecipitation with histone modification-specific antibodies are regularly ≤1% input. It is important to remember that the % input is a relative measure of immunoprecipitated target locus DNA relative to the total amount of the target locus present in the cleared chromatin sample and not an absolute quantification of the interaction between the histone modification of interest and the target genomic locus.

Time Considerations

Cells from exposure experiments can be fixed, frozen as cell pellets, and stored at −80°C (as indicated in steps 1–15 of Basic Protocol 1) at any time prior to completing the remainder of the ChIP protocol. We typically begin the fragmentation and immunoprecipitation (steps 16–25 of Basic Protocol 1) in the afternoon of "day 1" in such a manner that step 24 is completed immediately before leaving for the day. The bead-binding step (steps 26–27 of Basic Protocol 1) is then started in the morning on day 2 after preparing all of the required reagents and buffers. Washing and elution (steps 28–34 of Basic Protocol 1) are completed by late morning on day 2, and the cross-linking reversal/proteinase K digestion (steps 35–37 of

Basic Protocol 1) is completed by early to midafternoon. The DNA extraction (steps 38–41 of Basic Protocol 1) is completed by late afternoon, and the DNA precipitation (step 42) is left overnight. The precipitated DNA is pelleted, washed, and resuspended (steps 43–49 of Basic Protocol 1) the following morning, after which the purified DNA is ready to be used in downstream analyses.

Considerations for ChIP-Sequencing

High-throughput or next-generation sequencing (NGS) is a desirable secondary step following ChIP (ChIP-seq), as it allows for untargeted interrogation of the genome. This systems biology-based approach allows for the investigation of effects of a chemical exposure unconstrained by a prior hypothesis about affected targets, which allows for the discovery of novel toxicant-epigenome-gene interactions. Additionally, it allows for rapid measurement of a large number of targets following an exposure, which may be prohibitive when using qPCR. However, these experiments require careful planning at various stages to ensure data collected will be of use. In this section, we detail the stages at which investigators will have to make critical decisions in planning an NGS sequencing experiment including library construction, sequence depth, and considerations for computational analysis.

Library construction is the first critical decision made during a sequencing experiment (Head et al., 2014). This process can be summarized in six steps: fragmentation, end repair, phosphorylation of the 5′ end, A-tailing of the 3′ end, ligation of the adapters, and PCR to enrich product that has adaptors ligated to both ends. The size of target DNA fragments during the fragmentation stage is determined by the instrumentation and specific application. In the case of ChIP-seq, there is often low input and thus require a large number of cells as starting material. In-depth description of the specific protocols for these steps is beyond the scope of this chapter; however, detailed protocols are available in the literature (Brind'Amour et al., 2015; Ryan et al., 2015; Siegel et al., 2009).

Sequence depth is one of the primary considerations in designing ChIP-seq experiments (Bailey et al., 2013). Often, the assumption is the deeper the better, but deeper sequencing experiments can be costly and computationally laborious. Instead, sequence depth should be determined on the size of the genome and the number and size of protein binding sites. Library complexity, as discussed above, should also be assessed prior to determining depth. While numerous tools exist for this, the R package *preseq* is one of the best annotated and most commonly used (Daley and Smith, 2013). Histone modifications are relatively localized and thus as few as 20 million reads may be adequate to interrogate the human genome. The adequacy of a chosen sequencing depth can be determined by including a saturation analysis, specifically during peak calling and read mapping, as a component of the overall data analysis workflow. If this analysis suggests inadequate reads, technical replicate reads should be pooled before continuing.

Once the sequencing is completed, computational analysis is performed to answer the question or questions posed during the experiment (Bailey et al., 2013). Computational analysis is often treated as a black box by many investigators who rely on bioinformaticians for processing; however, understanding this process can lead to better communication with bioinformatic professionals and a better interpretation of the data. There are several available pipelines, but they all follow the procedure described here (Golosova et al., 2014; Han et al., 2015; Qin et al., 2016; Yan et al., 2014; Zhang et al., 2014). They may be implemented in a number of software environments including Bash, C++, Python, Perl, and R. The first step in data processing is filtering using a quality cutoff. Specifically, raw data are often analyzed by software similar to the R package *fastqcr* to generate a quality score that identifies sequencing error or bias. Next, data are mapped to the genome so as to describe to their location relative to gene regions. Available methods include Bowtie, MAQ/BWA, and SOAP (Reinert et al., 2015). Bowtie and Bowtie2, a newer version of the Bowtie pipeline, are the most commonly used due to their small computation footprint. However, they may fail to align reads with valid mapping when configured for maximum speed. Mapping is then followed by peak calling (Steinhauser et al., 2016; Thomas et al., 2017). The sensitivity and specificity of these steps is dependent upon the algorithm and normalization methods chosen. Several have been designed well for more broadly spreading weaker pattern characteristic of most histone marks. SICER, CCAT, ZINBA, and RSEG have all been designed to work specifically with ChIP-seq data, but more general packages such as SPP, MACS (version 2), and PeakRanger can also be used. This should then be followed by a reproducibility analysis (Hansen et al., 2015). Finally, differential binding sites can be identified, which can be done in either a qualitative or quantitative manner. The qualitative approach test whether overlapping peak regions look different between treatment conditions. The quantitative approach relies on whether read counties at a particular genomic location are different between treatment groups (Wu et al., 2015). Lastly, these peaks can be mapped to relevant genomic regions such as transcriptional start sites and gene promoters. Of note, there is a newly available software package to facilitate the analysis of CHIP-seq data called BioWardrobe. It integrates the analyses described earlier in a convenient graphic user interface that is accessible to biologists (Vallabh et al., 2018). Choosing how to map and analyze data should be considered carefully depending upon the question of interest. For example, studies attempting to identify de novo changes in histones associated with treatments should carefully consider mapping strategies but may be more likely to use a qualitative approach in identifying differential PTM regions. However, those that are looking to confirm a hypothesis regarding particular gene regions may focus more on the quantitative assessment of peaks during the hypothesis testing phase.

APPENDIX

50X cOmplete protease inhibitor solution

Dissolve one cOmplete protease inhibitor tablet in 1 mL of water
Store at −20°C in 250 μL single-use aliquots for up to 1 month

500 mM EDTA, pH 8.0

93.06 g EDTA disodium salt dehydrate
Water to 500 mL
Adjust pH to 8.0 with NaOH
Store at room temperature
Note: EDTA does not dissolve at this concentration below pH 8.0

10 mg/mL ethidium bromide (EtBr)

100 mg Ethidium bromide
Water to 10 mL

Store at room temperature
Note: light sensitive, cover the tube in aluminum foil

2.5 M glycine

9.38 g Glycine
Water to 50 mL
Store at room temperature

10 mg/mL glycogen

Add water directly to the original manufacturer's container to give a final glycogen concentration of 10 mg/mL, and then, prepare 500 μL aliquots and store at −20°C.

10% IGEPAL CA-630

5 mL IGEPAL CA-630
45 mL Water
Filter sterilize (0.22 μm pore)
Store at room temperature

5 M lithium chloride (LiCl)

105.98 g LiCl
Water to 500 mL
Store at room temperature

100 mM phenylmethylsulfonyl fluoride (PMSF)

174.2 mg PMSF
Methanol to 10 mL
Store at −20°C for up to 3 days

3 M sodium acetate (NaOAc)

61.52 g Sodium acetate
Water to 250 mL
Adjust the pH to 5.3 with glacial acetic acid
Store at room temperature

1.0 M sodium butyrate

2.75 g Sodium butyrate
Water to 25 mL
Filter sterilize (0.22 μm pore)
Prepare 250 μL aliquots and store at −20°C

500 mM sodium bicarbonate

10.5 g Sodium bicarbonate
Water to 250 mL
Filter sterilize (0.22 μm pore)
Store at room temperature

5 M sodium chloride

146.1 g Sodium chloride
Water to 500 mL
Store at room temperature

10% sodium deoxycholate

25 g Sodium deoxycholate
Water to 250 mL
Filter sterilize (0.22 μm pore)
Store at room temperature

10% sodium dodecysulfate (SDS)

25 g SDS
Water to 250 mL
Filter sterilize (0.22 μm pore)
Store at room temperature

500 mM sodium fluoride

525 mg Sodium fluoride
Water to 25 mL
Filter sterilize (0.22 μm pore)
Prepare 250 μL aliquots and store at −20°C

100 mM sodium pyrophosphate

665 mg Sodium pyrophosphate
Water to 25 mL
Filter sterilize (0.22 μm pore)
Prepare 250 μL aliquots and store at −20 °C

Soybean trypsin inhibitor (30X stock)

Dissolve powder in water at 30 mg/mL
Freeze in 1 mL aliquots at −80 °C

1.0 M Tris, pH 8.1

60.57 g Tris base
Water to 500 mL
Adjust pH to 8.1 with HCl

10% Triton X-100

5 mL Triton X-100
45 mL Water
Filter sterilize (0.22 μm pore)
Store at room temperature

Supplier Appendix

100 bp DNA ladder (NEB #N3231L)
6X Purple gel loading dye (NEB #B7025S)
Agarose (Sigma #A6013)
Ammonium chloride (Sigma #213330)
Boric acid (Fisher #BP168)
Bovine serum albumin (Sigma #A7906)
Cell strainer (Falcon #352350)
Chloroform (Sigma #650498)
cOmplete protease inhibitor tablets (Roche #11844600)
Dulbecco's phosphate-buffered saline (Life Technologies #14190-144)
Ethanol (Sigma #459844)
Ethidium bromide (Sigma #E8751)
Ethylenediaminetetraacetic acid disodium salt dihydrate (EDTA; Sigma #E5134)
Glycine (Sigma #G8878)
Glycogen (Sigma #G8751)
Hank's balanced salt solution (Life Technologies #14025-092)
IGEPAL CA-630 (Sigma #I3021)
Isoamyl alcohol (Sigma #320021)
Lithium chloride (Sigma #62476)
Paraformaldehyde (16%; Electron Microscopy Sciences #15710)
Phenol/chloroform/isoamyl alcohol (25:24:1; Life Technologies #15593-031)
Phenylmethylsulfonyl fluoride (PMSF; Sigma #78830)
Potassium bicarbonate (Sigma #P9144)
Protein A/G acrylamide beads (Thermo #53133)
Proteinase K (Sigma #P2308)
RNaseA (Life Technologies #12091-021)
Sodium acetate (Sigma #S2889)
Sodium bicarbonate (Fisher #S233-500)
Sodium butyrate (Sigma #B5887)
Sodium chloride (Sigma #S7653)
Sodium deoxycholate (Sigma #D6750)
Sodium dodecyl sulfate (Bio-Rad #161-0301)
Sodium fluoride (Sigma #S7920)
Sodium pyrophosphate tetrabasic (Sigma #P8010)
Soybean trypsin inhibitor (Sigma #T6522)
Tris base (Fisher #BP152)
Triton X-100 (Sigma #T8787)
0.25% Trypsin (Life Technologies #25200-056)

Acknowledgments

The material in this article has been reviewed by the Environmental Protection Agency and approved for publication. The contents of this article do not necessarily represent agency policy, nor does mention of trade names or commercial products constitute endorsement or recommendations for use. This work was supported by US Environmental Protection Agency intramural research funds. Support was also provided by National Institutes of Health Curriculum in Toxicology Training Grant (T32ES007126-34).

References

Baccarelli, A., Bollati, V., 2009. Epigenetics and environmental chemicals. Curr. Opin. Pediatr. 21, 243–251.

Bailey, T., Krajewski, P., Ladunga, I., Lefebvre, C., Li, Q., Liu, T., Madrigal, P., Taslim, C., Zhang, J., 2013. Practical guidelines for the comprehensive analysis of ChIP-seq data. PLoS Comput. Biol. 9(11): e1003326.

Brind'Amour, J., Liu, S., Hudson, M., Chen, C., Karimi, M.M., Lorincz, M.C., 2015. An ultra-low-input native ChIP-seq protocol for genome-wide profiling of rare cell populations. Nat. Commun. 6, 6033.

Brutlag, D., Schlehuber, C., Bonner, J., 1969. Properties of formaldehyde-treated nucleohistone. Biochemistry 8, 3214–3218.

Burris, H.H., Baccarelli, A., 2014. Environmental epigenetics: from novelty to scientific discipline. J. Appl. Toxicol. 34, 113–116.

Carey, M.F., Peterson, C.L., Smale, S.T., 2009. Chromatin immunoprecipitation (ChIP). Cold Spring Harb Protoc 4, e1–e8.

Cedar, H., Bergman, Y., 2009. Linking DNA methylation and histone modification: patterns and paradigms. Nat. Rev. Genet. 10, 295–304.

Dahl, J.A., Collas, P., 2008. A rapid micro chromatin immunoprecipitation assay (μChIP). Nat. Protoc. 3, 1032–1045.

Daley, T., Smith, A.D., 2013. Predicting the molecular complexity of sequencing libraries. Nat. Methods 10 (4), 325–327.

Gilmour, D.S., Lis, J.T., 1984. Detecting protein-DNA interactions in vivo: distribution of RNA polymerase in specific bacterial genes. Proc. Natl. Acad. Sci. U. S. A. 81, 4275–4279.

Gilmour, D.S., Lis, J.T., 1985. In vivo interactions of RNA polymerase II with genes of *Drosophila melanogaster*. Mol. Cell. Biol. 5, 2009–2018.

Gilmour, D.S., Lis, J.T., 1986. RNA polymerase II interacts with the promoter region of the noninduced *hsp70* gene in *Drosophila melanogaster* cells. Mol. Cell. Biol. 6, 3984–3989.

Golosova, O., Henderson, R., Vaskin, Y., Gabrielian, A., Grekhov, G., Nagarajan, V., Oler, A.J., Quinones, M., Hurt, D., Fursov, M., Huyen, Y., 2014. Unipro UGENE NGS pipelines and components for variant calling, RNA-seq and ChIP-seq data analyses. PeerJ.. 2e644.

Han, B.W., Wang, W., Zamore, P.D., Weng, Z., 2015. piPipes: a set of pipelines for piRNA and transposon analysis via small RNA-seq, RNA-seq, degradome- and CAGE-seq, ChIP-seq and genomic DNA sequencing. Bioinformatics 31 (4), 593–595.

Hansen, P., Hecht, J., Ibrahim, D.M., Krannich, A., Truss, M., Robinson, P.N., 2015. Saturation analysis of ChIP-seq data for reproducible identification of binding peaks. Genome Res. 25 (9), 1391–1400.

Head, S.R., Komori, H.K., LaMere, S.A., Whisenant, T., Van Nieuwerburgh, F., Salomon, D.R., Ordoukhanian, P., 2014. Library construction for next-generation sequencing: overviews and challenges. Biotechniques 56 (2). 61–64, 66, 68, passim.

Hecht, A., Strahl-Bolsinger, S., Grunstein, M., 1996. Spreading of transcriptional repressor SIR3 from telomeric heterochromatin. Nature 383, 92–96.

Hoffman, E.A., Frey, B.L., Smith, L.M., Auble, D.T., 2015. Formaldehyde crosslinking: a tool for the study of chromatin complexes. J. Biol. Chem. 290, 26404–26411.

Ilyin, Y.V., Georgiev, G.P., 1969. Heterogeneity of deoxynucleoprotein particles as evidenced by ultracentrifugation in cesium chloride density gradient. J. Mol. Biol. 41, 299–303.

Jackson, V., 1978. Studies on histone organization in the nucleosome using formaldehyde as a reversible cross-linking agent. Cell 15, 945–954.

Jackson, V., Chalkley, R., 1981. A new method for the isolation of replicative chromatin: selective deposition of histone on both new and old DNA. Cell 23, 121–134.

Jenuwein, T., Allis, C.D., 2001. Translating the histone code. Science 293, 1074–1080.

Kouzarides, T., 2007. Chromatin modifications and their function. Cell 128, 693–705.

McCullough, S.D., Bowers, E.C., On, D.M., Morgan, D.S., Dailey, L.A., Hines, R.N., Devlin, R.B., Diaz-Sanchez, D., 2016. Baseline chromatin modification levels may predict interindividual variability in ozone-induce gene expression. Toxicol. Sci. 150, 216–224.

Milne, T.A., Zhao, K., Hess, J.L., 2009. Chromatin immunoprecipitation (ChIP) for analysis of histone modifications and chromatin-associated proteins. Methods Mol. Biol. 538, 409–423.

O'Neill, L.P., Turner, B.M., 2003. Immunoprecipitation of native chromatin: NChIP. Methods 31, 76–82.

Qin, Q., Mei, S., Wu, Q., et al., 2016. ChiLin: a comprehensive ChIP-seq and DNase-seq quality control and analysis pipeline. BMC Bioinformatics 17, 404. https://doi.org/10.1186/s12859-016-1274-4.

Reinert, K., Langmead, B., Weese, D., Evers, D.J., 2015. Alignment of next-generation sequencing reads. Annu. Rev. Genomics Hum. Genet. 16, 133–151.

Ryan, R.J., Drier, Y., Whitton, H., Cotton, M.J., Kaur, J., Issner, R., Gillespie, S., Epstein, C.B., Nardi, V., Sohani, A.R., Hochberg, E.P., Bernstein, B.E., 2015. Detection of enhancer-associated rearrangements reveals mechanisms of oncogene dysregulation in B-cell lymphoma. Cancer Discov. 5 (10), 1058–1071.

Siegel, T.N., Hekstra, D.R., Kemp, L.E., Figueiredo, L.M., Lowell, J.E., Fenyo, D., Wang, X., Dewell, S., Cross, G.A.M., 2009. Four histone variants mark the boundaries of polycistronic transcription units in Trypanosoma brucei. Genes Dev. 23, 1063–1076.

Solomon, M.J., Varshavsky, A., 1985. Formaldehyde-mediated DNA-protein crosslinking: a probe for in vivo chromatin structures. Proc. Natl. Acad. Sci. U. S. A. 82, 6470–6474.

Solomon, M.J., Larsen, P.L., Varshavsky, A., 1988. Mapping protein-DNA interactions in vivo with formaldehyde: evidence that histone H4 is retained on a highly transcribed gene. Cell 53, 937–947.

Steinhauser, S., Kurzawa, N., Eils, R., Herrmann, C., 2016. A comprehensive comparison of tools for differential ChIP-seq analysis. Brief. Bioinform. 17 (6), 953–966.

Strahl, B.D., Allis, C.D., 2000. The language of covalent histone modifications. Nature 403, 41–45.

Tan, M., Luo, H., Lee, S., Jin, F., Yang, J.S., Montellier, E., Buchou, T., Cheng, Z., Rousseaux, S., Rajagopal, N., Lu, Z., Ye, Z., Zhu, Q., Wysocka, J., Ye, Y., Khochbin, S., Ren, B., Zhao, Y., 2011. Identification of 67 histone marks and histone lysine crotonylation as a new type of histone modification. Cell 146, 1016–1028.

Thomas, R., Thomas, S., Holloway, A.K., Pollard, K.S., 2017. Features that define the best ChIP-seq peak calling algorithms. Brief. Bioinform. 18 (3), 441–450.

Vallabh, S., Kartashov, A.V., Barski, A., 2018. Analysis of ChIP-Seq and RNA-Seq data with BioWardrobe. Methods Mol. Biol. 1783, 343–360.

Van Lente, F., Jackson, J.F., Weintraub, H., 1975. Identification of specific crosslinked histones after treatment of chromatin with formaldehyde. Cell 5, 45–50.

Wu, Q., Won, K.J., Li, H., 2015. Nonparametric tests for differential histone enrichment with ChIP-Seq data. Cancer Inform. 14 (Suppl. 1), 11–22.

Yan, H., Evans, J., Kalmbach, M., Moore, R., Middha, S., Luban, S., Wang, L., Bhagwate, A., Li, Y., Sun, Z., Chen, X., Kocher, J.P., 2014. HiChIP: a high-throughput pipeline for integrative analysis of ChIP-Seq data. BMC Bioinformatics 15, 280.

Zhang, Y., Sun, Y., Cole, J.R., 2014. A scalable and accurate targeted gene assembly tool (SAT-assembler) for next-generation sequencing data. PLoS Comput. Biol. 10 (8), e1003737. https://doi.org/10.1371/journal.pcbi.1003737.

Further Reading

Roadmap Epigenomics Consortium, Kundaje, A., Meuleman, W., Ernst, J., Bilenky, M., Yen, A., Heravi Moussavi, A., Kheradpour, P., Zhang, Z., Wang, J., Ziller, M.J., Amin, V., Whitaker, J.W., Schultz, M.D., Ward, L.D., Sarkar, A., Quon, G., Sandstrom, R.S., Eaton, M.L., Wu, Y.-C., Pfenning, A.R., Wang, X., Claussnitzer, M., Liu, Y., Coarfa, C., Harris, R.A., Shoresh, N., Epstein, C.B., Gjoneska, E., Leung, D., Xie, W., Hawkins, R.D., Lister, R., Hong, C., Gascard, P., Mungall, A.J., Moore, R., Chuah, E., Tam, A., Canfield, T.K., Hansen, R.S., Kaul, R., Sabo, P.J., Bansal, M.S., Carles, A., Dixon, J.R., Farh, K.-H., Feizi, S., Karlic, R., Kim, A.-R., Kulkarni, A., Li, D., Lowdon, R., Elliott, G., Mercer, T.R., Neph, S.J., Onuchic, V., Polak, P., Rajagopal, N., Ray, P., Sallari, R.C., Siebenthall, K.T., Sinnott-Armstrong, N.A., Stevens, M., Thurman, R.E., Wu, J., Zhang, B., Zhou, X., Beaudet, A.E., Boyer, L.A., De Jager, P.L., Farnham, P.J., Fisher, S.J., Haussler, D., Jones, S.J.M., Li, W., Marra, M.A., McManus, M.T., Sunyaev, S., Thomson, J.A., Tlsty, T.D., Tsai, L.-H., Wang, W., Waterland, R.A., Zhang, M.Q., Chadwick, L.H., Bernstein, B.E., Costello, J.F., Ecker, J.R., Hirst, M., Meissner, A., Milosavljevic, A., Ren, B., Stamatoyannopoulos, J.A., Wang, T., Kellis, M., 2015. Integrative analysis of 111 reference human epigenomes. Nature 518, 317–330.

Methods for Analysis of DNA Methylation

Karilyn E. Sant*, Jaclyn M. Goodrich†

*San Diego State University School of Public Health, San Diego, CA, United States
†University of Michigan School of Public Health, Ann Arbor, MI, United States

OUTLINE

Toxicoepigenetics
https://doi.org/10.1016/B978-0-12-812433-8.00015-0

347

INTRODUCTION

The epigenome, including the DNA methylome, is the dynamic regulatory framework that controls the use of genomic information to help define cellular identity and govern the response of cells, tissues, organs, and individuals to their environment. Researchers have demonstrated that individual epigenetic patterns are modified by the environment, heritable, and contribute to growth, development, and disease risk (Faulk and Dolinoy, 2011). DNA methylation is particularly susceptible to environmental perturbation in early life, evident by changes to DNA methylation associated with prenatal exposures to toxicants (e.g., lead, arsenic, mercury, bisphenol A, and cigarette smoke) (Koestler et al., 2013; Goodrich et al., 2015; Nahar et al., 2014; Joubert et al., 2012; Cardenas et al., 2015b), maternal BMI (Liu et al., 2014), gestational diabetes (Finer et al., 2015), maternal stress (Vidal et al., 2014), and more. DNA methylation occurs when DNA methyltransferases covalently attach a methyl group to the carbon-5 position of cytosine bases, producing 5-methylcytosine (5mC). In mammals, methylation is most common at cytosine-guanine dinucleotides, which are referred to as CpG sites (Dodge et al., 2002). CpG islands are genomic regions rich in CpG sites and are often located in the promoter region of genes. Hypermethylation of promoter CpG islands typically leads to the silencing of transcription, though this can vary for other regulatory regions. Recent studies find evidence for partial methylation of non-CpG cytosines especially in pluripotent embryonic stem cells, indicating the role of novel epigenetic mechanisms during differentiation and development (Lister et al., 2009). Furthermore, understanding the environmental influence on methylation levels in early life and as individuals age (Kochmanski et al., 2017) is an important part of risk assessment for the barrage of toxicants individuals encounter on a daily basis.

DNA methylation is an epigenetic modification of high interest, as methylation state is dynamic during development and these changes help to direct the processes of differentiation and organogenesis. Epigenetic modifications, including CpG methylation, are generally stable in somatic cells; however, during at least two developmental time periods, the epigenome undergoes extensive reprogramming. These critical windows of development include gametogenesis (Hajkova et al., 2002) and early preimplantation (Reik et al., 2001). While environmental perturbation to DNA methylation during this early gestational period is fairly well characterized, there is also evidence that exposures in adolescence and adulthood can alter the epigenome. For example, DNA methylation is associated with adolescent and/or adulthood exposures to toxicants such as phthalates (Goodrich et al., 2016), lead (Wright et al., 2010; Hanna et al., 2012), particulate matter (Panni et al., 2016; Dai et al., 2017), nitrogen dioxide (de F. C. Lichtenfels et al., 2018), and PAHs (Kim et al., 2016). DNA methylation changes with age in a predictable manner at specific genes and regions (Jung and Pfeifer, 2015), as well as stochastically. In 2013, Horvath et al. proposed an "epigenetic clock" calculated from the DNA methylation levels of 353 specific CpG sites that can be reliably used to predict biological age from multiple tissue types (Horvath, 2013). Since then, accelerated epigenetic aging—when the biological age estimate from the epigenetic clock is greater than chronological age—has been associated with all-cause mortality and cancer (Marioni et al., 2015; Perna et al., 2016; Durso et al., 2017; Ambatipudi et al., 2017), heart disease risk factors, metabolic syndrome, obesity, and age of menopause (Horvath et al., 2014, 2016; Nevalainen et al., 2017; Levine et al., 2016; Quach et al., 2017). Environmental factors influence the rate of epigenetic aging (Quach et al., 2017) (Nwanaji-Enwerem et al., 2017; Ward-Caviness et al., 2016), a phenomenon referred to as "environmental deflection" (Kochmanski et al., 2017). Characterizing the relationship between environmental exposures and the epigenetic clock, as well as the underlying mechanisms, is a rapidly emerging interest within the field of toxicoepigenetics with the potential to improve chemical risk assessment and develop more approaches to mitigate the effects of harmful chemical exposures.

Given its importance in early development and aging, DNA methylation is a crucial epigenetic modification to profile, and numerous methods are available to interrogate global methylation levels and gene-specific and epigenome-wide measures. This chapter serves as an introduction to methods for DNA methylation analysis. Section "Other DNA Epigenetic Modifications" provides an introduction to DNA epigenetic modifications other than 5mC, which should be considered when selecting appropriate methods. Section "Bisulfite Conversion" gives an introduction to bisulfite conversion and how it is used in epigenetic research. Section "Global DNA Methylation" provides an introduction to global methylation approaches. Section "Gene-Specific Methylation" includes information about methodologies for gene-specific quantitative approaches. Sections "Epigenome-Wide DNA Methylation Arrays" and "Next Generation Sequencing Approaches" describe epigenome-wide approaches using array-based technologies and next-generation sequencing, respectively. Overall, this chapter allows readers to select an appropriate method for DNA methylation screening for toxicoepigenetic studies in humans or model organisms.

OTHER DNA EPIGENETIC MODIFICATIONS

While 5mC is the most common epigenetic modification to DNA, several other covalent modifications exist. Once 5mC is produced, these marks are fairly stable and heritable. However, recent research has highlighted the importance of the DNA demethylation pathway, whose activity is mediated by ten-eleven translocation (TET) enzymes (Kohli and Zhang, 2013; Ito et al., 2011). These enzymes can initiate the demethylation of 5mC into 5-hydroxymethylcytosine, 5-formylcytosine, and 5-carboxylcytosine, and ultimately, unmethylated cytosine can be produced (Fig. 1). These other modifications are less frequently interrogated, yet they are becoming increasingly examined with improved technology. A fundamental understanding of the structures of these compounds is essential to choose the DNA methylation analyses most appropriate for research, since several technologies include or exclude these altered structures in their outputs indiscriminately.

Hydroxymethylcytosine (5hmC)

Other than 5mC, 5hmC is the most commonly studied DNA epigenetic modification. 5hmC is an intermediate in the demethylation pathway. Demethylation can be passive or active, the latter of which occurs when TET enzymes oxidize 5mC (Wu and Zhang, 2017). 5hmC is important to the toxicoepigenetics as this modification has been implicated in stem cell differentiation, regulation of gene expression in the brain, and levels that have been associated with exposures such as lead (Kriaucionis and Heintz, 2009; Tahiliani et al., 2009;

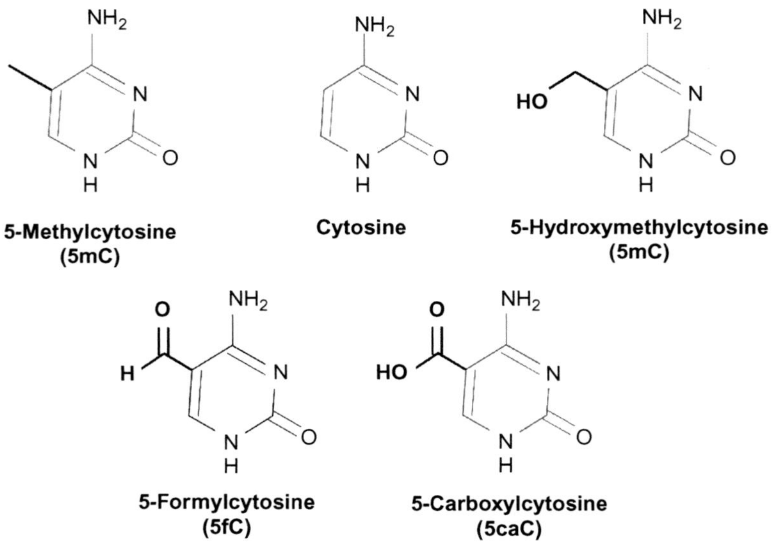

FIG. 1   Structures of modified cytosines. R groups are highlighted in bold type.

Wen et al., 2014; Ficz et al., 2011; Sen et al., 2015). Many of the methods that have been described in this chapter fail to distinguish between 5mC and 5hmC. Traditional bisulfite sequencing analysis provides a measure of total methylation that includes 5mC and 5hmC.

Fortunately, there are several methods that can distinguish between 5hmC and 5mC. At a global scale, ELISA methods are available to estimate total 5hmC or total levels of other covalent DNA modifications from samples. At the epigenome-wide scale, affinity-enrichment methods utilize an anti-5hmC antibody to capture and sequence the hydroxymethylome (hMeDIP-seq). Likewise, hMeDIP prior to bisulfite conversion and array hybridization has been utilized to generate percent 5hmC levels from the Infinium 450K platform (Sen et al., 2015). In addition to affinity-enrichment methods, bisulfite-sequencing-based methods can be adapted to generate 5hmC profiles by pretreating DNA prior to bisulfite conversion. One such method specifically labels 5hmC with a modified glucose molecule through the use of a beta glucosyltransferase. Labeled 5hmC is then biotinylated and captured using streptavidin (Song et al., 2011) followed by library preparation and sequencing. 5hmC enrichment at specific regions is calculated over input DNA. Another method, oxidative bisulfite sequencing (oxBs-seq), oxidizes DNA by pretreating with KRuO$_4$, which converts 5hmC to 5-formylcytosine (5fC) (Booth et al., 2013). Because 5fC is not protected from bisulfite conversion, this creates a sequence difference between the oxBS DNA and the comparative unpretreated DNA. In addition, Tet-assisted bisulfite sequencing (TAB-seq) (Yu et al., 2012), a method that, in combination with whole-genome bisulfite sequencing (WGBS), offers base-pair resolution for both 5mC and 5hmC, is an option. In this method, 5hmC is first protected from reacting to sodium bisulfite by labeling with beta glucosyltransferase, and then, the sample is exposed first to TET1 oxidation followed by sodium bisulfite treatment. This results in the ultimate conversion of all remaining cytosines and 5mC to uracil, thus distinguishing them from the protected 5hmCs. Direct comparison of sequencing results from the TAB-seq to WGBS allows for correct distinction between 5mC and 5hmC on a genome-wide basis. The suite of methods available to profile 5hmC continues to grow, though the cost and use of sample are nearly double for many of these methods that require profiling and comparing total 5mC or input genomic DNA along with the 5hmC-specific fraction.

5-Formylcytosine (5fC)

Following the production of 5hmC, 5fC can be generated by TET enzymes. Physically, 5fC can directly affect DNA and nucleosome structure and ultimately the chromatin state of localized DNA (Li et al., 2017; Raiber et al., 2015, 2017). Little is currently known about the function of 5fC. Though commonly an intermediate in the DNA demethylation pathway, 5fC can be a stable modification (Bachman et al., 2015). 5fC can be converted by TET enzymes to 5-carboxylcytosine or directly converted to unmodified cytosine by terminal deoxynucleotidyl transferase (TdT) or thymine DNA glycosylase (TDG) and resulting base excision repair (Zhang et al., 2012). Several studies have shown that 5fC is dynamic during development and aging, may have tissue-specific signatures, and is increasingly found in enhancer and intragenic regions of genes (Iurlaro et al., 2016; Qian et al., 2015; Bachman et al., 2015). Furthermore, 5fC increased C/EBPα and C/EBPβ DNA binding at regulatory CpG sites, suggesting an important epigenetic function for cellular differentiation and disease states (Khund Sayeed et al., 2015).

5-Carboxylcytosine (5caC)

Like 5hmC and 5fC, 5caC is an intermediate of the DNA demethylation pathway mediated by TET enzymes. 5caC can also be converted to unmodified cytosine by terminal deoxynucleotidyl transferase (TdT) or thymine DNA glycosylase (TDG) and resulting base excision repair (Zhang et al., 2012). Also, like 5fC, little is known about the functional consequences of 5caC. However, studies are increasingly associating elevated 5caC with cellular differentiation and carcinogenesis and finding that 5caC patterns may be age-, tissue-, and disease-state-specific (Lewis et al., 2017; Eleftheriou et al., 2015; Inoue et al., 2011).

BISULFITE CONVERSION

Bisulfite conversion of DNA is an essential first step for many gene-specific and epigenome-wide analyses. Since current sequencing methods cannot distinguish between methylated and unmethylated cytosines, different methylation-specific sequences must be created. Treatment of DNA with sodium bisulfite converts unmethylated cytosine into uracil via three steps: sulfonation, hydrolytic deamination, and desulfonation (Hayatsu et al., 1970; Clark et al., 2006). Following conversion, PCR amplification instead incorporates thymine, creating a stable sequence difference between methylated (cytosine) and unmethylated (thymine) cytosine residues throughout the genome (Fig. 2). It is important to note that bisulfite conversion does not distinguish between 5mC and 5hmC, and therefore, there is

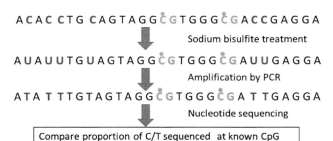

A C A C C T G C A G T A G G Č G T G G G Č G A C C G A G G A

⬇ Sodium bisulfite treatment

A U A U U T G U A G T A G G Č G T G G G Č G A U U G A G G A

⬇ Amplification by PCR

A T A T T T G T A G T A G G Č G T G G G Č G A T T G A G G A

⬇ Nucleotide sequencing

Compare proportion of C/T sequenced at known CpG
sites to quantify % of methylated cells in original sample

FIG. 2 Estimating DNA methylation follow-ing bisulfite conversion. Treatment with sodium bisulfite converts unmethylated cyto-sine to uracil. Following amplification and sequencing of the treated DNA, comparison of the proportion of cytosines and thymines at expected CpG sites enables quantification of the percent of methylated cells at each site in a given sample.

ambiguity as to the specific modification that occurs at each unconverted CpG locus following conversion (Huang et al., 2010b). However, several protocols exist that can distinguish between 5mC and 5hmC (Booth et al., 2013).

A detailed description of the bisulfite conversion protocol has been published (Clark et al., 2006). However, several commercially available kits are available to perform bisulfite conver-sion, including the EpiTect® Bisulfite Kit (Qiagen) and the EZ DNA Methylation™ Kit (Zymo Research). These kits contain already buffered solutions to enhance the reaction and often contain spin columns to maximize DNA output and purity. These kits have well-established and widely utilized protocols, which maintain DNA integrity and yield while still ensuring complete bisulfite conversion (Huang et al., 2010a). However, several control procedures can be utilized to confirm bisulfite conversion efficiency. Analytic standards for humans and mice can be purchased containing universally methylated or unmethylated DNA for comparison to the region or sequence of interest (Zymo Research and Promega). By producing a gradient of controls using these templates, a standard curve can be constructed for comparison to the experimental DNA for more confident quantification (Tse et al., 2011). When using several platforms for sequencing the bisulfite-converted DNA (such as PyroMark, discussed in Section "Repetitive Elements"), software examines the percentage of non-CpG cytosines. Because only 5mC in a CpG site should still remain as cytosine following bisulfite treatment, the percentage of remaining non-CpG cytosines typically is inversely proportional to the con-version efficiency. Therefore, examining any remaining C at these non-CpG sites is known as a bisulfite treatment control. Though studies have confirmed that the bisulfite conversion kits are highly efficient (>97%), these easy extra control steps help to increase data rigor and reproducibility (Worm Ørntoft et al., 2017; Holmes et al., 2014).

GLOBAL DNA METHYLATION

Global DNA methylation assays are used to approximate total genomic 5mC concentra-tions. Though no inference about specific gene expression can be made from global DNA methylation measures, approximations of total genomic methylation are widely utilized in studies as a biosensor for biochemical changes such as nutrition, age, or chemical exposures. Global methylation is more resistant to change than locus-specific measures of DNA meth-ylation, so it is most often used to examine changes during and following dynamic periods of epigenomic change such as embryonic programming (Reik, 2007), to compare healthy and diseased states (i.e., cancer) or longitudinally to observe epigenetic drift (Fraga et al., 2005).

Most approaches for global interrogation of DNA methylation are available for almost all species, and these methods offer researchers across the life sciences the opportunity to approximate total genomic methylation percentages. Global DNA methylation percentages can vary by cell, tissue, and age. DNA methylation is dynamic during gametogenesis and organogenesis, and epigenetic drift can occur with aging and cellular senescence (Fraga et al., 2005). These temporal relationships can also be modulated by toxicant exposures throughout the life course (Kochmanski et al., 2017). Furthermore, DNA methylation can vary from cell to cell within a tissue, and epigenomes can vary even further between different cell and tissue types. Because each of the global methods is designed to examine very distinct regions (i.e., repetitive elements) or sequences in the genome, global methylation methodologies do not yield 100% concordant results with one another (Knothe et al., 2016; Lisanti et al., 2013).

Examining global methylation of whole organs will probe highly variable genomes and provide an estimated average readout of all cell populations in the organ. Therefore, selection of tissue section during in vivo studies or cell-type selection for in vitro experiments is an important consideration for epigenomics—though the dawn of single-cell resolution technologies has improved these outcomes. Though more homogeneity is expected from in vitro experiments, this heterogeneity should also be considered when using stem cells, coculture, and microphysiological systems. Here, we present commonly utilized methods for the quantification of global DNA methylation and highlight important considerations during study design for their use. See Table 1 for an overview of these methods.

LUminometric Methylation Assay (LUMA)

The LUMA is used to approximate the total genomic percent methylation (Karimi et al., 2006a,b). Numerous studies have utilized LUMA to assess changes in global DNA methylation in response to toxicant exposures (Sant et al., 2016; Nahar et al., 2015), nutritional alterations

TABLE 1 Characteristics of Methods Used to Interrogate Global DNA Methylation

Method	Genomic Coverage (%), *Humans*	Input DNA Required	Method Similar for all Species?	Bisulfite Conversion Required?	Can Distinguish Between 5mc and Other Modifications?
LUMA	~8%	>250 ng	Yes	No	No
Repetitive elements (Alu/LINE-1)	~10%–15% each	>250 ng	No	Yes	No
Endogenous retroviruses (IAP)	~8%	>250 ng	No	Yes	No
Mass spectrometry	100%	>10 ng	Yes	No	Yes
ELISA	100%	>20 ng	No	No	Yes, separate kits available

5. PROTOCOLS FOR TOXICOEPIGENETICS RESEARCH

(Sant et al., 2013; Jousse et al., 2011; McKay et al., 2012), and ontogenically throughout the life course (Bjornsson et al., 2008; Thompson et al., 2010). We have previously published a detailed LUMA protocol (Sant et al., 2012). LUMA uses methylation-specific restriction enzyme digestion followed by pyrosequencing to quantify percent methylation. Both the *MspI* and *HpaII* restriction enzymes cleave DNA at the approximately 2.3 million (in humans) 5′-CCGG-3′ sites (Kinney et al., 2011; Ball et al., 2009). Though these sites account for only ~8% of CpG sites in the genome, these sites are often found within CpG islands that account for the majority of genomic methylation, and therefore, interrogation of CCGG sites is one of the most comprehensive ways to estimate global DNA methylation (Ball et al., 2009). *MspI* is methylation-insensitive and will therefore cleave at methylated, unmethylated, and hydroxymethylated 5′-CCGG-3′ sites (though it cannot cleave at sites with other 5′ modifications) (Ito et al., 2011). *HpaII*, on the other hand, is sensitive and can only cleave unmodified sites. Thus, the ratio of nucleotide incorporation for unmethylated/total cleavage will yield the percent of unmethylated sites. These results are typically normalized to *EcoRI* digestion as an internal control, since *EcoRI* sites are widely prevalent throughout the genome and are not methylation-sensitive. The global methylation percentage is then easily calculated from the nucleotide incorporation data using the following formula: $1 - [(HpaII(G)/EcoRI(T))/(MspI(G)/EcoRI(T))] \times 100$ (for more detailed information and methods, please see Sant et al., 2012).

LUMA has several major advantages. The assay requires minimal input DNA (250–500 ng of genomic DNA) and does not require bisulfite conversion and sequencing. LUMA is readily accessible to all model organisms because it does not require a reference genome and has been used in ecological studies to examine species ranging from fish to mammals (Head et al., 2014). *EcoRI* digestion also provides an internal standard to control for slightly variable amounts of input DNA per sample. Additionally, LUMA is high-throughput, with a very short run time; for example, a 96-well plate can run in under 20 min. However, LUMA may also have some disadvantages. Data suggest that global DNA methylation detected by LUMA may be variable based upon the method of DNA isolation utilized (Soriano-Tárraga et al., 2013). Therefore, the type of sample and the extraction method should be considered prior to selecting a DNA isolation method. Because LUMA only interrogates 5′-CCGG-3′ sites, this selective bias may influence results. Though 5′-CCGG-3′ sites are universal across organisms, these sites may not be distributed evenly throughout the genome and do not represent all genomic CpG sites. Furthermore, the ability of *MspI* to also cleave at hydroxymethylated sites suggests that global methylation percentages estimated from the LUMA assay are actually the total percentage of both methylated and hydroxymethylated DNA.

Repetitive Elements

Examination of CpG methylation within repetitive elements throughout the genome is one of the most popular methods to estimate global DNA methylation. Several classes of repetitive elements have substantial CpG composition, including long interspersed nuclear elements (LINEs; 21% of the human genome) and *Alu* repeats (11%–14% of the human genome) (Batzer and Deininger, 2002; Treangen and Salzberg, 2011; Luo et al., 2014). While LINEs occur across vertebrate species, *Alu* is specific to primates but represents almost a quarter of genomic CpG sites (Britten et al., 1988). For both *Alu* and LINE-1, both the nucleotide sequence and copy number throughout the genome can vary between species and strains. For example, LINE-1 is estimated to represent ~20% of the murine genome, vary in sequence by ~3% across strains in species, and vary in abundance by as much as 40% (Mouse Genome Sequencing Consortium, 2002; Rebuzzini et al., 2009; Sookdeo et al., 2013). Furthermore, the GC content of LINE-1 is variable between species (Boissinot and Sookdeo, 2016). For example, the mammalian LINE-1 contains approximately 43% GC content, while the zebra fish LINE-1 contains merely 37%. CpG content also varies across the *Alu* subfamilies. While the *AluY* class is the most CpG-rich, the *AluS* subfamily represents the largest percent of the genome. Methylation of *Alu* and LINE-1 is commonly used to estimate global DNA methylation because they comprise such a large percentage of total genome (~10% and 15% of the human genome, respectively) and genomic CpG sites (representing ~25% and 10% of CpG sites, respectively), and methylation of both repeat elements has been frequently shown to be sensitive to toxicological modification (Faulk et al., 2016; Rusiecki et al., 2008; Bollati et al., 2007; Tarantini et al., 2009; Goodrich et al., 2015; Lander et al., 2001; Zheng et al., 2017).

To assess percent methylation at these sites, DNA first undergoes sodium bisulfite conversion to create nucleotide differences between unmethylated and methylated CpGs throughout the genome, as described in Section "Bisulfite Conversion" (Clark et al., 1994). Each repetitive element is then amplified by PCR and sequenced in order to determine the percent methylation at each CpG site (Yang et al., 2004). Sequencing software examines the percent of C versus T incorporation at each site by calculating the peak heights of each (proportional to read counts) and creating a ratio of methylated/unmethylated + methylated sites. These methods are highly reproducible and accessible regardless of preferred sequencing methodology (Lisanti et al., 2013). Several sequencing platforms are built on quality control measures to ensure data integrity. For example, building C and T sequentially into the analytic sequence directly probes C versus T integration at a specific site and can then be used to directly compare site-specific methylation. Bisulfite treatment controls (Section "Bisulfite Conversion") should also be added to ensure bisulfite conversion efficiency. Drawbacks to using repetitive element methylation to approximate total genomic methylation include the fact that methylation across these regions may be more or less sensitive to toxicological or nutritional modulation than across the entire genome and may not be an accurate estimation of total methylation. Furthermore, these methods are not accessible to all model species, and sequences can vary.

Endogenous Retroviruses

Intracisternal A particles (IAPs) are endogenous retroviral transposable elements, with the ability to insert throughout the genome. IAPs are believed to be mouse-specific and can disrupt gene expression via insertion into gene regulatory regions. When IAPs contain CpG sites that can be differentially methylated and therefore influence gene expression, these genes can produce metastable epialleles (Rakyan et al., 2002). Metastable epialleles are defined as alleles that can be differentially expressed between genetically identical individuals or cell populations (Dolinoy et al., 2007a; Rakyan et al., 2002). Metastable epialleles are indicative of epigenetically labile loci, which can be manipulated by environmental (nongenetic) factors including exposures and diet. On a grand scale, these minute changes have the potential to produce phenotypic and physiological effects. Several IAP-generated metastable epialleles have been discovered throughout the murine genome, including the *CDK5-activator-binding protein (Cabp^{IAP})*, *axin fused (Axin^{fu})*, and *viable yellow agouti (A^{vy})*. The *Axin^{fu}* and *A^{vy}* mouse models allow for direct visualization of methylation at each respective locus due to phenotypic traits, namely, tail kink (*Axin^{fu}*) and coat color (*A^{vy}*). The agouti mouse model in particular has been widely utilized as a biosensor for methylation in studies examining the effects of diet and toxicant exposures on DNA methylation (Dolinoy et al., 2007b, 2006). Mice with hypomethylated *A^{vy}* IAP loci have more yellow coats, while hypermethylation yields more pseudoagouti (brown) coats (Fig. 3). While the agouti mouse model does not represent total genomic methylation, it does provide a visual indicator of IAP methylation. Because these regions are inherently labile, a visual readout of IAP methylation can be used to characterize widespread epigenetic changes—especially with regard to other metastable epialleles and labile loci. Similarly to *Alu* and LINE-1, these regions can also be subjected to bisulfite sequencing to confirm that phenotypic changes are indeed associated with methylation at these loci and across the genome.

Mass Spectrometry

The "gold standard" for the assessment of global DNA methylation is mass spectrometry. Mass spectrometry relies on mass differences rather than sequencing or nucleotide incorporation to quantify DNA methylation. Samples are injected into a column and separated by mass, yielding a chromatograph from which abundance can be quantified by calculating the area under the curve. These readouts can quantify even subtle mass differences, such as methylation, between DNA fragments. The original method, high-performance liquid chromatography (HPLC), requires large amounts of input DNA (measured on the milligram scale), but as several mass spectrometric methodologies emerged, these methods became more efficient (can be run on as little as 10 ng input DNA) and higher-throughput. Global methylation analysis using mass spectrometry is highly accurate, sensitive, more reproducible, and less biased than other methods since it probes across the entire genome regardless of locus or sequence. Additionally, this method can easily distinguish between 5mC and other DNA epigenetic modifications. However, the technical expertise required to perform these methods is greater than other methods, making this method more costly and inaccessible to those without training in analytic chemistry. Detailed protocols for mass spectrometric analysis of global DNA methylation have been previously published (Berdasco et al., 2009).

Enzyme-Linked Immunosorbent Assay (ELISA)

ELISA has been widely utilized as an easy, high-throughput analysis of DNA methylation. These assays detect 5mC using specific fluorescent antibodies or traditional colorimetric detection, which can then be quantified using a plate reader. While ELISA does interrogate 5mC throughout the entire genome, these assays can be species-specific and may not be available for researchers using alternative (viz., nonmammalian) models. Several companies have developed commercially available kits, requiring variable amounts of input DNA ranging from 20 to 1000 ng. ELISA methods not only can be more costly but also are easy to perform

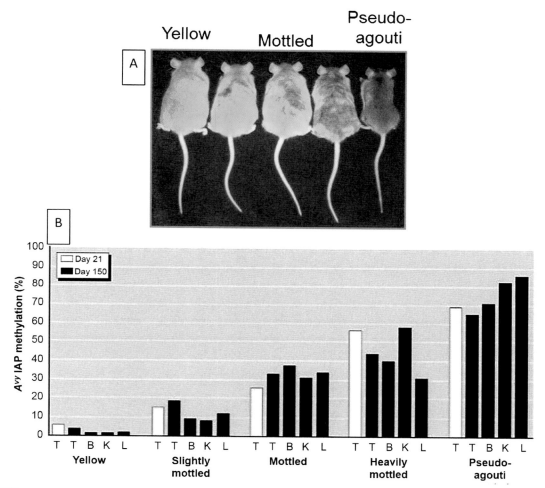

FIG. 3 The agouti mouse model: DNA methylation with corresponding phenotypic variation. (A) Coat color distribution, ranging from all yellow to brown (pseudoagouti) of yellow viable agouti (A^{vy}) mice. (B) Coat color corresponds to DNA methylation at the A^{vy} IAP. Average A^{vy} IAP methylation across nine CpG sites is shown in mice of each coat-color group at two postnatal ages in tail (T) and at day 150 in the brain (B), kidney (K), and liver (L). *Reprinted with permission from the author Dr. Dana Dolinoy. Photo and graph first appeared in Dolinoy D.C., Weidman, J.R., Waterland, R.A., Jirtle, R.L., 2006. Maternal genistein alters coat color and protects A(vy) mouse offspring from obesity by modifying the fetal epigenome. Environ. Health Perspect. 114, 567–572.*

and very high-throughput. Several commercially available kits are at market, including the MethylFlash Global DNA Methylation (5-mC) ELISA Easy Kit (Colorimetric) (EpiGentek). These kits are also available for the detection of 5hmC and 5fC. Because these assays are based upon relative quantification, controls are crucial. Positive (completely methylated) and negative (completely demethylated) DNA controls should be used (and are often included), and all samples are quantified using a standard curve prepared from dilutions of these controls.

GENE-SPECIFIC METHYLATION

Gene-specific analysis of DNA methylation can provide keen insight into the more subtle and, potentially, specifically targeted effects of DNA methylation. While the investigation of global DNA methylation provides information concerning systemic DNA methylation alterations, global methods lack the resolution to define the effects of environmental factors on discrete genes and processes within cells, tissues, and organs. Additional detail can be obtained by analyzing DNA methylation within specific gene regulatory regions or CpG loci using several gene- or locus-specific methods.

There are several major considerations for research investigating gene-specific methylation. First, there may be higher degrees of variability between tissues or even between cells for gene-specific methylation. These methylation patterns can help to direct gene expression for important factors in signal transduction and cellular communication, which ultimately govern processes such as tissue patterning, proliferation, and cell polarity. It is important to consider the number of targets, as gene-specific analyses with many loci can become costly and arduous. As high-throughput epigenomic analyses become more accessible and affordable, they may be more appropriate for broader applications; however, these gene-specific analyses are crucial for interrogating differences in DNA methylation at a small number of loci and validating the findings of large-scale epigenomic approaches.

Pyrosequencing

Pyrosequencing is one of the most widely employed methods for DNA sequencing. This technique quantifies the incorporation of specific nucleotides added into wells containing complementary DNA template by measuring luminescence generated via luciferase activity. Polymerization of nascent DNA strands leads to the release of one pyrophosphate (PPi) molecule per incorporated deoxynucleotide, which is used by ATP sulfurylase in the presence of adenosine $5'$-phosphosulfate (APS) to produce ATP. The presence of ATP allows for the conversion of luciferin to oxyluciferin, which releases light, in the presence of luciferase and oxygen (O_2). The intensity of emitted light is quantified temporally after each nucleotide addition to determine DNA sequence. This intensity of the resulting peak is increased proportionally to the number of nucleotide additions at a given time, and this peak amplitude depends on the amount of template in each well. This technology can be utilized for DNA methylation analysis by utilizing bisulfite-converted DNA and ratiometrically comparing C/T incorporation at CpG sites (see Fig. 2 for an example of how the sequence would look). At any target CpG site, PyroMark software analyzes the nucleotide incorporation of both C and T (separately, in succession), so that the ratio of methylated/unmethylated template can be quantified. Pyrosequencing has several strengths and weaknesses. As with many sequencing technologies, "slippage" can occur at homopolymer tracks as sequential nucleotide incorporation can occur during the same addition step. Because pyrosequencing is highly mechanized, run failure due to low template availability or physical jarring of instruments can disrupt data collection. However, pyrosequencing provides excellent resolution, requires less time, and is cost-efficient compared with other sequencing methods. Because only primers differ between all assays, reagents are affordable and can be repeatedly used.

Because primer design is integral to the success of pyrosequencing, primer design software (e.g., PyroMark Assay Design, Qiagen) is readily available, and validated primers are commercially available. This software has several quality control measures in place to improve data rigor, including a bisulfite treatment control (Section "Bisulfite Conversion"). Reference peaks, the spiking of a known concentration, can also be built into the assay template, enabling direct 1:1 comparison between samples and wells. PyroMark software accounts for all of these built-in controls and provides users with an easy-to-understand quality score to improve data reproducibility.

Methylation-Specific qPCR

Methylation-specific quantitative PCR (e.g., EpiTect MethyLight PCR Kit, Qiagen) is a practical and popular strategy for gene-specific methylation analysis (Eads et al., 2000). This strategy employs the use of strategically designed sets of primer pairs for the same region, which differ in sequence to target either methylated or unmethylated sequences of DNA. This approach subjects bisulfite-converted DNA to a modified TaqMan qPCR approach that pairs a single primer pair with two differentially labeled probes. One probe specifically anneals to the target region following bisulfite conversion, while the other only anneals to the target region if the cytosine was protected from bisulfite conversion due to methylation (Qiagen, 2011). The relative abundance of the protected and converted cytosine within the target region is determined by TaqMan qPCR. A limitation of the MethyLight assay is its inability to quantify locus-specific methylation in dense CpG regions. If the probe-annealing region has multiple CpGs, the readout of the assay would show if that region was hypo- or hypermethylated in only a semiquantitative manner. Otherwise, multiple primers could be created within the same region of interest to specifically interrogate each CpG, but this redundancy may be costly. However, the MethyLight assay is affordable and high-throughput for the assessment of a few CpGs and can utilize equipment already found in most life science laboratories. As with many other strategies, use of methylated and unmethylated control DNA should be utilized as quality controls for these experiments. Furthermore, constructing a standard curve from these known quantity and known methylation controls can be used to more rigorously quantify sample methylation.

Mass Spectrometry

Mass spectrometry methods, such as the EpiTYPER® assay (Agena Bioscience), have been widely utilized in gene-specific DNA methylation analyses. This assay amplifies the regions of interest in bisulfite-converted DNA using T7 promoter-tagged primers (Suchiman et al., 2015). Following transcription, RNA is cleaved at uracil sites, creating fragments of differing mass based upon methylation status. Fragments are analyzed using matrix-assisted laser desorption/ionization time-of-flight (MALDI-TOF) mass spectrometry on the MassARRAY® (Agena Bioscience) system, and fragment lengths are aligned to the region of interest to quantify methylation at each locus in the amplicon. The EpiTYPER platform has several strengths, including high reproducibility and the ability to interrogate amplicons of up to 600 bp. Mass spectrometry methods, such as EpiTYPER, require technical expertise in mass spectrometry or the adoption of commercial platforms such as MassARRAY that can be expensive.

Cloning & Sequencing

Cloning and Sanger sequencing are the original "gold standard" methods for gene-specific analysis of DNA methylation. Detailed protocols for performing these methods have been previously published (Zhang et al., 2009). This procedure also utilizes bisulfite-converted DNA. Regions of interest are amplified and inserted into vectors that have been engineered to indicate the insertion of the amplicon. The resulting products are then sequenced, and site-specific methylation can be quantified. This method is more time-intensive than modern methodologies and also may require more input DNA to produce a larger clone population. Like the other methods mentioned above, homopolymer tracks can introduce read "slippage" during sequencing, introducing error. Furthermore, multiple copies of the same cloned sequence can introduce bias, so proper control methods must be used. However, Sanger sequencing allows for reading of longer and denser regions of interest and therefore remains an important DNA methylation research tool.

EPIGENOME-WIDE DNA METHYLATION ARRAYS

In the past two decades, novel approaches have been introduced and refined for the analysis of the DNA methylome. The rapid advancement of high-throughput and relatively affordable arrays for screening DNA methylation at thousands to millions of CpG sites is accelerating the fields of not only basic biology but also medical applications and toxicoepigenetics with a discovery-based approach to finding and refining targets (Ambatipudi et al., 2017; Shenker et al., 2013; Joubert et al., 2016; Figueroa et al., 2009, 2010; Cardenas et al., 2015a; Kim et al., 2014). These approaches combine DNA treatments such as bisulfite conversion, digestion with methylation-sensitive enzymes, or anti-5-methylcytosine antibodies, with probe-based arrays and/or deep sequencing (see Section "Next Generation Sequencing Approaches") (Huang et al., 2010a; Laird, 2010). Cost efficiency, high-throughput capacity, accuracy (quantitative vs. relative), and resolution vary by technology and applicability to various species. Epigenome-wide platforms can be classified as either biased or unbiased approaches. An unbiased approach reveals the full regulatory network at the level of the whole genome (such as WGBS, in Section "Sequencing Bisulfite Converted DNA"), while a biased analysis is limited to certain loci or regions of the genome, such as promoter regions or heavily methylated sequences.

Microarray technology, well known in the gene expression world, is also used for DNA methylation analysis as it provides a high-throughput method to interrogate DNA methylation at thousands of loci. In general, microarray technologies are hybridization-based assays that analyze thousands of sequences simultaneously without requiring a large sample volume. Fluorescently labeled nucleic acids are hybridized to reporter molecules, such as oligonucleotides ("probes"), which are built onto a solid surface or microscopic polystyrene beads (Huang et al., 2010a). After an initial wash to reduce nonspecific hybridization, the microarrays are scanned under a confocal fluorescent microscope at wavelengths appropriate for the given fluorescent labels. Fluorescence is emitted when labeled nucleic acids hybridize to and are captured by the complementary probe and the total signal. The subsequent fluorescence intensity is measured at each spot on the array, which corresponds to each sequence-specific probe, and the strength of this signal depends on the amount of sample binding. These raw intensity data can be compared with control probes included on the array (i.e., probes that are not expected to have any sample binding) to estimate background signal (Ball and Sherlock, 2007). Microarray raw data require processing prior to interpretation to reduce technical variability such as differences in background signal and scanned intensity that vary by batch (i.e., between individual chips/arrays run at the same time or between arrays run on different days) or even by position within a chip or array. Fortunately, bioinformatics tools and pipelines have been established to identify and reduce these types of technical variability (see example pipelines for the Infinium BeadChips (Fortin et al., 2017; Pidsley et al., 2013)).

Bisulfite Conversion Based Arrays

The ability to quantify DNA methylation following bisulfite conversion not only is used in candidate gene analyses (Section "Global DNA Methylation") but also has been adapted to interrogate CpG sites throughout the epigenome in biased or partially biased array technologies. Illumina offers a variety of platforms for epigenome-wide analysis including the Infinium BeadChip and the GoldenGate methylation assay based on bisulfite-converted DNA genotyping (Bock et al., 2010). Bisulfite-converted DNA is measured by two probes, one that recognizes methylated cytosines and another that recognizes unmethylated cytosines. Single-base-pair extension allows for the incorporation of fluorescently labeled nucleotides, and the ratio of fluorescence signal from adenine/thymine versus guanine/cytosine is used to estimate percent methylation at a given CpG site (Weisenberger et al., 2008).

Infinium BeadChips

The most common tools used in epigenome-wide association studies (EWAS) in human populations have been the Illumina Infinium series of probe-based arrays (Fig. 4). The Infinium series provides DNA methylation quantification at single-CpG-site resolution with a low sample input requirement of 500 ng bisulfite-converted DNA. The first two versions of this platform, the Infinium HumanMethylation27 ("27K") and HumanMethylation450 ("450K") BeadChips, provided coverage at >27,000 and > 480,000 CpG sites spanning 14,495 and all designable RefSeq human genes, respectively. The 27K was semibiased toward promoter regions and CpG islands, while the 450K added coverage to all known human genes and to regions outside of CpG islands such as CpG shores, shelves, and "open sea." Hundreds of epidemiology studies have been published with epigenomic data from the 27K and 450K platforms including toxicoepigenetic studies reporting DNA methylation profiles associated with exposures to maternal smoke, arsenic, lead, bisphenol A, air pollution, phthalates, and more (Joubert et al., 2016; Wu et al., 2017; Cardenas et al., 2015a; Kim et al., 2013; Solomon et al., 2017; Gruzieva et al., 2017). In December 2015, the latest version of the Infinium array, the MethylationEPIC BeadChip, was released. The EPIC array includes 90% of the CpG sites

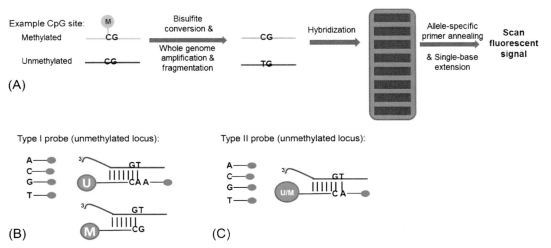

FIG. 4 DNA methylation analysis via Infinium BeadChips. (A) Following bisulfite conversion, DNA is amplified and fragmented prior to hybridization to the BeadChip. The Infinium MethylationEPIC BeadChips contain wells for eight samples, and each well contains probes to capture methylated or unmethylated targets (> 850,000 target loci). When sample DNA binds to the complementary probes, single-base extension occurs, the labeled nucleotides fluoresce, and the signal is scanned at each spot. (B) 450K and EPIC utilize two probe types. In this example of a type I probe set, two probes exist for a given locus: a probe that will match the unmethylated target (U) and one that will match the methylated target (M). Here, the sample DNA is unmethylated and matches the U probe. Upon annealing, single-base extension occurs, and the incorporated nucleotide releases a fluorescent signal. This does not occur at the M probe. (C) Type II probes contain only a signal probe for each locus that can hybridize to unmethylated or methylated DNA at that locus. Single-base extension occurs at the 5' C/T position of the target CpG. The color of the fluorescent signal depends on which nucleotide was incorporated, signifying whether the CpG site was originally methylated or not.

from the 450K and provides comprehensive quantification at >850,000 CpG sites across all known genes and in intergenic and key regulatory regions (e.g., gene promoters and enhancers (Moran et al., 2016)). By design, the majority of new probes are added to the upgrade target enhancer regions.

For all Infinium arrays, CpG methylation is calculated as the ratio between the methylated probe signal relative to the total (methylated and unmethylated) probe signal at every locus. The resulting value, known as the beta (β) value, ranges from 0 to 1, reflecting the proportion of DNA methylation at a given CpG site in the sample with 0 being completely unmethylated and 1 being completely methylated (Weisenberger et al., 2008). The beta value is also often expressed as a percentage by multiplying by 100. Occasionally, the M-value is also used that is defined as the $\log_2$ transformation of methylated to unmethylated probe intensities and can have better statistical properties. Even so, the beta value is more interpretable and far more widely utilized in the literature (Du et al., 2010). The Infinium arrays have many advantages. DNA methylation is always quantified at the same CpG sites enabling meta-analysis across studies or validation in additional cohorts. Bioinformatics pipelines for annotating and normalizing data are well developed and freely accessible through R packages (Fortin et al., 2017; Morris et al., 2014). Control probes are built into the arrays that can be used for quality assurance and quality control as well as to correct for batch effects both within and across studies (Fortin et al., 2014).

The 450K and EPIC arrays are also uniquely poised to address a common limitation of epidemiological epigenetic studies—the issue of cell-type heterogeneity in the most commonly surveyed sample types (DNA from blood leukocytes or umbilical cord blood leukocytes). Due to accessibility, blood DNA is often used to survey for epigenetic changes in human populations though blood is a heterogeneous mixture of cell types whose proportions can be influenced by disease, exposure, and age, leading to differing epigenomic profiles. Blood-cell-type-specific differentially methylated CpG sites have been identified on the 450K and EPIC, and methylation levels at these CpG sites can be used to estimate cell-type proportions of blood or cord blood samples using a validated method based on data from sorted cells (Houseman et al., 2012; Bakulski et al., 2016). A reference-free method has also been developed to estimate cell proportions in other tissue types (e.g., placenta or tumors) (Houseman et al., 2014). Limitations of the Infinium arrays include coverage at only a portion of the total CpG sites in the human epigenome and technical issues with the array such as probe-type biases in fluorescent signal and SNPs located within a subset of probes (most of which can be corrected through the bioinformatics pipeline (Teschendorff et al., 2013; Daca-Roszak et al., 2015)). Additionally, this platform is only available for human research though recently researchers successfully utilized the EPIC array to quantify mouse DNA methylation at a subset of >19,000 probes that also uniquely aligned to bisulfite-converted murine DNA (Needhamsen et al., 2017).

GoldenGate Assay for Methylation

The Illumina GoldenGate assay for methylation applies hybridization technology to provide highly sensitive and accurate DNA methylation quantification at up to 1532 CpG sites in a biased approach (Huang et al., 2010a). Up to 96 samples can be analyzed at once with as little as 250 ng of genomic DNA following bisulfite conversion. The coverage level has advantages and disadvantages. Predefined (e.g., the cancer methylation panel) or custom sets of CpG sites can be interrogated including only genetic pathways of most interest to a given study while reducing the need to adjust for thousands and thousands of multiple comparisons in statistical analysis. Unfortunately, this platform is only available for the human epigenome. Due to the limited coverage, this technology is in some ways more similar to the gene-region-based approaches in Section "Global DNA Methylation" than genome-wide approaches, though the interrogated sites can be distributed anywhere in the genome.

Restriction-Enzyme or Affinity Based Arrays

Restriction-enzyme- and affinity-based methods are combined with array technology to compare methylation levels between samples at thousands to millions of CpG sites. Platforms range from biased (e.g., focused on promoter regions) to unbiased (whole genome), and DNA methylation data range from semiquantitative to quantitative. While protocols vary widely by platform, the general method involves shearing genomic DNA, treating the DNA to generate an enriched fraction (e.g., enriched fraction of methylated DNA following restriction enzyme digestion or immunoprecipitation), labeling the enriched and input genomic DNA fractions separately, and cohybridizing the fractions to an array. After the arrays are scanned and the data are annotated, differentially methylated regions can be identified by probe (high resolution) or in genomic blocks (by smoothing the data across probes). Commonly used treatments include methylation-sensitive and methylation-insensitive restriction enzyme digestion and immunoprecipitation with antimethylcytosine antibody (MeDIP). Some of these methods can be applied to DNA from any species with a mapped genome, while others are restricted to species for which arrays are available (typically human, mouse, and rat). Cost and magnitude of generated data vary by array. One limitation is a relatively large input of required genomic DNA and, in the case of immunoprecipitation, the reliance on high-quality antibodies to produce reliable results (Bulyk, 2006; Minard et al., 2009).

Promoter Tiling Arrays

Tiling arrays use photolithographic technology in which short probes span the genome with high specificity and high resolution. This type of tiling array is commercially available through NimbleGen and Affymetrix, but may not be available for many model organisms. Since short probes may be subject to increased random signals compared with longer oligonucleotides (Zilberman and Henikoff, 2007; Huang et al., 2010a; Bock et al., 2010), NimbleGen tiling arrays utilize longer 60-mer probes. Most of the commercially available tiling arrays for humans, rats, and mice are biased with in-depth coverage of promoter regions and annotated CpG islands. DNA treatment prior to array hybridization includes methylated DNA immunoprecipitation (MeDIP) or methyl-CpG-binding domain-based capture (MBD-cap). Using MBD-cap followed by hybridization to the NimbleGen Mouse DNA Methylation 3X720K CpG Island Plus RefSeq Promoter Array, we previously reported differentially methylated genes in adult mice with prenatal BPA exposure that had developed hepatic tumors (Weinhouse et al., 2016).

CHARM

Comprehensive high-throughput arrays for relative methylation (CHARM) is an array-based method used to obtain highly quantitative DNA methylation data following McrBC restriction-enzyme-based fractionation of methylated and unmethylated DNA (Irizarry et al., 2008; Ladd-Acosta et al., 2010). Following hybridization to the Roche NimbleGen CHARM tiling array, data can be generated at up to 5 million CpG sites covered by 2 million probes throughout the genome. The data analysis pipeline for CHARM involves a novel normalization and smoothing algorithm that allows for the identification of differentially methylated regions as opposed to specific CpG sites. Unlike some other array platforms, CHARM can be used for mouse, rat, and human studies (Lee et al., 2011).

NEXT GENERATION SEQUENCING APPROACHES

Next-generation sequencing (NGS) technologies rapidly produce large amounts of sequence data at relatively low costs. These technologies emerged with the advent of the Human Genome Project and have steadily advanced ever since. Biases created by specific probes, allele-specific differences, and amplifications that appear in microarray technology are bypassed with sequencing technology (Laird, 2010). NGS-based methods can be used to obtain comprehensive DNA methylation data for any species with a mapped genome. Given that the methylation status is analyzed at every cytosine, deep sequencing provides great resolution for methylation profiles. However, the high cost of total sequencing runs and heavy reliance on computational analysis limit the widespread use in toxicological and medical research (Zilberman and Henikoff, 2007).

Sequencing Bisulfite Converted DNA

NGS of bisulfite-converted DNA provides quantitative base-pair resolution mapping of DNA methylation. Whole-genome (WGBS) (Lister et al., 2009), reduced representation (RRBS) (Meissner et al., 2008), and enhanced reduced representation (ERRBS) (Akalin et al., 2012b) bisulfite sequencing all quantify DNA methylation at mapped CpG sites but vary by CpG sites covered and cost per sample. In general, following bisulfite conversion of genomic DNA and library preparation, samples are sequenced (e.g., 50-base-pair reads on an Illumina HiSeq 2000 or HiSeq 4000), and reads are mapped to a reference genome. The reads are then used to quantify the proportion of cytosine versus thymine at a given CpG site to calculate percent of ratio of methylated to unmethylated cytosine at each CpG locus. Replicates are highly reproducible at both the CpG ($R^2 = 0.95$–0.97) and CpG island levels ($R^2 = 0.99$). WGBS is the quantitative platform that other epigenome-wide technologies are often compared with as it is highly reproducible and accurate and can cover all CpG sites in the genome. Library preparation reagents are available that are optimized to work with low starting amounts of DNA (<10 ng).

With adequate sequencing coverage ($\sim$30 times), WGBS generates quantitative data on the methylation status of every CpG site in the genome. This level of coverage is fairly unrealistic for most studies as it involves sequencing $\sim$800 million aligned reads, and lower levels of sequencing coverage have been shown to be adequate to detect differentially methylated regions between samples or experimental groups (Ziller et al., 2015). Publicly available reference DNA methylomes are available for more than 100 human cell types from the National Institutes of Health Epigenomics Roadmap Project (Kundaje et al., 2015). Unfortunately, this high-quality analysis is relatively low-throughput and comes at a high cost as four lanes are required on a sequencer such as the Illumina HiSeq 4000 for one sample to achieve >10 times coverage and generate data at every single CpG site in the human genome. Fortunately, RRBS and ERRBS reduce cost and increase sample throughput at the expense of lower coverage. For example, RRBS or ERRBS with one sample per sequencing lane can produce data at nearly 3 million CpG sites (Harris et al., 2010). RRBS involves enrichment of genomic DNA for CpG-rich regions via MspI digestion prior to bisulfite conversion and sequencing (Gu et al., 2011). MspI is a restriction enzyme that cleaves between the two cytosine residues in 5'-CCGG-3' sequences whether or not the second cytosine is methylated. This digestion leaves fragments with CG at each end that are eventually bisulfite-converted and sequenced, thus enriching for CpG-rich regions of the genome such as CpG islands and promoters. ERRBS is a refinement of RRBS that increases coverage in areas of the genome beyond CpG islands and promoters (Garrett-Bakelman et al., 2015). The ERRBS protocol includes modifications to the library preparation (including expansion of the size selection step after enzyme digestion from the typical 40–220 to 70–320 bp) and the bioinformatics pipeline to increase coverage to up to 10% of CpGs including those in CpG islands and shores (Akalin et al., 2012a).

RRBS and ERRBS have the advantage of lower cost and a smaller dataset focused on functionally relevant regions. However, unlike in WGBS (all possible CpG sites) and array technologies (known CpG sites), there is no guarantee that the same CpG sites will be sequenced in every sample or experiment that can hinder comparison across experiments.

Sequencing Affinity-Enriched DNA

Sequencing preceded by MeDIP or MBD-cap (MeDIP-seq or MBD-cap) is an unbiased approach to survey the DNA methylome of any species. Sheared DNA fragments are subject to antibody binding for the enrichment of methylated cytosine regions. These methylated fragments are purified and amplified for higher DNA yields (Weber et al., 2005; Huang et al., 2010a) and sequenced and mapped to a reference genome. These methods rely on a high-quality α-5mC antibody with high specificity, low variation between batches (i.e., different lots of the antibody), and low cross-reactivity that generates minimal background signal. While resolution is CpG-site-specific, generated data are relative enrichment of DNA methylation as opposed to percent methylation. Protocols using as little as 1 ng of genomic DNA have been developed and validated. Further, these approaches can be used with diverse sample types such as DNA from archived blood spots (Staunstrup et al., 2016; Zhao et al., 2014).

CONCLUSIONS AND CONSIDERATIONS

Going forward, the methods described in this chapter can be applied to toxicoepigenetic studies to understand the impact environmental factors have on DNA methylation and hydroxymethylation at vulnerable periods in the life span and ultimately how this contributes to the manifestation of toxicity and disease. Investigators must carefully consider the advantages and limitations of each method along with study hypotheses, sample source (species, tissue type, quality, and quantity of DNA), available laboratory resources and technical expertise, and budget when selecting a method.

Investigators working with epigenome-wide methods should plan for additional considerations unique to high-dimensional and array data such as data storage, bioinformatics pipelines, statistical power, and the need to validate findings. With regard to statistical power, it is important to consider that effect sizes are often small when comparing DNA methylation between exposed and nonexposed individuals or even between cases and controls for some diseases (Breton et al., 2017). While small sample sizes can still be used in relatively homogenous and well-controlled experimental models (e.g., control vs. exposed isogenic mice), heterogeneous human studies should be adequately powered to detect small differences in DNA methylation after accounting for thousands to millions of multiple comparisons. Furthermore, top differentially methylated genes or genomic regions should be validated using a gene-specific method such as pyrosequencing of EpiTYPER in some of the same samples (technical validation) and in other samples (replication cohort) to increase confidence in results, especially those from epigenome-wide arrays (Rakyan et al., 2011; Chadwick et al., 2015). Functional validation of differentially methylated regions should also be examined whenever possible. For example, the expression of the corresponding gene should be assessed, and the relationship between DNA methylation and gene expression should be evaluated. Casual evidence for an epigenetic effect can also be evaluated using methods such as Mendelian randomization. Researchers should also consider if genome-wide methylation assays should be paired with analyses to identify differentially methylated regions, such as Bumphunter, comb-p, or DMRcate, which can improve interpretability and statistical power (Peters et al., 2015; Jaffe et al., 2012; Pedersen et al., 2012). The growth of publicly available high-dimensional data and the formation of multicohort consortia such as the Pregnancy and Childhood Epigenetics (PACE) (Joubert et al., 2016; Felix et al., 2018) and Cohorts for Heart and Aging Research in Genetic Epidemiology (CHARGE) (Psaty et al., 2009; Ligthart et al., 2016; Joehanes et al., 2016) for humans and the National Institutes of Health Toxicant Exposures and Responses by Genomic and Epigenomic Regulators of Transcription (NIH TaRGET II) consortium of mice (Wang et al., 2018) with epigenome-wide data will help to facilitate collaborations to replicate findings in the future.

Acknowledgments

The authors acknowledge Muna Nahar and Dana Dolinoy for past insights into this topic. Funding for KES was provided by the National Institutes of Health (NIH; F32 ES028085). JMG was supported by NIH (grant numbers P01 ES02284401 and 1U2CES026553) and the US Environmental Protection Agency (EPA; RD 83543601). The contents of this publication are solely the responsibility of the authors and do not necessarily represent the official views of the US EPA or the NIH. Further, the US EPA does not endorse the purchase of any commercial products or services mentioned in this chapter.

References

Akalin, A., Garrett-Bakelman, F.E., Kormaksson, M., Busuttil, J., Zhang, L., Khrebtukova, I., Milne, T.A., Huang, Y., Biswas, D., Hess, J.L., Allis, C.D., Roeder, R.G., Valk, P.J., Lowenberg, B., Delwel, R., Fernandez, H.F., Paietta, E., Tallman, M.S., Schroth, G.P., Mason, C.E., Melnick, A., Figueroa, M.E., 2012a. Base-pair resolution DNA methylation sequencing reveals profoundly divergent epigenetic landscapes in acute myeloid leukemia. PLoS Genet. 8: e1002781.

Akalin, A., Garrett-Bakelman, F.E., Kormaksson, M., Busuttil, J., Zhang, L., Khrebtukova, I., Milne, T.A., Huang, Y., Biswas, D., Hess, J.L., Allis, C.D., Roeder, R.G., Valk, P.J., Lowenberg, B., Delwel, R., Fernandez, H.F., Paietta, E., Tallman, M.S., Schroth, G.P., Mason, C.E., Melnick, A., Figueroa, M.E., 2012b. Base-pair resolution DNA methylation sequencing reveals profoundly divergent epigenetic landscapes in acute myeloid leukemia. PLoS Genet. 8: e1002781.

Ambatipudi, S., Horvath, S., Perrier, F., Cuenin, C., Hernandez-Vargas, H., Le Calvez-Kelm, F., Durand, G., Byrnes, G., Ferrari, P., Bouaoun, L., Sklias, A., Chajes, V., Overvad, K., Severi, G., Baglietto, L., Clavel-Chapelon, F., Kaaks, R., Barrdahl, M., Boeing, H., Trichopoulou, A., Lagiou, P., Naska, A., Masala, G., Agnoli, C., Polidoro, S., Tumino, R., Panico, S., Dolle, M., Peeters, P.H., Onland-Moret, N.C., Sandanger, T.M., Nost, T.H., Weiderpass, E., Quiros, J.R., Agudo, A., Rodriguez-Barranco, M., Huerta Castano, J.M., Barricarte, A., Fernandez, A.M., Travis, R.C., Vineis, P., Muller, D.C., Riboli, E., Gunter, M., Romieu, I., Herceg, Z., 2017. DNA methylome analysis identifies accelerated epigenetic ageing associated with postmenopausal breast cancer susceptibility. Eur. J. Cancer 75, 299–307.

Bachman, M., Uribe-Lewis, S., Yang, X., Burgess, H.E., Iurlaro, M., Reik, W., Murrell, A., Balasubramanian, S., 2015. 5-Formylcytosine can be a stable DNA modification in mammals. Nat. Chem. Biol. 11, 555–557.

Bakulski, K.M., Feinberg, J.I., Andrews, S.V., Yang, J., Brown, S., Mckenney, S., Witter, F., Walston, J., Feinberg, A.P., Fallin, M.D., 2016. DNA methylation of cord blood cell types: applications for mixed cell birth studies. Epigenetics 11, 354–362.

Ball, C.A., Sherlock, G., 2007. What are MicroArrays? An introduction to microarray methods for measuring the transcriptome. In: Barnes, M.R., Wiley, I. (Eds.), Bioinformatics for Geneticists: A Bioinformatics Primer for the Analysis of Genetic Data. Wiley, Chichester/Hoboken, NJ.

Ball, M.P., Li, J.B., Gao, Y., Lee, J.-H., Leproust, E., Park, I.-H., Xie, B., Daley, G.Q., Church, G.M., 2009. Targeted and genome-scale methylomics reveals gene body signatures in human cell lines. Nat. Biotechnol. 27, 361–368.

Batzer, M.A., Deininger, P.L., 2002. Alu repeats and human genomic diversity. Nat. Rev. Genet. 3, 370–379.

Berdasco, M.A., Fraga, M.F., Esteller, M., 2009. Quantification of global DNA methylation by capillary electrophoresis and mass spectrometry. In: Tost, J. (Ed.), DNA Methylation: Methods and Protocols. Humana Press, Totowa, NJ.

Bjornsson, H.T., Sigurdsson, M.I., Fallin, M., et al., 2008. Intra-individual change over time in DNA methylation with familial clustering. JAMA 299, 2877–2883.

Bock, C., Tomazou, E.M., Brinkman, A.B., Muller, F., Simmer, F., Gu, H., Jager, N., Gnirke, A., Stunnenberg, H.G., Meissner, A., 2010. Quantitative comparison of genome-wide DNA methylation mapping technologies. Nat. Biotechnol. 28, 1106–1114.

Boissinot, S., Sookdeo, A., 2016. The evolution of LINE-1 in vertebrates. Genome Biol. Evol. 8, 3485–3507.

Bollati, V., Baccarelli, A., Hou, L., Bonzini, M., Fustinoni, S., Cavallo, D., Byun, H.-M., Jiang, J., Marinelli, B., Pesatori, A.C., Bertazzi, P.A., Yang, A.S., 2007. Changes in DNA methylation patterns in subjects exposed to low-dose benzene. Cancer Res. 67, 876–880.

Booth, M.J., Ost, T.W.B., Beraldi, D., Bell, N.M., Branco, M.R., Reik, W., Balasubramanian, S., 2013. Oxidative bisulfite sequencing of 5-methylcytosine and 5-hydroxymethylcytosine. Nat. Protoc. 8, 1841–1851.

Breton, C.V., Marsit, C.J., Faustman, E., Nadeau, K., Goodrich, J.M., Dolinoy, D.C., Herbstman, J., Holland, N., Lasalle, J.M., Schmidt, K., Yousefi, P., Perera, F., Joubert, B.R., Wiemels, J., Taylor, M., Yang, I.V., Chen, R., Hew, K.M., Freeland, D.M., Miller, R., Murphy, S.K., 2017. Small-magnitude effect sizes in epigenetic end points are important in children's environmental health studies: the Children's Environmental Health and Disease Prevention Research Center's Epigenetics Working Group. Environ. Health Perspect. 125, 511–526.

Britten, R.J., Baron, W.F., Stout, D.B., Davidson, E.H., 1988. Sources and evolution of human Alu repeated sequences. Proc. Natl. Acad. Sci. U. S. A. 85, 4770–4774.

Bulyk, M., 2006. DNA microarray technologies for measuring protein-DNA interactions. Curr. Opin. Biotechnol. 17, 422–430.

Cardenas, A., Houseman, E.A., Baccarelli, A.A., Quamruzzaman, Q., Rahman, M., Mostofa, G., Wright, R.O., Christiani, D.C., Kile, M.L., 2015a. In utero arsenic exposure and epigenome-wide associations in placenta, umbilical artery, and human umbilical vein endothelial cells. Epigenetics 10, 1054–1063.

Cardenas, A., Koestler, D.C., Houseman, E.A., Jackson, B.P., Kile, M.L., Karagas, M.R., Marsit, C.J., 2015b. Differential DNA methylation in umbilical cord blood of infants exposed to mercury and arsenic in utero. Epigenetics 10 (6), 1–8.

Chadwick, L.H., Sawa, A., Yang, I.V., Baccarelli, A., Breakefield, X.O., Deng, H.-W., Dolinoy, D.C., Fallin, M.D., Holland, N.T., Houseman, E.A., Lomvardas, S., Rao, M., Satterlee, J.S., Tyson, F.L., Vijayanand, P., Greally, J.M., 2015. New insights and updated guidelines for epigenome-wide association studies. Neuroepigenetics 1, 14–19.

Clark, S.J., Harrison, J., Paul, C.L., Frommer, M., 1994. High sensitivity mapping of methylated cytosines. Nucleic Acids Res. 22, 2990–2997.

Clark, S.J., Statham, A., Stirzaker, C., Molloy, P.L., Frommer, M., 2006. DNA methylation: bisulphite modification and analysis. Nat. Protoc. 1, 2353.

Daca-Roszak, P., Pfeifer, A., Zebracka-Gala, J., Rusinek, D., Szybinska, A., Jarzab, B., Witt, M., Zietkiewicz, E., 2015. Impact of SNPs on methylation readouts by Illumina Infinium HumanMethylation450 BeadChip Array: implications for comparative population studies. BMC Genomics 16, 1003.

Dai, L., Mehta, A., Mordukhovich, I., Just, A.C., Shen, J., Hou, L., Koutrakis, P., Sparrow, D., Vokonas, P.S., Baccarelli, A.A., Schwartz, J.D., 2017. Differential DNA methylation and PM2.5 species in a 450K epigenome-wide association study. Epigenetics 12, 139–148.

de F. C. Lichtenfels, A.J., van der Plaat, D.A., de Jong, K., van Diemen, C.C., Postma, D.S., Nedeljkovic, I., van Duijn, C.M., Amin, N., la Bastide-van Gemert, S., de Vries, M., Ward-Caviness, C.K., Wolf, K., Waldenberger, M., Peters, A., Stolk, R.P., Brunekreef, B., Boezen, H.M., Vonk, J.M., 2018. Long-term air pollution exposure, genome-wide DNA methylation and lung function in the lifelines cohort study. Environ. Health Perspect. 126:027004.

Dodge, J.E., Ramsahoye, B.H., Wo, Z.G., Okano, M., Li, E., 2002. De novo methylation of MMLV provirus in embryonic stem cells: CpG versus non-CpG methylation. Gene 289, 41–48.

Dolinoy, D.C., Weidman, J.R., Waterland, R.A., Jirtle, R.L., 2006. Maternal genistein alters coat color and protects A(vy) mouse offspring from obesity by modifying the fetal epigenome. Environ. Health Perspect. 114, 567–572.

Dolinoy, D.C., Das, R., Weidman, J.R., Jirtle, R.L., 2007a. Metastable epialleles, imprinting, and the fetal origins of adult diseases. Pediatr. Res. 61, 30R.

Dolinoy, D.C., Huang, D., Jirtle, R.L., 2007b. Maternal nutrient supplementation counteracts bisphenol A-induced DNA hypomethylation in early development. Proc. Natl. Acad. Sci. 104, 13056–13061.

Du, P., Zhang, X., Huang, C.C., Jafari, N., Kibbe, W.A., Hou, L., Lin, S.M., 2010. Comparison of beta-value and M-value methods for quantifying methylation levels by microarray analysis. BMC Bioinf. 11, 587.

Durso, D.F., Bacalini, M.G., Sala, C., Pirazzini, C., Marasco, E., Bonafe, M., do Valle, I.F., Gentilini, D., Castellani, G., Faria, A.M.C., Franceschi, C., Garagnani, P., Nardini, C., 2017. Acceleration of leukocytes' epigenetic age as an early tumor and sex-specific marker of breast and colorectal cancer. Oncotarget 8, 23237–23245.

Eads, C.A., Danenberg, K.D., Kawakami, K., Saltz, L.B., Blake, C., Shibata, D., Danenberg, P.V., Laird, P.W., 2000. MethyLight: a high-throughput assay to measure DNA methylation. Nucleic Acids Res. 28, e32.

Eleftheriou, M., Pascual, A.J., Wheldon, L.M., Perry, C., Abakir, A., Arora, A., Johnson, A.D., Auer, D.T., Ellis, I.O., Madhusudan, S., Ruzov, A., 2015. 5-Carboxylcytosine levels are elevated in human breast cancers and gliomas. Clin. Epigenetics 7, 88.

Faulk, C., Dolinoy, D.C., 2011. Timing is everything: the when and how of environmentally induced changes in the epigenome of animals. Epigenetics 6, 791–797.

Faulk, C., Kim, J.H., Anderson, O.S., Nahar, M.S., Jones, T.R., Sartor, M.A., Dolinoy, D.C., 2016. Detection of differential DNA methylation in repetitive DNA of mice and humans perinatally exposed to bisphenol A. Epigenetics 11, 489–500.

Felix, J.F., Joubert, B.R., Baccarelli, A.A., Sharp, G.C., Almqvist, C., Annesi-Maesano, I., Arshad, H., Baiz, N., Bakermans-Kranenburg, M.J., Bakulski, K.M., Binder, E.B., Bouchard, L., Breton, C.V., Brunekreef, B., Brunst, K.J., Burchard, E.G., Bustamante, M., Chatzi, L., Cheng Munthe-Kaas, M., Corpeleijn, E., Czamara, D., Dabelea, D., Davey Smith, G., de Boever, P., Duijts, L., Dwyer, T., Eng, C., Eskenazi, B., Everson, T.M., Falahi, F., Fallin, M.D., Farchi, S., Fernandez, M.F., Gao, L., Gaunt, T.R., Ghantous, A., Gillman, M.W., Gonseth, S., Grote, V., Gruzieva, O., Haberg, S.E., Herceg, Z., Hivert, M.F., Holland, N., Holloway, J.W., Hoyo, C., Hu, D., Huang, R.C., Huen, K., Jarvelin, M.R., Jima, D.D., Just, A.C., Karagas, M.R., Karlsson, R., Karmaus, W., Kechris, K.J., Kere, J., Kogevinas, M., Koletzko, B., Koppelman, G.H., Kupers, L.K., Ladd-Acosta, C., Lahti, J., Lambrechts, N., Langie, S.A.S., Lie, R.T., Liu, A.H., Magnus, M.C., Magnus, P., Maguire, R.L., Marsit, C.J., Mcardle, W., Melen, E., Melton, P., Murphy, S.K., Nawrot, T.S., Nistico, L., Nohr, E.A., Nordlund, B., Nystad, W., Oh, S.S., Oken, E., Page, C.M., Perron, P., Pershagen, G., Pizzi, C., Plusquin, M., Raikkonen, K., Reese, S.E., Reischl, E., Richiardi, L., Ring, S., Roy, R.P., Rzehak, P., Schoeters, G., Schwartz, D.A., Sebert, S., Snieder, H., Sorensen, T.I.A., Starling, A.P., et al., 2018. Cohort profile: pregnancy and childhood epigenetics (PACE) consortium. Int. J. Epidemiol. 47 (1), 22–23u. https://doi.org/10.1093/ije/dyx190.

Ficz, G., Branco, M.R., Seisenberger, S., Santos, F., Krueger, F., Hore, T.A., Marques, C.J., Andrews, S., Reik, W., 2011. Dynamic regulation of 5-hydroxymethylcytosine in mouse ES cells and during differentiation. Nature 473, 398–402.

Figueroa, M.E., Skrabanek, L., Li, Y., Jiemjit, A., Fandy, T.E., Paietta, E., Fernandez, H., Tallman, M.S., Greally, J.M., Carraway, H., Licht, J.D., Gore, S.D., Melnick, A., 2009. MDS and secondary AML display unique patterns and abundance of aberrant DNA methylation. Blood 114, 3448–3458.

Figueroa, M.E., Lugthart, S., Li, Y., Erpelinck-Verschueren, C., Deng, X., Christos, P.J., Schifano, E., Booth, J., van Putten, W., Skrabanek, L., Campagne, F., Mazumdar, M., Greally, J.M., Valk, P.J., Lowenberg, B., Delwel, R., Melnick, A., 2010. DNA methylation signatures identify biologically distinct subtypes in acute myeloid leukemia. Cancer Cell 17, 13–27.

Finer, S., Mathews, C., Lowe, R., Smart, M., Hillman, S., Foo, L., Sinha, A., Williams, D., Rakyan, V.K., Hitman, G.A., 2015. Maternal gestational diabetes is associated with genome-wide DNA methylation variation in placenta and cord blood of exposed offspring. Hum. Mol. Genet. 24, 3021–3029.

Fortin, J.P., Labbe, A., Lemire, M., Zanke, B.W., Hudson, T.J., Fertig, E.J., Greenwood, C.M., Hansen, K.D., 2014. Functional normalization of 450k methylation array data improves replication in large cancer studies. Genome Biol. 15, 503.

Fortin, J.P., Triche Jr., T.J., Hansen, K.D., 2017. Preprocessing, normalization and integration of the Illumina HumanMethylationEPIC array with minfi. Bioinformatics 33, 558–560.

Fraga, M.F., Ballestar, E., Paz, M.F., Ropero, S., Setien, F., Ballestar, M.L., Heine-Suñer, D., Cigudosa, J.C., Urioste, M., Benitez, J., Boix-Chornet, M., Sanchez-Aguilera, A., Ling, C., Carlsson, E., Poulsen, P., Vaag, A., Stephan, Z., Spector, T.D., Wu, Y.-Z., Plass, C., Esteller, M., 2005. Epigenetic differences arise during the lifetime of monozygotic twins. Proc. Natl. Acad. Sci. U. S. A. 102, 10604–10609.

Garrett-Bakelman, F.E., Sheridan, C.K., Kacmarczyk, T.J., Ishii, J., Betel, D., Alonso, A., Mason, C.E., Figueroa, M.E., Melnick, A.M., 2015. Enhanced reduced representation bisulfite sequencing for assessment of DNA methylation at base pair resolution. J. Vis. Exp. 96, e52246. https://doi.org/10.3791/52246.

Goodrich, J.M., Sanchez, B.N., Dolinoy, D.C., Zhang, Z., Hernandez-Avila, M., Hu, H., Peterson, K.E., Tellez-Rojo, M.M., 2015. Quality control and statistical modeling for environmental epigenetics: a study on in utero lead exposure and DNA methylation at birth. Epigenetics 10, 19–30.

Goodrich, J.M., Dolinoy, D.C., Sánchez, B.N., Zhang, Z., Meeker, J.D., Mercado-Garcia, A., Solano-González, M., Hu, H., Téllez-Rojo, M.M., Peterson, K.E., 2016. Adolescent epigenetic profiles and environmental exposures from early life through peri-adolescence. Environ. Epigenet. 2:dvw018.

Gruzieva, O., Xu, C.J., Breton, C.V., Annesi-Maesano, I., Anto, J.M., Auffray, C., Ballereau, S., Bellander, T., Bousquet, J., Bustamante, M., Charles, M.A., de Kluizenaar, Y., den Dekker, H.T., Duijts, L., Felix, J.F., Gehring, U., Guxens, M., Jaddoe, V.V., Jankipersadsing, S.A., Merid, S.K., Kere, J., Kumar, A., Lemonnier, N., Lepeule, J., Nystad, W., Page, C.M., Panasevich, S., Postma, D., Slama, R., Sunyer, J., Soderhall, C., Yao, J., London, S.J., Pershagen, G., Koppelman, G.H., Melen, E., 2017. Epigenome-wide meta-analysis of methylation in children related to prenatal NO2 air pollution exposure. Environ. Health Perspect. 125 (1), 104–110. https://doi.org/10.1289/EHP36.

Gu, H., Smith, Z.D., Bock, C., Boyle, P., Gnirke, A., Meissner, A., 2011. Preparation of reduced representation bisulfite sequencing libraries for genome-scale DNA methylation profiling. Nat. Protoc. 6, 468–481.

Hajkova, P., Erhardt, S., Lane, N., Haaf, T., El-Maarri, O., Reik, W., Walter, J., Surani, M.A., 2002. Epigenetic reprogramming in mouse primordial germ cells. Mech. Dev. 117, 15–23.

Hanna, C.W., Bloom, M.S., Robinson, W.P., Kim, D., Parsons, P.J., Vom Saal, F.S., Taylor, J.A., Steuerwald, A.J., Fujimoto, V.Y., 2012. DNA methylation changes in whole blood is associated with exposure to the environmental contaminants, mercury, lead, cadmium and bisphenol A, in women undergoing ovarian stimulation for IVF. Hum. Reprod. 27, 1401–1410.

Harris, R.A., Wang, T., Coarfa, C., Nagarajan, R.P., Hong, C., Downey, S.L., Johnson, B.E., Fouse, S.D., Delaney, A., Zhao, Y., Olshen, A., Ballinger, T., Zhou, X., Forsberg, K.J., Gu, J., Echipare, L., O'Geen, H., Lister, R., Pelizzola, M., Xi, Y., Epstein, C.B., Bernstein, B.E., Hawkins, R.D., Ren, B., Chung, W.Y., Gu, H., Bock, C., Gnirke, A., Zhang, M.Q., Haussler, D., Ecker, J.R., Li, W., Farnham, P.J., Waterland, R.A., Meissner, A., Marra, M.A., Hirst, M., Milosavljevic, A., Costello, J.F., 2010. Comparison of sequencing-based methods to profile DNA methylation and identification of monoallelic epigenetic modifications. Nat. Biotechnol. 28, 1097–1105.

Hayatsu, H., Wataya, Y., Kai, K., Iida, S., 1970. Reaction of sodium bisulfite with uracil, cytosine, and their derivatives. Biochemistry 9, 2858–2865.

Head, J.A., Mittal, K., Basu, N., 2014. Application of the LUminometric Methylation Assay to ecological species: tissue quality requirements and a survey of DNA methylation levels in animals. Mol. Ecol. Resour. 14, 943–952.

Holmes, E.E., Jung, M., Meller, S., Leisse, A., Sailer, V., Zech, J., Mengdehl, M., Garbe, L.-A., Uhl, B., Kristiansen, G., Dietrich, D., 2014. Performance evaluation of kits for bisulfite-conversion of DNA from tissues, cell lines, FFPE tissues, aspirates, lavages, effusions, plasma, serum, and urine. PLoS ONE. 9:e93933.

Horvath, S., 2013. DNA methylation age of human tissues and cell types. Genome Biol. 14, R115.

Horvath, S., Erhart, W., Brosch, M., Ammerpohl, O., von Schonfels, W., Ahrens, M., Heits, N., Bell, J.T., Tsai, P.C., Spector, T.D., Deloukas, P., Siebert, R., Sipos, B., Becker, T., Rocken, C., Schafmayer, C., Hampe, J., 2014. Obesity accelerates epigenetic aging of human liver. Proc. Natl. Acad. Sci. U. S. A. 111, 15538–15543.

Horvath, S., Gurven, M., Levine, M.E., Trumble, B.C., Kaplan, H., Allayee, H., Ritz, B.R., Chen, B., Lu, A.T., Rickabaugh, T.M., Jamieson, B.D., Sun, D., Li, S., Chen, W., Quintana-Murci, L., Fagny, M., Kobor, M.S., Tsao, P.S., Reiner, A.P., Edlefsen, K.L., Absher, D., Assimes, T.L., 2016. An epigenetic clock analysis of race/ethnicity, sex, and coronary heart disease. Genome Biol. 17, 171.

Houseman, E.A., Accomando, W.P., Koestler, D.C., Christensen, B.C., Marsit, C.J., Nelson, H.H., Wiencke, J.K., Kelsey, K.T., 2012. DNA methylation arrays as surrogate measures of cell mixture distribution. BMC Bioinformatics 13, 86.

Houseman, E.A., Molitor, J., Marsit, C.J., 2014. Reference-free cell mixture adjustments in analysis of DNA methylation data. Bioinformatics 30, 1431–1439.

Huang, Y.-W., Huang, T.H.M., Wang, L.-S., 2010a. Profiling DNA methylomes from microarray to genome-scale sequencing. Technol. Cancer Res. Treat. 9, 139–147.

Huang, Y., Pastor, W.A., Shen, Y., Tahiliani, M., Liu, D.R., Rao, A., 2010b. The behaviour of 5-hydroxymethylcytosine in bisulfite sequencing. PLoS ONE. 5:e8888.

Inoue, A., Shen, L., Dai, Q., He, C., Zhang, Y., 2011. Generation and replication-dependent dilution of 5fC and 5caC during mouse preimplantation development. Cell Res. 21, 1670–1676.

Irizarry, R.A., Ladd-Acosta, C., Carvalho, B., Wu, H., Brandenburg, S.A., Jeddeloh, J.A., Wen, B., Feinberg, A.P., 2008. Comprehensive high-throughput arrays for relative methylation (Charm). Genome Res. 18, 780–790.

Ito, S., Shen, L., Dai, Q., Wu, S.C., Collins, L.B., Swenberg, J.A., He, C., Zhang, Y., 2011. Tet proteins can convert 5-methylcytosine to 5-formylcytosine and 5-carboxylcytosine. Science (New York, N.Y.) 333, 1300–1303.

Iurlaro, M., Mcinroy, G.R., Burgess, H.E., Dean, W., Raiber, E.-A., Bachman, M., Beraldi, D., Balasubramanian, S., Reik, W., 2016. In vivo genome-wide profiling reveals a tissue-specific role for 5-formylcytosine. Genome Biol. 17, 141.

Jaffe, A.E., Murakami, P., Lee, H., Leek, J.T., Fallin, M.D., Feinberg, A.P., Irizarry, R.A., 2012. Bump hunting to identify differentially methylated regions in epigenetic epidemiology studies. Int. J. Epidemiol. 41, 200–209.

Joehanes, R., Just, A.C., Marioni, R.E., Pilling, L.C., Reynolds, L.M., Mandaviya, P.R., Guan, W., Xu, T., Elks, C.E., Aslibekyan, S., Moreno-Macias, H., Smith, J.A., Brody, J.A., Dhingra, R., Yousefi, P., Pankow, J.S., Kunze, S., Shah, S.H., Mcrae, A.F., Lohman, K., Sha, J., Absher, D.M., Ferrucci, L., Zhao, W., Demerath, E.W., Bressler, J., Grove, M.L., Huan, T., Liu, C., Mendelson, M.M., Yao, C., Kiel, D.P., Peters, A., Wang-Sattler, R., Visscher, P.M., Wray, N.R., Starr, J.M., Ding, J., Rodriguez, C.J., Wareham, N.J., Irvin, M.R., Zhi, D., Barrdahl, M., Vineis, P., Ambatipudi, S., Uitterlinden, A.G., Hofman, A., Schwartz, J., Colicino, E., Hou, L., Vokonas, P.S., Hernandez, D.G., Singleton, A.B., Bandinelli, S., Turner, S.T., Ware, E.B., Smith, A.K., Klengel, T., Binder, E.B., Psaty, B.M., Taylor, K.D., Gharib, S.A., Swenson, B.R., Liang, L., Demeo, D.L., O'connor, G.T., Herceg, Z., Ressler, K.J., Conneely, K.N., Sotoodehnia, N., Kardia, S.L., Melzer, D., Baccarelli, A.A., van Meurs, J.B., Romieu, I., Arnett, D.K., Ong, K.K., Liu, Y., Waldenberger, M., Deary, I.J., Fornage, M., Levy, D., London, S.J., 2016. Epigenetic signatures of cigarette smoking. Circ. Cardiovasc. Genet. 9, 436–447.

Joubert, B.R., Haberg, S.E., Nilsen, R.M., Wang, X., Vollset, S.E., Murphy, S.K., Huang, Z., Hoyo, C., Midttun, O., Cupul-Uicab, L.A., Ueland, P.M., Wu, M.C., Nystad, W., Bell, D.A., Peddada, S.D., London, S.J., 2012. 450K epigenome-wide scan identifies differential DNA methylation in newborns related to maternal smoking during pregnancy. Environ. Health Perspect. 120, 1425–1431.

Joubert, B.R., Felix, J.F., Yousefi, P., Bakulski, K.M., Just, A.C., Breton, C., Reese, S.E., Markunas, C.A., Richmond, R.C., Xu, C.J., Kupers, L.K., Oh, S.S., Hoyo, C., Gruzieva, O., Soderhall, C., Salas, L.A., Baiz, N., Zhang, H., Lepeule, J., Ruiz, C., Ligthart, S., Wang, T., Taylor, J.A., Duijts, L., Sharp, G.C., Jankipersadsing, S.A., Nilsen, R.M., Vaez, A., Fallin, M.D., Hu, D., Litonjua, A.A., Fuemmeler, B.F., Huen, K., Kere, J., Kull, I., Munthe-Kaas, M.C., Gehring, U., Bustamante, M., Saurel-Coubizolles, M.J., Quraishi, B.M., Ren, J., Tost, J., Gonzalez, J.R., Peters, M.J., Haberg, S.E., Xu, Z., van Meurs, J.B., Gaunt, T.R., Kerkhof, M., Corpeleijn, E., Feinberg, A.P., Eng, C., Baccarelli, A.A., Benjamin Neelon, S.E., Bradman, A., Merid, S.K., Bergstrom, A., Herceg, Z., Hernandez-Vargas, H., Brunekreef, B., Pinart, M., Heude, B., Ewart, S., Yao, J., Lemonnier, N., Franco, O.H., Wu, M.C., Hofman, A., Mcardle, W., van der Vlies, P., Falahi, F., Gillman, M.W., Barcellos, L.F., Kumar, A., Wickman, M., Guerra, S., Charles, M.A., Holloway, J., Auffray, C., Tiemeier, H.W., Smith, G.D., Postma, D., Hivert, M.F., Eskenazi, B., Vrijheid, M., Arshad, H., Anto, J.M., Dehghan, A., Karmaus, W., Annesi-Maesano, I., Sunyer, J., Ghantous, A., Pershagen, G., Holland, N., Murphy, S.K., Demeo, D.L., Burchard, E.G., Ladd-Acosta, C., Snieder, H., Nystad, W., et al., 2016. DNA methylation in newborns and maternal smoking in pregnancy: genome-wide consortium meta-analysis. Am. J. Hum. Genet. 98, 680–696.

Jousse, C., Parry, L., Lambert-Langlais, S., Maurin, A.-C., Averous, J., Bruhat, A., Carraro, V., Tost, J., Letteron, P., Chen, P., Jockers, R., Launay, J.-M., Mallet, J., Fafournoux, P., 2011. Perinatal undernutrition affects the methylation and expression of the leptin gene in adults: implication for the understanding of metabolic syndrome. FASEB J. 25, 3271–3278.

Jung, M., Pfeifer, G.P., 2015. Aging and DNA methylation. BMC Biol. 13, 7.

Karimi, M., Johansson, S., Ekström, T., 2006a. Using LUMA: a luminometric-based assay for global DNA-methylation. Epigenetics 1, 45–48.

Karimi, M., Johansson, S., Stach, D., Corcoran, M., Grandér, D., Schalling, M., Bakalkin, G., Lyko, F., Larsson, C., Ekström, T.J., 2006b. LUMA (LUminometric Methylation Assay)—a high throughput method to the analysis of genomic DNA methylation. Exp. Cell Res. 312, 1989–1995.

Khund Sayeed, S., Zhao, J., Sathyanarayana, B.K., Golla, J.P., Vinson, C., 2015. C/EBPβ (Cebpb) protein binding to the C/EBP|CRE DNA 8-mer TTGC|GTCA is inhibited by 5hmC and enhanced by 5mC, 5fC, and 5caC in the CG dinucleotide. Biochim. Biophys. Acta 1849, 583–589.

Kim, J.H., Rozek, L.S., Soliman, A.S., Sartor, M.A., Hablas, A., Seifeldin, I.A., Colacino, J.A., Weinhouse, C., Nahar, M.S., Dolinoy, D.C., 2013. Bisphenol A-associated epigenomic changes in prepubescent girls: a cross-sectional study in Gharbiah, Egypt. Environ. Health 12, 33.

Kim, J.H., Sartor, M.A., Rozek, L.S., Faulk, C., Anderson, O.S., Jones, T.R., Nahar, M.S., Dolinoy, D.C., 2014. Perinatal bisphenol A exposure promotes dose-dependent alterations of the mouse methylome. BMC Genomics 15, 30.

Kim, Y.H., Lee, Y.S., Lee, D.H., Kim, D.S., 2016. Polycyclic aromatic hydrocarbons are associated with insulin receptor substrate 2 methylation in adipose tissues of Korean women. Environ. Res. 150, 47–51.

Kinney, S.M., Chin, H.G., Vaisvila, R., Bitinaite, J., Zheng, Y., Estève, P.-O., Feng, S., Stroud, H., Jacobsen, S.E., Pradhan, S., 2011. Tissue-specific distribution and dynamic changes of 5-hydroxymethylcytosine in mammalian genomes. J. Biol. Chem. 286, 24685–24693.

Knothe, C., Shiratori, H., Resch, E., Ultsch, A., Geisslinger, G., Doehring, A., Lötsch, J., 2016. Disagreement between two common biomarkers of global DNA methylation. Clin. Epigenetics 8, 60.

Kochmanski, J., Montrose, L., Goodrich, J.M., Dolinoy, D.C., 2017. Environmental deflection: the impact of toxicant exposures on the aging epigenome. Toxicol. Sci. 156, 325–335.

Koestler, D.C., Avissar-Whiting, M., Houseman, E.A., Karagas, M.R., Marsit, C.J., 2013. Differential DNA methylation in umbilical cord blood of infants exposed to low levels of arsenic. Environ. Health Perspect. 121, 971–977.

Kohli, R.M., Zhang, Y., 2013. TET enzymes, TDG and the dynamics of DNA demethylation. Nature 502, 472–479.

Kriaucionis, S., Heintz, N., 2009. The nuclear DNA base 5-hydroxymethylcytosine is present in purkinje neurons and the brain. Science 324, 929–930.

Kundaje, A., Meuleman, W., Ernst, J., Bilenky, M., Yen, A., Heravi-Moussavi, A., Kheradpour, P., Zhang, Z., Wang, J., Ziller, M.J., Amin, V., Whitaker, J.W., Schultz, M.D., Ward, L.D., Sarkar, A., Quon, G., Sandstrom, R.S., Eaton, M.L., Wu, Y.C., Pfenning, A.R., Wang, X., Claussnitzer, M., Liu, Y., Coarfa, C., Harris, R.A., Shoresh, N., Epstein, C.B., Gjoneska, E., Leung, D., Xie, W., Hawkins, R.D., Lister, R., Hong, C., Gascard, P., Mungall, A.J., Moore, R., Chuah, E., Tam, A., Canfield, T.K., Hansen, R.S., Kaul, R., Sabo, P.J., Bansal, M.S., Carles, A., Dixon, J.R., Farh, K.H., Feizi, S., Karlic, R., Kim, A.R., Kulkarni, A., Li, D., Lowdon, R., Elliott, G., Mercer, T.R., Neph, S.J., Onuchic, V., Polak, P., Rajagopal, N., Ray, P., Sallari, R.C., Siebenthall, K.T., Sinnott-Armstrong, N.A., Stevens, M., Thurman, R.E., Wu, J., Zhang, B., Zhou, X., Beaudet, A.E., Boyer, L.A., de Jager, P.L., Farnham, P.J., Fisher, S.J., Haussler, D., Jones, S.J., Li, W., Marra, M.A., Mcmanus, M.T., Sunyaev, S., Thomson, J.A., Tlsty, T.D., Tsai, L.H., Wang, W., Waterland, R.A., Zhang, M.Q., Chadwick, L.H., Bernstein, B.E., Costello, J.F., Ecker, J.R., Hirst, M., Meissner, A., Milosavljevic, A., Ren, B., Stamatoyannopoulos, J.A., Wang, T., Kellis, M., 2015. Integrative analysis of 111 reference human epigenomes. Nature 518, 317–330.

Ladd-Acosta, C., Aryee, M.J., Ordway, J.M., Feinberg, A.P., 2010. Comprehensive high-throughput arrays for relative methylation (CHARM). In: Current Protocols in Human Genetics. Chapter 20, Unit 20.1.1-19.

Laird, P.W., 2010. Principles and challenges of genome-wide DNA methylation analysis. Nat. Rev. Genet. 11, 191–203.

Lander, E.S., Linton, L.M., Birren, B., Nusbaum, C., Zody, M.C., Baldwin, J., Devon, K., Dewar, K., Doyle, M., Fitzhugh, W., Funke, R., Gage, D., Harris, K., Heaford, A., Howland, J., Kann, L., Lehoczky, J., Levine, R., Mcewan, P., Mckernan, K., Meldrim, J., Mesirov, J.P., Miranda, C., Morris, W., Naylor, J., Raymond, C., Rosetti, M., Santos, R., Sheridan, A., Sougnez, C., Stange-Thomann, Y., Stojanovic, N., Subramanian, A., Wyman, D., Rogers, J., Sulston, J., Ainscough, R., Beck, S., Bentley, D., Burton, J., Clee, C., Carter, N., Coulson, A., Deadman, R., Deloukas, P., Dunham, A., Dunham, I., Durbin, R., French, L., Grafham, D., Gregory, S., Hubbard, T., Humphray, S., Hunt, A., Jones, M., Lloyd, C., Mcmurray, A., Matthews, L., Mercer, S., Milne, S., Mullikin, J.C., Mungall, A., Plumb, R., Ross, M., Shownkeen, R., Sims, S., Waterston, R.H., Wilson, R.K., Hillier, L.W., Mcpherson, J.D., Marra, M.A., Mardis, E.R., Fulton, L.A., Chinwalla, A.T., Pepin, K.H., Gish, W.R., Chissoe, S.L., Wendl, M.C., Delehaunty, K.D., Miner, T.L., Delehaunty, A., Kramer, J.B., Cook, L.L., Fulton, R.S., Johnson, D.L., Minx, P.J., Clifton, S.W., Hawkins, T., Branscomb, E., Predki, P., Richardson, P., Wenning, S., Slezak, T., Doggett, N., Cheng, J.F., Olsen, A., Lucas, S.,

Elkin, C., Uberbacher, E., Frazier, M., et al., 2001. Initial sequencing and analysis of the human genome. Nature 409, 860–921.

Lee, R.S., Tamashiro, K.L., Aryee, M.J., Murakami, P., Seifuddin, F., Herb, B., Huo, Y., Rongione, M., Feinberg, A.P., Moran, T.H., Potash, J.B., 2011. Adaptation of the Charm DNA methylation platform for the rat genome reveals novel brain region-specific differences. Epigenetics 6, 1378–1390.

Levine, M.E., Lu, A.T., Chen, B.H., Hernandez, D.G., Singleton, A.B., Ferrucci, L., Bandinelli, S., Salfati, E., Manson, J.E., Quach, A., Kusters, C.D., Kuh, D., Wong, A., Teschendorff, A.E., Widschwendter, M., Ritz, B.R., Absher, D., Assimes, T.L., Horvath, S., 2016. Menopause accelerates biological aging. Proc. Natl. Acad. Sci. U. S. A. 113, 9327–9332.

Lewis, L.C., Lo, P.C.K., Foster, J.M., Dai, N., Corrêa, I.R., Durczak, P.M., Duncan, G., Ramsawhook, A., Aithal, G.P., Denning, C., Hannan, N.R.F., Ruzov, A., 2017. Dynamics of 5-carboxylcytosine during hepatic differentiation: potential general role for active demethylation by DNA repair in lineage specification. Epigenetics 12, 277–286.

Li, F., Zhang, Y., Bai, J., Greenberg, M.M., Xi, Z., Zhou, C., 2017. 5-Formylcytosine yields DNA-protein cross-links in nucleosome core particles. J. Am. Chem. Soc. 139, 10617–10620.

Ligthart, S., Marzi, C., Aslibekyan, S., Mendelson, M.M., Conneely, K.N., Tanaka, T., Colicino, E., Waite, L.L., Joehanes, R., Guan, W., Brody, J.A., Elks, C., Marioni, R., Jhun, M.A., Agha, G., Bressler, J., Ward-Caviness, C.K., Chen, B.H., Huan, T., Bakulski, K., Salfati, E.L., Fiorito, G., Wahl, S., Schramm, K., Sha, J., Hernandez, D.G., Just, A.C., Smith, J.A., Sotoodehnia, N., Pilling, L.C., Pankow, J.S., Tsao, P.S., Liu, C., Zhao, W., Guarrera, S., Michopoulos, V.J., Smith, A.K., Peters, M.J., Melzer, D., Vokonas, P., Fornage, M., Prokisch, H., Bis, J.C., Chu, A.Y., Herder, C., Grallert, H., Yao, C., Shah, S., Mcrae, A.F., Lin, H., Horvath, S., Fallin, D., Hofman, A., Wareham, N.J., Wiggins, K.L., Feinberg, A.P., Starr, J.M., Visscher, P.M., Murabito, J.M., Kardia, S.L., Absher, D.M., Binder, E.B., Singleton, A.B., Bandinelli, S., Peters, A., Waldenberger, M., Matullo, G., Schwartz, J.D., Demerath, E.W., Uitterlinden, A.G., van Meurs, J.B., Franco, O.H., Chen, Y.I., Levy, D., Turner, S.T., Deary, I.J., Ressler, K.J., Dupuis, J., Ferrucci, L., Ong, K.K., Assimes, T.L., Boerwinkle, E., Koenig, W., Arnett, D.K., Baccarelli, A.A., Benjamin, E.J., Dehghan, A., 2016. DNA methylation signatures of chronic low-grade inflammation are associated with complex diseases. Genome Biol. 17, 255.

Lisanti, S., Omar, W.A.W., Tomaszewski, B., de Prins, S., Jacobs, G., Koppen, G., Mathers, J.C., Langie, S.A.S., 2013. Comparison of methods for quantification of global DNA methylation in human cells and tissues. PLoS ONE. 8: e79044.

Lister, R., Pelizzola, M., Dowen, R., Hawkins, R.D., Hon, G., Tonti-Filippini, J., Nery, J., Lee, L., Ye, Z., Ngo, Q.-M., Edsall, L., Antosiewicz-Bourget, J., Stewart, R., Ruotti, V., Millar, A.H., Thomson, J., Ren, B., Ecker, J., 2009. Human DNA methylomes at base resolution show widespread epigenomic differences. Nature 462, 315–322.

Liu, X., Chen, Q., Tsai, H.-J., Wang, G., Hong, X., Zhou, Y., Zhang, C., Liu, C., Liu, R., Wang, H., Zhang, S., Yu, Y., Mestan, K.K., Pearson, C., Otlans, P., Zuckerman, B., Wang, X., 2014. Maternal preconception body mass index and offspring cord blood DNA methylation: exploration of early life origins of disease. Environ. Mol. Mutagen. 55, 223–230.

Luo, Y., Lu, X., Xie, H., 2014. Dynamic Alu methylation during normal development, aging, and tumorigenesis. Biomed. Res. Int. 2014, 12.

Marioni, R.E., Shah, S., Mcrae, A.F., Chen, B.H., Colicino, E., Harris, S.E., Gibson, J., Henders, A.K., Redmond, P., Cox, S.R., Pattie, A., Corley, J., Murphy, L., Martin, N.G., Montgomery, G.W., Feinberg, A.P., Fallin, M.D., Multhaup, M.L., Jaffe, A.E., Joehanes, R., Schwartz, J., Just, A.C., Lunetta, K.L., Murabito, J.M., Starr, J.M., Horvath, S., Baccarelli, A.A., Levy, D., Visscher, P.M., Wray, N.R., Deary, I.J., 2015. DNA methylation age of blood predicts all-cause mortality in later life. Genome Biol. 16, 25.

McKay, J.A., Groom, A., Potter, C., Coneyworth, L.J., Ford, D., Mathers, J.C., Relton, C.L., 2012. Genetic and non-genetic influences during pregnancy on infant global and site specific DNA methylation: role for folate gene variants and vitamin B12. PLoS ONE. 7:e33290.

Meissner, A., Mikkelsen, T.S., Gu, H., Wernig, M., Hanna, J., Sivachenko, A., Zhang, X., Bernstein, B.E., Nusbaum, C., Jaffe, D.B., Gnirke, A., Jaenisch, R., Lander, E.S., 2008. Genome-scale DNA methylation maps of pluripotent and differentiated cells. Nature 454, 766–770.

Minard, M., Jain, A., Barton, M., 2009. Analysis of epigenetic alterations to chromatin during development. Genesis 47, 559–572.

Moran, S., Arribas, C., Esteller, M., 2016. Validation of a DNA methylation microarray for 850,000 CpG sites of the human genome enriched in enhancer sequences. Epigenomics 8, 389–399.

Morris, T.J., Butcher, L.M., Feber, A., Teschendorff, A.E., Chakravarthy, A.R., Wojdacz, T.K., Beck, S., 2014. ChAMP: 450k chip analysis methylation pipeline. Bioinformatics 30, 428–430.

Mouse Genome Sequencing Consortium, 2002. Initial sequencing and comparative analysis of the mouse genome. Nature 420, 520.

Nahar, M.S., Kim, J.H., Sartor, M.A., Dolinoy, D.C., 2014. Bisphenol A-associated alterations in the expression and epigenetic regulation of genes encoding xenobiotic metabolizing enzymes in human fetal liver. Environ. Mol. Mutagen. 55, 184–195.

Nahar, M.S., Liao, C., Kannan, K., Harris, C., Dolinoy, D.C., 2015. In utero bisphenol A concentration, metabolism, and global DNA methylation across matched placenta, kidney, and liver in the human fetus. Chemosphere 124, 54–60.

Needhamsen, M., Ewing, E., Lund, H., Gomez-Cabrero, D., Harris, R.A., Kular, L., Jagodic, M., 2017. Usability of human Infinium MethylationEPIC BeadChip for mouse DNA methylation studies. BMC Bioinformatics 18, 486.

Nevalainen, T., Kananen, L., Marttila, S., Jylhava, J., Mononen, N., Kahonen, M., Raitakari, O.T., Hervonen, A., Jylha, M., Lehtimaki, T., Hurme, M., 2017. Obesity accelerates epigenetic aging in middle-aged but not in elderly individuals. Clin. Epigenetics 9, 20.

Nwanaji-Enwerem, J.C., Dai, L., Colicino, E., Oulhote, Y., Di, Q., Kloog, I., Just, A.C., Hou, L., Vokonas, P., Baccarelli, A.A., Weisskopf, M.G., Schwartz, J.D., 2017. Associations between long-term exposure to PM2.5 component species and blood DNA methylation age in the elderly: the VA normative aging study. Environ. Int. 102, 57–65.

Panni, T., Mehta, A.J., Schwartz, J.D., Baccarelli, A.A., Just, A.C., Wolf, K., Wahl, S., Cyrys, J., Kunze, S., Strauch, K., Waldenberger, M., Peters, A., 2016. Genome-wide analysis of DNA methylation and fine particulate matter air pollution in three study populations: KORA F3, KORA F4, and the normative aging study. Environ. Health Perspect. 124, 983–990.

Pedersen, B.S., Schwartz, D.A., Yang, I.V., Kechris, K.J., 2012. Comb-p: software for combining, analyzing, grouping and correcting spatially correlated P-values. Bioinformatics 28, 2986–2988.

Perna, L., Zhang, Y., Mons, U., Holleczek, B., Saum, K.U., Brenner, H., 2016. Epigenetic age acceleration predicts cancer, cardiovascular, and all-cause mortality in a German case cohort. Clin. Epigenetics 8, 64.

Peters, T.J., Buckley, M.J., Statham, A.L., Pidsley, R., Samaras, K., R, V.L., Clark, S.J., Molloy, P.L., 2015. De novo identification of differentially methylated regions in the human genome. Epigenetics Chromatin 8, 6.

Pidsley, R., Y Wong, C.C., Volta, M., Lunnon, K., Mill, J., Schalkwyk, L.C., 2013. A data-driven approach to preprocessing Illumina 450K methylation array data. BMC Genomics 14, 293.

Psaty, B.M., O'donnell, C.J., Gudnason, V., Lunetta, K.L., Folsom, A.R., Rotter, J.I., Uitterlinden, A.G., Harris, T.B., Witteman, J.C., Boerwinkle, E., 2009. Cohorts for heart and aging research in genomic epidemiology (Charge) consortium: design of prospective meta-analyses of genome-wide association studies from 5 cohorts. Circ. Cardiovasc. Genet. 2, 73–80.

Qiagen, 2011. EpiTect® MethyLight PCR Handbook. Available from: https://www.qiagen.com/us/shop/epigenetics/epitect-methylight-pcr-kits/ [Accessed 20 September 2017].

Qian, Y., Tu, J., Tang, N.L.S., Kong, G.W.S., Chung, J.P.W., Chan, W.-Y., Lee, T.-L., 2015. Dynamic changes of DNA epigenetic marks in mouse oocytes during natural and accelerated aging. Int. J. Biochem. Cell Biol. 67, 121–127.

Quach, A., Levine, M.E., Tanaka, T., Lu, A.T., Chen, B.H., Ferrucci, L., Ritz, B., Bandinelli, S., Neuhouser, M.L., Beasley, J.M., Snetselaar, L., Wallace, R.B., Tsao, P.S., Absher, D., Assimes, T.L., Stewart, J.D., Li, Y., Hou, L., Baccarelli, A.A., Whitsel, E.A., Horvath, S., 2017. Epigenetic clock analysis of diet, exercise, education, and lifestyle factors. Aging (Albany NY) 9, 419–446.

Raiber, E.-A., Murat, P., Chirgadze, D.Y., Beraldi, D., Luisi, B.F., Balasubramanian, S., 2015. 5-Formylcytosine alters the structure of the DNA double helix. Nat. Struct. Mol. Biol. 22, 44–49.

Raiber, E.-A., Portella, G., Martinez Cuesta, S., Hardisty, R., Murat, P., Li, Z., Iurlaro, M., Dean, W., Spindel, J., Beraldi, D., Dawson, M., Reik, W., Balasubramanian, S., 2017. 5-Formylcytosine controls nucleosome positioning through covalent histone-DNA interaction. bioRxiv. https://doi.org/10.1101/224444.

Rakyan, V.K., Blewitt, M.E., Druker, R., Preis, J.I., Whitelaw, E., 2002. Metastable epialleles in mammals. Trends Genet. 18, 348–351.

Rakyan, V.K., Down, T.A., Balding, D.J., Beck, S., 2011. Epigenome-wide association studies for common human diseases. Nat. Rev. Genet. 12, 529–541.

Rebuzzini, P., Castiglia, R., Nergadze, S.G., Mitsainas, G., Munclinger, P., Zuccotti, M., Capanna, E., Redi, C.A., Garagna, S., 2009. Quantitative variation of LINE-1 sequences in five species and three subspecies of the subgenus Mus and in five Robertsonian races of Mus musculus domesticus. Chromosom. Res. 17, 65–76.

Reik, W., 2007. Stability and flexibility of epigenetic gene regulation in mammalian development. Nature 447, 425–432.

Reik, W., Dean, W., Walter, J., 2001. Epigenetic reprogramming in mammalian development. Science 293, 1089–1093.

Rusiecki, J.A., Baccarelli, A., Bollati, V., Tarantini, L., Moore, L.E., Bonefeld-Jorgensen, E.C., 2008. Global DNA hypomethylation is associated with high serum-persistent organic pollutants in greenlandic inuit. Environ. Health Perspect. 116, 1547–1552.

Sant, K., Nahar, M., Dolinoy, D., 2012. DNA methylation screening and analysis. In: Harris, C., Hansen, J.M. (Eds.), Developmental Toxicology. Humana Press, New York, NY.

Sant, K.E., Dolinoy, D.C., Nahar, M.S., Harris, C., 2013. Inhibition of proteolysis in histiotrophic nutrition pathways alters DNA methylation and one-carbon metabolism in the organogenesis-stage rat conceptus. J. Nutr. Biochem. 24, 1479–1487.

Sant, K.E., Dolinoy, D.C., Jilek, J.L., Shay, B.J., Harris, C., 2016. Mono-2-ethylhexyl phthalate (MEHP) alters histiotrophic nutrition pathways and epigenetic processes in the developing conceptus. J. Nutr. Biochem. 27, 211–218.

Sen, A., Cingolani, P., Senut, M.C., Land, S., Mercado-Garcia, A., Tellez-Rojo, M.M., Baccarelli, A.A., Wright, R.O., Ruden, D.M., 2015. Lead exposure induces changes in 5-hydroxymethylcytosine clusters in CpG islands in human embryonic stem cells and umbilical cord blood. Epigenetics 10 (7), 607–621. https://doi.org/10.1080/15592294. 2015.1050172.

Shenker, N.S., Polidoro, S., van Veldhoven, K., Sacerdote, C., Ricceri, F., Birrell, M.A., Belvisi, M.G., Brown, R., Vineis, P., Flanagan, J.M., 2013. Epigenome-wide association study in the European Prospective Investigation into Cancer and Nutrition (EPIC-Turin) identifies novel genetic loci associated with smoking. Hum. Mol. Genet. 22, 843–851.

Solomon, O., Yousefi, P., Huen, K., Gunier, R.B., Escudero-Fung, M., Barcellos, L.F., Eskenazi, B., Holland, N., 2017. Prenatal phthalate exposure and altered patterns of DNA methylation in cord blood. Environ. Mol. Mutagen. 58 (6), 398–410. https://doi.org/10.1002/em.22095.

Song, C.-X., Szulwach, K.E., Fu, Y., Dai, Q., Yi, C., Li, X., Li, Y., Chen, C.-H., Zhang, W., Jian, X., Wang, J., Zhang, L., Looney, T.J., Zhang, B., Godley, L.A., Hicks, L.M., Lahn, B.T., Jin, P., He, C., 2011. Selective chemical labeling reveals the genome-wide distribution of 5-hydroxymethylcytosine. Nat. Biotechnol. 29, 68–72.

Sookdeo, A., Hepp, C.M., Mcclure, M.A., Boissinot, S., 2013. Revisiting the evolution of mouse LINE-1 in the genomic era. Mob. DNA 4, 3.

Soriano-Tárraga, C., Jiménez-Conde, J., Giralt-Steinhauer, E., Ois, Á., Rodríguez-Campello, A., Cuadrado-Godia, E., Fernández-Cadenas, I., Montaner, J., Lucas, G., Elosua, R., Roquer, J., 2013. DNA isolation method is a source of global DNA methylation variability measured with LUMA. Experimental analysis and a systematic review. PLoS ONE. 8e60750.

Staunstrup, N.H., Starnawska, A., Nyegaard, M., Christiansen, L., Nielsen, A.L., Borglum, A., Mors, O., 2016. Genome-wide DNA methylation profiling with MeDIP-seq using archived dried blood spots. Clin. Epigenetics 8, 81.

Suchiman, H.E.D., Slieker, R.C., Kremer, D., Slagboom, P.E., Heijmans, B.T., Tobi, E.W., 2015. Design, measurement and processing of region-specific DNA methylation assays: the mass spectrometry-based method Epityper. Front. Genet. 6, 287.

Tahiliani, M., Koh, K.P., Shen, Y., Pastor, W.A., Bandukwala, H., Brudno, Y., Agarwal, S., Iyer, L.M., Liu, D.R., Aravind, L., Rao, A., 2009. Conversion of 5-methylcytosine to 5-hydroxymethylcytosine in mammalian DNA by Mll partner TET1. Science (New York, N.Y.) 324, 930–935.

Tarantini, L., Bonzini, M., Apostoli, P., Pegoraro, V., Bollati, V., Marinelli, B., Cantone, L., Rizzo, G., Hou, L., Schwartz, J., Bertazzi, P.A., Baccarelli, A., 2009. Effects of particulate matter on genomic DNA methylation content and iNOS promoter methylation. Environ. Health Perspect. 117, 217–222.

Teschendorff, A.E., Marabita, F., Lechner, M., Bartlett, T., Tegner, J., Gomez-Cabrero, D., Beck, S., 2013. A beta-mixture quantile normalization method for correcting probe design bias in Illumina Infinium 450 k DNA methylation data. Bioinformatics 29, 189–196.

Thompson, R.F., Atzmon, G., Gheorghe, C., Liang, H.Q., Lowes, C., Greally, J.M., Barzilai, N., 2010. Tissue-specific dysregulation of DNA methylation in aging. Aging Cell 9, 506–518.

Treangen, T.J., Salzberg, S.L., 2011. Repetitive DNA and next-generation sequencing: computational challenges and solutions. Nat. Rev. Genet. 13, 36–46.

Tse, M.Y., Ashbury, J.E., Zwingerman, N., King, W.D., Taylor, S.A.M., Pang, S.C., 2011. A refined, rapid and reproducible high resolution melt (HRM)-based method suitable for quantification of global LINE-1 repetitive element methylation. BMC. Res. Notes 4, 565.

Vidal, A.C., Benjamin Neelon, S.E., Liu, Y., Tuli, A.M., Fuemmeler, B.F., Hoyo, C., Murtha, A.P., Huang, Z., Schildkraut, J., Overcash, F., Kurtzberg, J., Jirtle, R.L., Iversen, E.S., Murphy, S.K., 2014. Maternal stress, preterm birth, and DNA methylation at imprint regulatory sequences in humans. Genet. Epigenet. 6, 37–44.

Wang, T., Pehrsson, E.C., Purushotham, D., Li, D., Zhuo, X., Zhang, B., Lawson, H.A., Province, M.A., Krapp, C., Lan, Y., Coarfa, C., Katz, T.A., Tang, W.Y., Wang, Z., Biswal, S., Rajagopalan, S., Colacino, J.A., Tsai, Z.T.-Y., Sartor, M.A., Neier, K., Dolinoy, D.C., Pinto, J., Hamanaka, R.B., Mutlu, G.M., Patisaul, H.B., Aylor, D.L., Crawford, G.E., Wiltshire, T., Chadwick, L.H., Duncan, C.G., Garton, A.E., Mcallister, K.A., Ta, R.I.I.C., Bartolomei, M.S., Walker, C.L., Tyson, F.L., 2018. The Niehs TaRGET II Consortium and environmental epigenomics. Nat. Biotechnol. 36, 225.

Ward-Caviness, C.K., Nwanaji-Enwerem, J.C., Wolf, K., Wahl, S., Colicino, E., Trevisi, L., Kloog, I., Just, A.C., Vokonas, P., Cyrys, J., Gieger, C., Schwartz, J., Baccarelli, A.A., Schneider, A., Peters, A., 2016. Long-term exposure to air pollution is associated with biological aging. Oncotarget 7, 74510–74525.

Weber, M., Davies, J., Wittig, D., Oakeley, E., Haase, M., Lam, W., Schbeler, D., 2005. Chromosome-wide and promoter-specific analyses identify sites of differential DNA methylation in normal and transformed human cells. Nat. Genet. 37, 853–862.

Weinhouse, C., Sartor, M.A., Faulk, C., Anderson, O.S., Sant, K.E., Harris, C., Dolinoy, D.C., 2016. Epigenome-wide DNA methylation analysis implicates neuronal and inflammatory signaling pathways in adult murine hepatic tumorigenesis following perinatal exposure to bisphenol A. Environ. Mol. Mutagen. 57 (6), 435–446. https://doi.org/10.1002/em.22024.

Weisenberger, D.J., van den Berg, D., Pan, F., Berman, B.P., Laird, P.W., 2008. Comprehensive DNA Methylation Analysis on the Illumina Infinium Assay Platform. Application Note: Illumina Epigenetic Analysis, Illumina.

Wen, L., Li, X., Yan, L., Tan, Y., Li, R., Zhao, Y., Wang, Y., Xie, J., Zhang, Y., Song, C., Yu, M., Liu, X., Zhu, P., Li, X., Hou, Y., Guo, H., Wu, X., He, C., Li, R., Tang, F., Qiao, J., 2014. Whole-genome analysis of 5-hydroxymethylcytosine and 5-methylcytosine at base resolution in the human brain. Genome Biol. 15, R49.

Worm Ørntoft, M.-B., Jensen, S.Ø., Hansen, T.B., Bramsen, J.B., Andersen, C.L., 2017. Comparative analysis of 12 different kits for bisulfite conversion of circulating cell-free DNA. Epigenetics 12, 626–636.

Wright, R.O., Schwartz, J., Wright, R.J., Bollati, V., Tarantini, L., Park, S.K., Hu, H., Sparrow, D., Vokonas, P., Baccarelli, A., 2010. Biomarkers of lead exposure and DNA methylation within retrotransposons. Environ. Health Perspect. 118, 790–795.

Wu, X., Zhang, Y., 2017. TET-mediated active DNA demethylation: mechanism, function and beyond. Nat. Rev. Genet. 18 (9), 517–534. https://doi.org/10.1038/nrg.2017.33.

Wu, S., Hivert, M.F., Cardenas, A., Zhong, J., Rifas-Shiman, S.L., Agha, G., Colicino, E., Just, A.C., Amarasiriwardena, C., Lin, X., Litonjua, A.A., Demeo, D.L., Gillman, M.W., Wright, R.O., Oken, E., Baccarelli, A.A., 2017. Exposure to low levels of lead in utero and umbilical cord blood DNA methylation in project viva: an Epigenome-Wide Association Study. Environ. Health Perspect. 125:087019.

Yang, A.S., Estécio, M.R.H., Doshi, K., Kondo, Y., Tajara, E.H., Issa, J.-P.J., 2004. A simple method for estimating global DNA methylation using bisulfite PCR of repetitive DNA elements. Nucleic Acids Res. 32, e38.

Yu, M., Hon, G.C., Szulwach, K.E., Song, C.X., Zhang, L., Kim, A., Li, X., Dai, Q., Shen, Y., Park, B., Min, J.H., Jin, P., Ren, B., He, C., 2012. Base-resolution analysis of 5-hydroxymethylcytosine in the mammalian genome. Cell 149, 1368–1380.

Zhang, Y., Rohde, C., Tierling, S., Stamerjohanns, H., Reinhardt, R., Walter, J., Jeltsch, A., 2009. DNA methylation analysis by bisulfite conversion, cloning, and sequencing of individual clones. In: Tost, J. (Ed.), DNA Methylation: Methods and Protocols. Humana Press, Totowa, NJ.

Zhang, L., Lu, X., Lu, J., Liang, H., Dai, Q., Xu, G.-L., Luo, C., Jiang, H., He, C., 2012. Thymine DNA glycosylase specifically recognizes 5-carboxylcytosine-modified DNA. Nat. Chem. Biol. 8, 328–330.

Zhao, M.T., Whyte, J.J., Hopkins, G.M., Kirk, M.D., Prather, R.S., 2014. Methylated DNA immunoprecipitation and high-throughput sequencing (MeDIP-seq) using low amounts of genomic DNA. Cell Rep. 16, 175–184.

Zheng, Y., Joyce, B.T., Liu, L., Zhang, Z., Kibbe, W.A., Zhang, W., Hou, L., 2017. Prediction of genome-wide DNA methylation in repetitive elements. Nucleic Acids Res. 45, 8697–8711.

Zilberman, D., Henikoff, S., 2007. Genome-wide analysis of DNA methylation patterns. Development 134, 3959–3965.

Ziller, M.J., Hansen, K.D., Meissner, A., Aryee, M.J., 2015. Coverage recommendations for methylation analysis by whole genome bisulfite sequencing. Nat. Methods 12, 230–232.

Further Reading

Bakulski, K.M., Lee, H., Feinberg, J.I., Wells, E.M., Brown, S., Herbstman, J.B., Witter, F.R., Halden, R.U., Caldwell, K., Mortensen, M.E., Jaffe, A.E., Moye Jr., J., Caulfield, L.E., Pan, Y., Goldman, L.R., Feinberg, A.P., Fallin, M.D., 2015. Prenatal mercury concentration is associated with changes in DNA methylation at Tceanc2 in newborns. Int. J. Epidemiol. 44, 1249–1262.

Methods for Analyzing miRNA Expression

Valentina Bollati, Laura Dioni

EPIGET (Epidemiology, Epigenetics and Toxicology) Lab, Department of Clinical Sciences and Community Health, University of Milan, Milan, Italy

O U T L I N E

INTRODUCTION

MicroRNAs (miRNAs) are short (~22 nucleotides) noncoding RNA molecules. The mature and biologically effective transcript is produced through a multistep process involving incremental cleavages of a much longer primary transcript (pri-miRNA) of hundreds or thousands of nucleotides in length to give a pre-miRNA (70- to 100-nucleotide hairpin precursor), finally affording a functional and mature miRNA.

The interest in this class of regulatory molecules has grown exponentially in the past few years, and it is becoming increasingly clear that miRNAs have a central role in controlling pivotal biological processes, such as cell proliferation and cell death. In fact, they are able to regulate gene expression by forming a complex with the corresponding messenger RNA (mRNA), thus suppressing mRNA translation or causing mRNA degradation.

According to the current knowledge, each single miRNA regulates hundreds or thousands of mRNAs, based on its sequence. Although much research has been conducted so far in order to clarify the complex relationships regulating miRNA-target interactions, the kinetics behind their response to changes in miRNA concentrations are not entirely clear. In addition, it is still under debate whether changes in the expression of one of these target mRNAs affect the expression of other mRNAs, thus causing a higher availability of the common targeting miRNA.

Considering the strong impact miRNAs exert on molecular pathways, aberrant miRNA expression (both in target tissues and in biofluids) has been proposed as a biomarker of disease (e.g., neurodegenerative disorders, diabetes, cardiovascular disease, immune system disorders, and cancer). In addition, miRNAs can serve as a therapeutic target, as they can be either modulated by antagonists (e.g., anti-miRs) or replaced by synthetic oligos (Bader et al., 2010; Fig. 1).

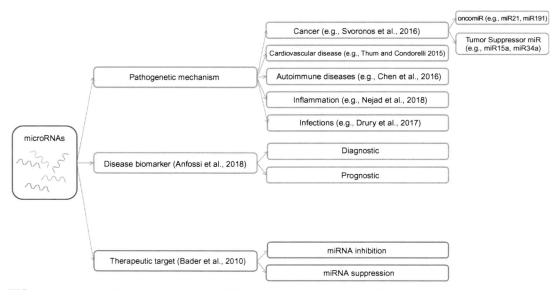

FIG. 1 Main studies about three miRNAs roles: pathogenic mechanism, diseases biomarker, and therapeutic target.

As such, in the past few years, many different approaches have been developed in order to allow a quantitative or at least qualitative profiling of miRNAs and a precise validation of specific miRNAs.

This type of analysis, although very promising, suffers from some technical limitations due to the nature of miRNAs, which are lowly abundant in biological samples: miRNAs, in fact, represent ~0.01% of the mass of a total RNA sample; moreover, each cell type expresses a specific signature of miRNAs. In the attempt of better understanding tissue specificity of miRNA expression, several consortia are working toward the identification of specific miRNA signatures in different tissues. For example, the recent work conducted by Aaron Smith and colleagues allowed to identify tissue-specific miRNAs in 23 different tissues of rats (Smith et al., 2016).

Currently (miRBase v22, accessed 5 May 2018), 38,589 precursor miRNAs and 48,885 mature miRNA sequences have been identified in 271 different species. During profiling analysis, pri-miRNAs and pre-miRNAs need to be distinguished from mature transcripts, generating an additional pitfall due to the high sequence similarity. It is important to underline that several mature miRNAs can be generated from a single pre-miRNA, and often, these transcripts are produced in clusters that may regulate pathways involved in diverse biological functions.

Moreover, miRNAs belonging to the same family might differ by only a single nucleotide, making efficient discrimination among members difficult. Taking into account these different aspects characterizing the challenges of miRNA analysis, we will discuss the pros and cons of the techniques currently available to measure and characterize miRNA expression and abundance.

miRNA ORIGIN

miRNAs are relatively stable molecules that can be easily investigated in different sources, such as tissue, blood, serum, and formalin-fixed paraffin-embedded (FFPE) tissues (Osman, 2012). Their stability makes them easy to investigate, even in samples and tissues that were stored prior to the setup of specific protocols for miRNA analysis.

However, miRNA expression is, at least in part, tissue-specific, and miRNA analysis might be impacted by cellular heterogeneity of the biological sample investigated. Indeed, several phenomena might modify RNA expression (including miRNAs), while, at the same time, they can cause a shift in cell composition. For example, whenever an inflammatory reaction is present, an increased number of inflammatory cells also are produced, which can contribute to the total mRNA and miRNA obtained from the sample.

Besides the traditional investigation of miRNAs in tissues and cells, the study of circulating miRNAs is attracting increasing interest. Circulating miRNAs have been used in epidemiological studies (Vrijens et al., 2015; Motta et al., 2017), because they can be employed as noninvasive biomarkers for a variety of indications. However, experimental conditions can deeply impact the identification of specific miRNAs. In fact, miRNAs "circulating" in blood either can be derived from circulating cells (i.e., neutrophils, eosinophils, basophils, lymphocytes, monocytes, platelets, or erythrocytes) or can be extracellular and released by either passive or active release processes such as miRNAs encapsulated in exosomes or microvesicles, Ago-2-miRNA complexes, high-density lipoprotein-miRNA complexes, or miRNAs bound to other protective proteins. The complete separation of one of the abovementioned components is impossible to achieve today, but several protocols are available that can at least preferentially enrich for one or a few of the components. An alternative approach might be to track tissue specificity using miRNAs (e.g., the ones reported in the miRNA atlas projects currently available) as they are known to be expressed just in a particular tissue or cell type.

miRNA NOMENCLATURE

miRNA nomenclature is evolving rapidly, and the numbers of miRNA species are sequential in order of discovery. In general, the first three-letter prefix (not always reported) refers to the species in which the miRNA has been identified (e.g., hsa for human or mmu for mouse). Homologues in different organisms receive the same name that is based on the similarity of the ~22 nt sequence to previously identified miRNAs (Griffiths-Jones, 2004). The maturity of the molecule is also indicated in the name: miR indicates a mature miRNA, whereas pri-miR or pre-miR indicates the primary transcript or the precursor form of the miRNA, respectively. Moreover, the suffix -3p or -5p specifies whether the mature miRNA is produced from the 3' or the 5' arm of the precursor.

As some miRNA families exist and are characterized by related but not identical sequences, members of the same family are named with the same number followed by a lowercase letter (e.g., hsa-miR-148a and hsa-miR-148b). On the contrary, miRNAs that share the same identical sequence as the mature miRNA but originate in different locus of the genome are indicated by a numerical suffix (e.g., hsa-miR-16-1 and hsa-miR-16-2). Some recent studies have also discovered isomiRs, variant miRNA sequences that derive from the same gene but vary in sequence due to posttranscriptional modifications; they may have different targets and thus different cell functions (Zhou et al., 2012).

In addition, the symbol * is sometimes reported beside a miRNA name. In this case, as commercially available platforms are not always fully up to date in terms of miRNA nomenclature, * is used to point toward the "minor form" between two mature miRNAs produced from the 5' and 3' arms. It is currently accepted that both the dominant and the minor forms can be functional, and sometimes, the * form can even be more actively transcribed than the dominant one, making this nomenclature inaccurate. For this reason, it is now suggested to use the more recent convention of adding -3p or -5p to the miR name.

Recent deep-sequencing experiments and powerful bioinformatics study have led to an important increase of the number of known miRNA sequences and have revealed the diversity and variability of mature and functional short noncoding RNAs, including their genomic origins, biogenesis pathways, and sequence variability (Desvignes et al., 2015). Detailed information on the correct nomenclature and function of identified miRNAs can be retrieved through the available databases (such as miRBase, ZooMir, and miRNEST). The current version of miRBase (Kozomara and Griffiths-Jones, 2014), the most commonly used reference microRNA database, is available at www.mirbase.org and lists a total of 38,589 entries representing hairpin precursor miRNAs, corresponding to 48,885 mature miRNA products, in 271 species (release 22 March 2018). In addition, miRBase has an interface to data from deep sequencing, in the NCBI gene expression omnibus (GEO), to provide information about predicted precursor hairpin sequences and experimentally identified mature miRNA sequences and links to miRNA-target prediction databases (Pritchard et al., 2012).

PREANALYTICAL VARIABLES

Preanalytic variables might significantly impact miRNA analysis; therefore, all specimen types should have standard and careful workflows in each study. The small size of miRNAs and their ubiquitous presence in all tissue types deeply affect their analysis, so methodologies for miRNA detection are still under development (Pritchard et al., 2012).

The three main preanalytic variables are classified as biological, environmental, and technical and are shared by both extracellular and cellular miRNAs. According to the current opinion of the scientific community, both external and patient-intrinsic factors such as age, sex, race, body mass index, diet, fasting state, underlying illness/organ dysfunction, exercise, vitamin supplementation, medications, smoking, circadian variation, chemical exposure experienced, and even altitude can alter miRNA concentrations and are often difficult to standardize across studies (Khan et al., 2017). Technical variables include specimen collection, transport, processing, and storage as well as the miRNA extraction method, and a strong effort is needed in order to standardize these conditions (Butz and Patocs, 2015). In addition, blood cell contamination and hemolysis are particularly relevant for extracellular miRNAs (Kirschner et al., 2013) and for highly vascularized tissue specimens; multiple studies have now focused on assembling lists of miRNAs that a researcher can use to determine whether a miRNA of interest may be affected by contamination of blood cells such as platelets or hemolysate (Nair et al., 2014). For example, based on the variability in measured amounts of miR-451 and miR-16 (typically expressed in red blood cells), some authors suggest measuring free hemoglobin at an absorbance of 414 nm with a cutoff of 0.2 or higher to identify hemolyzed specimens before screening miRNA levels (Kirschner et al., 2011).

miRNAs are stable in a variety of biological fluids such as blood, lymph, cerebrospinal liquid, bronchoalveolar lavage, urine, saliva, and tears (Hanson et al., 2009). In addition, they can be engaged with proteins (Arroyo et al., 2011); lipoproteins (Vickers et al., 2011); and membrane-coated vesicles such as microparticles, exosomes, and apoptotic bodies (Raposo and Stoorvogel, 2013). Measurement of circulating miRNAs is a good noninvasive tool for epigenetic studies, but further recommendations for the preanalytic treatment of specimens are required, such as minimization of freeze/thaw cycles (Zhao et al., 2014) and storage up to 6 years at −20 or −80°C (Grasedieck et al., 2012) to significantly reduce RNase activity. The current recommendations for blood collection, the most widely studied source, suggest using a 22-gauge needle, discarding the first small amount of blood drawn (Nair et al., 2014; Witwer et al., 2013), using EDTA as the anticoagulant (Tiberio et al., 2015), and avoiding other anticoagulants such as citrate, which may trigger hemolysis, and heparin, which inhibits enzymes used in downstream miRNA quantification assays.

Clinical tissue samples, formalin-fixed-paraffin-embedded (FFPE) or fresh-frozen, provide an important biospecimen for in situ quantitative analysis of miRNAs. In order to prepare these specimens correctly, time is critical, especially the "ischemic" time window occurring from sample excision to stabilization, which has many effects on the downstream recovery of nucleic acids and protein antigens (Masuda et al., 1999) because of autolysis. Although miRNAs tend to remain stable after tissue collection, the ischemic time (spent either at room temperature or at 4°C) and the loss of blood during this period (Chafin et al., 2013) can potentially change gene expression (including miRNAs) or, to a lesser degree, cause sample degradation (Khan et al., 2017). The "warm ischemic time" should be < 1 h, and the "cold ischemic time" should be < 24 h. Relatively to FFPE tissue only, due to the harsh nature of the fixation and embedding procedures, the obtained RNAs are heavily fragmented and chemically cross-linked by formalin (Nagy et al., 2016), raising doubts on their stability. Recently, through whole transcriptome sequencing (RNA-seq), the stability of total RNA has been assessed, showing a strong quantitative concordance between total RNA from FFPE stored for <2 years and high-quality fresh-frozen total RNA samples (Hester et al., 2016). As miRNAs are more stable than mRNA, an additional evidence, provided by other studies, shows a high concordance of miRNAs recovery between fresh-frozen and FFPE tissue specimens up to 9 years from collection (Xi et al., 2007; Meng et al., 2013).

miRNA ISOLATION METHODS

To ensure RNA stability, it is important that biological samples are immediately frozen in liquid nitrogen and stored at −70°C or immediately covered in a RNA stabilization reagent or disrupted and homogenized in the presence of RNase-inhibiting or RNase-denaturing reagents.

At the beginning of every RNA isolation procedure, a complete disruption of plasma membranes of cells and organelles is absolutely required to release all the RNA contained in the sample, and an efficient homogenization is necessary to reduce the viscosity of the lysates produced by disruption. These two steps can be performed differentially according to different starting material: for animal isolated cells, the use of lysis buffer and "vortexing" is recommended; for animal tissues, the disruption by rapid agitation with beads, by mortar and pestle, or by syringe and needle followed by an incubation with lysis buffer is recommended. In particular, for blood and urine cells, these steps can be performed after an appropriate separation, by centrifugation, of different components of these body fluids.

In general, the miRNA extraction methods are very similar to the ones used for the isolation of total RNA, except for some modifications to retain or enrich the small RNA fraction. The numerous RNA preparation technologies available can be classified into three general techniques: organic extraction methods, filter-based "spin basket" protocols, and magnetic particle methods. Since the yield depends on the sample type, on the tissue type, on the cellular differentiation level, on the starting material quantity, and/or on the isolation method, it's widely variable and can be of 0.2–5 μg of total RNA from 1 mg sample and 1–30 pg of total RNA for cell (Vomelova et al., 2009).

Organic Extraction Methods

Organic extraction methods are generally considered the "gold standard" for RNA isolation. According to this method, the specimen is homogenized in a phenol-containing solution and then is centrifuged with subsequent stratification into three phases: a lower organic phase; a middle phase containing denatured proteins and genomic DNA; and an upper aqueous phase with RNA, which is recovered and subsequently collected by alcohol precipitation and rehydration (Sambrook, 2012; Chomczynski and Sacchi, 2006). Organic methods are labor-intensive but can provide maximum possible yield; therefore, they are suggested in studies with a limited number of samples with expected low yields.

Filter-Based Extraction Methods

Filter-based, "spin basket" formats utilize membranes (usually a glass fiber, silica derivative, or ion exchange membrane) that are located at the bottom of a small plastic basket. Specimens are lysed in a buffer with RNase inhibitors (usually guanidine thiocyanate). Then, RNA molecules are bound to the membrane by passing the lysate through the membrane during centrifugation. Washing solutions are successively passed through the membrane and discarded. An appropriate elution solution is applied to the membrane, and the mobilized RNA is collected into a tube by centrifugation or by vacuum, using small laboratory pumps (Moret et al., 2013). Filter-based methods are laborsaving and more expensive than previous organic methods, but they are particularly suitable for high-throughput study, and final elution can be performed in a small volume, which is important for downstream miRNA analyses requiring more concentrated samples.

Magnetic Particle-Based Extraction Methods

Magnetic particle-based extraction methods use small particles (with a diameter of 0.5–1 µm) with a paramagnetic core and a coating modified to bind to RNA. Paramagnetic particles migrate when exposed to a magnetic field but retain minimal magnetic memory once the field is removed. In this isolation procedure, samples are lysed in a solution containing RNase inhibitors, and the small particles interact with RNA molecules; subsequently, the magnetic particles can be collected using an external magnetic field and resuspended once the field is removed. After several rounds of release, resuspension in wash solutions, and recapture, the RNA is released into an elution solution, and the magnetic particles are finally removed (Dioni et al., 2017). Magnetic beads offer many benefits compared with other technologies for isolating RNAs. Beads bind RNA more efficiently than glass fiber filters, resulting in a higher and more consistent RNA yield. Additionally, there is no risk of filter clogging due to cellular debris in high-sample-input studies; on the other hand, in these methods, it's crucial to completely remove beads at the end of the procedure, as they could inhibit downstream miRNA analyses.

Magnetic particle-based isolation procedures aren't very laborious and can be automated easier than other methods; however, these systems are rigidly designed and don't allow many ad hoc variations for miRNA-enrichment strategies.

Hybrid Extraction Methods

Hybrid extraction methods that combine the effectiveness of organic extraction with the easiness of "spin basket" formats are also available. Several miRNA studies support the good performances of these hybrid methods, which often allow enrichment of the small RNA (<200 nucleotides) fraction and/or separation of the large RNA (>200 nucleotides) component; in fact, one of the principal concerns for circulating miRNA studies is their relatively low concentration (Tan et al., 2015), especially in biological fluids (Duy et al., 2015; Guo et al., 2017; Andreasen et al., 2010; Blondal et al., 2013). Enrichment of small RNAs may be beneficial for downstream applications where the presence of large RNAs is generally allowed but could increase the background noise.

Optimization of Extraction Methods

The optimization and evaluation of different methodologies of small RNA purification from various biofluids are still ongoing (El-Khoury et al., 2016; Hammerle-Fickinger et al., 2010). Some authors have suggested the use of carriers, such as glycogen, yeast transfer RNA, or linear acrylamide, during the final RNA precipitation step to improve the isolation efficiency (McAlexander et al., 2013) because these carriers are insoluble in isopropyl alcohol and form a precipitate that traps RNA molecules, including miRNAs (Ban et al., 2017). Where assessment of miRNA isolation efficiency is crucial (e.g., in biofluids and in extracellular vesicles), a fixed amount of a synthetic miRNA that is not expressed in the examined biological matrix may be "spiked in" at an early stage of the isolation protocol to be used as an efficiency extraction control.

QUALITY AND QUANTITY OF miRNA

The two main critical aspects in miRNA studies are the possibility to quantify very small amounts of RNAs isolated from biofluids and assessment of the presence of the small RNA component. The concentration of total RNAs should be determined by measuring the absorbance at 260 nm ($A260$) using a spectrophotometer, but this analysis does not allow the discrimination between small and large RNA components. An absorbance of one unit at 260 nm is equivalent to 44 µg of total RNA/mL; this relationship is valid only for measurements conducted at a neutral pH. The ratio of the readings at wavelengths of 260 and 280 nm ($A260/A280$) provides an estimate of the RNA purity with respect to contaminants that absorb in the UV spectrum, such as proteins. For accurate values, it is recommended to measure the absorbance in 10 mM Tris-Cl and pH 7.5, where pure RNA has an $A260/A280$ ratio of 1.9–2.1. However, quantification of small amounts of RNA (typically <2 ng/µL) is almost impossible using spectrophotometric instruments. For all these reasons, miRNA quantity and quality should be analyzed using either capillary electrophoresis (CE) or fluorimetry. As discussed below, the ideal approach is the combination of both these methodologies because CE gives a better qualitative value, while dual-channel fluorimetry gives a better quantitative value.

CE

CE is electrophoresis performed in a capillary tube, where the separating force is the difference in the charge-to-size ratio. Not a flow through the column, but an electric field moves the separation, as the capillary is filled with a conductive fluid (at a certain pH value). In order to analyze a sample, RNA is introduced in the capillary, either by a pressure injection or by an electrokinetic injection. A high voltage is generated over the capillary, and due to this electric field (up to >300 V/cm), the sample components move through the capillary at different speeds, which are related to the number of nucleotides in the miRNA, to the opposite-charged electrode. The obtained data (generally called an electropherogram) can provide insight on the characteristic RNA patterns (Harrington et al., 2009), distinguishing small and large components (small, 18S, and 28S, if present). RNA integrity number (RIN) was developed to remove individual interpretation in RNA quality control. The RIN software algorithm takes the entire electrophoretic trace into account and allows for the classification of total RNA, based on a numbering system from 1 to 10, with 1 being the most degraded profile and 10 being the most intact; samples with RIN > 7.0 are considered acceptable. However, in case of isolation of the enriched small RNA fraction, the RIN is not applicable. Nonetheless, a visual inspection of the output electropherogram allows to verify small RNA presence, defined by a well-defined peak at about 25 nucleotides.

Dual-Channel Fluorimetry

Dual-channel fluorimetry, designed to provide sensitive fluorescent detection, is widely used in miRNA studies (Deben et al., 2013) because it can provide the most accurate quantification of miRNA content, even at concentrations <2 ng/µL (Garcia-Elias et al., 2017). A dual-channel fluorimeter is equipped with two excitation filters and two emission filters, and it is easy to use and suitable for standard RNA study applications (Landolt et al., 2016; Grabmuller et al., 2015). Quantitating unknown RNAs require instrument calibration using a blank and a single standard sample (generally provided by the manufacturer) that should be appropriate for the expected range of miRNA concentrations.

miRNA DETECTION TECHNIQUES

Although many techniques that were originally designed to profile mRNAs can, in principle, be applied to miRNA analysis, miRNAs show some peculiar properties, thus creating an additional challenge to their investigation. The first one is represented by their short size (~22 nucleotides for the mature form), which is not sufficient to permit primer annealing in conventional polymerase chain reaction (PCR). Moreover, they do not have a poly-A tail, thereby prohibiting the use of a universal primer for reverse transcription. In addition, several members of the same family often only differ by 1–2 nucleotides, making the discrimination among them very demanding. Furthermore, several experimental designs require not only a single miRNA quantification but also the screening of hundreds of miRNAs at the same time. Despite these factors, the currently available techniques are discussed below so that researchers can choose the best approach for a given experimental design. In order to give

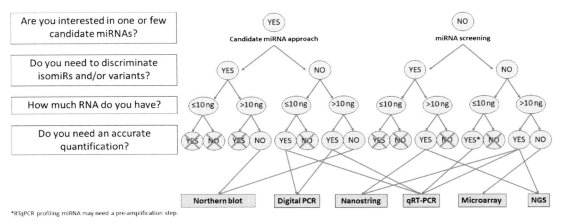

*RTqPCR profiling miRNA may need a pre-amplification step.

FIG. 2   Choosing a technique in "single candidate" or "screening" miRNA study.

support in this choice, Fig. 2 depicts a decision tree, aimed at assisting the researcher in the choice of analytic method best suited for a specific study.

Northern Blotting

The first approach that was applied to assess the accumulation of target miRNAs is northern blot analysis (Lee et al., 1993). This technique combines polyacrylamide gel electrophoresis of RNA samples with a transfer to a positively charged nylon membrane as the first step and hybridization with a labeled probe. The main advantage of this technique is that it does not require any special equipment. Moreover, it grants quantification of the expression level and separation according to the miRNA size. This size separation step allows the discrimination of pri-miRNA from pre-miRNA and miR. On the contrary, the main problem with using DNA oligonucleotide probes is their very low sensitivity, especially when investigating miRNAs that are not abundant, making the application of this protocol very problematic whenever a cell or tissue source is limited. The standard input RNA requirement is typically 3–50 μg of total RNA per sample.

To avoid this problem, a new method based on locked nucleic acid (LNA)-modified oligonucleotide probes has been reported. This method is able to increase the sensitivity of detection by ~10-fold, compared with standard DNA probes. LNAs are a class of high-affinity RNA analogues in which the ribose ring is "locked" in the ideal conformation for Watson-Crick binding, resulting in a higher thermal stability when hybridized to a complementary strand. In fact, for each incorporated LNA monomer, the melting temperature of the duplex increases by 2–8°C. LNA nucleotides have an additional bridge composed by 4'C and 2'O atoms. The hybrid LNA/DNA probes may improve miRNA detection because of short hybridization time, high efficiency, discriminatory power, and high melting temperature. The minimal length of the LNA/DNA probe was determined to be 12 nucleotides, and these probes usually contain 30% LNA nucleotides (Kloosterman et al., 2006). LNA probes are commercially available (from Exiqon, Vedbaek, Denmark) for many other expression studies, but their price is >50-fold over the cost of traditional probes applied in standard RT-PCR.

Northern blotting is a technique suitable for single assay mature miRNA studies (Varallyay et al., 2007), but not for profiling, as it is still not possible to perform multiplex reaction with a large number of miRNA assays.

In Situ Hybridization

A traditional method that allows both the quantification and localization in single cells is in situ hybridization (ISH), which has been modified for a more efficient detection of a limited number of small RNAs.

Depending on the detection method, ISH can be divided into chromogenic enzyme-based in situ hybridization and fluorescent in situ hybridization (FISH) (Urbanek et al., 2015). Some variations propose different probe types as LNA probe (Exiqon, Denmark) previously described or 2′-fluoro-modified RNA (2′-F RNA) that ensures increased binding to the target and better nuclease resistance (Li et al., 2013) or 2′-O-methyl (2′-O-Me) RNA modification that has faster hybridization kinetics and greater ability to bind structured targets than DNA probes (Majlessi et al., 1998). Commercial assays for the detection of several miRNAs using in situ hybridization and fluorescence in situ hybridization are also available (Lin et al., 2008). Some authors introduced 1-ethyl-3-(3-dimethylaminopropyl)carbodiimide (EDC) in combination with traditional formaldehyde fixation to improve miRNA recovery, in the first step of ISH method (Pena et al., 2009). Others improved "permeabilization step" with longer time treatment or with several commercial buffers, especially in the presence of chemically modified probes that need a stronger permeabilization treatment compared with short probes (Ortega et al., 2015).

The "washing step" of an ISH protocol has been optimized to preserve the probe-target complex but at the same time eliminate off-target binding; at this purpose, tetramethylammonium chloride (TMAC) is used in combination with RNase A, also adjusting the washing temperature to provide specificity of binding while preserving the signal strength (Deo et al., 2006).

The "detection step" in the first ISH experiments used radiolabeled probes with many critical issues, so in the following years, nonradioactive probes have been introduced that used enzyme-based detection. Some nonradioactive haptens, combined with probes, are detected by histochemical enzymatic reactions, after the application of enzyme-conjugated antihapten antibodies. Alkaline phosphatase (AP) is most commonly used with nitro blue tetrazolium/5-bromo-4-chloro-3-indolyl phosphate (NBT/BCIP) (Jorgensen et al., 2010).

Recently, nanotechnology was introduced in ISH method to increase lifetime of the organic fluorophores, improving miRNA studies. Metal nanoshells, composed of silica spheres with encapsulated $Ru(bpy)3^{2+}$ complexes as cores and thin silver layers as shells, have been used for the detection of low-abundant miRNAs. This system needs to be optimized to provide better penetration of the nanoshells through the cell membrane, mobility in the cells, and high specificity toward the miRNA target (Zhang et al., 2010).

qRT-PCR Analysis

qRT-PCR is a highly specific and reproducible method to measure transcripts or gene copies (Heid et al., 1996). In addition, probed-based quantitative PCR is a highly sensitive detection method for quantification of relative miRNA expression in biological samples (Shi and Chiang, 2005; Livak and Schmittgen, 2001). Since other methods require a larger amount of RNA sample, RT-PCR is the gold standard for those applications that are characterized by a low RNA availability; in miRNA studies, a minimum of 1 ng of sample is required for profiling and even <1 ng for single assays.

Some strategies have been developed in order to solve some issues typical of RT-PCR analysis of miRNAs (Dellett and Simpson, 2016), such as their small size (they have the same length of conventional PCR primers), their highly variable GC content (which complicates the optimization of a unique protocol in a profiling assay), the presence of pre-miRNA and pri-miRNA, and the highly similar sequences of mature miRNAs belonging to the same family (Guo and Chen, 2014). Some authors performed a miRNA absolute quantification protocol using a standard curve generated with a synthetic miRNA spike-in that was previously inserted in all biological samples and diluted serially, from a fixed known concentration. These dilutions are carried out in parallel with qPCR of samples, and output data are used to calculate the normalization factor for each miRNA of interest (Kroh et al., 2010). The miRNA RT-PCR methods can be classified into two categories, which are based on "universal" and "specific" reverse transcription (RT) reactions, respectively. In addition, a third probe-ligation method that does not require an RT reaction is also available (Jin et al., 2016). In the "universal RT" methods, all miRNA sequences are elongated by an identical tail used to activate the RT reaction to cDNA, for example, the addition of a poly-A tail to the 3' end (Shi and Chiang, 2005), polyuridylation with poly(U) polymerase (Mei et al., 2012; Benes et al., 2015), and recently combination of linker ligation and end tailing; all the RT reactions are located in the same tube, but they suffer from high background noise, and the polyadenylation and ligation steps also introduce bias. The "specific RT" methods include the use of a linear primer (Raymond et al., 2005; Sharbati-Tehrani et al., 2008); "princer" probes (Huang et al., 2015); and, very often, "stem-loop RT primers" (st-primers) developed by Chen et al. (2005) (TaqMan), which selectively target mature miRNAs and easily cover a dynamic range of linear quantification of seven $\log_{10}$ units; this strategy is considered the "gold standard" RT-PCR method and is used for validation of other incoming techniques (Benes and Castoldi, 2010), for example, one of the most recent novel methods providing a "two-tailed qRT-PCR" with two hemiprobes complementary to two different parts of the target miRNA (Androvic et al., 2017).

To date, the most widespread qPCR probes use TaqMan Chemistry; in miRNA qPCR, "TaqMan small RNA assays" utilize st-primers and minor groove binder (MGB) probe. The st-primers are composed of (1) a short single-stranded sequence at the 3' end that anneals to the 3' end of the target miRNA, (2) a double-stranded segment (stem), and (3) a loop. This st-primer forms a specific RNA/DNA duplex, thus preventing binding with pri-miRNA and pre-miRNA and allowing the elongation of each miRNA. The TaqMan MGB probes contain (1) a reporter dye (FAM dye) linked to the 5' end of the probe; (2) an MGB at the 3' end of the probe, which increases the melting temperature without increasing the probe length; and (3) a nonfluorescent quencher at the 3' end of the probe to measure the reporter dye contributions more accurately. During PCR, the TaqMan MGB probe anneals specifically to a complementary sequence between the forward and reverse primer sites. When the probe is intact, the proximity of the reporter dye to the quencher dye results in suppression of the reporter signal.

The DNA polymerase cleaves only probes that are hybridized to the target and separates the reporter dye from the quencher dye; this separation results in increased fluorescence by the reporter.

These assays are available in several formats; individual miRNA assays are useful for validation of target-specific workflows, while miRNA prespotted 384-well plates or cards and TaqMan OpenArray MicroRNA Panels are ideal for profiling hundreds of miRNAs. The panels cover most relevant human miRNAs aligned with Sanger miRBase v21, including 754 assays, and can generate a global miRNA expression profile in a single workday, using as little as 1 ng of RNA. This is a good strategy for large epidemiological studies (Pergoli et al., 2017; Cavalleri et al., 2017) in which sample throughput is critical. A typical tool is the Applied Biosystems QuantStudio 12K Flex Real-Time PCR System, which performs three samples per panel and three panels per run, allowing the processing of a total of nine samples in parallel and up to 36 samples per 8 h workday.

Combining Megaplex Primer Pools to streamline the conversion of miRNA to cDNA and preamplification prior to quantification, the OpenArray technology is a broadly applicable nanoliter fluidics platform for low-volume (33 nL) RT-PCR reactions. It utilizes a microscope slide-sized plate with 3072 through-holes, with each plate containing 48 subarrays with 64 through-holes; the holes measure 300 μm in diameter and 300 μm in depth and have hydrophilic (inner) and hydrophobic (outside) coatings, where the reagents are retained via surface tension.

Comparative analysis of high-throughput platforms of qPCR indicates that the better choice for circulating miRNA signatures is inversely related to the qPCR reaction volume (Farr et al., 2015). The OpenArray system, with a 33 nL reaction volume, was therefore found to be the most reproducible high-throughput platform tested (Morrison et al., 2006).

Microarray Analysis

Microarray technologies have been widely applied to gene expression studies. As most research groups have access to microarray facilities with experienced personnel, they are now easily adapting this approach to miRNA analysis. Microarray currently represents a widely used high-throughput method for measuring miRNA levels, because they can quantify large numbers of miRNAs simultaneously in a single experiment, are more cost-effective than a next-generation sequencing, but have a very limited potential for discovery of novel miRNA sequences and not discriminate between a mature miRNA and its precursor.

Microarrays for miRNA screening are usually based on hybridization between a target molecule to be analyzed and a probe, which occurs on a glass slide carrying short spotted or in situ synthetized oligonucleotides. Each oligonucleotide (i.e., DNA, RNA, or LNA) can hybridize to a single miRNA. Microarrays may be made in-house or can be purchased from several commercial suppliers, for example, Agilent, Exiqon, Illumina, Genosensor, Affymetrix, and LC Biosciences (Callari et al., 2012). The coverage of each platform is different. For example, Agilent (Agilent, Santa Clara, California, the United States) assures coverage of the entire human miRNome as listed in the most recent version of miRBase, whereas Exiqon (Exiqon, Vedbaek, Denmark) provides microarrays based on LNA that can simultaneously measure human, mouse, and rat miRNA expression on the same glass slide.

The procedure requires that the miRNA purified is first dephosphorylated at the 3′ terminus with calf phosphatase and labeled with a fluorophore bound by the T4 RNA ligase. Then, miRNA sample is hybridized to the glass slide for several hours up to days; it is washed to remove any unbound labeled miRNA and then scanned for fluorescence.

Approximately 50–150 ng of total RNA is required for each experiment ranging across different commercial platforms, which is considered a low input for those techniques that do not require a PCR amplification. This method is well suited for comparing two experimental conditions or evaluating the relative abundance of miRNAs in two different states (e.g., cases vs controls) but cannot give an absolute quantification. It represents an ideal approach for small nonhuman species where the RNA yield (especially in biofluids) is expected to be very low.

It is a general consensus that the initial results of microarray screening need to be validated with another technique such as real-time quantitative reverse transcription-PCR (qRT-PCR) analysis.

Nanostring Analysis

NanoString nCounter platform (NanoString Technologies, Seattle, Washington, the United States) is a relatively recent technique applicable to miRNA profiling. It is based on hybridization and allows a direct measurement of the miRNA amount without the need for reverse transcription or cDNA amplification or labeling of the starting material. This platform is characterized from novel technologies of direct molecular barcoding and digital detection of targets; for each miRNA of interest, two specific probes are designed: the first one is called the "capture probe," which contains a biotin molecule and anneals at the 3′ end. The biotin component will then allow the molecule to be absorbed to a solid phase through streptavidin binding. The second probe is called the "reporter probe," which contains a specific color-coded sequence and anneals at the 5′ end, and it carries a fluorescent signal. After hybridization of the probes, the probe-miRNA complex is aligned and immobilized on a cartridge then placed in the nCounter Digital Analyzer for data collection. High specificity is ensured by a stepwise control of hybridization and ligation temperatures. About 800 pairs of probes for 800 miRNAs of interest may be performed in parallel in a single reaction, so it is possible to detect several miRNAs recognizing similar variants of a sequence. The data obtained are highly reliable if the amount of starting material is sufficient (>10 ng) and of good quality. This newly emergent technique has been used in case-control cancer study (Bailey, 2015), neurodegeneration study (Kumar et al., 2013), and bowel syndrome (Fourie et al., 2014). However, it is an expensive technology, present in research laboratories or facilities, but not in routine clinical laboratories (Kappel and Keller, 2017). NanoString technology has some advantages, it is faster than microarray, it avoids amplification and measures absolute quantification reducing bias problem, and it provides also a better dynamic range from quite low starting material (Waggott et al., 2012).

Digital qPCR (DqPCR) Analysis

DqPCR analysis, a recently developed technology for miRNA analysis, is characterized by the ability to obtain an absolute quantification, without the need for internal controls and a subsequent normalization step. DqPCR is based on the concept that each template molecule is partitioned into a single microreaction, in which interference from other nucleic acids is minimal, leading to a much greater sensitivity and accuracy compared with conventional qPCR methods. These partitions are made either by separating copies into physical chambers via microfluidics or by creating water-in-oil immersion droplets that hydrostatically separate targets. The respective technologies are named chamber dPCR (Borzi et al., 2017) and droplet dPCR (Hindson et al., 2013). Given a few basic quantitative assumptions, such as (1) targets independently segregate, (2) targets are fully accessible to the probes and primers, (3) the droplet volume is known, and (4) each partition contains a limited number of targets, then, the absolute copy number of genomic material can be calculated by applying the Poisson correction (Redshaw et al., 2013).

In the dPCR method for miRNA measurements, some of the original limitations still exist, which creates the same measurement uncertainty as evidenced with qRT-PCR. For example, miRNA has such short target sequences that primer and probe optimization strategies, such as melting temperature, length, and guanine-cytosine content, might be more difficult (Redshaw et al., 2013). An absolute quantification of circulating miRNAs can also be applied to dPCR without the need of a standard curve (Ferracin et al., 2015). Moreover, dPCR proved to be more tolerant than qPCR to the presence of inhibitors in the amplification reaction (Dingle et al., 2013).

Next-Generation Sequencing (NGS)

NGS has recently become a widely accepted approach for detecting miRNA sequences. The major advantage of this technique is that it provides the possibility of quantifying known miRNAs and identifying new unknown miRNA sequences. With this technique, it is possible to discover new miRNAs that may have a diagnostic or prognostic value. This approach is based on the preparation of a library containing small RNA cDNAs prepared from the sample to be analyzed. These sequences are then massively sequenced in parallel, meaning that millions of cDNA molecules can be sequenced in a single run.

Among the NGS technologies currently available and extensively reviewed (Metzker, 2010), it is worth mentioning that "sequencing by synthesis," commercialized by Illumina, is one of the most used approaches. Purified high-quality miRNAs are ligated to adaptors at the 3′ and 5′ ends and then converted to cDNAs. This reverse transcription is followed by an amplification step, which includes index sequences in the cDNAs and makes them suitable for tagging the cDNA library and later sequencing several samples in a single reaction. cDNA libraries are then applied to a flow cell, where single cDNAs are captured and specifically amplified, giving a clonal copy spot for every single cDNA. Each spot is then sequenced using the fluorescence-labeled dye terminators and a charge-coupled device camera.

Although we mentioned the undeniable potential of NGS, which is able to detect novel miRNA sequences, it should also be noted that only a minority of putative miRNA sequences identified by NGS are then confirmed as new miRNAs. Each NGS reaction can in fact identify many unknown small RNAs, but not all of them can be classified as bona fide miRNAs. Stringent criteria have been identified to annotate a putative small miRNA sequence as miRNA, including a length of ~22 nucleotides, the presence in the -3p and -5p sequencing version, the conservation across species, and a genomic sequence that predicts a possible precursor that can be folded into the duplex structure (Yang and Qu, 2012). An advantage of the sequencing platform for miRNA profiling is the large dynamic range provided by this technique if enough input is available (Willenbrock et al., 2009). Obstacles in the use of NGS as the leading methodology for miRNA analysis include the high costs (even though they are decreasing), the very complex computational analysis required, and the difficulty to standardize the data coming from different experiments, as the library preparation step introduces a bias that preferentially captures a different pull down of miRNAs.

STRATEGIES FOR NORMALIZATION

The normalization stage refers to all of the technical adjustments made to the measured value, which help to reduce the contribution exerted by experimental factors (e.g., the variation in the RNA amount pipetted in a single reaction) not directly related to biological differences among samples. Although an exhaustive list of all the normalization strategies applied to every single experimental approach is outside the goals of this chapter, we need to acknowledge that the choice of the best strategy is strictly dependent on the main experimental question, the platform being used, the number of miRNAs investigated, and the specimen type.

In general, the most common strategy for normalization (especially for RT-PCR) that has been consistently applied to tissue/cell miRNAs is the use of a "reference" housekeeping small nucleolar RNA (e.g., RNU6, RNU6B, RNU44, and RNU48 for human) (Schwarzenbach et al., 2015). However, this approach is made more complex by the need to confirm that this reference is not influenced by the conditions under study. In order to normalize for variations during miRNA extraction, as previously mentioned, it is also possible to add a fixed quantity of one or more spike-in controls during the extraction. This method eliminates the experimental procedure bias but not the deviation in sampling and quality of tissue. The combined normalization of an endogenous and exogenous reference miRNA has been proposed to compensate both the bias (Sourvinou et al., 2013). Whenever normalization is required to standardize the profiling of miRNA data, it is possible to use the "global mean" (either geometric or arithmetic mean) as a normalization factor. This strategy allows normalization for a global tendency rather than choosing a specific reference control that could be altered and then influence the final result, as it uses the average expression level of all miRNAs detected in each sample as a normalization factor (Mestdagh et al., 2009). Global mean strategy is especially suitable for circulating miRNA quantification, where data normalization is even more complex. In this case, validated housekeeping RNAs are not available and the reference housekeeping gene used to normalize the cellular miRNA signature might be not present in the extracellular environment.

Circulating miRNA normalization poses an additional difficulty: miRNAs in biofluids can be easily contaminated by surrounding tissues and/or can be influenced by the dilution of the liquid matrix, from which have been isolated. Proposed possible solution to overlap the problem is to normalize through quantification of creatinine, which is a product of normal muscle metabolism and is produced and removed through the kidney at a relatively constant rate. Blood and urine creatinine concentrations could be therefore used to normalize plasma and urine miRNA measurements (Wolenski et al., 2017). To date, several normalization strategies have been used for circulating miRNA, but no consensus exists on the best strategy for normalization (Kok et al., 2015). In order to assist the researcher in the process of choosing the best normalization strategy, some tools have been proposed, such as the NormFinder (Andersen et al., 2004) and geNorm (Vandesompele et al., 2002) algorithms, which calculate the ratio of each housekeeping target or normalization factors to another, assuming that this ratio is constant among different samples. These algorithms are equally suitable for circulating and tissue miRNAs.

CHALLENGES

Several of the above-described methodologies are high-throughput assays and thus produce large amounts of data, making a proper computational approach essential. An exponential growth of bioinformatics tools, specifically designed to analyze miRNA data, has been observed. A recent review has reported a comprehensive list of the tools available to date (Akhtar et al., 2016) and describes the purpose for which they have been developed, such as miRNA discovery, miRNA-target prediction, validated miRNA targets, miRNA expression analysis, identification of miRNA regulatory networks, analysis of miRNA metabolic and signaling pathways, investigation of miRNA and transcription factor interactions, and linking miRNAs to diseases.

It is important to underline that the results obtained with the various tools might be different, due to the specific analytic strategy applied. Moreover, many tools are unable to discriminate among isoforms or to distinguish if the effect of a specific stimulus/disease is associated with a downregulation or an upregulation of a specific miRNA or, in the case of biofluid based biomarkers, whether there is an increase in total abundance and not in relative expression. In addition, many tools are not capable of performing statistical analysis to discover differentially expressed miRNAs.

As miRNA function does not require a perfect match with the potential target mRNA, correlating miRNA expression with mRNA targets poses several challenges. First of all, we still do not have a clear understanding of the definition of a target region within an mRNA, even if most bioinformatics approaches still rely on the 3′ UTR matching to the 5′ terminus of the putative miR (Marin et al., 2013). Moreover, a single mRNA is regulated by several miRNAs, and each miRNA is able to target several mRNAs, thus making pathway analysis very complex. Given the fact that computational data analysis is a key step that can generate contradictory results using different tools and that these results can have an impact on the final interpretation of the biological meaning of an altered expression, caution is highly recommended.

CONCLUSIONS

In this chapter, we focused on the main techniques available to date for miRNA analysis. This field of research is of increasing interest, and many attempts are still ongoing to bypass the limitations of the approaches described above and further summarized in Table 1.

TABLE 1 MicroRNA Analysis Technologies, Advantages and Disadvantages

Method	Analysis Type	Advantages	Disadvantages	Main Assay/Platform	Vendor	RNA Quantity	References
Northern blotting	Single assay	Established method. Easily adapted to existing workflow. Low cost	Low-throughput method. Not accurate quantification	32P-labeled miRNA probes	Sigma	3–50 μg	Lee and Ambros (2001)
				Dig-labeled miRNA probes	Biorad		Ramkissoon et al. (2006)
				miRcury LNA oligonucleotide probes	Qiagen		Valoczi et al. (2004)
In situ hybridization	Single assay	Established method. Allows localization	Low-throughput method. Not accurate quantification	LNA RNA probes	Exiqon	5–10 μm-thick tissue sections	Yamamichi et al. (2009)
				2'-O-methyl LNA RNA probes	Exiqon		Soe et al. (2011)
				2F DNA/RNA probes	Exonbio Lab		Li et al. (2013)
				NBT/BCIP detection system	Roche		Chaudhuri et al. (2013)
Quantitative real-time PCR	Single assay	Sensitive and specific method. Accurate quantification. Low sample input	Not able to identify novel miRNAs	TaqMan Small RNA Assays	ThermoFisher	<10 ng	Chen et al. (2005)
				miRCURY LNA qPCR primers/probe	Exiqon		Tabrizi et al. (2015)
	Profiling			TaqMan TLDA Card	ThermoFisher		Kodani et al. (2011)
				TaqMan OpenArray	ThermoFisher		Hudson et al. (2013)
				Biomark HD System	Fluidigm		Petriv et al. (2010)
				SmartChip Human MicroRNA	Wafergen		De Wilde et al. (2014)
				miScript miRNA PCR Array	SABiosence/Qiagen		Hu et al. (2013)
Microarray	Profiling	Established method. Easily adapted to existing workflow	Not able to identify novel miRNAs. Not accurate quantification	Geniom Biochip miRNA	CBC	50–150 ng	Lange (2010)

Method	Type	Advantages	Disadvantages	Product	Company	Amount	References
NanoString nCounter	Profiling	Sensitive, accurate, and specific method. Can discriminate isomiRs and/or variants	Cannot identify novel miRNAs	miRcury LNA microRNA Array	Exiqon		Sewer et al. (2014)
				µParaFlo Biochip Array	LC Biosciences		Kong et al. (2014)
				MicroRNA Microarray	Agilent		Wang et al. (2007)
				GeneChip miRNA Array	Affymetrix		Chen et al. (2008)
				Sentrix Array Matrix BeadChips	Illumina		Siegrist et al. (2009)
				GenoExplorer microRNA chips	Genosensor		Park et al. (2011)
				nCounter Human miRNA Expression Assay	Nanostring	10–100 ng	Wyman et al. (2011)
Digital quantitative PCR	Single assay	Sensitive and specific method. Accurate quantification. No need for normalization	Not able to identify novel miRNAs	QS 3D Digital PCR	ThermoFisher	< 10 ng	Conte et al. (2015)
				Droplet Digital PCR	Biorad		Hindson et al. (2013)
Next-generation sequencing	Profiling	Accurate and specific methods. Can detect novel miRNAs. Can discriminate isomiRs and/or variants	Significant computational support needed for data analysis	*High-throughput NGS platform*		10 ng	
				HiSeq 2000 Genome Analyzer IIX	Illumina		Kato et al. (2011)
				SOLiD System	ABI		Ramsingh et al. (2010)
				Gs FLX 454 System	Roche		Margulies et al. (2005)
				Smaller-scale NGS platform			
				Ion Torrent	ThermoFisher		Schageman et al. (2013)
				MiSeq System	Illumina		Murakami et al. (2014)
				GS Junior Seq System	Roche		Trachtenberg and Holcomb (2013)

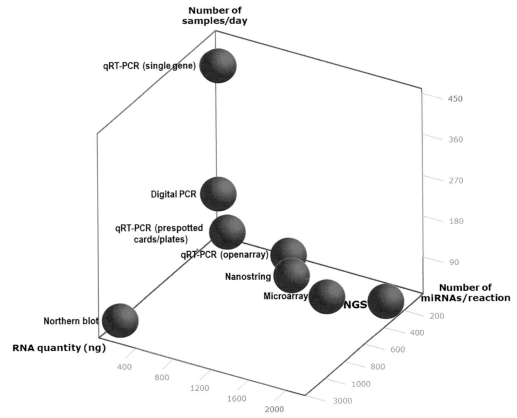

FIG. 3 Applications pattern in specific conditions of RNA input, number of samples/day and number of miRNAs/reaction in human studies.

In order to fully appreciate the potential of miRNAs in routine practice, it is especially important to develop methodologies that are easy to use and that can be highly standardized to ensure reproducibility among different diagnostic laboratories. However, each method is suitable for a specific application in a specific condition of input RNA. An example, focused on human studies, is illustrated in Fig. 3.

Finally, bioinformatics and statistical data analysis are fundamental steps, and their difficulty cannot be underestimated because the quality of these steps will have an impact on the final results obtained.

References

Akhtar, M.M., Micolucci, L., Islam, M.S., Olivieri, F., Procopio, A.D., 2016. Bioinformatic tools for microRNA dissection. Nucleic Acids Res. 44, 24–44.

Andersen, C.L., Jensen, J.L., Orntoft, T.F., 2004. Normalization of real-time quantitative reverse transcription-PCR data: a model-based variance estimation approach to identify genes suited for normalization, applied to bladder and colon cancer data sets. Cancer Res. 64, 5245–5250.

Andreasen, D., Fog, J.U., Biggs, W., Salomon, J., Dahslveen, I.K., Baker, A., Mouritzen, P., 2010. Improved microRNA quantification in total RNA from clinical samples. Methods 50, S6–S9.

Androvic, P., Valihrach, L., Elling, J., Sjoback, R., Kubista, M., 2017. Two-tailed RT-qPCR: a novel method for highly accurate miRNA quantification. Nucleic Acids Res. 45, e144.

Arroyo, J.D., Chevillet, J.R., Kroh, E.M., Ruf, I.K., Pritchard, C.C., Gibson, D.F., Mitchell, P.S., Bennett, C.F., Pogosova-Agadjanyan, E.L., Stirewalt, D.L., Tait, J.F., Tewari, M., 2011. Argonaute2 complexes carry a population of circulating microRNAs independent of vesicles in human plasma. Proc. Natl. Acad. Sci. USA 108, 5003–5008.

Bader, A.G., Brown, D., Winkler, M., 2010. The promise of microRNA replacement therapy. Cancer Res. 70, 7027–7030.

Bailey, S.T., Westerling, T., Brown, M., 2015. Loss of estrogen-regulated microRNA expression increases HER2 signaling and is prognostic of poor outcome in luminal breast cancer. Cancer Res. 75, 436–445.

Ban, E., Chae, D.K., Yoo, Y.S., Song, E.J., 2017. An improvement of miRNA extraction efficiency in human plasma. Anal. Bioanal. Chem. 409, 6397–6404.

Benes, V., Castoldi, M., 2010. Expression profiling of microRNA using real-time quantitative PCR, how to use it and what is available. Methods 50, 244–249.

Benes, V., Collier, P., Kordes, C., Stolte, J., Rausch, T., Muckentaler, M.U., Haussinger, D., Castoldi, M., 2015. Identification of cytokine-induced modulation of microRNA expression and secretion as measured by a novel microRNA specific qPCR assay. Sci. Rep. 5, 11590.

Blondal, T., Jensby Nielsen, S., Baker, A., Andreasen, D., Mouritzen, P., Wrang Teilum, M., Dahlsveen, I.K., 2013. Assessing sample and miRNA profile quality in serum and plasma or other biofluids. Methods 59, S1–S6.

Borzi, C., Calzolari, L., Conte, D., Sozzi, G., Fortunato, O., 2017. Detection of microRNAs using chip-based QuantStudio 3D digital PCR. Methods Mol. Biol. 1580, 239–247.

Butz, H., Patocs, A., 2015. Technical aspects related to the analysis of circulating microRNAs. EXS 106, 55–71.

Callari, M., Dugo, M., Musella, V., Marchesi, E., Chiorino, G., Grand, M.M., Pierotti, M.A., Daidone, M.G., Canevari, S., De Cecco, L., 2012. Comparison of microarray platforms for measuring differential microRNA expression in paired normal/cancer colon tissues. PLoS One 7, e45105.

Cavalleri, T., Angelici, L., Favero, C., Dioni, L., Mensi, C., Bareggi, C., Palleschi, A., Rimessi, A., Consonni, D., Bordini, L., Todaro, A., Bollati, V., Pesatori, A.C., 2017. Plasmatic extracellular vesicle microRNAs in malignant pleural mesothelioma and asbestos-exposed subjects suggest a 2-miRNA signature as potential biomarker of disease. PLoS One 12, e0176680.

Chafin, D., Theiss, A., Roberts, E., Borlee, G., Otter, M., Baird, G.S., 2013. Rapid two-temperature formalin fixation. PLoS One 8, e54138.

Chaudhuri, A.D., Yelamanchili, S.V., Fox, H.S., 2013. Combined fluorescent in situ hybridization for detection of microRNAs and immunofluorescent labeling for cell-type markers. Front. Cell. Neurosci. 7, 160.

Chen, C., Ridzon, D.A., Broomer, A.J., Zhou, Z., Lee, D.H., Nguyen, J.T., Barbisin, M., Xu, N.L., Mahuvakar, V.R., Andersen, M.R., Lao, K.Q., Livak, K.J., Guegler, K.J., 2005. Real-time quantification of microRNAs by stem-loop-RT-PCR. Nucleic Acids Res. 33, e179.

Chen, J., Lozach, J., Garcia, E.W., Barnes, B., Luo, S., Mikoulitch, I., Zhou, L., Schroth, G., Fan, J.B., 2008. Highly sensitive and specific microRNA expression profiling using BeadArray technology. Nucleic Acids Res. 36, e87.

Chomczynski, P., Sacchi, N., 2006. The single-step method of RNA isolation by acid guanidinium thiocyanate-phenol-chloroform extraction: twenty-something years on. Nat. Protoc. 1, 581–585.

Conte, D., Verri, C., Borzi, C., Suatoni, P., Pastorino, U., Sozzi, G., Fortunato, O., 2015. Novel method to detect microRNAs using chip-based QuantStudio 3D digital PCR. BMC Genomics 16, 849.

De Wilde, B., Lefever, S., Dong, W., Dunne, J., Husain, S., Derveaux, S., Hellemans, J., Vandesompele, J., 2014. Target enrichment using parallel nanoliter quantitative PCR amplification. BMC Genomics 15, 184.

Deben, C., Zwaenepoel, K., Boeckx, C., Wouters, A., Pauwels, P., Peeters, M., Lardon, F., Baay, M., Deschoolmeester, V., 2013. Expression analysis on archival material revisited: isolation and quantification of RNA extracted from FFPE samples. Diagn. Mol. Pathol. 22, 59–64.

Dellett, M., Simpson, D.A., 2016. Considerations for optimization of microRNA PCR assays for molecular diagnosis. Expert. Rev. Mol. Diagn. 16, 407–414.

Deo, M., Yu, J.Y., Chung, K.H., Tippens, M., Turner, D.L., 2006. Detection of mammalian microRNA expression by in situ hybridization with RNA oligonucleotides. Dev. Dyn. 235, 2538–2548.

Desvignes, T., Batzel, P., Berezikov, E., Eilbeck, K., Eppig, J.T., McAndrews, M.S., Singer, A., Postlethwait, J.H., 2015. miRNA nomenclature: a view incorporating genetic origins, biosynthetic pathways, and sequence variants. Trends Genet. 31, 613–626.

Dingle, T.C., Sedlak, R.H., Cook, L., Jerome, K.R., 2013. Tolerance of droplet-digital PCR vs real-time quantitative PCR to inhibitory substances. Clin. Chem. 59, 1670–1672.

Dioni, L., Sucato, S., Motta, V., Iodice, S., Angelici, L., Favero, C., Cavalleri, T., Vigna, L., Albetti, B., Fustinoni, S., Bertazzi, P., Pesatori, A., Bollati, V., 2017. Urinary chromium is associated with changes in leukocyte miRNA expression in obese subjects. Eur. J. Clin. Nutr. 71, 142–148.

Duy, J., Koehler, J.W., Honko, A.N., Minogue, T.D., 2015. Optimized microRNA purification from TRIzol-treated plasma. BMC Genomics 16, 95.

El-Khoury, V., Pierson, S., Kaoma, T., Bernardin, F., Berchem, G., 2016. Assessing cellular and circulating miRNA recovery: the impact of the RNA isolation method and the quantity of input material. Sci. Rep. 6, 19529.

Farr, R.J., Januszewski, A.S., Joglekar, M.V., Liang, H., Mcaulley, A.K., Hewitt, A.W., Thomas, H.E., Loudovaris, T., Kay, T.W., Jenkins, A., Hardikar, A.A., 2015. A comparative analysis of high-throughput platforms for validation of a circulating microRNA signature in diabetic retinopathy. Sci. Rep. 5, 10375.

Ferracin, M., Lupini, L., Salamon, I., Saccenti, E., Zanzi, M.V., Rocchi, A., Da Ros, L., Zagatti, B., Musa, G., Bassi, C., Mangolini, A., Cavallesco, G., Frassoldati, A., Volpato, S., Carcoforo, P., Hollingsworth, A.B., Negrini, M., 2015. Absolute quantification of cell-free microRNAs in cancer patients. Oncotarget 6, 14545–14555.

Fourie, N.H., Peace, R.M., Abey, S.K., Sherwin, L.B., Rahim-Williams, B., Smyser, P.A., Wiley, J.W., Henderson, W.A., 2014. Elevated circulating miR-150 and miR-342-3p in patients with irritable bowel syndrome. Exp. Mol. Pathol. 96, 422–425.

Garcia-Elias, A., Alloza, L., Puigdecanet, E., Nonell, L., Tajes, M., Curado, J., Enjuanes, C., Diaz, O., Bruguera, J., Marti-Almor, J., Comin-Colet, J., Benito, B., 2017. Defining quantification methods and optimizing protocols for microarray hybridization of circulating microRNAs. Sci. Rep. 7, 7725.

Grabmuller, M., Madea, B., Courts, C., 2015. Comparative evaluation of different extraction and quantification methods for forensic RNA analysis. Forensic. Sci. Int. Genet. 16, 195–202.

Grasedieck, S., Scholer, N., Bommer, M., Niess, J.H., Tumani, H., Rouhi, A., Bloehdorn, J., Liebisch, P., Mertens, D., Dohner, H., Buske, C., Langer, C., Kuchenbauer, F., 2012. Impact of serum storage conditions on microRNA stability. Leukemia 26, 2414–2416.

Griffiths-Jones, S., 2004. The microRNA registry. Nucleic Acids Res. 32, D109–D111.

Guo, L., Chen, F., 2014. A challenge for miRNA: multiple isomiRs in miRNAomics. Gene 544, 1–7.

Guo, Y., Vickers, K., Xiong, Y., Zhao, S., Sheng, Q., Zhang, P., Zhou, W., Flynn, C.R., 2017. Comprehensive evaluation of extracellular small RNA isolation methods from serum in high throughput sequencing. BMC Genomics 18, 50.

Hammerle-Fickinger, A., Riedmaier, I., Becker, C., Meyer, H.H., Pfaffl, M.W., Ulbrich, S.E., 2010. Validation of extraction methods for total RNA and miRNA from bovine blood prior to quantitative gene expression analyses. Biotechnol. Lett. 32, 35–44.

Hanson, E.K., Lubenow, H., Ballantyne, J., 2009. Identification of forensically relevant body fluids using a panel of differentially expressed microRNAs. Anal. Biochem. 387, 303–314.

Harrington, C.A., Winther, M., Garred, M.M., 2009. Use of bioanalyzer electropherograms for quality control and target evaluation in microarray expression profiling studies of ocular tissues. J. Ocul. Biol. Dis. Infor. 2, 243–249.

Heid, C.A., Stevens, J., Livak, K.J., Williams, P.M., 1996. Real time quantitative PCR. Genome Res. 6, 986–994.

Hester, S.D., Bhat, V., Chorley, B.N., Carswell, G., Jones, W., Wehmas, L.C., Wood, C.E., 2016. Editor's highlight: dose-response analysis of RNA-seq profiles in archival formalin-fixed paraffin-embedded samples. Toxicol. Sci. 154, 202–213.

Hindson, C.M., Chevillet, J.R., Briggs, H.A., Gallichotte, E.N., Ruf, I.K., Hindson, B.J., Vessella, R.L., Tewari, M., 2013. Absolute quantification by droplet digital PCR versus analog real-time PCR. Nat. Methods 10, 1003–1005.

Hu, F., Meng, X., Tong, Q., Liang, L., Xiang, R., Zhu, T., Yang, S., 2013. BMP-6 inhibits cell proliferation by targeting microRNA-192 in breast cancer. Biochim. Biophys. Acta 1832, 2379–2390.

Huang, T., Yang, J., Liu, G., Jin, W., Liu, Z., Zhao, S., Yao, M., 2015. Quantification of mature microRNAs using pincer probes and real-time PCR amplification. PLoS One 10, e0120160.

Hudson, J., Duncavage, E., Tamburrino, A., Salerno, P., Xi, L., Raffeld, M., Moley, J., Chernock, R.D., 2013. Overexpression of miR-10a and miR-375 and downregulation of YAP1 in medullary thyroid carcinoma. Exp. Mol. Pathol. 95, 62–67.

Jin, J., Vaud, S., Zhelkovsky, A.M., Posfai, J., Mcreynolds, L.A., 2016. Sensitive and specific miRNA detection method using SplintR ligase. Nucleic Acids Res. 44, e116.

Jorgensen, S., Baker, A., Moller, S., Nielsen, B.S., 2010. Robust one-day in situ hybridization protocol for detection of microRNAs in paraffin samples using LNA probes. Methods 52, 375–381.

Kappel, A., Keller, A., 2017. miRNA assays in the clinical laboratory: workflow, detection technologies and automation aspects. Clin. Chem. Lab. Med. 55, 636–647.

Kato, M., Chen, X., Inukai, S., Zhao, H., Slack, F.J., 2011. Age-associated changes in expression of small, noncoding RNAs, including microRNAs, in C. elegans. RNA 17, 1804–1820.

Khan, J., Lieberman, J.A., Lockwood, C.M., 2017. Variability in, variability out: best practice recommendations to standardize pre-analytical variables in the detection of circulating and tissue microRNAs. Clin. Chem. Lab. Med. 55, 608–621.

Kirschner, M.B., Kao, S.C., Edelman, J.J., Armstrong, N.J., Vallely, M.P., Van Zandwijk, N., Reid, G., 2011. Haemolysis during sample preparation alters microRNA content of plasma. PLoS One 6, e24145.

Kirschner, M.B., Edelman, J.J., Kao, S.C., Vallely, M.P., Van Zandwijk, N., Reid, G., 2013. The impact of hemolysis on cell-free microRNA biomarkers. Front. Genet. 4, 94.

Kloosterman, W.P., Wienholds, E., de Bruijn, E., Kauppinen, S., Plasterk, R.H., 2006. In situ detection of miRNAs in animal embryos using LNA-modified oligonucleotide probes. Nat. Methods 3, 27–29.

Kodani, M., Yang, G., Conklin, L.M., Travis, T.C., Whitney, C.G., Anderson, L.J., Schrag, S.J., Taylor Jr., T.H., Beall, B.W., Breiman, R.F., Feikin, D.R., Njenga, M.K., Mayer, L.W., Oberste, M.S., Tondella, M.L., Winchell, J.M., Lindstrom, S.L., Erdman, D.D., Fields, B.S., 2011. Application of TaqMan low-density arrays for simultaneous detection of multiple respiratory pathogens. J. Clin. Microbiol. 49, 2175–2182.

Kok, M.G., Halliani, A., Moerland, P.D., Meijers, J.C., Creemers, E.E., Pinto-Sietsma, S.J., 2015. Normalization panels for the reliable quantification of circulating microRNAs by RT-qPCR. FASEB J. 29, 3853–3862.

Kong, Y., Wu, J., Yuan, L., 2014. MicroRNA expression analysis of adult-onset Drosophila Alzheimer's disease model. Curr. Alzheimer Res. 11, 882–891.

Kozomara, A., Griffiths-Jones, S., 2014. miRBase: annotating high confidence microRNAs using deep sequencing data. Nucleic Acids Res. 42, D68–D73.

Kroh, E.M., Parkin, R.K., Mitchell, P.S., Tewari, M., 2010. Analysis of circulating microRNA biomarkers in plasma and serum using quantitative reverse transcription-PCR (qRT-PCR). Methods 50, 298–301.

Kumar, P., Dezso, Z., MacKenzie, C., Oestreicher, J., Agoulnik, S., Byrne, M., Bernier, F., Yanagimachi, M., Aoshima, K., Oda, Y., 2013. Circulating miRNA biomarkers for Alzheimer's disease. PLoS One 8, e69807.

Landolt, L., Marti, H.P., Beisland, C., Flatberg, A., Eikrem, O.S., 2016. RNA extraction for RNA sequencing of archival renal tissues. Scand. J. Clin. Lab. Invest. 76, 426–434.

Lange, J., 2010. microRNA profiling on automated biochip platform reveals biomarker signatures from blood samples. Nat. Methods 7, 162

Lee, R.C., Ambros, V., 2001. An extensive class of small RNAs in Caenorhabditis elegans. Science 294, 862–864.

Lee, R.C., Feinbaum, R.L., Ambros, V., 1993. The C. elegans heterochronic gene lin-4 encodes small RNAs with antisense complementarity to lin-14. Cell 75, 843–854.

Li, J., Li, X., Li, Y., Yang, H., Wang, L., Qin, Y., Liu, H., Fu, L., Guan, X.Y., 2013. Cell-specific detection of miR-375 downregulation for predicting the prognosis of esophageal squamous cell carcinoma by miRNA in situ hybridization. PLoS One 8, e53582.

Lin, S.L., Chiang, A., Chang, D., Ying, S.Y., 2008. Loss of mir-146a function in hormone-refractory prostate cancer. RNA 14, 417–424.

Livak, K.J., Schmittgen, T.D., 2001. Analysis of relative gene expression data using real-time quantitative PCR and the 2(-Delta Delta C(T)) method. Methods 25, 402–408.

Majlessi, M., Nelson, N.C., Becker, M.M., 1998. Advantages of 2'-O-methyl oligoribonucleotide probes for detecting RNA targets. Nucleic Acids Res. 26, 2224–2229.

Margulies, M., Egholm, M., Altman, W.E., Attiya, S., Bader, J.S., Bemben, L.A., Berka, J., Braverman, M.S., Chen, Y.J., Chen, Z., Dewell, S.B., Du, L., Fierro, J.M., Gomes, X.V., Godwin, B.C., He, W., Helgesen, S., Ho, C.H., Irzyk, G.P., Jando, S.C., Alenquer, M.L., Jarvie, T.P., Jirage, K.B., Kim, J.B., Knight, J.R., Lanza, J.R., Leamon, J.H., Lefkowitz, S.M., Lei, M., Li, J., Lohman, K.L., Lu, H., Makhijani, V.B., Mcdade, K.E., Mckenna, M.P., Myers, E.W., Nickerson, E., Nobile, J.R., Plant, R., Puc, B.P., Ronan, M.T., Roth, G.T., Sarkis, G.J., Simons, J.F., Simpson, J.W., Srinivasan, M., Tartaro, K.R., Tomasz, A., Vogt, K.A., Volkmer, G.A., Wang, S.H., Wang, Y., Weiner, M.P., Yu, P., Begley, R.F., Rothberg, J.M., 2005. Genome sequencing in microfabricated high-density picolitre reactors. Nature 437, 376–380.

Marin, R.M., Sulc, M., Vanicek, J., 2013. Searching the coding region for microRNA targets. RNA 19, 467–474.

Masuda, N., Ohnishi, T., Kawamoto, S., Monden, M., Okubo, K., 1999. Analysis of chemical modification of RNA from formalin-fixed samples and optimization of molecular biology applications for such samples. Nucleic Acids Res. 27, 4436–4443.

McAlexander, M.A., Phillips, M.J., Witwer, K.W., 2013. Comparison of methods for miRNA extraction from plasma and quantitative recovery of RNA from cerebrospinal fluid. Front. Genet. 4, 83.

Mei, Q., Li, X., Meng, Y., Wu, Z., Guo, M., Zhao, Y., Fu, X., Han, W., 2012. A facile and specific assay for quantifying microRNA by an optimized RT-qPCR approach. PLoS One 7, e46890.

Meng, W., McElroy, J.P., Volinia, S., Palatini, J., Warner, S., Ayers, L.W., Palanichamy, K., Chakravarti, A., Lautenschlaeger, T., 2013. Comparison of microRNA deep sequencing of matched formalin-fixed paraffin-embedded and fresh frozen cancer tissues. PLOS ONE 8(5), e64393.

Mestdagh, P., Van Vlierberghe, P., De Weer, A., Muth, D., Westermann, F., Speleman, F., Vandesompele, J., 2009. A novel and universal method for microRNA RT-qPCR data normalization. Genome Biol. 10, R64.

Metzker, M.L., 2010. Sequencing technologies—the next generation. Nat. Rev. Genet. 11, 31–46.

Moret, I., Sanchez-Izquierdo, D., Iborra, M., Tortosa, L., Navarro-Puche, A., Nos, P., Cervera, J., Beltran, B., 2013. Assessing an improved protocol for plasma microRNA extraction. PLoS One 8, e82753.

Morrison, T., Hurley, J., Garcia, J., Yoder, K., Katz, A., Roberts, D., Cho, J., Kanigan, T., Ilyin, S.E., Horowitz, D., Dixon, J.M., Brenan, C.J., 2006. Nanoliter high throughput quantitative PCR. Nucleic Acids Res. 34, e123.

Motta, V., Bonzini, M., Grevendonk, L., Iodice, S., Bollati, V., 2017. Epigenetics applied to epidemiology: investigating environmental factors and lifestyle influence on human health. Med. Lav. 108, 10–23.

Murakami, Y., Tanahashi, T., Okada, R., Toyoda, H., Kumada, T., Enomoto, M., Tamori, A., Kawada, N., Taguchi, Y.H., Azuma, T., 2014. Comparison of hepatocellular carcinoma miRNA expression profiling as evaluated by next generation sequencing and microarray. PLoS One 9, e106314.

Nagy, Z.B., Wichmann, B., Kalmar, A., Bartak, B.K., Tulassay, Z., Molnar, B., 2016. miRNA isolation from FFPET specimen: a technical comparison of miRNA and Total RNA isolation methods. Pathol. Oncol. Res. 22, 505–513.

Nair, V.S., Pritchard, C.C., Tewari, M., Ioannidis, J.P., 2014. Design and analysis for studying microRNAs in human disease: a primer on -omic technologies. Am. J. Epidemiol. 180, 140–152.

Ortega, F.G., Lorente, J.A., Garcia Puche, J.L., Ruiz, M.P., Sanchez-Martin, R.M., de Miguel-Perez, D., Diaz-Mochon, J.J., Serrano, M.J., 2015. miRNA in situ hybridization in circulating tumor cells—MishCTC. Sci. Rep. 5, 9207.

Osman, A., 2012. MicroRNAs in health and disease—basic science and clinical applications. Clin. Lab. 58, 393–402.

Park, J.H., Ahn, J., Kim, S., Kwon, D.Y., Ha, T.Y., 2011. Murine hepatic miRNAs expression and regulation of gene expression in diet-induced obese mice. Mol. Cell 31, 33–38.

Pena, J.T., Sohn-Lee, C., Rouhanifard, S.H., Ludwig, J., Hafner, M., Mihailovic, A., Lim, C., Holoch, D., Berninger, P., Zavolan, M., Tuschl, T., 2009. miRNA in situ hybridization in formaldehyde and EDC-fixed tissues. Nat. Methods 6, 139–141.

Pergoli, L., Cantone, L., Favero, C., Angelici, L., Iodice, S., Pinatel, E., Hoxha, M., Dioni, L., Letizia, M., Albetti, B., Tarantini, L., Rota, F., Bertazzi, P.A., Tirelli, A.S., Dolo, V., Cattaneo, A., Vigna, L., Battaglia, C., Carugno, M., Bonzini, M., Pesatori, A.C., Bollati, V., 2017. Extracellular vesicle-packaged miRNA release after short-term exposure to particulate matter is associated with increased coagulation. Part. Fibre Toxicol. 14, 32.

Petriv, O.I., Kuchenbauer, F., Delaney, A.D., Lecault, V., White, A., Kent, D., Marmolejo, L., Heuser, M., Berg, T., Copley, M., Ruschmann, J., Sekulovic, S., Benz, C., Kuroda, E., Ho, V., Antignano, F., Halim, T., Giambra, V., Krystal, G., Takei, C.J., Weng, A.P., Piret, J., Eaves, C., Marra, M.A., Humphries, R.K., Hansen, C.L., 2010. Comprehensive microRNA expression profiling of the hematopoietic hierarchy. Proc. Natl. Acad. Sci. USA 107, 15443–15448.

Pritchard, C.C., Cheng, H.H., Tewari, M., 2012. MicroRNA profiling: approaches and considerations. Nat. Rev. Genet. 13, 358–369.

Ramkissoon, S.H., Mainwaring, L.A., Sloand, E.M., Young, N.S., Kajigaya, S., 2006. Nonisotopic detection of microRNA using digoxigenin labeled RNA probes. Mol. Cell. Probes 20, 1–4.

Ramsingh, G., Koboldt, D.C., Trissal, M., Chiappinelli, K.B., Wylie, T., Koul, S., Chang, L.W., Nagarajan, R., Fehniger, T.A., Goodfellow, P., Magrini, V., Wilson, R.K., Ding, L., Ley, T.J., Mardis, E.R., Link, D.C., 2010. Complete characterization of the microRNAome in a patient with acute myeloid leukemia. Blood 116, 5316–5326.

Raposo, G., Stoorvogel, W., 2013. Extracellular vesicles: exosomes, microvesicles, and friends. J. Cell Biol. 200, 373–383.

Raymond, C.K., Roberts, B.S., Garrett-Engele, P., Lim, L.P., Johnson, J.M., 2005. Simple, quantitative primer-extension PCR assay for direct monitoring of microRNAs and short-interfering RNAs. RNA 11, 1737–1744.

Redshaw, N., Wilkes, T., Whale, A., Cowen, S., Huggett, J., Foy, C.A., 2013. A comparison of miRNA isolation and RT-qPCR technologies and their effects on quantification accuracy and repeatability. BioTechniques 54, 155–164.

Sambrook, J., 2012. Molecular Cloning: A Laboratory Manual, fourth ed. Cold Spring Harbor, New York.

Schageman, J., Zeringer, E., Li, M., Barta, T., Lea, K., Gu, J., Magdaleno, S., Setterquist, R., Vlassov, A.V., 2013. The complete exosome workflow solution: from isolation to characterization of RNA cargo. Biomed. Res. Int. 2013, 253957.

Schwarzenbach, H., da Silva, A.M., Calin, G., Pantel, K., 2015. Data normalization strategies for microRNA quantification. Clin. Chem. 61, 1333–1342.

Sewer, A., Gubian, S., Kogel, U., Veljkovic, E., Han, W., Hengstermann, A., Peitsch, M.C., Hoeng, J., 2014. Assessment of a novel multi-array normalization method based on spike-in control probes suitable for microRNA datasets with global decreases in expression. BMC Res. Notes 7, 302.

Sharbati-Tehrani, S., Kutz-Lohroff, B., Bergbauer, R., Scholven, J., Einspanier, R., 2008. miR-Q: a novel quantitative RT-PCR approach for the expression profiling of small RNA molecules such as miRNAs in a complex sample. BMC Mol. Biol. 9, 34.

Shi, R., Chiang, V.L., 2005. Facile means for quantifying microRNA expression by real-time PCR. BioTechniques 39, 519–525.

Siegrist, F., Singer, T., Certa, U., 2009. MicroRNA expression profiling by bead array technology in human tumor cell lines treated with interferon-alpha-2a. Biol. Proced. Online 11, 113–129.

Smith, A., Calley, J., Mathur, S., Qian, H.R., Wu, H., Farmen, M., Caiment, F., Bushel, P.R., Li, J., Fisher, C., Kirby, P., Koenig, E., Hall, D.G., Watson, D.E., 2016. The rat microRNA body atlas; evaluation of the microRNA content of rat organs through deep sequencing and characterization of pancreas enriched miRNAs as biomarkers of pancreatic toxicity in the rat and dog. BMC Genomics 17, 694.

Soe, M.J., Moller, T., Dufva, M., Holmstrom, K., 2011. A sensitive alternative for microRNA in situ hybridizations using probes of 2'-O-methyl RNA + LNA. J. Histochem. Cytochem. 59, 661–672.

Sourvinou, I.S., Markou, A., Lianidou, E.S., 2013. Quantification of circulating miRNAs in plasma: effect of preanalytical and analytical parameters on their isolation and stability. J. Mol. Diagn. 15, 827–834.

Tabrizi, M., Khalili, M., Vasei, M., Nouraei, N., Mansour Samaei, N., Khavanin, A., Khajehei, M., Mowla, S.J., 2015. Evaluating the miR-302b and miR-145 expression in formalin-fixed paraffin-embedded samples of esophageal squamous cell carcinoma. Arch Iran Med 18, 173–178.

Tan, G.W., Khoo, A.S., Tan, L.P., 2015. Evaluation of extraction kits and RT-qPCR systems adapted to high-throughput platform for circulating miRNAs. Sci. Rep. 5, 9430.

Tiberio, P., Callari, M., Angeloni, V., Daidone, M.G., Appierto, V., 2015. Challenges in using circulating miRNAs as cancer biomarkers. Biomed. Res. Int. 2015, 731479.

Trachtenberg, E.A., Holcomb, C.L., 2013. Next-generation HLA sequencing using the 454 GS FLX system. Methods Mol. Biol. 1034, 197–219.

Urbanek, M.O., Nawrocka, A.U., Krzyzosiak, W.J., 2015. Small RNA detection by in situ hybridization methods. Int. J. Mol. Sci. 16, 13259–13286.

Valoczi, A., Hornyik, C., Varga, N., Burgyan, J., Kauppinen, S., Havelda, Z., 2004. Sensitive and specific detection of microRNAs by northern blot analysis using LNA-modified oligonucleotide probes. Nucleic Acids Res. 32, e175.

Vandesompele, J., De Preter, K., Pattyn, F., Poppe, B., Van Roy, N., De Paepe, A., Speleman, F., 2002. Accurate normalization of real-time quantitative RT-PCR data by geometric averaging of multiple internal control genes. Genome Biol 3. RESEARCH0034.

Varallyay, E., Burgyan, J., Havelda, Z., 2007. Detection of microRNAs by Northern blot analyses using LNA probes. Methods 43, 140–145.

Vickers, K.C., Palmisano, B.T., Shoucri, B.M., Shamburek, R.D., Remaley, A.T., 2011. MicroRNAs are transported in plasma and delivered to recipient cells by high-density lipoproteins. Nat. Cell Biol. 13, 423–433.

Vomelova, I., Vanickova, Z., Sedo, A., 2009. Methods of RNA purification. All ways (should) lead to Rome. Folia Biol. (Praha) 55, 243–251.

Vrijens, K., Bollati, V., Nawrot, T.S., 2015. MicroRNAs as potential signatures of environmental exposure or effect: a systematic review. Environ. Health Perspect. 123, 399–411.

Waggott, D., Chu, K., Yin, S., Wouters, B.G., Liu, F.F., Boutros, P.C., 2012. NanoStringNorm: an extensible R package for the pre-processing of NanoString mRNA and miRNA data. Bioinformatics 28, 1546–1548.

Wang, H., Ach, R.A., Curry, B., 2007. Direct and sensitive miRNA profiling from low-input total RNA. RNA 13, 151–159.

Willenbrock, H., Salomon, J., Sokilde, R., Barken, K.B., Hansen, T.N., Nielsen, F.C., Moller, S., Litman, T., 2009. Quantitative miRNA expression analysis: comparing microarrays with next-generation sequencing. RNA 15, 2028–2034.

Witwer, K.W., Buzas, E.I., Bemis, L.T., Bora, A., Lasser, C., Lotvall, J., Nolte-'T Hoen, E.N., Piper, M.G., Sivaraman, S., Skog, J., Thery, C., Wauben, M.H., Hochberg, F., 2013. Standardization of sample collection, isolation and analysis methods in extracellular vesicle research. J. Extracell. Vesicles 2, 20360.

Wolenski, F.S., Shah, P., Sano, T., Shinozawa, T., Bernard, H., Gallacher, M.J., Wyllie, S.D., Varrone, G., Cicia, L.A., Carsillo, M.E., Fisher, C.D., Ottinger, S.E., Koenig, E., Kirby, P.J., 2017. Identification of microRNA biomarker candidates in urine and plasma from rats with kidney or liver damage. J. Appl. Toxicol. 37, 278–286.

Wyman, S.K., Knouf, E.C., Parkin, R.K., Fritz, B.R., Lin, D.W., Dennis, L.M., Krouse, M.A., Webster, P.J., Tewari, M., 2011. Post-transcriptional generation of miRNA variants by multiple nucleotidyl transferases contributes to miRNA transcriptome complexity. Genome Res. 21, 1450–1461.

Xi, Y., Nakajima, G., Gavin, E., Morris, C.G., Kudo, K., Hayashi, K., Ju, J., 2007. Systematic analysis of microRNA expression of RNA extracted from fresh frozen and formalin-fixed paraffin-embedded samples. RNA 13, 1668–1674.

Yamamichi, N., Shimomura, R., Inada, K., Sakurai, K., Haraguchi, T., Ozaki, Y., Fujita, S., Mizutani, T., Furukawa, C., Fujishiro, M., Ichinose, M., Shiogama, K., Tsutsumi, Y., Omata, M., Iba, H., 2009. Locked nucleic acid in situ hybridization analysis of miR-21 expression during colorectal cancer development. Clin. Cancer Res. 15, 4009–4016.

Yang, J.H., Qu, L.H., 2012. DeepBase: annotation and discovery of microRNAs and other noncoding RNAs from deep-sequencing data. Methods Mol. Biol. 822, 233–248.

Zhang, J., Fu, Y., Mei, Y., Jiang, F., Lakowicz, J.R., 2010. Fluorescent metal nanoshell probe to detect single miRNA in lung cancer cell. Anal. Chem. 82, 4464–4471.

Zhao, H., Shen, J., Hu, Q., Davis, W., Medico, L., Wang, D., Yan, L., Guo, Y., Liu, B., Qin, M., Nesline, M., Zhu, Q., Yao, S., Ambrosone, C.B., Liu, S., 2014. Effects of preanalytic variables on circulating microRNAs in whole blood. Cancer Epidemiol. Biomark. Prev. 23, 2643–2648.

Zhou, H., Arcila, M.L., Li, Z., Lee, E.J., Henzler, C., Liu, J., Rana, T.M., Kosik, K.S., 2012. Deep annotation of mouse isomiR and iso-moR variation. Nucleic Acids Res. 40, 5864–5875.

Further Reading

Anfossi, S., Babayan, A., Pantel, K., Calin, G.A., 2018. Clinical utility of circulating non-coding RNAs—an update. Nat. Rev. Clin. Oncol. https://doi.org/10.1038/s41571-018-0035-x [Epub ahead of print].

Boisen, M.K., Dehlendorff, C., Linnemann, D., Schultz, N.A., Jensen, B.V., Hogdall, E.V., Johansen, J.S., 2015. MicroRNA expression in formalin-fixed paraffin-embedded cancer tissue: identifying reference microRNAs and variability. BMC Cancer 15, 1024.

Chen, J.Q., Papp, G., Szodoray, P., Zeher, M., 2016. The role of microRNAs in the pathogenesis of autoimmune diseases. Autoimmun. Rev. 15, 1171–1180.

Drury, R.E., O'Connor, D., Pollard, A.J., 2017. The clinical application of microRNAs in infectious disease. Front. Immunol. 8, 1182.

Kakimoto, Y., Kamiguchi, H., Ochiai, E., Satoh, F., Osawa, M., 2015. MicroRNA stability in postmortem FFPE tissues: quantitative analysis using autoptic samples from acute myocardial infarction patients. PLoS One 10, e0129338.

Mestdagh, P., Hartmann, N., Baeriswyl, L., Andreasen, D., Bernard, N., Chen, C., Cheo, D., D'Andrade, P., Demayo, M., Dennis, L., Derveaux, S., Feng, Y., Fulmer-Smentek, S., Gerstmayer, B., Gouffon, J., Grimley, C., Lader, E., Lee, K.Y., Luo, S., Mouritzen, P., Narayanan, A., Patel, S., Peiffer, S., Ruberg, S., Schroth, G., Schuster, D., Shaffer, J.M., Shelton, E.J., Silveria, S., Ulmanella, U., Veeramachaneni, V., Staedtler, F., Peters, T., Guettouche, T., Wong, L., Vandesompele, J., 2014. Evaluation of quantitative miRNA expression platforms in the microRNA quality control (miRQC) study. Nat. Methods 11, 809–815.

Nejad, C., Stunden, H.J., Gantier, M.P., 2018. A guide to miRNAs in inflammation and innate immune responses. FEBS J. https://doi.org/10.1111/febs.14482 [Epub ahead of print].

Peiro-Chova, L., Pena-Chilet, M., Lopez-Guerrero, J.A., Garcia-Gimenez, J.L., Alonso-Yuste, E., Burgues, O., Lluch, A., Ferrer-Lozano, J., Ribas, G., 2013. High stability of microRNAs in tissue samples of compromised quality. Virchows Arch. 463, 765–774.

Svoronos, A.A., Engelman, D.M., Slack, F.J., 2016. OncomiR or tumor suppressor? The duplicity of microRNAs in cancer. Cancer Res. 76, 3666–3670.

Thum, T., Condorelli, G., 2015. Long noncoding RNAs and microRNAs in cardiovascular pathophysiology. Circ. Res. 116, 751–762.

Index

Note: Page numbers followed by *f* indicate figures and *t* indicate tables.

Made in the USA
Las Vegas, NV
19 February 2022